Application of Compact Heat Exchangers for Combined Cycle Driven Efficiency in Next Generation Nuclear Power Plants

Bahman Zohuri

Application of Compact Heat Exchangers for Combined Cycle Driven Efficiency in Next Generation Nuclear Power Plants

A Novel Approach

Springer

Bahman Zohuri
Galaxy Advanced Engineering, Inc.
Albuquerque, NM, USA

ISBN 978-3-319-23536-3 ISBN 978-3-319-23537-0 (eBook)
DOI 10.1007/978-3-319-23537-0

Library of Congress Control Number: 2015949382

Springer Cham Heidelberg New York Dordrecht London

Printed on acid-free paper

Springer International Publishing AG Switzerland is part of Springer Science+Business Media (www.springer.com)

This book is dedicated to my daughter, Natalie.
She has always encouraged me with my publications.

Bahman Zohuri

Preface

Today's global energy market places many demands on power generation technology, including high thermal efficiency, low cost, rapid installation, reliability, environmental compliance, and operational flexibility.

The demand for clean, non-fossil-based electricity is growing; therefore, the world needs to develop new nuclear reactors with higher thermal efficiency in order to increase electricity generation and decrease the detrimental effects on the environment. The current fleet of nuclear power plants is classified as Generation III or less. However, these models are not as energy efficient as they should be because the operating temperatures are relatively low. Currently, a group of countries has initiated an international collaboration to develop the next generation of nuclear reactors, called Generation IV. The ultimate goal of developing such reactors is to increase the thermal efficiency from the current range of 30–35 % to 45–50 %. This increase in thermal efficiency would result in a higher production of electricity compared to current Pressurized Water Reactor (PWR) or Boiling Water Reactor (BWR) technologies.

A number of technologies are being investigated for the Next Generation Nuclear Plant (NGNP) that will produce heated fluids at significantly higher temperatures than current generation power plants. The higher temperatures offer the opportunity to significantly improve the thermodynamic efficiency of the energy conversion cycle. One of the concepts currently under study is the Molten Salt Reactor. The coolant from the Molten Salt Reactor may be available at temperatures as high as 800–1000 °C.

At these temperatures, an open Brayton cycle combined with a Rankine bottoming cycle appears to have some strong advantages over any other combinations of other cycles.

Combined-cycle thermal efficiency increases as gas turbine specific power increases. The gas turbine firing temperature is the primary determinant of specific power.

Gas turbine engines, both aircraft and industrial power generation, represent one of the most aggressive applications for structural materials. With ever-growing demands for increasing performance and efficiencies, all classes of materials are being pushed to higher temperature capabilities. These materials must also satisfy stringent durability and reliability criteria. As materials are developed to meet these demanding requirements, the processing of these materials often becomes very complicated and expensive. As a result, the cost of materials and processes has become a much larger consideration in the design and application of high performance materials. Both the aircraft engine and power generation industries are highly cost competitive, and market advantage today relies on reducing cost as well as increasing performance and efficiency.

Given the rise of the distributed power generation market, and the renewed attention to nuclear power plants (in particular new generation and small modular reactors), one of the most economical solutions today is to generate power through small gas turbine systems in the form of Brayton cycles, combined with these new and small modular reactors (SMRs). These gas turbines arbitrarily can be categorized as micro-turbines with output of (5–200 kW) and mini-turbines with output of (200–500 kW). The thermal efficiency of such micro-turbines is about 20% or less if no recuperator is used in the system. Using a recuperator (a regenerator can also be considered but has a number of problems) operating at 87% effectiveness, the efficiency of the gas turbine system increases to about 30%, a substantial performance improvement. However, the cost of the recuperator is about 25–30% of the total power plant; therefore total cost of ownership and return on investments are not very well justified. This necessitates the use of all prime surface heat exchangers with no brazing. Thus the quest for novel designs to enable a new generation of compact heat exchangers in support of combined cycles is underway, and an understanding of such innovative approaches among engineers and scientists is rising rapidly.

In this book, after providing the necessary concise information on all aspects of this innovative approach such as combined cycle and associated turbines such as micro-turbines combined, then goes to the discussion on various types of compact heat exchanger surfaces and novel designs that can be considered for the costeffective heat exchangers and packaging in the system.

The simple Brayton cycle is modified to include a recuperator (which will transfer heat from the turbine exhaust to preheat compressed high pressure air before going to the combustion chamber); it will require less fuel to obtain the desired turbine inlet temperature of compressed air and also the optimum pressure ratio (either for compressor or turbine) is reduced to typically 3–4. This improves the thermal efficiency of the cycle. Alternatively, a regenerator can also be used replacing a recuperator.

The development of high temperature/high strength materials, corrosion-resistant coatings, and improved cooling technology has led to increases in gas turbine firing temperatures. This increase in firing temperature is the primary development that has led to increases in Combined Cycle Gas Turbine (CCGT) thermal efficiencies. The improvements in combined-cycle thermal efficiencies and the commercial development of combined-cycle power plants have proceeded in parallel with advances in gas turbine technologies.

The Generation IV International Forum (GIF) Program has narrowed design options of the nuclear reactors to six concepts. These concepts are Gas-cooled Fast Reactor (GFR), Very High Temperature Reactor (VHTR), Sodium-cooled Fast Reactor (SFR), Lead-cooled Fast Reactor (LFR), Molten Salt Reactor (MSR), and Supercritical Water-cooled Reactor (SCWR). These nuclear-reactor concepts differ in their design in aspects such as the neutron spectrum, coolant, moderator, and operating temperature and pressure.

There are many different types of power reactors. What is common to them all is that they produce thermal energy that can be used for its own sake or converted into mechanical energy and ultimately, in the vast majority of cases, into electrical energy. Thermal-hydraulic issues related to both operating and advanced reactors are presented. Further thermal-hydraulics research and development is continuing in both experimental and computational areas for operating reactors, reactors under construction or ready for near-term deployment, and advanced Generation-IV reactors. As the computing power increases, the fine-scale multi-physics computational models, coupled with the systems analysis code, are expected to provide answers to many challenging problems in both operating and advanced reactor designs.

Compact heat-exchangers, filters, turbines, and other components in integrated Next-Generation Nuclear Power Plant combined cycle system must withstand demanding conditions of high temperatures and pressure differentials. Under the highly sulfiding conditions of the high temperature such as inlet hot steam or other related environmental effects, the performance of components degrade significantly with time unless expensive high alloy materials are used. Deposition of a suitable coating on a low-cost alloy may improve its resistance to such sulfidation attack and decrease capital and operating costs. A review of the literature indicates that the corrosion reaction is the competition between oxidation and sulfidation reactions. The Fe- and Ni-based high-temperature alloys are susceptible to sulfidation attack unless they are fortified with high levels of Cr, Al, and Si. To impart corrosion resistance, these elements need not be in the bulk of the alloy and need only be present at the surface layers.

Those that practice the art of Nuclear Engineering must have a physical and intuitive understanding of the mechanisms and balances of forces, which control the transport of heat and mass in all physical systems. This understanding starts at the molecular level, with intermolecular forces and the motion of molecules, and continues to the macroscopic level where gradients of velocity, temperature, and concentration drive the diffusion of momentum, heat, and mass, and the forces of pressure, inertia, and buoyancy balance to drive convection of fluids.

This text covers the fundamentals of combined cycles required to understand electrical power generation systems and related efficiency issues. It then covers the application of these principles to nuclear reactor power systems. It is a general approach to a Brayton combined cycle text, aimed at explaining combined cycles as deployed with compact heat exchangers in the loop, and applying them to the challenges facing actual nuclear power systems. It is written at an undergraduate level, but should also be useful to practicing engineers and scientists as well.

Chapter 1 provides the basic definitions and principles behind a typical nuclear reactor power plant layout and the scope of the thermodynamics behind it. Chapters 2 and 3 cover the cost of generating electricity and the rise in demand for energy sources, laying out the groundwork for justification of new-generation nuclear power plants. Chapter 4 concentrates on the fundamentals of a new approach to energy conversion technology, covering power conversion systems and their components, demonstrating how we can recover waste heat from a power plant in order to put it into use, thereby driving overall output efficiency higher, and achieving better revenue for day-to-day operations. Chapter 5 includes detailed consideration of air brayton cycles as deployed in nuclear power plants, and after covering heat exchangers in general, Chapter 6 then covers compact heat exchangers specifically, demonstrating how they are the most efficient and cost-effective option for application in the innovative approaches described in Chapter 7.

Taken as a whole the last few chapters of the book cover current and projected reactor systems and how the novel design of these compact heat exchangers from thermal design perspective in principle is applied to their innovative designs, operation, and safety analyses.

Detailed appendices cover metric and English system units and conversions, detailed steam and gas tables, heat transfer properties, and nuclear reactor system descriptions, as well as providing a holistic approach to the understanding of nuclear power plants and each generation.

Acknowledgments

The author would like to acknowledge all the individuals for their help, encouragement, and support. We have decided not to name them all since some of them may not be around to see the end result of their encouragement, but we hope they can at least read this acknowledgment wherever they may be.

Last but not least, special thanks to my parents, wife, children, and friends for providing constant encouragement, without which this book could not have been written. We especially appreciate their patience with pure frequent absence from home and long hours in front of the computer during the preparation of this book.

My sincere appreciation goes to Professor Patrick McDaniel of the Department of Nuclear Engineering at the University of New Mexico Albuquerque, New Mexico, who provided me the knowledge that I have now and continues teaching me what I need to know to go forward. He is a true gentleman.

I am also indebted to my another teacher, mentor, and now a true friend, Professor Dimiter N. Petsev of the University of New Mexico, Chemical Engineering Department, whom I have a lot of respect as well.

Contents

1 Definitions and Basic Principles 1
1.1 Typical Pressurized Water Reactor 1
1.2 Scope of Thermodynamics 3
1.3 Units 5
1.3.1 Fundamental Units 5
1.3.2 Thermal Energy Units 6
1.3.3 Unit Conversion 6
1.4 Classical Thermodynamics 7
1.5 Open and Closed Systems 8
1.6 System Properties 10
1.6.1 Density 10
1.6.2 Pressure 11
1.6.3 Temperature 13
1.7 Properties of the Atmosphere 14
1.8 The Laws of Thermodynamics 15
References 16

2 Electricity, an Essential Necessity in Our Life 17
2.1 Cost of Generating Electricity Today 17
2.2 Nuclear Power Plants 20
2.3 Cost of Electricity from New Nuclear Power Plants Stations 20
2.3.1 Pros and Cons of New Nuclear Power Plants 21
2.4 Is Nuclear Power a Global warming Solution? 25
2.5 Prediction of Energy Consumption Worldwide 26
2.6 Current Energy Consumption by Capita 29
2.7 The Next Nuclear Age: Can Safe Nuclear Power Work for America or the World? 31
References 35

3 Energy Resources and the Role of Nuclear Energy 37
3.1 The World's Energy Resources 37
3.2 Today's Global Energy Market 38
3.3 End of Cheap Oil and the Future of Energy 39
3.4 What To Do About Coal 42
3.5 The Future of Energy 43
3.6 Nuclear Reactors for Power Production 45
3.7 Future Nuclear Power Plant System 46
3.8 Next Generation of Nuclear Power Reactors for Power Production 47
3.9 Goals for Generation IV Nuclear Energy Systems 49
3.10 A Technology Roadmap for Generation IV Nuclear Energy Systems 50
3.11 The Description of the Six Most Promising Nuclear Power Systems 51
Reference 57

4 A New Approach to Energy Conversion Technology 59
4.1 Power Conversion Study and Technology Options Assessment 59
4.2 Waste Heat Recovery 64
4.2.1 Advantages and Disadvantages of Waste Heat Recovery 64
4.3 Power Conversion System Components 65
4.3.1 Heat Exchangers 65
4.3.2 Compact Heat Exchangers 79
4.4 Development of Gas Turbine 81
4.5 Turbomachinery 83
4.6 Heat Transfer Analysis 85
4.7 Combined Cycle Power Plant 86
4.8 Advanced Computational Materials Proposed for Gen IV Systems 89
4.9 Material Classes Proposed for Gen IV Systems 91
4.10 Generation IV Materials Challenges 91
4.11 Generation IV Materials Fundamental Issues 93
4.12 Capital Cost of Proposed Gen IV Reactors 94
4.12.1 Economic and Technical of Combined Cycle Performance 96
4.12.2 Economic Evaluation Technique 97
4.12.3 Output Enhancement 98
References 101

5 Air Brayton Cycles for Nuclear Power Plants 103
5.1 Background 103
5.2 Typical Cycles 107

5.3 Methodology 107
5.4 Combined Cycle Code Equations 110
5.5 Computer Code Flowchart 111
5.6 Validation of Methodology 112
5.7 Nominal Analysis Parameters 113
5.8 Combined Cycle System Baseline 114
5.9 Recuperated Cycle System Baseline 114
5.10 Nominal Results for Combined Cycle Model 115
5.11 Nominal Results for Recuperated Cycle Model 118
5.12 Extension of Results Versus Peak Turbine Temperatures 121
5.13 Conclusions 122
References 123

6 Heat Exchangers 125
6.1 Heat Exchanger Types 125
6.2 Classification According to Transfer Processes 128
6.2.1 Indirect Contact Type Heat Exchangers 128
6.2.2 Direct Contact Type Heat Exchangers 128
6.3 Classification of Heat Exchanger by Construction Type 130
6.3.1 Tubular Heat Exchangers 130
6.3.2 Plate Heat Exchangers 131
6.3.3 Plate Fin Heat Exchangers 132
6.3.4 Tube Fin Heat Exchangers 132
6.3.5 Regenerative Heat Exchangers 133
6.4 Condensers 133
6.5 Boilers 134
6.6 Classification According to Compactness 135
6.7 Types of Applications 135
6.8 Cooling Towers 135
6.9 Regenerators and Recuperators 137
6.10 Heat Exchanger Analysis: Use of the LMTD 142
6.11 Effectiveness-NTU Method for Heat Exchanger Design 149
6.12 Special Operating Conditions 155
6.13 Compact Heat Exchangers and Their Classifications 155
References 160

7 Effective Design of Compact Heat Exchangers for NGNP 161
7.1 Introduction 161
7.2 Classification of Heat Exchangers 164
7.3 Compact Heat Exchanger-Driven Efficiencies in Brayton Cycle 167
7.4 Thermal Energy Transfer for Process Heat Application in Enhanced Mode 176
7.5 Design Criteria for Process Heat Exchangers 186
7.6 Thermal and Hydraulic Design 191
7.6.1 Equations and Parameters 192

7.7 The Overall Heat Exchanger Design Process 209
7.7.1 Input Information Needed 210
7.8 Design Summary 212
7.9 Compact Heat Exchangers in Practice 222
7.10 Heat Exchanger Materials and Comparisons 222
7.11 Guide to Compact Heat Exchangers 224
7.11.1 Generic Advantages of Compact Design 225
References 225

Appendix A: Table and Graphs Compilations 229

Appendix B: Nuclear Power Plants 317

Nuclear Systems Acronyms 361

Index 365

About the Author

Bahman Zohuri is currently at the Galaxy Advanced Engineering, Inc., a consulting company that he started himself in 1991 when he left both semiconductor and defense industries after many years working as a chief scientist. After graduating from the University of Illinois in the field of Physics and Applied Mathematics, as well as the University of New Mexico from Nuclear Engineering Department, he joined Westinghouse Electric Corporation where he performed thermal hydraulic analysis and natural circulation for Inherent Shutdown Heat Removal System (ISHRS) in the core of a Liquid Metal Fast Breeder Reactor (LMFBR) as a secondary fully inherent shut system for secondary loop heat exchange. All these designs were used for Nuclear Safety and Reliability Engineering for Self-Actuated Shutdown System. He designed the Mercury Heat Pipe and Electromagnetic Pumps for Large Pool Concepts of LMFBR for heat rejection purpose for this reactor around 1978 where he received a patent for it. He later on was transferred to defense division of Westinghouse where he was responsible for the dynamic analysis and method of launch and handling of MX missile out of canister. The results are applied to MX launch seal performance and muzzle blast phenomena analysis (i.e., missile vibration and hydrodynamic shock formation). He also was involved in analytical calculation and computation in the study of Nonlinear Ion Wave in Rarefying Plasma. The results are applied to the propagation of "Soliton Wave" and the resulting charge collector traces, in the rarefactions characteristic of the corona of a laser irradiated target pellet. As part of his graduate research work at Argonne National Laboratory, he performed computation and programming of multi-exchange integral in surface physics and solid-state physics. He holds different patents in areas such as diffusion processes and design of diffusion furnace while he was senior process engineer working for different semiconductor industries such as Intel, Varian, and National Semiconductor corporations. Later on he joined Lockheed Missile and Aerospace Corporation as Senior Chief Scientist and was responsible for research and development (R&D) and the study of vulnerability, survivability, and both radiation and laser hardening of various components of the Strategic Defense Initiative, also known as Star Wars.

This included of payload (i.e., IR sensors) for the Defense Support Program, Boost Surveillance and Tracking Satellite, and Space Surveillance and Tracking Satellite against laser or nuclear threats. While there, he also studied and performed the analysis of characteristics of laser beam and nuclear radiation interaction with materials, Transient Radiation Effects in Electronics (TREE), Electromagnetic Pulse (EMP), System Generated Electromagnetic Pulse (SGEMP), Single-Event Upset (SEU), Nuclear Blast and, Thermomechanical, hardness assurance, maintenance, device technology.

He did few years of consulting under his company Galaxy Advanced Engineering with Sandia National Laboratories (SNL), where he supported development of operational hazard assessments for the Air Force Safety Center (AFSC) in connection with other interest parties. Intended use of the results was their eventual inclusion in Air Force Instructions (AFIs) specifically issued for Directed Energy Weapons (DEW) operational safety. He completed the first version of a comprehensive library of detailed laser tools for Airborne Laser (ABL), Advanced Tactical Laser (ATL), Tactical High Energy Laser (THEL), Mobile/Tactical High Energy Laser (M-THEL), etc.

He also oversaw SDI computer programs related to Battle Management C^3, artificial Intelligent, and autonomous systems. He is the author of several publications and holds various patents such as Laser Activated Radioactive Decay and Results of Thru-Bulkhead Initiation.

Recently, he has published two other books with CRC and Francis Taylor on the subject of:

1. Heat Pipe Design and Technology: A Practical Approach, Published by CRC Publishing Company.
2. Dimensional Analysis and Self-Similarity Methods for Engineering and Scientist, Published by Springer Publishing Company.
3. High Energy Laser (HEL): Tomorrow's Weapon in Directed Energy Weapons Volume I, Published by Trafford Publishing Company.
4. Thermodynamics in Nuclear Power Plant Systems, Published by Springer Publishing Company.
5. Thermal-Hydraulic Analysis of Nuclear Reactors, Published by Springer Publishing Company.

Chapter 1
Definitions and Basic Principles

Nuclear power plants currently generate better than 20 % of the central station electricity produced in the USA. The USA currently has 104 operating power producing reactors, with 9 more planned. France has 58 with 1 more planned. China has 13 with 43 planned. Japan has 54 with 3 more planned. In addition, Russia has 32 with 12 more planned. Nuclear generated electricity has certainly come into its own existence and is the safest, cleanest, and greenest form of electricity currently produced on this planet. However, many current thermodynamics texts ignore nuclear energy and use few examples of nuclear power systems. Nuclear energy presents some interesting thermodynamic challenges and it helps to introduce them at the fundamental level. Our goal here is to introduce thermodynamics as the energy conversion science that it is and apply it to nuclear systems. Certainly, there are many aspects of thermodynamics that are given little or no coverage. However, that is true for any textual introduction to this science; however by considering concrete systems, it is easier to give insight into the fundamental laws of the science and to provide an intuitive feeling for further study [1–6].

1.1 Typical Pressurized Water Reactor

By far the most widely built nuclear system is the pressurized water reactor (PWR). There are a number of reasons for this. Steam turbines have for many decades been the dominant means of generating mechanical energy to turn electrical generators. The temperatures reached in the thermodynamic cycle of a PWR are within the range of fairly common engineering materials. They were the first systems built and operated reliably to produce electricity. A typical PWR system is described in Fig. 1.1.

B. Zohuri, *Application of Compact Heat Exchangers For Combined Cycle Driven Efficiency In Next Generation Nuclear Power Plants*,
DOI 10.1007/978-3-319-23537-0_1

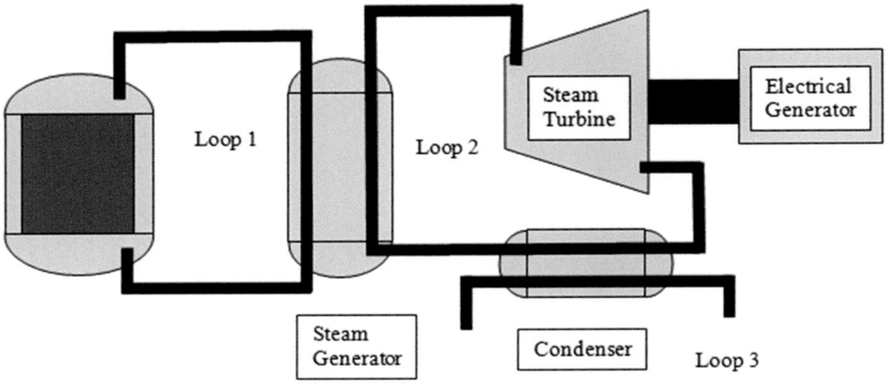

Fig. 1.1 Pressurized water reactor schematic

The basic PWR consists of five major components, the reactor core, steam generator(s), steam turbine, condenser, and electrical generator and three water/steam loops. Each loop requires a pump that is not shown to keep the diagram cleaner. The nuclear energy is converted to thermal energy in the reactor core. This thermal energy is then transported via the first loop to the steam generator where it is passed to the water in the second loop. The water in the second loop enters as a liquid and is turned to steam. The steam then passes to the turbine where the thermal energy is converted to mechanical energy to rotate the electrical generator. After the thermal energy has been converted to mechanical energy in the steam turbine, the low-pressure steam passes to the condenser to be cooled by the water in the third loop. The second law of thermodynamics tells us that we cannot simply expand the steam to a low enough energy state that it can return to the steam generator in its original liquid state. Therefore, we must extract more thermal energy from the low-pressure steam to return it to its liquid state where it can be pumped back into the steam generator. The third loop is called the circulating water system and it is open to the environment. There are multiple ways of providing this cooling water including intake and return to a river, or the ocean, intake and return to a cooling pond, or intake from a river and exhaust through a cooling tower. However, we are getting ahead of ourselves.

Consider for a minute why nuclear energy is so useful. A great deal of energy is produced by a very little mass.

Example Calculation: Calculate the U-235 consumed to produce 1 MW of thermal energy for 1 day. Note that a Megawatt is a unit of power, or energy per unit time,

$$1\,\text{MW} = 10^6\,\text{W} = 10^6\,\text{J/s} \quad 1\ \text{day} = 24\,\text{h} = 24 \times 3600\,\text{s}$$

The energy released in fission of a U-235 atom is ~200 MeV

$$1\,\text{eV} = 1.6 \times 10^{-19}\,\text{J} \quad 1\,\text{MeV} = 1.6 \times 10^{-13}\,\text{J} \quad 200\,\text{MeV} = 32\ \text{pJ}$$

Fissioning 1 atom of U-235 produces 3.2×10^{-11} J

To produce 10^6 J requires $106/3.2 \times 10^{-11}$ atoms $= 3.125 \times 10^{16}$ atoms

And for a duration of 8.64×10^4 s

The total number of atoms consumed will be $3.125 \times 8.64 \times 10^{20}$ atoms

Therefore 2.7×10^{21} atoms will be consumed

A gram mole of U-235 is 6.022×10^{23} atoms

So a gram is $6.022 \times 10^{23}/235 = 2.563 \times 10^{21}$ atoms/g

Therefore 1 MW-day of nuclear energy consumes 1.05 g of U-235

The fundamental thing to understand is that a PWR converts nuclear energy to electrical energy and it does this by converting the nuclear energy first to thermal energy and then converting the thermal energy to mechanical energy, which is finally converted to electrical energy. The science of thermodynamics deals with each of these conversion processes. To quantify how each of these processes takes place we must understand and apply the laws of thermodynamics.

1.2 Scope of Thermodynamics

Thermodynamics is the science that deals with energy production, storage, transfer, and conversion. It is a very broad subject that affects most fields of science including biology and microelectronics. The primary forms of energy considered in this text will be nuclear, thermal, chemical, mechanical, and electrical. Each of these can be converted to a different form with widely varying efficiencies. Predominantly thermodynamics is most interested in the conversion of energy from one form to another via thermal means. However, before addressing the details of thermal energy conversion, consider a more familiar example. Newtonian mechanics defines work as force acting through a distance on an object. Performing work is a way of generating mechanical energy. Work itself is not a form of energy, but a way of transferring energy to a mass. So when one mass gains energy, another mass, or field, must lose that energy.

Consider a simple example. A 65-kg woman decides to go over Niagara Falls in a 25-kg wooden barrel. (The first person to go over the fall in a barrel was a woman, Annie Taylor.) Niagara Falls has a vertical drop of 50 m and has the highest flow rate of any waterfall in the world. The force acting on the woman and the barrel is the force of gravity, which at the surface of the earth produces a force of 9.8 N for every kilogram of matter that it acts on. So we have:

$$W = F \times D \quad F = (65 + 25) \times 9.8 = 882.0\,\text{N} \quad D = 50\,\text{m}$$
$$W = 882.0 \times 50.0 = 44,100\,\text{N-m} = 44.1\,\text{kJ}$$

A Newton meter is a joule and 1000 J is a kilo-joule. Therefore, when the woman and barrel went over the falls, by the time they had reached the bottom, the force of gravity had performed 44.1 kJ of work on them. The gravitational field had 44.1 kJ of potential energy stored in it, when the woman and the barrel were at the top of the falls. This potential energy was converted to kinetic energy by the time the barrel reached the bottom of the falls. Kinetic energy is also measured in joules, as with all other forms of energy. However, we are usually most interested in velocities when we talk about kinetic energies, so let us extract the velocity with which she hit the waters of the inlet to Lake Ontario.

$$\Delta\mathrm{KE} = \Delta\mathrm{PE} = 44.1\,\mathrm{kJ} = 1/2\,\mathrm{mV}^2 = (90/2)\,\mathrm{kg} \times \mathrm{V}^2 \quad \mathrm{V}^2 = 44.1\,\mathrm{kJ}/(90/2)\,\mathrm{kg}$$

Now it is a matter of converting units. A joule is a Newton-meter. 1 N is defined as 1 kg accelerated at the rate of 1 m/s/s. So

$$\begin{aligned} 44.1\,\mathrm{kJ} &= 44,100\,\mathrm{N\text{-}m} \\ &= 44,100\,\mathrm{kg\,m/s/s\text{-}m} \\ &= 44,100\,\mathrm{kg}\ \ (\mathrm{m/s})^2 \end{aligned}$$

$$\begin{aligned} V^2 &= 44,100\,\mathrm{kg}\ \ (\mathrm{m/s})^2/(90/2)\,\mathrm{kg} \\ &= 490/(1/2) = 980\ \ (\mathrm{m/s})^2 \\ V &= 31.3\,\mathrm{m/s}\ \ (\sim 70\,\mathrm{m\,ph}) \end{aligned}$$

Needless to say she recommended that no one ever try that again. Of course, others have, some have made it, and some have drowned.

Before leaving this example, it is worth pointing out that when we went to calculate the velocity, it was unaffected by the mass of the object that had dropped the 50 m. So one-half the velocity squared represents what we will call a specific energy, or energy per kilogram. In addition, the potential energy at the top of the falls could be expressed as a specific potential energy relative to the waters below. The potential energy per pound mass would just be the acceleration of gravity times the height of the falls. Typically, we will use lower case letters to represent specific quantities and upper case letters to represent extensive quantities. Extensive quantities are dependent upon the amount of mass present. Specific quantities are also referred to as intensive variables, though there are some intensive variables that have no extensive counterpart, such as pressure or temperature:

$$\mathrm{p.e.} = \mathrm{mgh/m} = \mathrm{gh} = 9.8 \times 50 = 0.49\,\mathrm{kJ/kg}$$

It is also worth pointing out that Newton's law of gravity states that

$$F = G\frac{m_1 M_2}{R^2} \tag{1.1}$$

where m_1 is the smaller mass and M_2 is the mass of the Earth. We can find the specific force on an object by dividing the gravitational force by the mass of the object. For distances like 50 m on the surface of the Earth $(R = 6,378,140\,\text{m})$ we can treat R as constant, but if the distance the gravitational force acts through is comparable to the radius of the Earth, an integration would be required. Even on the top of Mount Everest, the gravitational potential is within 0.25 % of that at Sea Level, so gravity is essentially constant for all systems operating on the face of the Earth.

1.3 Units

In this section, we discuss the System International (SI) and English (E) Systems.

1.3.1 Fundamental Units

Before going any further, it will be a very good idea to discuss units for physical quantities and the conversion of units from one system to another. Unfortunately, the field of thermodynamics is beset with two popular systems of units. One is the System International (SI) consisting of the kilogram, meter, and second. The other is the English (E) system consisting of the pound-mass, foot, and second.

Starting with the SI system, the unit of force is the Newton. The unit of work or energy is the Joule, and the unit of pressure is the Pascal. We have:

$$\begin{aligned}1\,\text{N} &= 1\,\text{kg-m/s}^2\\1\,\text{J} &= 1\,\text{N-m}\\1\,\text{Pa} &= 1\,\text{N/m}^2\end{aligned}$$

Now the acceleration of gravity at Sea Level on Earth is 9.8066 m/s^2, so a 100 kg mass will weigh 980.66 N. Also when we want to avoid spelling out very large or small quantities, we will usually use the standard abbreviations for powers of 10 in units of 1000. We have:

$$\begin{aligned}\text{kilo} &= 10^3\\\text{mega} &= 10^6\\\text{giga} &= 10^9\\\text{deci} &= 10^{-1}\\\text{centi} &= 10^{-2}\\\text{milli} &= 10^{-3}\\\text{micro} &= 10^{-6}\\\text{nano} &= 10^{-9}\end{aligned}$$

For the English system, we have:

$$\begin{aligned}
&\text{lbm} \Rightarrow 1\,\text{lb f (at Sea Level)}\\
&1\,\text{ft-lb f} = 1\,\text{lb f} \times 1\,\text{ft}\\
&1\,\text{British Thermal Unit (BTU)} = 778\,\text{ft-lb f}\\
&1\,\text{psi} = 1\,\text{lb f/in}^2
\end{aligned}$$

Note that the fact that 1 lbf = 1 lbm at Sea Level on Earth means that a mass of 100 lbm will weigh 100 lbf at Sea Level on Earth. The acceleration of gravity at Sea Level on Earth is 32.174 ft/s^2. Thus we have 1 lbf/(1 lbm-ft/s) = 32.174. If we move to another planet where the acceleration of gravity is different, the statement that 1 lbm ⇒ 1 lbf does not hold.

Consider comparative weights on Mars. The acceleration of gravity on Mars is 38.5 % of the acceleration of gravity on Earth. So in the SI system we have:

$$W = 0.385 \times 9.8066\,\text{m/s}^2 \times 100\ \text{kg} = 377.7\,\text{N}$$

In the English system, we have:

$$W = 0.385 \times 100\ \text{lbm} = 38.5\ \text{lb f}$$

1.3.2 Thermal Energy Units

The British thermal unit (Btu) is defined to be the amount of heat that must be absorbed by a 1 lb-mass to raise its temperature by 1 °F. The calorie is the SI unit that is defined in a similar way. It is the amount of heat that must be absorbed by 1 g of water to raise its temperature by 1 °C. This raises the question as to how a calorie compares with a joule since both appear to be measures of energy in the SI system. James Prescott Joule spent a major part of his life proving that thermal energy was simply another form of energy like mechanical kinetic or potential energy. Eventually his hypothesis was accepted and the conversion factor between the calorie and joule has been defined by:

$$1\ \text{cal} = 4.1868\ \text{J}$$

The constant 4.1868 is called the mechanical equivalent of heat.

1.3.3 Unit Conversion

As long as one remains in either the SI system or the English system, calculations and designs are simple. However, that is no longer possible as different organizations and different individuals usually think and work in their favorite system. In order to communicate with an audience that uses both SI and English systems, it is important to be able to convert back and forth between the two systems. The basic conversion factors are,

$$\begin{aligned}
1 \ \text{kg} &= 2.20462 \ \text{lbm} \\
1 \ \text{lbm} &= 0.45359 \ \text{kg} \\
1 \ \text{m} &= 3.2808 \ \text{ft} \\
1 \ \text{ft} &= 0.3048 \ \text{m} \\
1 \ \text{J} &= 0.00094805 \ \text{Btu} \\
1 \ \text{Btu} &= 1055 \ \text{J} \\
1 \ \text{atm} &= 14.696 \ \text{psi} \\
1 \ \text{atm} &= 101325 \ \text{Pa} \\
1 \ \text{psi} &= 6894.7 \ \text{Pa} \\
1 \ \text{bar} &= 100,000.0 \ \text{Pa} \\
1 \ \text{bar} &= 14.504 \ \text{psi}
\end{aligned}$$

The bar unit is simply defined by rounding off Sea Level atmospheric pressure to the nearest 100 kPa. There are many more conversion factors defined in the Appendix, but they are all derived from this basic few.

1.4 Classical Thermodynamics

Classical thermodynamics was developed long before the atomic theory of matter was accepted. Therefore, it treats all materials as continuous and all derivatives well defined by a limiting process. Steam power and an ability to analyze it and optimize it was one of the main drivers for the development of thermodynamic theory. The fluids involved always looked continuous. A typical example would be the definition of the density of a substance at a point. We have,

$$\rho = \lim_{\Delta V \to 0} \frac{\Delta m}{\Delta V} \tag{1.2}$$

As long as ΔV does not get down to the size of an atom, this works. Since classical thermodynamics was developed, however, we have come to understand that all gases and liquids are composed of very small atoms or molecules and a limiting process that gets down to the atomic or molecular level will eventually become discontinuous and chaotic. Nevertheless, the continuous model still works well for the macroscopic systems that will be discussed in this text and Classical Thermodynamics is based on it.

At times, we will refer to an atomistic description of materials in order to develop a method of predicting specific thermodynamic variables that classical thermodynamics cannot predict. A typical example is the derivative that is called the constant volume specific heat. This variable is defined as the rate of change of the internal energy stored in a substance as a function of changes in its temperature. Classical thermodynamics demonstrates that this variable has to exist and makes great use of it, but it has no theory for calculating it from first principles. An atomistic view will allow us to make some theoretical estimates of its value. Therefore, at times we will deviate from the classical model and adopt an atomistic view that will improve our understanding of the subject.

Classical thermodynamics is also an equilibrium science. The laws of thermodynamics apply to objects or systems in equilibrium with themselves and their surroundings. By definition, a system in equilibrium is not likely to change. However, we are generally interested in how systems change as thermal energy is converted to and from other forms of energy. This presents a bit of a dilemma in that the fundamental laws are only good for a system in equilibrium and the parameters we want to predict are a result of thermal energy changes in the system. To get around this dilemma, we define what is called a quasi-equilibrium process. A quasi-equilibrium process is one that moves from one system state to another so slowly and so incrementally that it looks like a series of equilibrium states. This is a concept that classical thermodynamics had a great deal of difficulty clarifying and quantifying. Basically, a process was a quasi-equilibrium process if the laws of equilibrium thermodynamics could characterize it. This is sort of a circular definition, but once again, we will find that the atomistic view allows us to make some predictions and quantifications that identify a quasi-equilibrium process. Quasi-equilibrium processes can occur very rapidly on time scales typical of human observation. For example, the expansion of the hot gases out the nozzle of a rocket engine can be well described as a quasi-equilibrium process with classical thermodynamics.

1.5 Open and Closed Systems

In the transfer and conversion of thermal energy, we will be interested in separating the *entire universe* into a *system* and its *environment*. We will mainly be interested in the energy transfers and conversions that go on within the *system*, but in many cases we will need to consider its interactions with the rest of the world or its *environment*. Systems that consist of a *fixed amount of mass* that is contained within fixed boundaries are called *closed systems*. Systems that *pass the mass back and forth* to the environment will be called *open systems*. Both *open* and *closed systems* allow energy to flow across their borders, but the flow of mass determines whether they are *open* or *closed systems*. *Open systems* will also carry energy across their borders with the mass as it moves. Consider the simple compressed gas in the piston as a *closed* system (Fig. 1.2).

In analyzing the closed system, we will be concerned about the changes in the internal energy of the compressed gas as it interacts with its environment and the transfers of mechanical and thermal energies across its boundary.

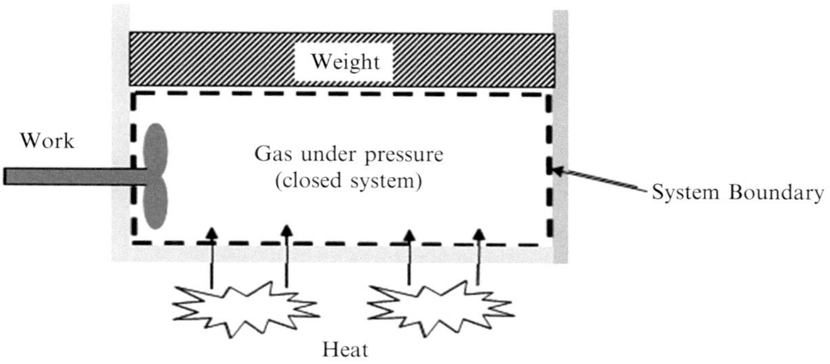

Fig. 1.2 A closed system

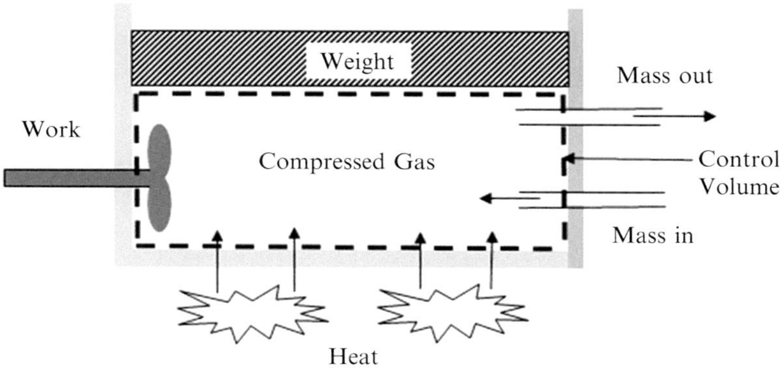

Fig. 1.3 An open system

In analyzing open systems, the concept of a *control volume* comes into play. The *control volume* is the boundary for the open system where the energy changes that we are interested in takes place. The thing separates the open system from its environment. Consider the following open system where we have now allowed mass to flow in and out of the piston of our closed system above (Fig. 1.3).

The *control volume* looks a lot like our system boundary from before, and it is. The only difference is that we now allow mass to flow in and out of our *control volume*. Thermal and mechanical energy can still flow across the boundary, or in and out of the *control volume*. The mass flowing in and out can also carry energy with it either way.

1.6 System Properties

In order to characterize a system we will have to identify its properties. Initially there are three main properties that we will be concerned with—density, pressure, and temperature all of which are *intensive* variables. We will use intensive properties to characterize the equilibrium states of a system. Systems will be composed of *pure substances* and *mixtures of pure substances*. A *pure substance* is a material that consists of only one type of atom, or one type of molecule. A *pure substance* can exist in multiple phases. Normally the phases of concern will be gas, liquid, and solid, though for many pure substances there can be several solid phases. Water is an example of a pure substance that can readily be observed in any of its three phases.

A solid phase is typically characterized as having a fixed volume and fixed shape. A solid is rigid and incompressible. A liquid has a fixed volume but no fixed shape. It deforms to fit the shape of the container that is in it. It is not rigid but is still relatively incompressible. A gas has no fixed shape and no fixed volume. It expands to fit the container that is in it. To characterize a system composed of one or more pure components and one or more phases, we will need to specify the correct number of intensive variables required to define a state. Gibbs Phase Rule named after J. Willard Gibbs who first derived it gives the correct number of intensive variables required to completely define an equilibrium state in a mixture of pure substances. It is

$$V = C - P + 2 \tag{1.3}$$

$V =$ number of variables required to define an equilibrium state.
$C =$ the number of pure components (substances) present.
$P =$ the number of phases present.

So for pure steam at Sea Level and above 100 °C, we have one component and one phase so the number of variables required to specify an equilibrium state is 2, typically temperature and pressure. However, temperature and density would also work. If we have a mixture of steam and liquid water in the system, we have one component and two phases, so only one variable is required to specify the state, either pressure or temperature would work. If we have a mixture like air that is composed of oxygen, nitrogen, and argon, we have three components and three phases (the gas phase for each component), we are back to requiring two variables. As we progress, we will introduce additional intensive variables that can be used to characterize the equilibrium states of a system in addition to density, pressure, and temperature.

1.6.1 Density

Density is defined as the mass per unit volume. The standard SI unit is kilograms per cubic meter (kg/m^3). The Standard English unit is pounds mass per cubic foot (lbm/ft^3). If the mass per unit volume is not constant in a system, it can be defined at a point by a suitable limiting process that converges for engineering purposes long before we get to the atomistic level. The inverse of density is specific volume. Specific volume is an intensive variable, whereas volume is an extensive variable. The standard unit for specific volume in the SI system is cubic meters per kilogram (m^3/kg). The standard unit in the English system is cubic feet per pound mass (ft^3/lbm).

1.6.2 *Pressure*

Pressure is defined as force per unit area. The standard unit for pressure in the SI system is the Newton per square meter or Pascal (Pa). This unit is fairly small for most engineering problems so pressures are more commonly expressed in kilo-Pascals (kPa) or mega-Pascals (MPa). The standard unit in the English system really does not exist. The most common unit is pounds force per square inch (psi). However, many other units exist and the appropriate conversion factors are provided in the Appendix.

Pressure as an intensive variable is constant in a closed system. It really is only relevant in liquid or gaseous systems. The force per unit area acts equally in all directions and on all surfaces for these phases. It acts normal to all surfaces that contain or exclude the fluid. (The term fluid includes both gases and liquids.) The same pressure is transmitted throughout the entire volume of liquid or gas at equilibrium (Pascal's law). This allows the amplification of force by a hydraulic piston. Consider the system in Fig. 1.4, where the force on the piston at B is greater than the force on the piston at A because the pressure on both is the same and the area of piston B is much larger.

In a gravity field, the pressure in a gas or liquid increases with the height of a column of the fluid. For instance, in a tube containing a liquid held vertically, the weight of all of the liquid above a point in the tube is pressing down on the liquid at that point. Consider Fig. 1.5, then:

$$\begin{aligned} \mathrm{d}p &= \rho g \ \mathrm{d}h \\ p(0) &= P(H) + \int_0^H \rho g \ \mathrm{d}h \end{aligned} \tag{1.4}$$

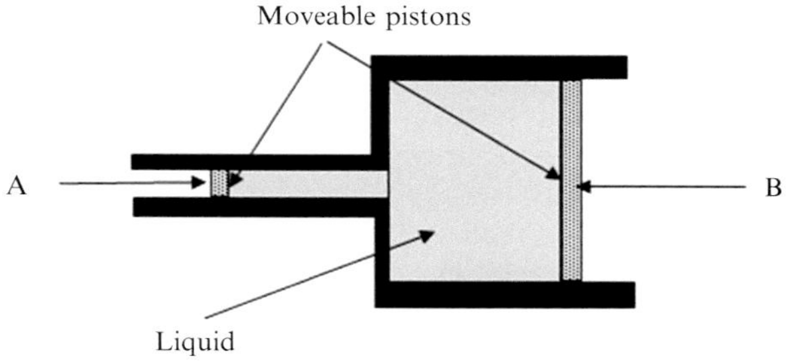

Fig. 1.4 A hydraulic amplifier

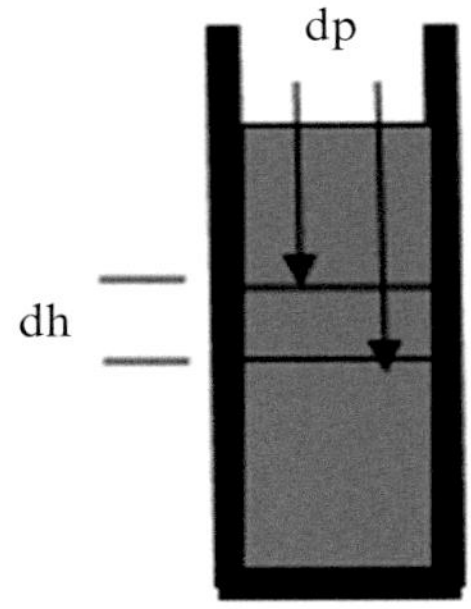

Fig. 1.5 Pressure in a liquid column

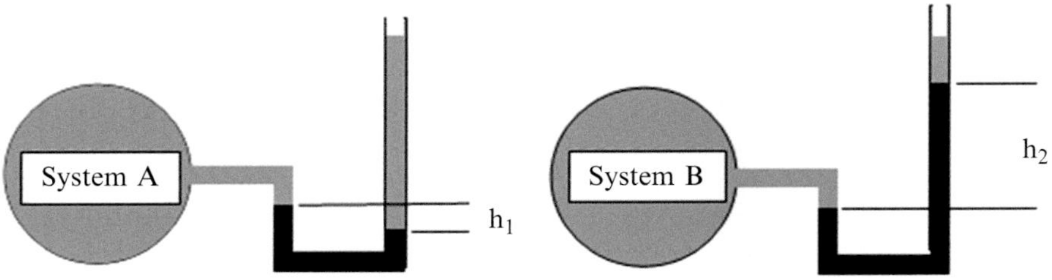

Fig. 1.6 Pressure measurement with manometers

Thus, the pressure at the bottom of the container is equal to the pressure on the top of the fluid in the container plus the integral of the weight of the fluid per unit area in the container.

This raises an interesting concept. Often it will be important to distinguish between *absolute pressure* and *gage pressure*. The preceding equation calculates the *absolute pressure*. The *gage pressure* is simply the pressure exerted by the weight of the column without the external pressure on the top surface of the liquid. It is certainly possible to have a negative gage pressure, but not possible to have a negative absolute pressure. A *vacuum pressure* occurs when the absolute pressure in a system is less than the pressure in the environment surrounding the system.

Using the setup in Fig. 1.6, a very common way of measuring pressure is an instrument called a manometer. A manometer works by measuring the difference in the height of a fluid in contact with two different pressures. A manometer can measure absolute pressure by filling a closed end tube with the liquid and then inverting it into a reservoir of liquid that is open to the pressure that is to be measured. Manometers can also measure a vacuum gage pressure. Consider Fig. 1.6.

The tall tubes on the right in each system are open to the atmosphere. System A is operating at a small negative pressure, or vacuum, relative to the atmosphere. System B is operating at a positive pressure relative to the atmosphere. The magnitude of the pressure in each case can be calculated by measuring the height difference between the fluids on the two sides of the U-tube and calculating its weight per unit area. This is the difference in the pressures inside Systems A or B and the atmospheric pressure pushing down on the open columns on the right.

1.6.3 Temperature

The other intensive variable to be considered at this point is the temperature. Almost everyone is familiar with temperature as a measure of coldness or hotness of a substance. As we continue our study of thermodynamics, we will greatly refine our concept of temperature but for now it is useful to discuss how a temperature scale is constructed. Traditionally the Fahrenheit scale was established by defining the freezing point of water at Sea Level pressure to be 32 °F and the boiling point of water to be 212 °F under the same conditions. A thermometer containing a fluid that expands readily as a function of temperature could be placed in contact with a system that contained ice and water vapor saturated air. The height of the fluid in the thermometer would be recorded as the 32 °F height. Then the same thermometer would be placed in a water container that was boiling and the height of the fluid in the thermometer marked as the 212 °F point. The difference in height between the two points would then be marked off in 180 divisions with each division representing 1 °F. The Celsius scale was defined in the same way by setting the freezing point of water at 0 °C and the boiling point at 100 °C. Water was chosen as the reference material because it was always available in most laboratories around the world.

When it became apparent that absolute temperatures were possibly more important than simply temperatures in the normal range of human experience, absolute temperature scales were defined. The freezing point of water was defined as 273.15 K and the boiling point was defined as 373.15 K, to match up with the Celsius scale. Note that the unit on the absolute scale is Kelvin, not degrees Kelvin. It was named in honor of Lord Kelvin who had a great deal to do with the development of temperature measurement and thermodynamics. The freezing point of water was further defined as the equilibrium of pure ice and air saturated water. However, it was difficult to attain this point because as ice melts it forms a layer of pure water around itself, which prevents direct contact of pure ice, and air-saturated water. Therefore, in 1954, the two-point method was abandoned and the triple point of water was chosen as a single standard. The triple point of water is 273.16 K, 0.01 K above the ice point for water at Sea Level pressure. A single point can be used to define the temperature scale if temperatures are measured with a constant volume, ideal gas thermometer. Basically, the ideal gas thermometer can measure the pressure exerted by a constant volume of gas in contact with the system to be measured. It can also measure the pressure exerted by the gas when in contact with a system at the triple point of water. The ratio of the two pressures gives the ratio of the measured absolute temperature to the absolute temperature of the triple point of water.

However, additional secondary standards are defined to simplify calibration over a broad range of temperatures. The International Practical Temperature Scale is defined as follows:

Triple point of equilibrium hydrogen	13.81 K
Boiling point of hydrogen at 33.33 kPa	17.042 K
Boiling point of hydrogen at 1 atm	20.28 K
Boiling point of neon	27.102 K
Triple point of oxygen	54.361 K
Boiling point of oxygen	90.188 K
Triple point of water	273.16 K
Boiling point of water	373.15 K
Freezing point of zinc	692.73 K
Freezing point of silver	1235.08 K
Freezing point of gold	1337.58 K

Once the absolute temperature scale in Kelvins was defined it became part of the SI system. An absolute scale matching the Fahrenheit scale between the freezing point of water and its boiling point has been defined for the English system. Since there are 180° between the freezing and boiling points in the Fahrenheit scale and 100° over the same range in the Kelvin scale, the absolute scale for the English system, where the unit of measurement is called a degree Rankine, is simply 1.8 times the number of Kelvins. So the freezing point of water on the Rankine scale is 491.67 °R and the boiling point is 671.67 °R. Absolute zero on the Rankine scale is −459.67 °F. To convert back and forth the following formulas apply:

$$\begin{aligned} T_{\mathrm{K}} &= T_{\mathrm{C}} + 273 \\ T_{\mathrm{C}} &= T_{\mathrm{K}} - 273 \\ T_{\mathrm{R}} &= T_{\mathrm{F}} + 460 \\ T_{\mathrm{F}} &= T_{\mathrm{R}} - 460 \end{aligned} \tag{1.5}$$

$$\begin{aligned} T_{\mathrm{R}} &= 1.8T_{\mathrm{K}} \\ T_{\mathrm{K}} &= \frac{5}{9}T_{\mathrm{R}} \\ T_{\mathrm{F}} &= 1.8T_{\mathrm{C}} + 32 \\ T_{\mathrm{C}} &= \frac{5}{9}(T_{\mathrm{F}} - 32) \end{aligned} \tag{1.6}$$

1.7 Properties of the Atmosphere

Before going further, it will be useful to have a model for the atmosphere that can be used for calculations. It is important to realize that the atmosphere at Sea Level supports a column of air that extends upward of 50 miles. Given the equation derived earlier for the pressure in a column of fluid, we have as always to begin at Sea Level.

$$\begin{aligned} &dp = -\rho g \ dh \\ &\text{Let} \rho = p/RT \\ &\text{Then} \\ &dp = -p\frac{g}{RT} \ dh \end{aligned} \tag{1.7a}$$

On integrating the last term of Eq. (1.7a), we obtain

$$p = p_{SL} e^{-\frac{g}{RT}h} \tag{1.7b}$$

To perform the integration, the above temperature has been assumed constant. This is not quite true as the standard lapse rate for the Troposphere up to about 40,000 ft is approximately 2 °C per 1000 ft or 3.6 °F per 1000 ft. This means that the air is denser than the exponential model predicts. However, it is approximately correct for the Troposphere particularly if only a limited range of elevations is considered and the average temperature is used. The initial values at Sea Level for the standard atmosphere are as follows:

Pressure	14.696 psi	101.325 kPa
Temperature	59 °F (519 °R)	15 °C (288 K)
Density	076,474 lbm/ft^3	1.225 kg/m^3

Composition	Mole fraction (%)
Nitrogen	78.08
Oxygen	20.95
Argon	0.93
Carbon Dioxide	0.03
Ne, He, CH_4 et al.	0.01

A more extensive model of the atmosphere as a function of altitude is provided in the Appendix. The relative composition is essentially constant up to the top of the Troposphere.

1.8 The Laws of Thermodynamics

It is useful at this time to state the Laws of Thermodynamics. Later chapters will elaborate on them, but realizing that there are four simple laws that all of the analysis is built around will provide some structure to guide the way forward.

Zeroth Law of Thermodynamics
Two bodies in thermal contact with a third body will be at the same temperature.

This provides a definition and method of defining temperatures, perhaps the most important intensive property of a system when dealing with thermal energy conversion problems.

First Law of Thermodynamics
Energy is always conserved when it is transformed from one form to another.

This is the most important law for analysis of most systems and the one that quantifies how thermal energy is transformed to other forms of energy.

Second Law of Thermodynamics
It is impossible to construct a device that operates on a cycle and whose sole effect is the transfer of heat from a cooler body to a hotter body.

Basically, this law states that it is impossible for heat to spontaneously flow from a cold body to a hot body. If heat could spontaneously flow from a cold body to a hot body, we could still conserve energy, so the First Law would hold. But every experiment that has ever been performed indicates that thermal energy always flows the other way. This law seems obvious enough but the implications are very significant, as we will see.

Third Law of Thermodynamics
It is impossible by means of any process, no matter how idealized, to reduce the temperature of a system to absolute zero in a finite number of steps.

This allows us to define a zero point for the thermal energy of a body be taken under consideration and subject of this matter is beyond the scope of this book.

References

1. Zohuri, B., & McDaniel, P. (2015). *Thermodynamics in nuclear power plant*. New York: Springer.
2. Cengel, Y. A., & Boles, M. A. (2008). *Thermodynamics: An engineering approach* (6th ed.). Boston: McGraw Hill.
3. Elliott, J. R., & Lira, C. T. (1999). *Introductory chemical engineering thermodynamics*. Upper Saddle River, NJ: Prentice Hall.
4. Hseih, J. S. (1975). *Principles of thermodynamics*. New York: McGraw-Hill.
5. Moran, M. J., & Shapiro, H. N. (2008). *Fundamentals of engineering thermodynamics* (6th ed.). New York: John Wiley & Sons.
6. Wylen, G. J., & Sonntag, R. E. (1978). *Fundamentals of classical thermodynamics, SI version 2e*. New York: John Wiley & Sons.

Chapter 2
Electricity, an Essential Necessity in Our Life

Early man relied on fire for the luxuries of light, heat, and cooking. Today, we take all these luxuries for granted. At the flick of a switch, a push of a button, or the turn of a knob, we can have instant power. Electricity plays a huge part in our everyday lives. Whether it is at home, school, the local shopping center, or our workplace, our daily routines rely heavily on the use of electricity. From the time we wake up in the morning until we hit the pillow at night, our daily life is dependent on electricity. The alarm we have to turn off each morning runs on electricity. The light in our bedroom, the hot shower we take before breakfast, Dad's electric razor. All these things need electricity in order to function. Even our first meal of the day is heavily dependent on electricity. The fridge that keeps all our food cool and fresh needs electricity to run or the grill that cooks your bacon and eggs also needs power to operate. This power generally (unless you have gas stove) comes from electricity. Electricity not only plays a big part in our daily lives at home, but it is extremely important for all the things that go on in the world around us in our modern life, such as industry that we depend on, communication as in the form of radio, television, e-mail, Internet, etc. Transport is another aspect of our daily life that depends on electricity to some degree.

2.1 Cost of Generating Electricity Today

One of the first questions that come to our mind about the necessity of electricity for our day-to-day life is this:

How much does it cost to generate electricity with different types of power plants?

The Annual Energy Outlook 2014 (AEO2014), prepared by the US Energy Information Administration (EIA), presents long-term annual projections of energy supply, demand, and prices focused on the USA through 2040, based on results

B. Zohuri, *Application of Compact Heat Exchangers For Combined Cycle Driven Efficiency In Next Generation Nuclear Power Plants*,
DOI 10.1007/978-3-319-23537-0_2

from EIA's National Energy Modeling System (NEMS). NEMS enables EIA to make projections under alternative, internally consistent sets of assumptions, the results of which are presented as cases. The analysis in AEO2014 focuses on five primary cases: a Reference case, Low and High Economic Growth cases, and Low and High Oil Price cases. Results from a number of other alternative cases also are presented, illustrating uncertainties associated with the Reference case projections. EIA published an Early Release version of the AEO2014 Reference case in December 2013. The projections in the US EIA's Annual Energy Outlook 2014 (AEO2014) focus on the factors that shape the US energy system over the long term.

EIA has historical data on the average annual operation, maintenance, and fuel costs for existing power plants by major fuel or energy source types in Table 2.1. Average Power Plant Operating Expenses for Major US Investor-Owned Electric Utilities, 2001–2012 (Mills per kilowatt-hour)[1] of the Electric Power Annual.

There are about 19,023 individual generators at about 6997 operational power plants in the USA with a nameplate generation capacity of at least 1 MW. A power plant can have one or more generators, and some generators may use more than one type of fuel.

There are currently 61 commercially operating nuclear power plants with 99 nuclear reactors in 30 states in the USA. Thirty-five of these plants have two or more reactors. The Palo Verde plant in Arizona has three reactors and had the largest combined net summer generating capacity of 3937 MW in 2012. Fort Calhoun in Nebraska with a single reactor had the smallest net summer capacity at 479 MW in 2012.

Four reactors were taken out of service in 2013: the Crystal River plant in Florida with one reactor in February; the Kewaunee plant in Wisconsin with one reactor in April; and the San Onofre plant in California with two reactors in June. The Vermont Yankee plant in Vermont, with a single reactor, was taken out of service in December 2014.

The role electricity plays in our lives by enhancing our productivity, comfort, safety, health, and economy is obvious. We live with the benefits of electricity every day. So much so that we take it for granted that whenever we plug our gadgets into the wall socket, the power will be there. While most people give little thought to where electricity comes from, there are many different ways to generate electricity—including coal, oil, gas, hydroelectric, nuclear, and solar. Each option inherits certain advantages that merit consideration whenever there is a need for a

[1] A mill is equal to 1/1000 of a US dollar, or 1/10 of 1 cent. Mills per kilo-watt-hour (kWh) equals dollars per mega-watt-hour (MWh). To convert mills per kWh to cents per kWh, divide mills per kWh by 10

1 mill/kWh = 0.1 cent/kWh

1 mill = 0.1 cents = 0.001 dollars

1 MW = 1000 kW

1 mill/kWh = 1 dollar/MWh.

Table 2.1 Average power plant operating expenses for major US investor-owned electric utilities, 2003–2013 (Mills per kilowatt-hour)

	Operation				Maintenance			
Year	Nuclear	Fossil steam	Hydroelectric	Gas Turbine and Small Scale	Nuclear	Fossil steam	Hydroelectric	Gas Turbine and Small Scale
2003	9.12	2.74	3.47	3.50	5.23	2.72	2.32	2.26
2004	8.97	3.13	3.88	4.27	5.38	2.96	2.76	2.14
2005	8.26	3.21	3.95	3.69	5.27	2.98	2.73	1.89
2006	9.03	3.57	3.76	3.51	5.69	3.19	2.70	2.16
2007	9.54	3.63	5.44	3.26	5.79	3.37	3.87	2.42
2008	9.89	3.72	5.78	3.77	6.20	3.59	3.89	2.72
2009	10.00	4.23	4.88	3.05	6.34	3.96	3.50	2.58
2010	10.50	4.04	5.33	2.79	6.80	3.99	3.81	2.73
2011	10.89	4.02	5.13	2.81	6.80	3.99	3.74	2.93
2012	12.49	4.38	6.71	2.46	7.32	4.48	4.63	2.75

	Fuel				Total			
Year	Nuclear	Fossil steam	Hydroelectric	Gas Turbine and Small Scale	Nuclear	Fossil steam	Hydroelectric	Gas Turbine and Small Scale
2003	4.60	17.29	–	43.89	18.95	22.75	5.79	49.66
2004	4.58	18.21	–	45.18	18.93	24.31	6.60	51.59
2005	4.63	21.69	–	55.52	18.15	27.88	6.88	61.10
2006	4.85	23.09	–	53.89	19.57	29.85	6.46	59.56
2007	4.99	23.88	–	58.75	20.32	30.88	9.32	64.43
2008	5.29	28.43	–	64.23	21.37	35.75	9.67	70.72
2009	5.35	32.30	–	51.93	21.69	40.48	8.38	57.55
2010	6.68	27.73	–	43.21	23.98	35.76	9.15	48.74
2011	7.01	27.08	–	38.80	24.70	35.09	8.88	44.54
2012	7.61	28.34	–	30.45	27.42	37.20	11.34	35.67

Hydroelectric category consists of both conventional hydroelectric and pumped storage
Gas Turbine and Small Scale category consists of gas turbine, internal combustion, photovoltaic, and wind plants
Notes: Expenses are average expenses weighted by net generation. A mill is a monetary cost and billing unit equal to 1/1000 of the US dollar (equivalent to 1/10 of 1 cent)
Total may not equal sum of components due to independent rounding
Sources: Federal Energy Regulatory Commission, FERC Form 1, "Annual Report of Major Electric Utilities, Licensees and Others via Ventyx Global Energy Velocity Suite"

new power plant. Nuclear generated electricity is unique in that it inherently addresses many of the shortcomings of the other means for power generation. The use of nuclear power provides answers for many problems in the areas of environment, safety, economics, reliability, sustainability, and even waste.

2.2 Nuclear Power Plants

Right now, nuclear energy provides about 20 % of the US electricity, a little bit less of the world's electricity. That works out to about 7 % of total energy we consume. There is a lot of opportunity for total energy fraction to go up, because nuclear energy can be used to produce transportation fuels. We can use it to produce hydrogen. We can use the heat to help with biofuel processing.

Nuclear generated electricity is not just produced in the USA. Most developed countries worldwide have nuclear power plants generating electricity for their citizens. Furthermore, nuclear power generation continues to grow annually. With concerns over the environmental effects of global warming and pollution from gases emitted from coal-fired plants, the demand for nuclear power is projected to continue to increase a great deal in the next decades.

Currently, 30 countries worldwide are operating 437 nuclear reactors for electricity generation and 67 new nuclear plants are under construction in 14 countries. Included in this number are 100 plants operating in 31 states.

While the USA can boast about having the most nuclear power plants, electrical power from these plants provides less than 20 % of all power supplied in the USA. Other countries are much more dependent on nuclear than the USA. The next figure ranks the per capita supply of nuclear power for the top 10 nuclear power generating countries. Currently, nuclear energy represents about 77 % of total electricity production in France, 54 % in Slovakia, 54 % in Belgium, 47 % in Ukraine, 43 % in Hungary, 42 % in Slovenia, 40 % in Switzerland, 40 % in Sweden, 35 % Korea Republic, and 33 % in Armenia.

2.3 Cost of Electricity from New Nuclear Power Plants Stations

Current discussions about possibilities to mitigate the effects of global warming have also opened discussions about a potential revival of nuclear power. In this context, it is often argued with very low cost of electricity from nuclear power plants. This seems to be one of the strongest arguments in favor of atomic energy. To determine the future cost of electricity from nuclear power, the cost from currently operating power stations is taken into account. However this is not correct.

In the abovementioned discussions about building new nuclear power stations, the cost for electricity from new and not from already existing nuclear power stations should be taken into account. This makes a huge difference as we will see further below. As a matter of fact, it is nearly impossible to estimate the cost of building new nuclear power stations. This is mainly a consequence of missing national and international safety standards. It is not clear which safety measures will have to be

Fig. 2.1 A typical structure view of nuclear power plant

applied and as a consequence the investment costs can barely be estimated. Figure 2.1 shows the structure of a typical nuclear power plant from outside.

Outside of USA, Finland is the only country in Europe, where a nuclear power plant is currently being built. In this situation, the best possible practice is to use the costs for the plant in Finland for cost comparisons with other technologies.

2.3.1 Pros and Cons of New Nuclear Power Plants

As a result of the current discussion how further global warming could be prevented or at least mitigated, the revival of nuclear power seems to be in everybody's—or at least in many politicians'—mind. It is interesting to see that in many suggestions to mitigate global warming, the focus is put on the advantages of nuclear power generation, and its disadvantages are rarely mentioned. With new generation of nuclear power plants known as GEN-IV, any disadvantages are playing very low key anyway. Bear in your mind that there is no perfect energy source. Each and every one has its own advantages and compromises.

Environmentally, nuclear power is once again considered a prominent alternative, despite the disregard it was met with in the 1970s. This is because it is now being touted as a more environmentally beneficial solution since it emits far fewer greenhouse gases during electricity generation than coal or other traditional power plants.

The environmental impact of any power generation station can be measured by quantifying the burden of fuel delivery, emissions of by-products and wastes, and the potential impact on the lives (human or otherwise) of those living nearby.

It is widely accepted as a somewhat dangerous, potentially problematic, but manageable source of generating electricity. Radiation is not easily dealt with, especially in nuclear waste and maintenance materials, and expensive solutions are needed to contain, control, and shield both people and the environment from its harm.

In contrast to fossil fuel plants (coal, oil, and gas), nuclear power plants do not produce any carbon dioxide or sulfur emissions, which are major contributors to the greenhouse effect and acid rain, respectively. According to the Nuclear Energy Institute, US nuclear power plants prevent 5.1 million tons of sulfur dioxide, 2.4 million tons of nitrogen oxide, and 164 million metric tons of carbon from entering the earth's atmosphere each year.[2]

Nuclear power reactors do contribute a measurable increase in radiation to the environment around a nuclear power plant. However, this increase is relatively small compared to natural background radiation and is less than the radioactivity released from a typical coal plant. Even with this increase in radiation, most employees of nuclear power plants receive exposures typically of workers in all occupations. In addition, no evidence exists that shows that small increases in radiation exposure have negative health effects.

Because nuclear power plants are relatively self-sufficient, plant sitting is more amenable to environmental concerns. Oil and gas plants must be sited close to major pipelines and hydroelectric plants must be sited on rivers. Impact to wildlife habitat and municipalities may have to be compromised in order to site these types of plants. For example, public outrage over fish kills from hydroelectric power plants have limited new plant construction and policy makers are seriously considering the removal of existing dams around the USA. Meanwhile, strict sitting regulation ensures that nuclear power plants have minimal impact to their surrounding areas.

The most pressing environmental concern facing the nuclear industry is the issue of waste disposal. All processes produce waste. Nuclear waste from a power plant is unique in that it can be highly radioactive. While highly radioactive waste is hazardous to all living beings, nuclear fuel is amenable to containment, treatment, reduction, and reprocessing (recycling). Processes have been developed to separate reusable fuel and the highly radioactive elements from used nuclear fuel. The waste products can then be made into a glass or ceramic waste pellet for disposal.

---

[2] http://www.nuclearconnect.org/know-nuclear/applications/electricity.

The hazard associated with this pellet has an expected duration of about 100 years. Considering that chemical hazards maintain their nature indefinitely, this waste form may be preferable. Currently, such a waste treatment process is not being utilized in the USA because of political resistance; however, research continues to find new solutions to this problem.

The dialogue about using nuclear power—and expanding it—centers on weighing these risks against the rewards, as well as the risks inherent in other forms of power generation. These are just some of the issues involved.[3]

1. PROS

- Lower carbon dioxide and other greenhouse released into the atmosphere in power generation.
- Low operating costs (relatively).
- Nuclear power generation does emit relatively low amounts of carbon dioxide (CO_2). The emissions of greenhouse gases and therefore the contribution of nuclear power plants to global warming is therefore relatively little.
- Known, developed technology "ready" for market, in particular, new generation of power plant (GEN-IV) and related research led by industries and universities on combined cycle such as Brayton that is a promising effort to drive the output efficiency of such reactors upward [1–8].
- This technology is readily available; it does not have to be developed first.
- It is possible to generate a high amount of electrical energy in one single plant.
- Large power-generating capacity able to meet industrial and city needs (as opposed to low-power technologies like solar that might meet only local, residential, or office needs but cannot generate power for heavy manufacturing).
- Existing and future nuclear waste can be reduced through waste recycling and reprocessing, similar to Japan and the EU (at added cost).

2. CONS

- High construction costs due to complex radiation containment systems and procedures.
- High subsidies needed for construction and operation, as well as loan guarantees.
- Subsidies and investment could be spent on other solutions (such as renewable energy systems).
- High-known risks in an accident.
- Unknown risks.
- Long construction time.
- Target for terrorism (as are all centralized power generation sources).

[3] http://www.triplepundit.com/special/nuclear-energy-pros-and-cons/.

- Waivers are required to limit liability of companies in the event of an accident. (This means that either no one will be responsible for physical, environmental, or health damages in the case of an accident or leakage over time from waste storage, or that the government will ultimately have to cover the cost of any damages.) Nuclear power plants as well as nuclear waste could be preferred targets for terrorist attacks. No atomic energy plant in the world could withstand an attack similar to 9/11 in New York. Such a terrorist act would have catastrophic effects for the whole world.
- Nuclear is a centralized power source requiring large infrastructure, investment, and coordination where decentralized sources (including solar and wind) can be more efficient, less costly, and more resilient. The time frame needed for formalities, planning and building of a new nuclear power generation plant is in the range of 20–30 years in the western democracies. In other words: It is an illusion to build new nuclear power plants in a short time.
- Uranium sources are just as finite as other fuel sources, such as coal, natural gas, etc., and are expensive to mine, refine, and transport, and produce considerable environmental waste (including greenhouse gases) during all of these processes.
- The majority of known uranium around the world lies under land controlled by tribes or indigenous peoples who do not support it being mined from the earth. The energy source for nuclear energy is Uranium. Uranium is a scarce resource; its supply is estimated to last only for the next 30–60 years depending on the actual demand.
- Shipping nuclear waste internationally poses an increased potential threat to interception to terrorism (though this has not happened yet with any of the waste shipped by other countries). Increasing the amount of waste shipped, particularly in less secure countries, is seen as a significant increase in risk to nuclear terrorism. The problem of radioactive waste is still an unsolved one. The waste from nuclear energy is extremely dangerous and it has to be carefully looked after for several thousand years (10,000 years according to US Environmental Protection Agency standards).

There is no doubt that the devastating earthquake, tsunami, and consequent multi-reactor damage in Japan will have a significant impact on the future use of nuclear energy, the nuclear industry, and the global nuclear order. The full impact will not be known for some time. Data about the incident unfolding at the Fukushima Daiichi nuclear power reactors were still being compiled and nobody for sure has any idea about the extended damage of this event.

To make wise choices about the future of nuclear power, we need improved knowledge of the safety, safeguards, and security features of both existing and new nuclear energy plants. Understanding the potential advantages and disadvantages of nuclear energy is critical for those stakeholders and decision-makers facing national energy challenges.

2.4 Is Nuclear Power a Global Warming Solution?

It is only possible to mitigate global warming if the worldwide consumption of fossil fuels can be drastically reduced in the next 10–15 years. There is simply no room for a scenario as it is predicted by the International Energy Agency (IEA).

It is also obvious that no combination of alternative technologies can replace the current usage of fossil fuels. There is simply not enough non-fossil fuel available for this. In order to mitigate global warming, we have to use the available energy much more efficiently. But this will not be enough either: We will have to change our behavior to reduce our personal energy consumption. We must change our current lifestyle and seriously strive for a sustainable living.

To answer the question of

Is Nuclear Power a Global warming Solution?

The IEA says that[4]:

The contribution of nuclear energy to reduce the cause of global warming is only 10 %

The IEA predicts a strong increase of the carbon dioxide emissions by the year 2030. Additionally, IEA investigated to which extent the abovementioned emissions of CO_2 could be prevented if politics applied rigorous measures.

From all measures investigated, nuclear energy was found to have the least effect (only 10 %). Almost 80 % of the desired effects are due to increasing the energy efficiency. This result is surprising, in particular if you think about how nuclear power is praised as solution to global warming by politicians like George W. Bush and Tony Blair. It seems like they would (again) head into the wrong direction. See Fig. 2.2.

> Instead of talking about measures to increase the energy efficiency, which accounts for 80 % of the effects, some politicians propagandize building nuclear power plants, which according to IEA can only account for 10 % of the desired effects. Here the focus is clearly on the wrong subject![5]

So, then the question comes to that: Do we have a serious problem in hand?

Thus as it was said earlier

Do we have to change our behavior to reduce our personal energy consumption?

We must change our current lifestyle and seriously strive for a sustainable living.

As part of an answer to the above question and statement, the following sections are presented for further argument on the matter of electricity generation cost and consumption worldwide.

[4] http://timeforchange.org/nuclear_power_articles?filter0%5B%5D=60.

[5] http://timeforchange.org/nuclear_power_articles?filter0%5B%5D=60.

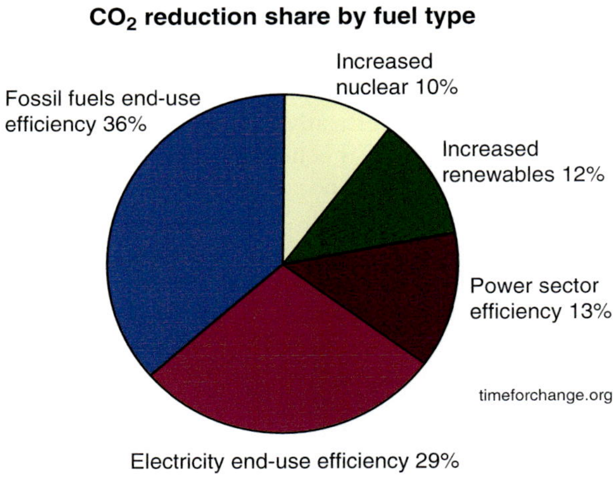

Fig. 2.2 CO_2 reduction share by fuel type

2.5 Prediction of Energy Consumption Worldwide

The other question is this:

How much energy will we consume in the future?

According to the American Energy Information Administration (EIA) and to the International Energy Agency (IEA), the worldwide energy consumption will on average continue to increase by 2 % per year. The graph below shows the actual values starting from 1980 until today in blue and the predictions of the energy consumption until the year 2030 in orange. See Fig. 2.3a and b.

More detailed data (in quadrillion BTU) of the actual and predicted energy consumption worldwide by geographic area is given by Table 2.2 in the following page.

The highest annual growth of energy consumption is predicted for Asia (3.7 %), non-OECD countries (3 %), and Central and South America (2.8 %). The lowest annual growth of energy consumption is predicted for Europe with 1 %.[6]

A yearly increase by 2 % leads to a doubling of the energy consumption every 35 years. This means the worldwide energy consumption is predicted to be twice as high in the year 2040 compared to today (2007).

The future of civilization will depend upon the indefinite supply of electricity. Clearly, there is a limit to the supply of fossil fuels. The most optimistic estimates have fossil fuel lasting no more than 100 years; however, they may become economically undesirable in much less time. Obviously wind, solar, and other renewable energy sources (such as ethanol) can sustain our world with power indefinitely. However, the power generation potential from even a small amount of uranium is so great, even nuclear fuel can be included on this list. In the right

[6] http://timeforchange.org/prediction-of-energy-consumption.

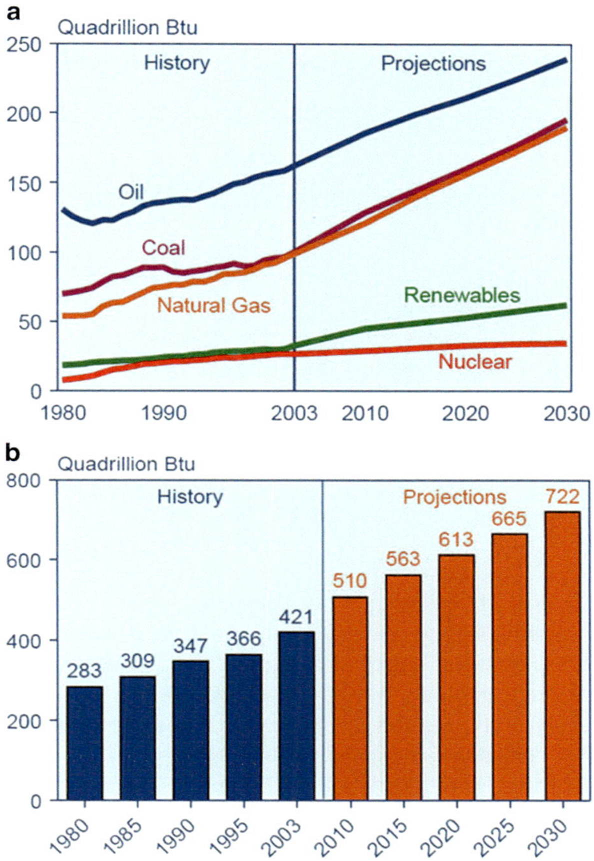

Fig. 2.3 (**a**) World marketed energy use by fuel type, 1980–2030. (**b**) World marketed energy consumption 1980–2030. *Source*: *History*: Energy Information Administration (EIA), International Energy Annual 2003 (May–July 2005), website www.eia.doe.gov/iea. *Projection*: EIA, System for the Analysis of Global Energy Market (2006)

configuration, nuclear power can provide electricity for generations. The right configuration is in the "Breeder Reactor." The design of the breeder reactor is such that even as fuel is consumed, new fuel is created as a by-product. Only a few breeder reactor plants have been built. Since plutonium—a material used in nuclear weapons—is created in these plants, governments have been hesitant to allow their construction. Nonetheless, applying the breeder concept can reduce fuel prices so low that even the extraction of uranium from the world's oceans would not be an overly expensive endeavor. In an article printed in the American Journal of Physics (vol. 51, Jan. 1983, B. Cohen), there is enough uranium in all the world's oceans and the earth's crust under the oceans to last 5 billion years (assuming that 6500 metric tons of uranium is removed annually). For all practical purposes, nuclear is a *reliable* and *sustainable* power source.

Nuclear power plants are one of the most economical forms of energy production. Nuclear fuel costs (as a function of power generation potential) represent only a fraction of the cost of fossil fuels. Including capital and non-fuel operating costs,

Table 2.2 Actual and predicted energy consumption worldwide by geographic area (http://timeforchange.org/prediction-of-energy-consumption)

Region	2003	2010	2015	2020	2025	2030	Average annual percent change, 2003–2030
OECD	234.3	256.1	269.9	281.6	294.5	308.8	1.0
North America	118.3	131.4	139.9	148.4	157.0	166.2	1.3
Europe	78.9	84.4	87.2	88.7	91.3	94.5	0.7
Asia	37.1	40.3	42.8	44.4	46.1	48.0	1.0
Non-OECD	186.4	253.6	293.5	331.5	371.0	412.8	3.0
Europe and Eurasia	48.5	56.5	62.8	68.7	74.0	79.0	1.8
Asia	83.1	126.2	149.4	172.8	197.1	223.6	3.7
Middle East	19.6	25.0	28.2	31.2	34.3	37.7	2.4
Africa	13.3	17.7	20.5	22.3	24.3	26.8	2.6
Central and South America	21.9	28.2	32.5	36.5	41.2	45.7	2.8
Total World	420.7	509.7	563.4	613.0	665.4	721.6	2.0

the cost of operating a nuclear power plant is roughly equivalent to fossil fuels. Recently, the average electricity production cost for nuclear energy was recognized as the cheapest source of electricity. As 2012, the average cost of power generation by nuclear plants was 2.40 cents/kW-h, for coal-fired plants 3.27 cents, for oil 22.48 cents, and for gas 3.40 cents. Costs for solar and wind are still well beyond that considered to be competitive to the public.

The cost of regulation and industry oversight of nuclear power generation is substantially more than that of other power generation sources; however, improvements in reliability and operational and maintenance efficiencies have contributed to reducing those costs. Currently, nuclear power plant capacity factors average over 85 %. This is competitive with those of fossil fired plants (average 50–60 %), or solar and wind which have capacity factors in the 30 % range, or even lower. Most plants are designed to operate in a base load configuration; that is, they run at full power regardless of the demand on electricity. Nuclear power plants are particularly well suited for this purpose since they are designed to produce large quantities of power and can sustain operation for up to 2 years without refueling.

The US nuclear energy plants can supply large amounts of predictable, reliable electricity through virtually every period of extreme heat and cold. During the 2014 Polar Vortex, nuclear energy generation saw no drop in output and on the coldest day operated at 95 % capacity. See Fig. 2.4.

From the fuel equivalency point of view, like fossil fuels, the nuclear fuel raw materials come from the Earth. Uranium, the primary fuel material, is mined. The environmental impact of mining is well known; however, the advantage of nuclear power comes from the amount of power that comes from a small amount of uranium. The power from 1 kg of uranium is approximately equivalent to 42 gal of oil, 1 ton of coal, or 17,000 cubic feet of natural gas. Therefore, as a function of

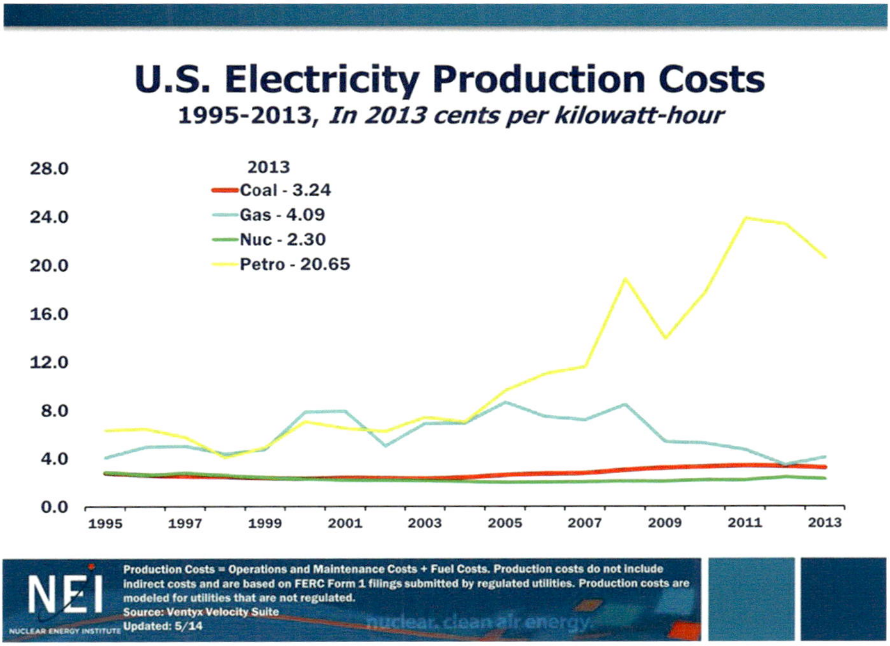

Fig. 2.4 US electricity production costs. http://www.nei.org/Knowledge-Center/Nuclear-Statistics/Costs-Fuel,-Operation,-Waste-Disposal-Life-Cycle/US-Electricity-Production-Costs

power consumption, very little uranium needs to be removed from the ground; hence, the environmental impact of uranium mines is much less compared with mining and drilling for fossil fuels.

Unlike oil or gas, nuclear fuel is solid; hence, nuclear fuel is immune to the environment problems posed by spillage during transportation to a power plant. Unused nuclear fuel is only slightly more radioactive that naturally occurring underground. Fuel delivery casks are designed with a high margin of safety to ensure that even in the event of a transportation accident, the environment remains free of contamination from the nuclear fuel.

2.6 Current Energy Consumption by Capita

The values are indicated as "kg oil equivalents" or kgoe. An example from the graph: People living in North America use per year and per person the energy equivalent to approx. 8000 kg oil, which is about 10,000 l of oil. To convert "kg oil equivalent" into kWh, multiply it with the factor 11.628. Example: 8000 kg oil is about 93,024 kWh (11.628×8000) or 93 MWh.

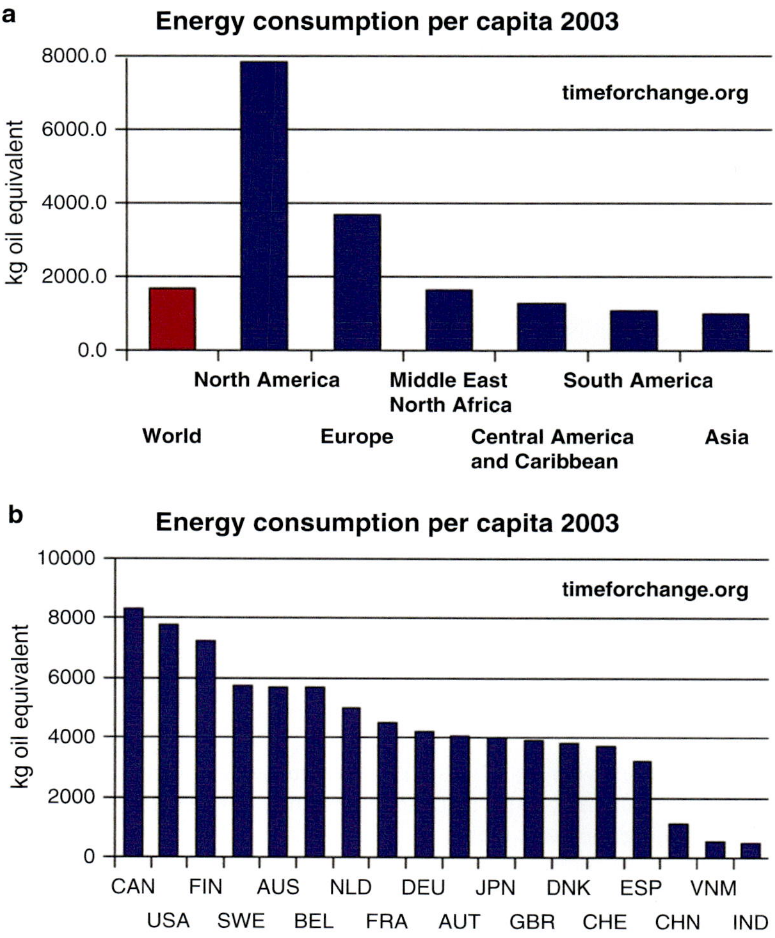

Fig. 2.5 (**a**) Energy consumption per capita 2003. (**b**) Energy consumption per capita 2003. *Note*: Most abbreviations for countries are self explaining, except perhaps NLD = Netherlands, DEU = Germany, AUT = Austria, GBR = UK, CHE = Switzerland, CHN = China, VNM = Vietnam, IND = India

There are huge differences between individual regions of the world. Data has been extracted from Earthtrends.wri.org, an excellent, very flexible source of data. This site is recommended if you want to go into more details. See Fig. 2.5a and b.

Here is an update of the energy consumption per capita and for the predicted energy consumption for the top energy-consuming countries (see Fig. 2.6).

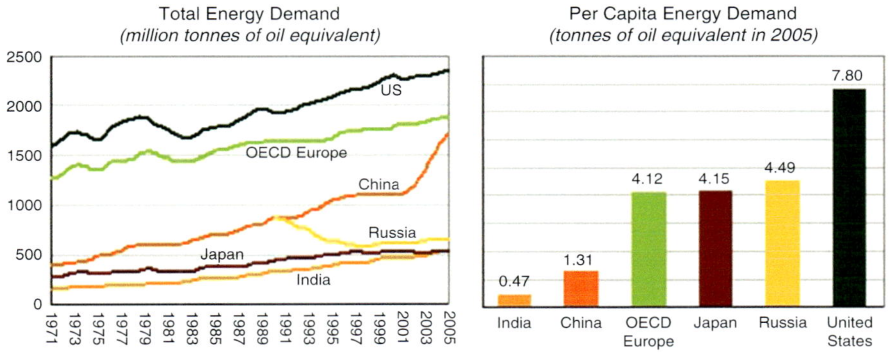

Fig. 2.6 Updated chart of energy consumption per capita

2.7 The Next Nuclear Age: Can Safe Nuclear Power Work for America or the World?

Before for us to be on our way and focus on the subject of this book, "Compact Heat Exchanger Design For Combined Cycle Driven Efficiency In New Generation of Nuclear Power Plants," we need to understand the basic rules and different generation of nuclear power from past to present and go the future aspect this source of generating electricity. This allows us to understand what would be the Total Cost of Ownership (TCO) and Return On Investment (ROI) for the owners of these power plants (i.e., utility companies) and how technology will help them to reduce TCO and increase ROI over the course of life cycle for each type and design of these plants, how safe they are, and which generation works for us better. For that we need to briefly describe each of these generations of nuclear power plants, then we can argue "*Can Safe Nuclear Power Work for America or the World*?"

Later on we can describe more details about each generation in particular GEN-IV which most of the research by this author and rest of the world are based on. Chapter 5 is devoted to such details and before that we have laid down the basic thermodynamic cycle in Chap. 3, in order to have better understanding of combined cycle and how the compact heat exchanger will play a role in order to drive the efficiency of such a system to a higher level in order to reduce the TCO and increase ROI.

Historically, the Department of Energy's (DOE) Advanced Test Reactor (ATR) located at Idaho National Laboratory (INL) through fission chain reaction started to produce 250 MW output energy—enough to power almost 200,000 homes. This reactor which is the epicenter of American nuclear energy research was a step toward commercialization nuclear power plants here in USA and around the world.

Over the past half century, 51 reactors have been built here, including first-generation prototypes of the 1950s; only three still operate. But it is among the relics of these early experiments that the country's energy future is taking shape.[7]

In recent years, the debate over nuclear power has moved to the front and center of quest for an alternative source of energy, spurred by concerns about foreign oil and the specter of global warming and rise of demand on production of more electricity for our day-to-day life, which is getting more and more dependency on it. But what many on both sides of the issue often fail to note is that America's 103 existing nuclear reactors are aging and coming to the end of their life cycle. Over the next few decades, they will have to be decommissioned—taking 20 % of the country's electrical supply with them.

In the Energy Policy Act of 2005, Congress approved up to $2.95 billion in incentives for new nuclear plants, and set aside another $1.25 billion for an experimental reactor to be built in the Idaho desert. The reactor will be the centerpiece of a modern-day of global life on earth with its population growing and its industry expanding, with scientists from around the world working together to revolutionize the production of nuclear power at reasonable cost and safer shape and form.

In principle, for nuclear reactors to produce energy in form of electricity, a chain reaction usually uranium as fuel takes place that turns into heat and fast-moving neutron with the core of reactor. A coolant (i.e., heat exchanger) takes away the heat and uses it to generate superheated steam and in return to spin a turbine to generate electivity, while a moderator within the core slows down the fast-moving neutron to keep the chain reaction under control. Hence, any material used in building a reactor needs to withstand the heat generated by this chain reaction as well as intense buildup of pressure and a constant barrage of neutrons for the reactor's projected life cycle. This is where reactors such ATR comes to play as a research reactor to pave the road for commercialization of nuclear power plant from concept of design to prototype and finally production stage.

The ATR uses 92 % enriched uranium (i.e., anything more than 20 % is considered weapons-grade) to generate a quadrillion neutrons per square centimeter per second—100 to 1000 times greater than commercial reactors. By cranking up the neutron dose, this reactor can simulate as much as 40 years of wear and tear on a new fuel or alloy in a single year.

This reactor is a simple water-cooled model built in 1967. But by tuning the pressure, temperature, and chemistry inside its core, scientists can use it to reproduce the conditions in just about any other type of reactor. Recently, folks in charge of ATR operation have tested chunks of graphite to see whether it is safe to extend the life of Britain's antiquated Magnox reactors. INL staffs are now gearing up for an even bigger challenge and that is testing parts for proposed Generation IV reactors, which would leap technologically two steps ahead of the Gen II designs operating commercially in the USA as of today.

[7] http://www.popularmechanics.com/science/energy/a869/3760347/.

Arguments by pro-nuclear power plants stand behind the cost-effectiveness of GEN-II plants, despite concerns about catastrophic accidents and radioactive waste disposal.

Rise of demand for production of more electricity at reduced total cost of ownership yet increase the return on investment has pushed a dozen utility companies around the country and the world once again to start the lengthy process of applying to Nuclear Regulatory Commission (NRC) for licensing to built new nuclear power plants and renew the operational license for their existing one. If all goes smoothly, they could produce power by the middle of the next decade. These reactors would be Generation III and III+ designs—evolutionary improvements on today's Generation II reactors, which use water in some form as both a coolant and a moderator (i.e., LWR, BWR, or PWR).

But, according to the DOE, what are really needed are even safer, cheaper reactors that produce less waste and use fuel that is not easily adapted for weapons production. To develop this kind of reactor, 10 countries, including the USA, joined forces in 2000 to launch the Generation IV International Forum. A committee of 100-plus scientists from participating countries evaluated more than 100 designs; after 2 years, they picked the six best. All of the final Gen IV concepts make a clean break from past designs. Some do not use a moderator, for instance. Others call for helium or molten lead to be used as coolants.

A top level view of Generation II and III reactors is depicted in Fig. 2.7 below and all 103 operational nuclear power plants today in USA are employing light water reactor (LWR) technology, which uses ordinary water as both a moderator and a coolant. The next wave of nuclear plants has taken these Generation II concepts to the next level, improving both safety and efficiency. Utilities plan to begin building Generation III reactors by the end of the decade.

In a Gen II Pressurized Water Reactor, water circulates through the core (*1*) where it is heated by the fuel's chain reaction. The hot water is then piped to a steam generator, and the steam spins a turbine (*2*) that produces electricity. The Gen III Evolutionary Pressurized Reactor improves upon this design primarily by enhancing safety features. Two separate 51-in.-thick concrete walls (*3*), the inner one lined

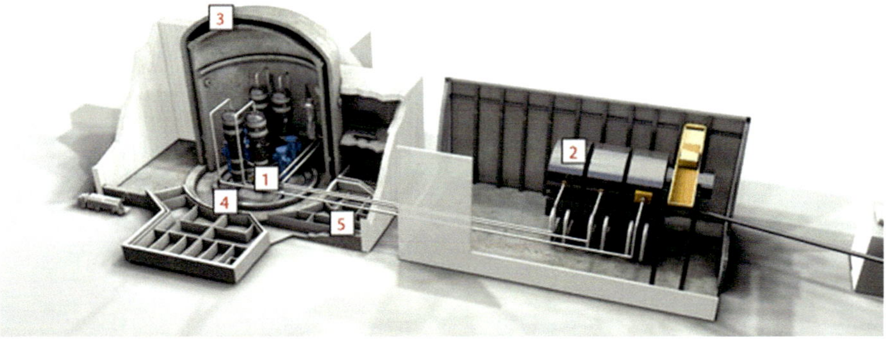

Fig. 2.7 High level depiction of GEN-II or GEN-III nuclear reactor plant (courtesy of Hearst Communications, Inc.)

with metal, are each strong enough to withstand the impact of a heavy commercial airplane. The reactor vessel sits on a 20-ft slab of concrete with a leak-tight "core catcher," (*4*) where the molten core would collect and cool in the event of a meltdown. There are also four safeguard buildings (*5*) with independent pressurizers and steam generators, each capable of providing emergency cooling of the reactor core.

Fourth-generation nuclear power plants differ radically from current reactors by replacing water coolants and moderators, reaching higher temperatures, and gaining the potential to create hydrogen, as well as electricity. Figure 2.8 is an illustration of high-level Generation IV of nuclear power plant, where six types of reactors are designated as GEN-IV as of today [1, 2].

One of the six Gen IV designs under consideration is the meltdown-proof pebble-bed reactor, which uses grains of uranium encased in balls of graphite as

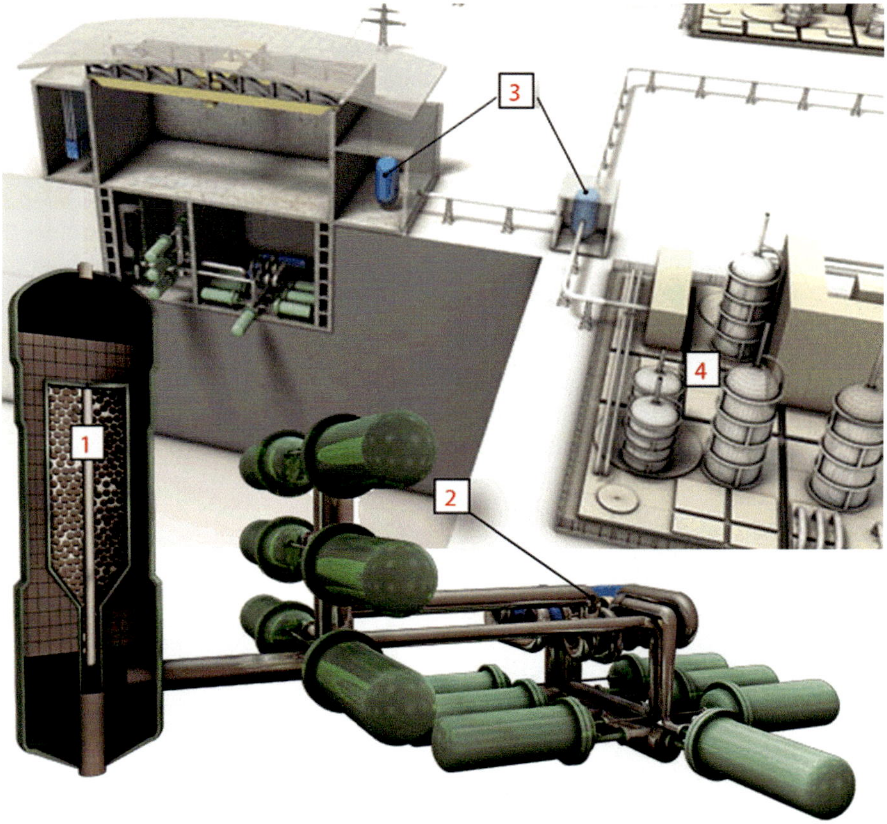

Fig. 2.8 High level depiction of GEN-IV nuclear reactor plant (courtesy of Hearst Communications, Inc.)

fuel. Helium gas is heated as it circulates through a vessel of these pebbles (*1*) and then powers a turbine (*2*) to generate electricity.

A heat exchanger (*3*) can transfer heat from the helium to adjacent facilities (*4*) for the production of hydrogen. The plant relies on "passive safety": If the cooling system fails, the nuclear reaction grinds to a halt on its own.

More details on these six reactors and the concept behind their design are given in Section 3.8 and they can also be found in Refs. [1, 2] or any search on the web as well.

References

1. Zohuri, B. (2015). *Combined cycle driven efficiency for next generation nuclear power plants: An innovative design approach*. New York: Springer.
2. Zohuri, B., & McDaniel, P. (2015). *Thermodynamics in nuclear power plant*. New York: Springer.
3. Zohuri, B. (2014). *Innovative open Air Brayton combined cycle systems for the next generation nuclear power plants*. Albuquerque, NM: University of New Mexico Publications.
4. Zohuri, B., McDaniel, P. & Cassiano R. De Olivera. (2015) Advanced Nuclear Open Air-Brayton Cycles for Highly Efficient Power Conversion *Nuclear Technology*. Volume 192, Number 1, Pages 48–60.
5. McDaniel, P. J., Zohuri, B., & de Oliveira, C. R. E. (2014, September). A combined cycle power conversion system for small modular LMFBRs. *ANS Transactions*.
6. Zohuri, B., McDaniel, P., & de Oliveira, C. R. E. (2014, June). A comparison of a recuperated open cycle (Air) Brayton power conversion system with the traditional steam rankine cycle for the next generation nuclear power plant. *ANS Transactions*.
7. McDaniel, P. J., de Oliveira, C. R. E., Zohuri, B., & Cole, J. (2012, November). A combined cycle power conversion system for the next generation nuclear power plant. *ANS Transactions*.
8. Forsberg, C., McDaniel, P., & Zohuri, B. (2015). Variable electricity and steam from salt, helium, and sodium cooled base-load reactors with gas turbines and heat storage. In *Proceedings of ICAPP 2015*, 03–06 May 2015. Nice (France) Paper 15115.

Chapter 3
Energy Resources and the Role of Nuclear Energy

Energy is broadly defined as the ability to produce a change from the existing conditions. Thus, the term energy implies that a capacity of action is present. The evaluation of energy is done by measuring certain effects that are classified by descriptive names, and these effects can be produced under controlled conditions. For example, a mass that is located at a certain position may have a potential energy or if the same mass is in motion then it may possess the kinetic energy or if its characteristics of composition such as temperature or pressure changes, then it is going through an energy process which is called internal energy. The internal energy can be measured by the release of an amount by change in potential energy experienced by external load.

3.1 The World's Energy Resources

For the past half century fossil fuels, namely, coal, oil, and natural gas, have supplied the major portion of the world's energy requirements. It has long been realized, however, that in the not too distant future these sources of energy will be largely exhausted. At present the total energy consumption, for all countries, is about 1×10^{17} Btu per year. Since the world's population is steadily growing and the power use per capita is increasing as well, the rate of energy utilization by the year 2020 could well be 5–10 times the current value. According to one estimate, the known coal, oil, gas, and oil shale which can be extracted at no more than twice the present cost would be equivalent to roughly 4×10^{19} Btu. This means that in about 100 years the world's economically useful reserves of fossil fuels may approach exhaustion.

B. Zohuri, *Application of Compact Heat Exchangers For Combined Cycle Driven Efficiency In Next Generation Nuclear Power Plants*,
DOI 10.1007/978-3-319-23537-0_3

3.2 Today's Global Energy Market

Today's global energy market places many demands on power generation technology including high thermal efficiency, low cost, rapid installation, reliability, environmental compliance, and operation flexibility. The conclusion, even considering some margin error in it, is inevitably that new sources of power must be found during the next 50 years or so if the earth is to support the growing population with some increase in living standards. Some consideration has been given to few such sources, for example, solar, wind energies, as well as nuclear energy. Although solar and wind energies are very attractive, developing large-scale processes along with large farms of such systems is still some years away; on the other hand nuclear energy has been made available and advanced in fission of heaviest elements or at the stage of still research using fusion of very light nuclei. Technology of fusion process in commercial use with controlled release of such energies using either magnetic confinement or laser drive pellet of deuterium and tritium, the two isotopes of hydrogen, is too far in advanced. Nuclear fission, on the other hand, has already been established as a practical means for production of energy and getting to be at the point of economically very competitive with energy produced from fossil fuels in very near future.

The total amount of basic raw materials as a source of fuel for fission power planet such as Uranium and Thorium, in the earth's crust, to a depth of 3 miles, is very large, possibly something like 10^{12} tons. However, much of this is present in minerals containing such a small proportion of the desired element that extraction would be very expensive and not very cost-effective in particular for high-grade ore reserves that are believed to be in the order of 2×10^6 tons, and therefore we need to reduce the cost of recovery from moderately low-grade ores to at least $100 or less per pound of metal with advancing technology in this matter.

Development of plant layout and modularization concepts requires an understanding of both primary and secondary systems.

General Electric's *ST*eam *A*nd *G*as (STAG)™ combined cycle power generation equipment has met these demands and surpassed them, taking power plant performance to unprecedented levels.

The development of steam and gas turbine combined cycles has paralleled gas turbine development, resulting in reliable combined cycle plants. Those incorporating GE's advanced gas turbine technology have achieved efficiency levels approaching 58 %, due primarily to the higher firing temperatures of advanced technology gas turbines. The MS9001H gas turbine will achieve 60 % efficiency in combined cycle application when it goes into full operation.

In addition to advances in gas turbine technology, steam turbine performance also has evolved. GE's STAG combined-cycle power generation product line includes steam cycle options that satisfy a wide range of economic considerations including fuel flexibility, fuel cost, duty cycle, and space limitations.

Heat-exchangers, filters, turbines, and other components in integrated coal gasification combined cycle system must withstand demanding conditions of high temperatures and pressure differentials. Under the highly sulfiding conditions of the high temperature coal gas, the performance of components degrades significantly with time unless expensive high alloy materials are used. Deposition of a suitable coating on a low-cost alloy may improve is resistance to such sulfidation attack and decrease capital and operating costs. A review of the literature indicates that the corrosion reaction is the competition between oxidation and sulfidation reactions. The Fe- and Ni-based high-temperature alloys are susceptible to sulfidation attack unless they are fortified with high levels of Cr, Al, and Si. To impart corrosion resistance, these elements need not be in the bulk of the alloy and need only be present at the surface layers.

3.3 End of Cheap Oil and the Future of Energy

Global production of conventional oil will begin to decline sooner than most people think, probably within 10 years. As we recall, two sudden price increases took place in 1973 and 1979 and rudely impacted the industrial world and made it to recognize its dependency on cheap crude oil. The first event in 1973 that caused an oil price increase took place in response to an Arab embargo during the Arab and Israeli war. The price tripled and then nearly doubled again when Iran's Shah was dethroned, sending the major economies into a spin. Just a few years earlier oil explorers had discovered enormous new oil reservoirs on the North Slope of Alaska and below the North Sea off the cost of Europe. The emotional and political reactions of most analysts predict a shortage of crude oil in the world due to these types of crises. Not having enough underground reservoirs for exploration of oil will put the future survival of the world economy on a critical path.

The five Middle Eastern nations who are members of Organization of Petroleum Exporting Countries (OPEC) were able to hike the price of crude oil not because oil was growing short but because they managed to control 36 % of the international market. Later, when due to pumped oil from Alaska and North Sea, the demand for crude oil sagged, and then prices of oil dropped and the OPEC's control of prices collapsed. The next oil crunch will not be so temporary. The exploration and discovery of oil fields, as well as production of it, around the world suggests that within the next decade, the supply of conventional oil will not support and cannot keep up with demand. Whether this conclusion is in contradiction with what oil companies are reporting is an open question. Today's oil production rate of about 23.6 GBO (Giga Barrel Oil) per year may suggest cheap crude oil for the next 43 years, or more, based on the official charts that show the reserves are growing. But there are three critical errors.

- *First*, it relies on distorted estimates of reserves.
- *A second mistake* is to pretend that production will remain constant.

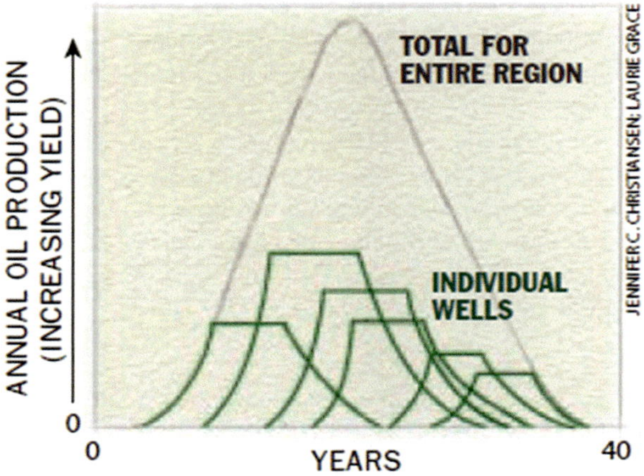

Fig. 3.1 FLOW OF OIL starts to fall from any large region when about half the crude is gone. Adding the output of fields of various sizes and ages (*green curves above*) usually yields a bell-shaped production curve for the region as a whole. M. King Hubbert, a geologist with Shell Oil, exploited this fact in 1956 to predict correctly that oil from the lower 48 American states would peak around 1969

- *Third* and the most important, conventional wisdom erroneously assume that the last bucket of oil can be pumped from the ground just as quickly as the barrels of oil gushing from wells today.

In fact, the rate at which any well—or any country—can produce oil always rises to a maximum and then, when about half the oil is gone, begins falling gradually back to zero.

From an economic perspective then, when the world runs completely out of oil is thus not directly relevant:

What matters is when production begins to taper off. Beyond that point, prices will rise unless demand declines commensurately. See Fig. 3.1.

Using several different techniques to estimate the current reserves of conventional oil and the amount still left to be discovered, many experts in the field concluded that the decline would begin before 2010.

In practice, companies and countries are often deliberately vague about the likelihood of the reserves they report, preferring instead to publicize whichever figure, within a P10–P90 range, best suits them. Exaggerated estimates can, for instance, raise the price of an oil company's stock. See Fig. 3.2.

The members of OPEC have faced an even greater temptation to inflate their reports because the higher their reserves, the more oil they are allowed to export. National companies, which have exclusive oil rights in the main OPEC countries, need not (and do not) release detailed statistics on each field that could be used to verify the country's total reserves. There is thus good reason to suspect that when, during the late 1980s, 6 of the 11 OPEC nations increased their reserve figures by colossal amounts, ranging from 42 % to 197 %, they did so only to boost their export quotas.

Meanwhile global demand for oil is currently rising at more than 2 % a year. Since 1985, energy use is up about 30 % in Latin America, 40 % in Africa, and 50 %

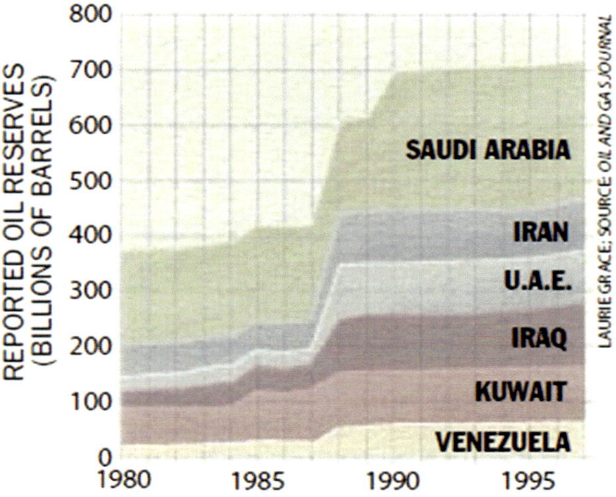

Fig. 3.2 SUSPICIOUS JUMP in reserves reported by six OPEC members added 300 billion barrels of oil to official reserve tallies yet followed no major discovery of new fields

in Asia. The Energy Information Administration forecasts that worldwide demand for oil will increase 60 % (to about 40 GBO a year) by 2020.

The switch from growth to decline in oil production will thus almost certainly create economic and political tension. Unless alternatives to crude oil quickly prove themselves, the market share of the OPEC states in the Middle East will rise rapidly. Within 2 years, these nations' share of the global oil business will pass 30 %, nearing the level reached during the oil-price shocks of the 1970s.

The world could thus see radical increases in oil prices. That alone might be sufficient to curb demand, flattening production for perhaps 10 years (demand fell more than 10 % after the 1979 shock and took 17 years to recover). Many Middle Eastern nations will soon themselves be past the midpoint. World production will then have to fall.

With sufficient preparation, however, the transition to the post-oil economy need not be traumatic. If advanced methods of producing liquid fuels from natural gas can be made profitable and scaled up quickly, gas could become the next source of transportation fuel (see "Liquid Fuels from Natural Gas," by Safaa A. Fouda, on page 92) [1]. Safer nuclear power, cheaper renewable energy, and oil conservation programs could all help postpone the inevitable decline of conventional oil.

Countries should begin planning and investing now. In November 2009, a panel of energy experts appointed by President Bill Clinton strongly urged the administration to increase funding for energy research by $1 billion over the next 5 years. That was a small step in the right direction, one that must be followed by giant leaps from the private sector.

The world is not running out of oil—at least not yet. What our society does face, and soon, is the end of the abundant and cheap oil on which all industrial nations depend.

3.4 What To Do About Coal

Cheap plentiful coal is expected to fuel power plants for the foreseeable future, but can we keep it from devastating the environment?

To deal with climate change means addressing the problems posed by emissions from coal-fired power plants. Unless we take prompt action to strictly limit the amount of carbon dioxide (CO_2) released into the atmosphere when consuming coal to make electricity, we have little chance of gaining control over global warming. The overview of burning coal's side effects are:

- Coal is widely burned for power plants to produce electricity, but it also produces large quantities of climate changing carbon dioxide.
- Compared with conventional power plants, new gasification facilities can more effectively and affordably extract CO_2 so it can be safely stored underground.
- The world must begin implementing carbon sequestration to stave off global warming.

Coal—the fuel that powered the Industrial Revolution—is a particularly worrisome source of energy, in part because burning it produces considerably more carbon dioxide per unit of electricity generated than burning either oil or natural gas does. In addition, coal is cheap and will remain abundant long after oil and natural gas have become very scarce. With coal plentiful and inexpensive, its use is expanding in the USA and elsewhere and is expected to continue rising in areas with abundant coal resources [1]. See Fig. 3.3.

Indeed, US power providers are expected to build the equivalent of nearly 280 500-MW, coal-fired electricity plants between 2003 and 2030. Meanwhile China is already constructing the equivalent of one large coal-fueled power station a week. Over their roughly 60-year life spans, the new generating facilities in operation by 2030 could collectively introduce into the atmosphere about as much carbon dioxide as was released by all the coal burned since the dawn of the Industrial Revolution [1].

Coal use can lead to a range of harmful consequences, including decapitated mountains, air pollution from acidic and toxic emissions, and water fouled with coal wastes. Extraction also endangers and can kill miners. Together such effects make coal production and conversion to useful energy one of the more destructive activities on the planet.

We need to find alternative answers to the above issues in order to deal with future demand. The only answer is to move as quickly as possible to alternative fuels—including natural gas and nuclear power, as well as solar, wind, and geothermal energy. "Running out of energy in the long run is not the problem, but the bind comes during the next 10 years and we need to get over our dependency on crude oil" [1].

Fig. 3.3 Iron hydroxide precipitate (*orange*) in a Missouri stream receiving acid drainage from surface coal mining. Image: US Geological Survey

3.5 The Future of Energy

The future of energy will be very different. For all the uncertainties highlighted in various reports by experts in the field, we can be certain that the energy world will look a lot different in 2030 than it does today. The world energy system will be transformed, but not necessarily in the way we would like to see. We can be confident of some of the trends highlighted in reports on current global trends in energy supply and consumption, environmentally, economically, and socially. But that can—and must—be altered when there is still time to change the road we are on. The growing weight of China, India, the Middle East, and other non-OECD regions in energy markets and in CO_2 emissions is something we need to take under consideration in order to deal with global warming. The rapidly increasing dominance of national oil companies and the emergence of low-carbon energy technologies seems one necessary solution to the problem in hand, but not sufficient enough. And while market imbalances could temporarily cause prices to fall back, it is becoming increasingly apparent that the era of cheap oil is over. But many of the key policy drivers (not to mention other, external factors) remain in doubt. It is within the power of all governments, of producing and consuming

Fig. 3.4 Typical nuclear plant in our backyard

countries alike, acting alone or together, to steer the world toward a cleaner, cleverer, and more competitive energy system. Time is running out and the time to act is now. So what we need to ask is that "*Can Nuclear Power Compete*."

A variety of companies that are in the energy production business say the answer may be yes. Manufacturers have submitted new designs to the Nuclear Regulatory Commission's safety engineers, and that agency has already approved some as ready for construction, if they are built on a previously approved site. Utilities, reactor manufacturers, and architecture/engineering firms have formed partnerships to build plants, pending final approvals. Swarms of students are enrolling in college-level nuclear engineering programs and rosy projections from industry and government predict a surge in construction. See Fig. 3.4.

Like another moon shot, the launch of new reactors after a 35-year hiatus in order is certainly possible, though not a sure bet. It would be easier this time, the experts say, because of technological progress over the intervening decades. But as with a project as large as a moon landing, there is another question: Would it be worthwhile?

In order to answer this question we need to at least satisfy the four unresolved problems associated with nuclear power that were brought up by an MIT report and they were mentioned at the beginning of this write-up. In order to discuss the first point, which is the cost of producing a nuclear power plant with its modern and today's technologies from total ownership and return on investment, we need to understand the nature of the beast from the day it was born in the basement of University of Chicago.

3.6 Nuclear Reactors for Power Production

In the USA, most reactor design and development for the generation of electrical power branched from early nuclear navy research, when it was realized that a compact nuclear power plant would have great advantages for submarine propulsion. Such a power plant would make possible long voyages across the oceans at high speeds without the necessity for resurfacing at frequent intervals. Argonne National Laboratory was assigned the task of designing such reactor. So the first generation of Pressurized Water Reactors (PWR) was born. It used highly enriched uranium as the fuel, and water under pressure as the moderator as well as coolant. The first prototype of this reactor named STR Mark 1 started operation at Arco, Idaho, in March 1953 and a production version of it was installed in the USS Nautilus, the first nuclear powered submarine in May, 1953.

As a result of the experience gained in successful operation of the submarine reactors, the first commercial version of a PWR was designed and installed at Shipping port, Pennsylvania, and went into operation on December 2, 1957, with a water pressure of 13.8 MPa (2000 psi). The steam produced in the heat exchanger was at a temperature of about 254 °C (490 °F) and a pressure of close to 4.14 MPa (600 psi). In order to make the reactor cost-effective and reduce the cost of the power produced, only a small number of the fuel elements were highly enriched in uranium-235 (U^{235}) as an alloy with zirconium. The remainder was of normal uranium dioxide. The change in core design required more real estate for the foot print of a commercialized PWR. This was not an issue for a land-based facility. The output power of this reactor was about 60 MW(e) and 230 MW(t). Further enhancement in core design increased the power to 150 MW(e) and 505 MW(t). Pressurized water reactors, using slightly (2–6 %) enriched uranium dioxide as the fuel, are now commonly used in the USA and other countries around the globe for commercial power generation. The most recent plants have electrical output in the neighborhood of 1000 MW(e) (3000 MW(t)).

Later on other reactor designs based on different fuel materials, moderators, and coolants with various electrical and thermal power outputs were born. Examples are the following:

- Boiling water reactor (BWR) initiated in 1953.
- Water-cooled graphite moderated in 1954.
- High temperature, gas-cooled reactor (HTGR).
- Liquid metal fast breeder reactors (LMFBR).

Basically, all commercial reactor power plants of present interest are systems for generating steam utilizing the heat of nuclear fission to boil water and produce steam for a turbine. They are often referred to as "*N*uclear *S*team *S*upply *S*ystems" or NSSS. The steam is expanded in a turbine which drives a generator to produce electricity in the conventional manner. The exhaust steam from the turbine passes on to a condenser where it is converted into liquid water and this is returned as feed water to the steam generator of the NSSS.

The proportion of the heat supplied in a power plant that is actually converted into electrical energy is called the thermal efficiency of the system; thus, in a nuclear installation:

$$\text{Thermal efficiency} = \frac{\text{Electrical energy generated}}{\text{Heat produced in the reactor}}$$

The maximum possible value of the thermal efficiency is the ideal *thermodynamic efficiency*, which is given by following relationship:

$$\text{Ideal thermodynamic efficiency} = \frac{T_2 - T_1}{T_2}$$

where

T_1 is the absolute temperature of the steam entering the turbine (K, Kelvin).
T_2 is the temperature at which heat is rejected to the condenser (K, Kelvin).

The ideal thermodynamic efficiency can be increased by having T_2 as high as possible and T_1 as low as possible. In practice, T_1 is more or less fixed by the ambient temperature; the thermal efficiency of a steam electric plant is then largely determined by the steam temperature, which should be as high as feasible.

Conditions in PWRs and BWRs are such that the steam temperature is lower than in modern fossil-fuel power plants, in which the heat is produced by burning coal, oil, or gas. The thermal efficiencies of these nuclear reactor plants are only about 33 %, compared with 40 % for the best fossil-fuel facilities. With the HTGRs and fast breeder reactors, however, the thermal efficiencies should equal to those of the best fossil-fuel plants, i.e., about 40 %.

3.7 Future Nuclear Power Plant System

In response to the difficulties in achieving suitability, a sufficiently high degree of safety and a competitive economic basis for nuclear power, the US Department of Energy initiated the Generation IV program in 1999. Generation IV refers to the broad division of nuclear designs into four categories as follows:

1. Early prototype reactor (Generation I).
2. The large central station nuclear power plants of today (Generation II).

3. The advanced light-water reactors and other systems with inherent safety features that have been designed in recent years (Generation III).
4. The next-generation system to be designed and built two decades from now (Generation IV).

By 2000, international interest in the Generation IV project had resulted in a nine-country coalition that includes:

1. Argentina
2. Brazil
3. Canada
4. France
5. Japan
6. South Africa
7. South Korea
8. The UK
9. The USA

Participants are mapping out and collaborating on the research and development of future nuclear energy systems.

Although the Generation IV program is exploring a wide variety of new systems, a few examples serve to illustrate the broad approaches to reactor designs that are developing to meet the objectives.

The next-generation systems are based on three general classes of reactors:

1. Gas-cooled
2. Water-cooled
3. Fast-spectrum

All these categories and their brief designs are discussed in the following sections.

3.8 Next Generation of Nuclear Power Reactors for Power Production

Experts are projecting worldwide electricity consumption will increase substantially in the coming decades, especially in the developing world. The accompanying economic growth and social progress will have a direct impact on rising electricity prices. This has focused fresh attention on nuclear power plants. New, safer, and more economical nuclear reactors could not only satisfy many of our future energy needs but could combat global warming as well. Today's existing nuclear power plants on line in the USA provide a fifth of the nation's total electrical output.

Taking into account the expected increase in energy demand worldwide and the growing awareness about global warming, climate change issues, and sustainable development, nuclear energy will be needed to meet future global energy demand.

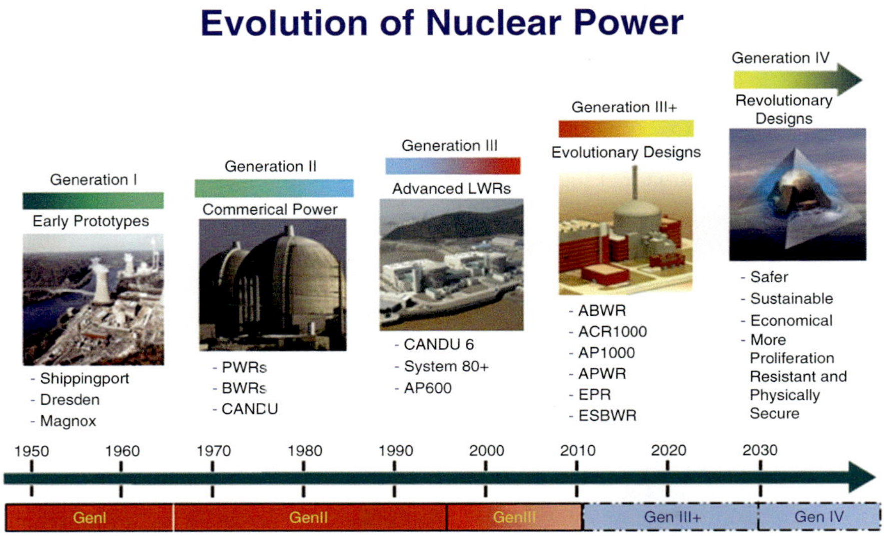

Fig. 3.5 Evolution of nuclear power plants

Nuclear power plant technology has evolved as distinct design generations as mentioned in the previous section and is briefly summarized here again as follows (see Fig. 3.5):

- First generation: prototypes, and first realizations (~1950–1970).
- Second generation: current operating plants (~1970–2030).
- Third generation: deployable improvements to current reactors (~2000 and on).
- Fourth generation: advanced and new reactor systems (2030 and beyond).

The *G*eneration IV *I*nternational *F*orum, or GIF, was chartered in July 2001 to lead the collaborative efforts of the world's leading nuclear technology nations to develop next generation nuclear energy systems to meet the world's future energy needs.

Eight technology goals have been defined for Generation IV systems in four broad areas:

1. Sustainability
2. Economics
3. Safety and reliability
4. Proliferation resistance and physical protection

A large number of countries share these ambitious goals as they aim at responding to economic, environmental, and social requirements of the twenty-first century. They establish a framework and identify concrete targets for focusing GIF R&D efforts.

3.9 Goals for Generation IV Nuclear Energy Systems

The next generation ("Generation IV") of nuclear energy systems is intended to meet the below goals (while being at least as effective as the "third" generation in terms of economic competitiveness, safety, and reliability) in order to provide a sustainable development of nuclear energy (Table 3.1).

In principle, the Generation IV systems should be marketable or deployable from 2030 onward. The systems should also offer a true potential for new applications compatible with an expanded use of nuclear energy, in particular in the fields of hydrogen or synthetic hydrocarbon production, seawater desalination, and process heat production.

It has been recognized that these objectives, widely and officially shared by a large number of countries, should be the basis of an internationally shared R&D program, which allows keeping open and consolidating the technical options, and avoiding any early or premature down selection.

In fact, because the next generation nuclear energy systems will address needed areas of improvement and offer great potential, many countries share a common interest in advanced R&D that will support their development. The international research community should explore such development benefits with the identification of promising research areas and collaborative efforts. The collaboration on R&D by many nations on the development of advanced next generation nuclear energy systems will in principle aid the progress toward the realization of such systems, by leveraging resources, providing synergistic opportunities, avoiding unnecessary duplication, and enhancing collaboration.

Table 3.1 Goals for Generation IV nuclear energy systems

Sustainability-1	Generation IV nuclear energy systems will provide sustainable energy generation that meets clean air objectives and provides long-term availability of systems and effective fuel utilization for worldwide energy production
Sustainability-2	Generation IV nuclear energy systems will minimize and manage their nuclear waste and notably reduce the long-term stewardship burden, thereby improving protection for the public health and the environment
Economics-1	Generation IV nuclear energy systems will have a clear life cycle cost advantage over other energy sources
Economics-2	Generation IV nuclear energy systems will have a level of financial risk comparable to other energy projects
Safety and reliability-1	Generation IV nuclear energy systems operations will excel in safety and reliability
Safety and reliability-2	Generation IV nuclear systems will have a very low likelihood and degree of reactor core damage
Safety and reliability-3	Generation IV nuclear energy systems will eliminate the need for offsite emergency response
Proliferation resistance and physical protection	Generation IV nuclear energy systems will increase the assurance that they are very unattractive and the least desirable route for diversion or theft of weapons usable materials, and provide increased physical protection against acts of terrorism

3.10 A Technology Roadmap for Generation IV Nuclear Energy Systems

The technology roadmap defines and plans the necessary research and development (R&D) to support the next generation of innovative nuclear energy systems known as Generation IV. The roadmap has been an international effort of ten countries, including Argentina, Brazil, Canada, France, Japan, Republic of Korea, South Africa, Switzerland, the UK, and the USA, the International Atomic Energy Agency, and the OECD Nuclear Energy Agency.

Beginning in 2001, over 100 experts from these countries and international organizations began work on defining the goals for new systems, identifying many promising concepts, and evaluating them, and defining the R&D needed for the most promising systems. By the end of 2002, the work resulted in a description of the six most promising systems and their associated R&D needs, and they are listed below.

1. Gas-cooled fast reactor (GFR): Features a fast-neutron-spectrum, helium-cooled reactor, and closed fuel cycle
2. Very-high-temperature reactor (VHTR): A graphite-moderated, helium-cooled reactor with a once-through uranium fuel cycle
3. Supercritical-water-cooled reactor (SCWR): A high-temperature, high-pressure, water-cooled reactor that operates above the thermodynamic critical point of water
4. Sodium-cooled fast reactor (SFR): Features a fast-spectrum, sodium-cooled reactor and closed fuel cycle for efficient management of actinides and conversion of fertile uranium
5. Lead-cooled fast reactor (LFR): Features a fast-spectrum, lead/bismuth eutectic liquid-metal-cooled reactor and a closed fuel cycle for efficient conversion of fertile uranium and management of actinides
6. Molten salt reactor (MSR): Produces fission power in a circulating molten salt fuel mixture with an epithermal-spectrum reactor and a full actinide recycling fuel cycle

These systems offer significant advances in sustainability, safety and reliability, economics, proliferation resistance, and physical protection. These six systems feature increased safety, improved economics for electricity production, and new products such as hydrogen for transportation applications, reduced nuclear wastes for disposal, and increased proliferation resistance.

In 2009, the Experts Group published an outlook on Generation IV R&D, to provide a view of what GIF members hope to achieve collectively in the period 2010–2014. All Generation IV systems have features aiming at performance improvement, new applications of nuclear energy, and/or more sustainable approaches to the management of nuclear materials. High-temperature systems offer the possibility of efficient process heat applications and eventually hydrogen production. Enhanced sustainability is achieved primarily through adoption of a

Table 3.2 Summary of the main characteristics of the six Generation IV systems

System	Neutron spectrum	Coolant	Temp. °C	Fuel cycle	Size (MWe)
VHTR (very high temperature gas reactor)	Thermal	Helium	900–1000	Open	250–300
SFR (sodium-cooled fast reactor)	Fast	Sodium	550	Closed	30–150 300–1500 1000–2000
SCWR (supercritical water-cooled reactor)	Thermal/fast	Water	510–625	Open/closed	300–700 1000–2000
GFR (gas-cooled fast reactor)	Fast	Helium	850	Closed	1200
LFR (lead-cooled fast reactor)	Fast	Lead	480–800	Closed	20–180 300–1200 600–1000
MSR (molten salt reactor)	Epithermal	Fluoride salt	700–800	Closed	1000

closed fuel cycle with reprocessing and recycling of plutonium, uranium, and minor actinides using fast reactors; this approach provides significant reduction in waste generation and uranium resource requirements.

Table 3.2 summarizes the main characteristics of the six Generation IV systems.

3.11 The Description of the Six Most Promising Nuclear Power Systems

A brief summary of each Generation-IV nuclear power system is as follows:

VHTR: The very-high temperature reactor is a next step in the evolutionary development of high-temperature reactors. The VHTR is a helium gas-cooled, graphite-moderated, thermal neutron spectrum reactor with a core outlet temperature greater than 900 °C, and a goal of 1000 °C, sufficient to support production of hydrogen by thermochemical processes. The reference reactor thermal power is set at a level that allows passive decay heat removal, currently estimated to be about 600 MWth. The VHTR is primarily dedicated to the cogeneration of electricity and hydrogen, as well as to other process heat applications. It can produce hydrogen from water by using thermochemical, electrochemical, or hybrid processes with reduced emission of CO_2 gases. At first, a once-through LEU (<20 % U^{235}) fuel cycle will be adopted, but a closed fuel cycle will be assessed, as well as potential

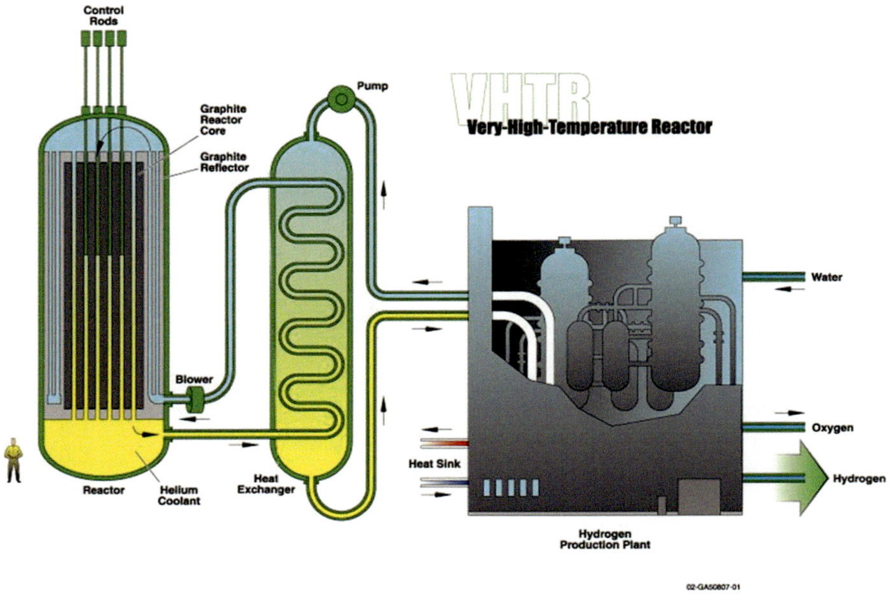

Fig. 3.6 Very high temperature reactor

symbiotic fuel cycles with other types of reactors (especially light-water reactors) for waste reduction. See Fig. 3.6.

SFR: The sodium-cooled fast reactor system uses liquid sodium as the reactor coolant, allowing high power density with low coolant volume fraction. The reactor can be arranged in a pool layout or a compact loop layout. Reactor size options under consideration range from small (50–300 MWe) modular reactors to larger reactors (up to 1500 MWe). The two primary fuel recycle technology options are advanced aqueous and pyrometallurgical processing. A variety of fuel options are being considered for the SFR, with mixed oxide preferred for advanced aqueous recycle and mixed metal alloy preferred for pyrometallurgical processing. Owing to the significant past experience accumulated with sodium-cooled reactors in several countries, the deployment availability of SFR systems is targeted for 2020. See Fig. 3.7.

SCWR: Supercritical-water-cooled reactors are a class of high-temperature, high-pressure water-cooled reactors operating with a direct energy conversion cycle and above the thermodynamic critical point of water (374 °C, 22.1 MPa). The higher thermodynamic efficiency and plant simplification opportunities afforded by a high-temperature, single-phase coolant translate into improved economics. A wide variety of options are currently considered: both thermal-neutron and fast-neutron spectra are envisaged and both pressure vessel and pressure tube

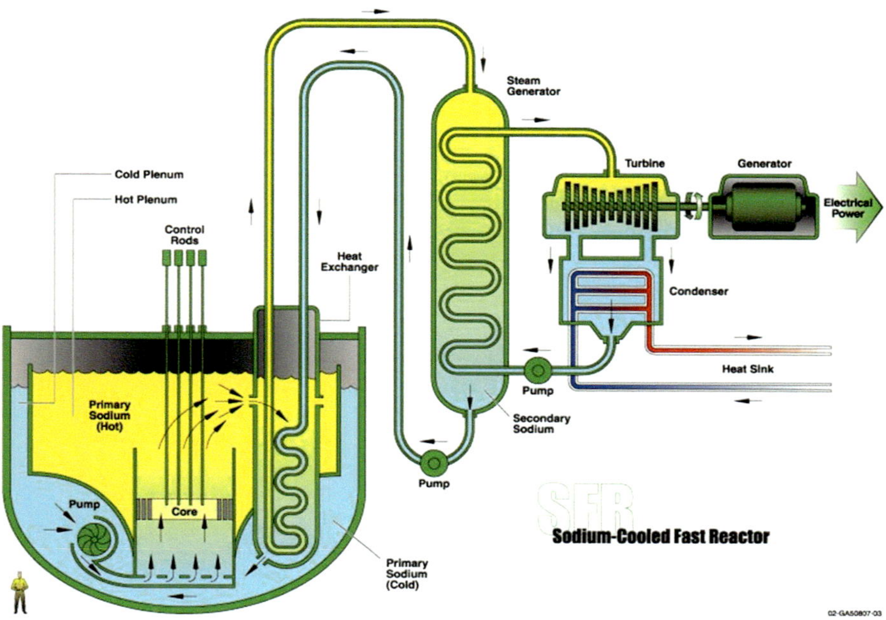

Fig. 3.7 Sodium-cooled fast reactor

configurations are considered. The operation of a 30–150 MWe technology demonstration is targeted for 2022. See Fig. 3.8.

GFR: The main characteristics of the gas-cooled fast reactor are fissile self-sufficient cores with fast neutron spectrum, robust refractory fuel, high operating temperature, high efficiency electricity production, energy conversion with a gas turbine, and full actinide recycling possibly associated with an integrated on-site fuel reprocessing facility. A technology demonstration reactor needed to qualify key technologies could be put into operation by 2020. See Fig. 3.9.

LFR: The lead-cooled fast reactor system is characterized by a fast-neutron spectrum and a closed fuel cycle with full actinide recycling, possibly in central or regional fuel cycle facilities. The coolant could be either lead (preferred option), or lead/bismuth eutectic. The LFR can be operated as a breeder; a burner of actinides from spent fuel, using inert matrix fuel; or a burner/breeder using thorium matrices. Two reactor size options are considered: a small transportable system of 50–150 MWe with a very long core life and a medium system of 300–600 MWe. In the long term a large system of 1200 MWe could be envisaged. The LFR system may be deployable by 2025. See Fig. 3.10.

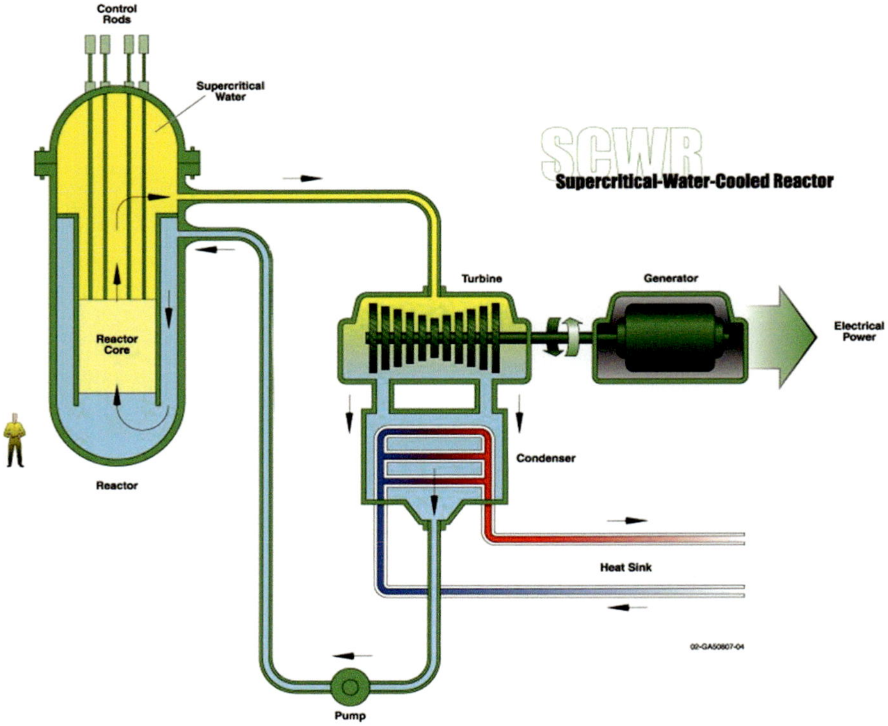

Fig. 3.8 Supercritical water-cooled temperature reactor

MSR: The molten salt reactor system embodies the very special feature of a liquid fuel. MSR concepts, which can be used as efficient burners of TRU from spent LWR fuel, have also a breeding capability in any kind of neutron spectrum ranging from thermal (with a thorium-based fuel cycle) to fast (with the U-Pu fuel cycle). Whether configured for burning or breeding, MSRs have considerable promise for the minimization of radiotoxic nuclear waste. See Fig. 3.11.

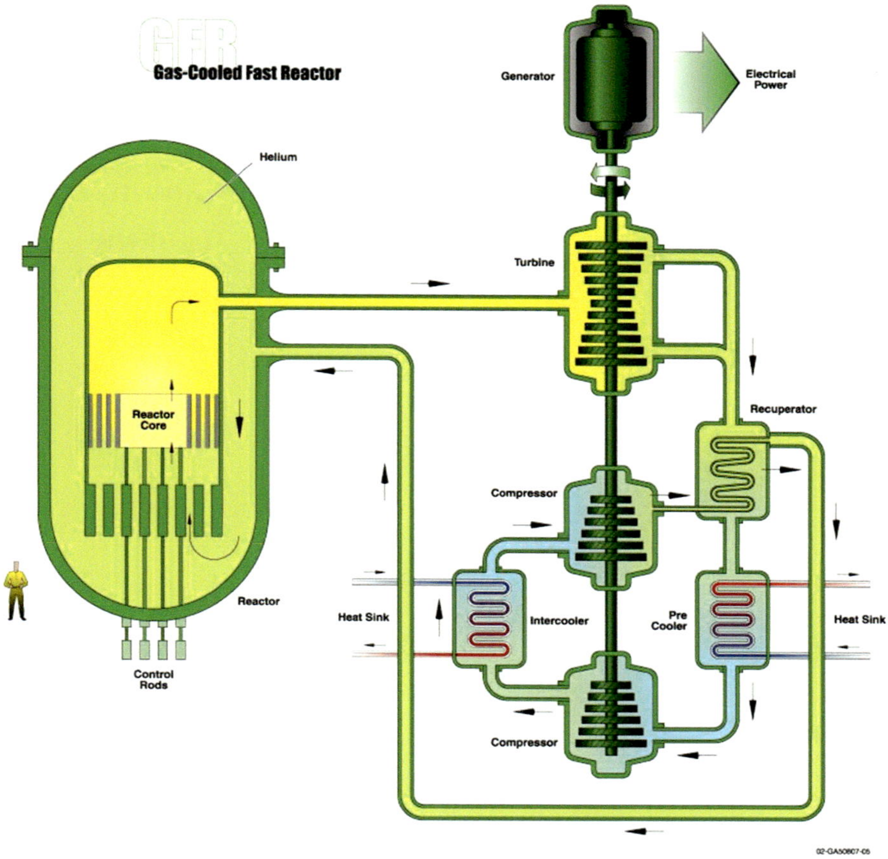

Fig. 3.9 Gas-cooled fast reactor

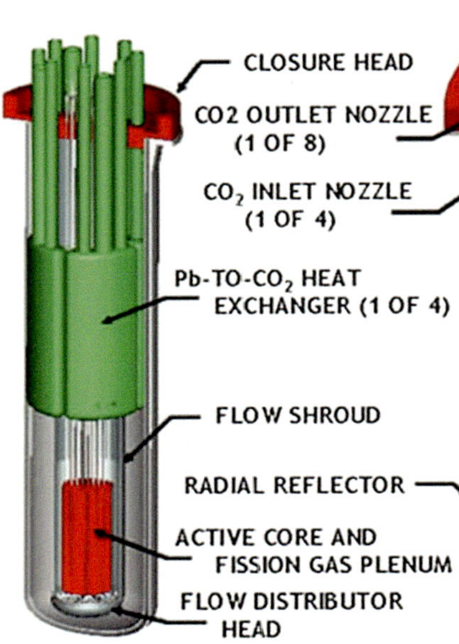

Fig. 3.10 Lead-cooled fast reactor

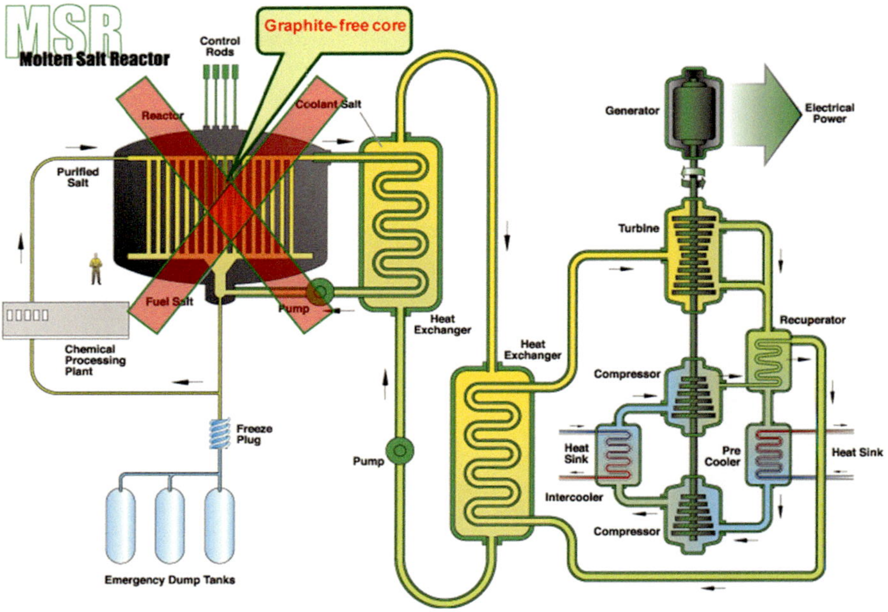

Fig. 3.11 Molten salt reactor

Reference

1. Scientific American. (2007). *Oil and the future of energy*. Guildford, CT: The Lyons Press.

Chapter 4
A New Approach to Energy Conversion Technology

A nuclear reactor produces and controls the release of energy from splitting the atoms of uranium. Uranium-fuelled nuclear power is a clean and efficient way of boiling water to make steam which drives turbine generators. Except for the reactor itself, a nuclear power station works like most coal or gas-fired power stations.

4.1 Power Conversion Study and Technology Options Assessment

In the USA, most reactors' design and development for the generation of electrical power was branched from early nuclear navy research, when it was realized that a compact nuclear power plant would have great advantages for submarine-driven nuclear propulsion system. To have such a power plant on board would make possible long voyages cross the oceans at high speeds without the necessity for resurfacing at frequent intervals.

Operating temperatures of conventional light water reactors, 280–320 °C limit power conversion systems to producing pressurized steam that drives a condensing steam turbine. After employing thermal recovery measures, nuclear plants using this Rankine cycle see a net plant efficiency of around 32–34 %. Comparatively, gas turbines with turbine inlet temperatures of up to and greater than 1400 °C have simple cycle efficiencies of around 40 % that can be boosted to around 60 % in a combined cycle. The ability of the combined cycle or Brayton with recuperator cycle to drastically improve net plant efficiency is an especially appealing feature to employ with a nuclear power source, given the very low fuel costs for nuclear energy, but has previously been technically infeasible given the high operating temperature requirements of a combined cycle. See Fig. 4.1.

One of the differentiating features of the fluoride salt high temperature reactors is the operating temperature range of the primary coolant loop, 600–700 °C (reactor

B. Zohuri, *Application of Compact Heat Exchangers For Combined Cycle Driven Efficiency In Next Generation Nuclear Power Plants*,
DOI 10.1007/978-3-319-23537-0_4

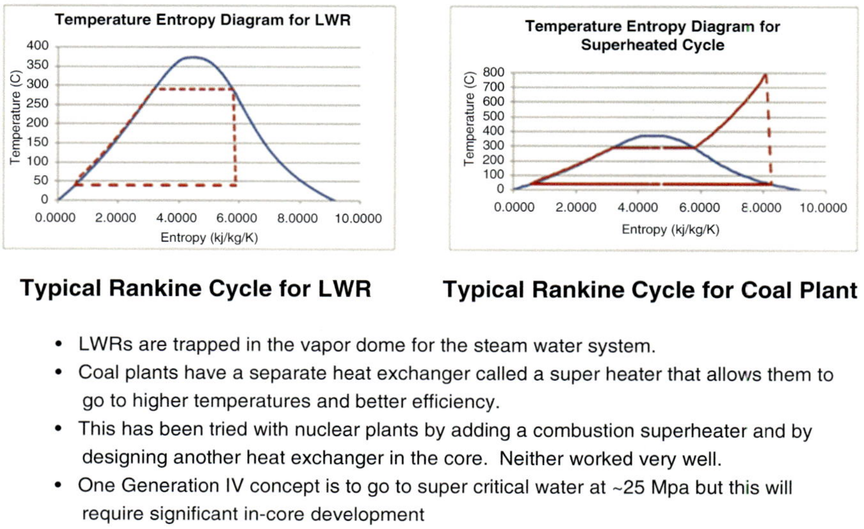

Fig. 4.1 Drawback of current Nuclear Power Plants (NPP) designs [1]

inlet and outlet temperatures, respectively). Although other advanced, high-temperature reactors have been developed, the high temperature characteristics of the lithium fluoride and beryllium fluoride eutectic (FLiBe) molten salt primary coolant used in fluoride salt high temperature reactors enables an operating temperature range that is uniquely suited to driving an open air combined cycle which sees proportional increases in efficiency and power generation with elevated turbine inlet temperatures.

A study was done in September of 2004 by a team of experts at the University of California, Nuclear Engineering Department. The executive summary shows that the electrical Power Conversion System (PCS) for the Next Generation Nuclear Plant (NGNP) will take advantage of a significantly higher reactor outlet temperature to provide greater efficiency than can be achieved by the current generation of light water reactors. In anticipation of the design, development and procurement of an advanced power conversion system for NGNP, the study was initiated to identify the major design and technology options and their trade-offs that must be considered in the evaluation of PCS options to support future research and procurement decisions. These PCS technology options affect cycle efficiency, capital cost, system reliability and maintainability and technical risk, and therefore the cost of electricity from Generation IV systems. A reliable evaluation and estimate of actual costs requires an optimized, integrated PCS design. At that early stage of the NGNP project, it was useful to identify the technology options that would be considered in the design of proposed PCS systems, identify the system performance and cost implications of these design options, and provide a general framework for evaluating the design choices and technology trade-offs.

The ultimate measure of the value of power conversion options is the cost of electricity produced, which is a function of capital and operating cost recovery and the system efficiency and reliability. Evaluating cost is difficult to do without detailed integrated designs, but it is possible to identify the factors that influence component and system performance, cost, and technical risk. In this study, several existing Brayton conversion system designs were studied to illustrate and evaluate the implications of the major design choices to assess performance against the Generation IV economics and sustainability goals, and to identify areas of technical incompleteness or weakness. Several reference system designs were considered to provide a semi-quantitative basis for performing comparisons. The reference systems included the GT-MHR, PBMR, GTHTR-300, Framatome indirect cycle design, and AHTR high temperature Brayton cycle designs. Where appropriate, Generation II, III, and III+ light water reactors (two 1970s designs, the EPR, and the ESBWR) were also considered.

The design choices and technology options considered relevant for the assessment of NGNP power conversion options included the cycle types and operational conditions, such as working fluid choices, direct vs. indirect, system pressure and interstage cooling and heating options. The cost and maintainability of the PCS is also influenced by the PCS layout and configuration including distributed vs. integrated PCS designs, single vs. multiple shafts, shaft orientation, and the implications for the pressure boundary.

From the summary Table 4.1, it is apparent that high temperature gas reactor power conversion design efforts to date have resulted in very different design choices based on project-specific requirements and performance or technical risk requirements.

In the review of existing designs and the evaluation of the major technology options, it immediately becomes apparent that the optimized design involves a complex trade-off of diverse factors, such as cost, efficiency, development time, maintainability, and technology growth path that must be considered in an integrated PCS system context before final evaluation. General observations derived from the review of the reference systems, including comparisons with light water reactor systems where applicable, include the following:

- There are key PCS design choices that can have large effects on PCS power density and nuclear island size, making careful and detailed analysis of design trade-offs important in the comparison of PCS options.
- Considering the major construction inputs for nuclear plants—steel and concrete—high temperature reactors appear to be able to break the economy-of-scale rules for LWRs, and achieve similar material-input performance at much smaller unit sizes.
- For HTRs, a much larger fraction of total construction inputs go into the nuclear island. To compete economically with LWRs, HTRs must find approaches to reduce the relative costs for nuclear-grade components and structures.

Table 4.1 Summary of PCS design features for representative gas reactor systems

Feature	PBMR (horizontal)	GT-MHR	GTHTR300	Framatome indirect	AHTR-IT
Thermal power (MWt)	400	600	600	600	2400
Direct vs. indirect cycle	Direct	Direct	Direct	Indirect	Indirect
Recuperated vs. combined cycle	Recuperated	Recuperated	Recuperated	Combined	Recuperated
Intercooled vs. non-intercooled	Intercooled	Intercooled	Non-intercooled	Intercooled	Intercooled/reheat
Integrated vs. distributed PCS	Distributed	Integrated	Distributed	Distributed	Distributed (modular)
Single vs. multiple TM shafts	Single (previously multiple)	Single	Single	Single	Multiple (modular)
Synchronous vs. asynchronous	Reduction to synchronous	Asynchronous	Synchronous	Synchronous	Synchronous
Vertical vs. horizontal TM	Horizontal	Vertical	Horizontal	Horizontal	Vertical
Submerged vs. external generator	External	Submerged	Submerged	External	Submerged

PCS technology options also include variations on the cycle operating conditions and the cycle type that can have an important impact on performance and cost. These options include:

- Working fluid choice—He, N_2, CO_2 or combinations have been considered. Working fluid physical characteristics influence cycle efficiency and component design.
- System pressure—Higher pressures lead to moderate efficiency increases and smaller PCS components, but increase the pressure boundary cost—particularly for the reactor vessel—which introduces a component-design, and a system cost and performance trade-off.
- Direct vs. indirect—Indirect cycles involve an intermediate heat exchanger (IHX) and resulting efficiency reduction, and more complex control requirements, but facilitate maintenance.
- Interstage cooling (or heating) results in higher efficiency but greater complexity.

Some of the observations from this assessment of these factors include:

- Differences between He vs. N_2 working fluids were not considered critical for turbomachinery design, because both involve similar differences from current combustion turbines, with the primary difference being in the heat exchanger size to compensate for the lower N_2 thermal conductivity.
- N_2 allows 3600-rpm compressor operation at thermal powers at and below 600 MW(t), while He compressors must operate at higher speeds requiring reduction gears, asynchronous generators, or multiple-shaft configurations. However, power uprating to approximately 800 MW(t) would permit 3600-rpm He compressor operation, providing a potentially attractive commercialization approach. Turbo machinery tolerances for He systems do not appear to be a key issue.
- Direct/indirect—Efficiency loss can be 2–4 %, depending on design, and the IHX becomes a critical component at high temperatures. Direct cycles have an extended nuclear grade pressure boundary. Maintainability is considered a key design issue for direct cycles.
- Interstage cooling, as well as bottoming cycles (Rankine), can result in significant efficiency improvements, but at a cost of complexity and lower temperature differences for heat rejection, affecting the potential for dry cooling and reduced environmental impact from heat rejection.

The PCS configuration and physical arrangement of the system components has important effects on the volume and material inputs into structures, on the pressure boundary volume and mass, on gas inventories and storage volume, on the uniformity of flow to heat exchangers, on pressure losses, and on maintainability. The major factors considered in this study included:

- Distributed vs. integrated PCS design approach—PCS components can be located inside a single pressure vessel (e.g., GT-MHR), or can be divided between multiple pressure vessels (e.g., PBMR, HTR-300). This is a major design choice, with important impacts in several areas of design and performance.

- Shaft orientation (vertical/horizontal)—Orientation affects the compactness of the system, the optimal design of ducting between turbo machinery and heat exchangers. Vertical turbo machinery provides a reduced PCS footprint area and building volume and can simplify the ducting arrangement to modular recuperator and intercooler heat exchangers.
- Single vs. multiple shafts—Single shafts may include flexible couplings or reduction gears. In multiple-shaft systems, turbo-compressors are separated from synchronous turbo-generators, allowing the compressors to operate at higher speed and reducing the number of compressor stages required. Multiple shafts and flexible couplings reduce the weight of the individual turbo machines that bearings must support.
- Pressure boundary design—The pressure vessels that contain the PCS typically have the largest mass of any PCS components and provide a significant (~33 %) contribution to the total PCS cost.

4.2 Waste Heat Recovery

Waste heat is heat generated in a process by way of fuel combustion or chemical reaction, which is then "dumped" into the environment and not reused for useful and economic purposes. The essential fact is not the amount of heat, but rather its "value." The mechanism to recover the unused heat depends on the temperature of the waste heat gases and the economics involved.

Large quantities of hot flue gases are generated from boilers, kilns, ovens, and furnaces. If some of the waste heat could be recovered then a considerable amount of primary fuel could be saved. The energy lost in waste gases cannot be fully recovered. However, much of the heat could be recovered and adopting the following measures as outlined in this chapter can minimize losses.

4.2.1 Advantages and Disadvantages of Waste Heat Recovery

Advantages:

These systems have many benefits which could be direct or indirect.

- Direct benefits:
  - The recovery process will add to the efficiency of the process and thus decrease the costs of fuel and energy consumption needed for that process.

- Indirect benefits:
  - Reduction in pollution: Thermal and air pollution will dramatically decrease since less flue gases of high temperature are emitted from the plant since most of the energy is recycled.
  - Reduction in the equipment sizes: As fuel consumption reduces, the control and security equipment for handling the fuel decreases. Also, filtering equipment for the gas is no longer needed in large sizes.
  - Reduction in auxiliary energy consumption: Reduction in equipment sizes means another reduction in the energy fed to those systems like pumps, filters, fans, etc.

Disadvantages:

- Capital cost: The capital cost to implement a waste heat recovery system may outweigh the benefit gained in heat recovered. It is necessary to put a cost to the heat being offset.
- Quality of heat: Often waste heat is of low quality (temperature). It can be difficult to efficiently utilize the quantity of low quality heat contained in a waste heat medium. Heat exchangers tend to be larger to recover significant quantities which increases capital cost.
- Maintenance of equipment: Additional equipment requires additional maintenance cost.

4.3 Power Conversion System Components

The effectiveness or efficiency of the major Power Conversion System (PCS) components, primarily the heat exchangers and turbo machinery, is clearly a major factor in system cost and performance. Over all summary of each of these components are given in the following sections.

4.3.1 Heat Exchangers

A *heat exchanger* is a device built for efficient heat transfer from one medium to another, whether the media are separated by a solid wall so that they never mix, or the media are in direct contact. They are widely used in space heating, refrigeration, air conditioning, power plants, chemical plants, petrochemical plants, petroleum refineries, and natural gas processing. One common example of a heat exchanger is the radiator in a car, in which a hot engine-cooling fluid, like antifreeze, transfers heat to air flowing through the radiator.

For efficiency, heat exchangers are designed to maximize the surface area of the wall between the two fluids, while minimizing resistance to fluid flow through the exchanger. The exchanger's performance can also be affected by the addition of fins or corrugations in one or both directions, which increase surface area and may channel fluid flow or induce turbulence.

A we mentioned the heat exchangers (HX) are transferring heat from one fluid to another and in particular at compact size is one of the important aspects of Power Conversion System (PCS) components, if we want to rely on combined cycles driven efficiency of next-generation nuclear power plants.

High heat transfer rates, modular construction, exotic alloys, low hold-up volume, and high internal turbulence are all combined to improve efficiencies, and process control, with longer maintenance cycles. The temperature difference at the pinch depends on the decision of heat exchanger; in general, the smaller the temperature difference the more expensive the heat exchanger [2].

Taking Fig. 4.2 under consideration, pinch point is defined as the point where the temperature difference is a minimum. Taking as an example a temperature difference of 20 K, say, then the combined cold stream can be moved from left to right on the diagram until the temperature difference at the pinch point is 20 K as shown in Fig. 4.2. It can then be seen that external heating of 90 kW and external cooling of 140 kW are required; all other energy changes can be achieved by heat exchangers between the various streams; the difference between the external heating and the external cooling is (90 − 140) = −50 kW, the same as in Table 4.1. Note that the process in a boiler or condenser would appear as a horizontal line on a diagram such as Fig. 4.2. The data is given in Table 4.2.

As we said compact heat exchangers have a vital role to produce higher efficiency if they are used in combined gas-steam power cycles vs. conventional shell-and-tube heat exchangers.

They are available with standard connections in various diameters and in different lengths. Compact heat exchangers can be designed in horizontal and

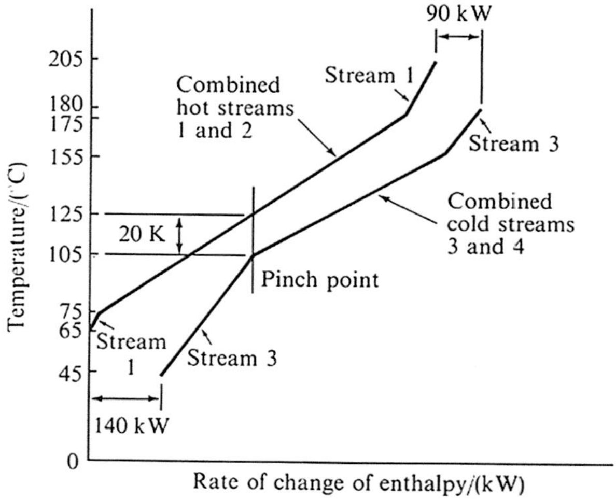

Fig. 4.2 Temperature against rate of enthalpy change for composite hot and cold streams

Table 4.2 Data for four fluid streams

Stream number	Initial temp. (°C)	Final temp. (°C)	Mass flow rate (kg/s)	Specific heat cap. (kJ/kg K)	Heat cap. rate (kW/K)	Rate of enthalpy increase (kW)
1	205	65	2.00	1.00	2.0	−280
2	175	75	3.20	1.25	4.0	−400
3	45	180	3.75	0.80	3.0	+405
4	105	155	3.00	1.50	4.5	+225
						−50

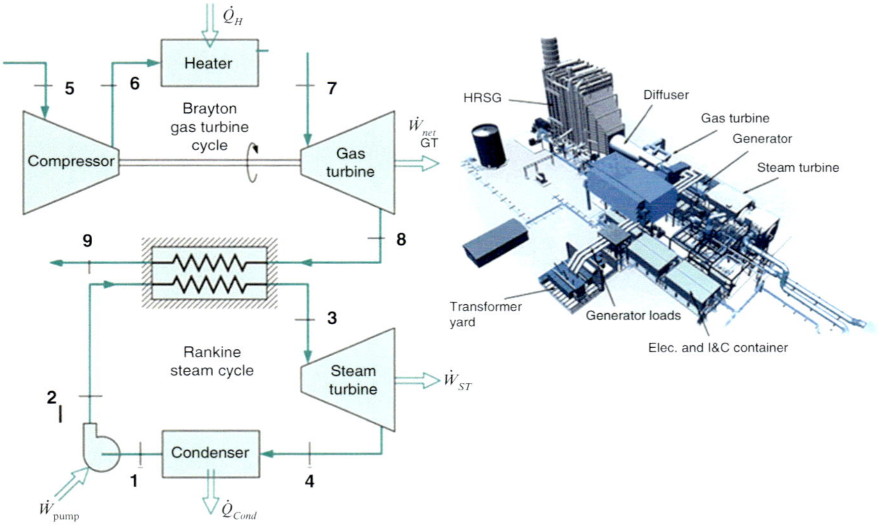

Fig. 4.3 Combined gas (Brayton)-steam (Rankine) cycle

vertical orientation and can be used for a wide variety of applications including among others the ones listed below:

- Compressed gas/water coolers
- Water/water coolers
- Oil/water coolers
- Preheaters
- Condensers and evaporators for chemical and technical processes of all kinds
- Machine coolers
- Oil coolers for hydraulic systems
- Oil and water coolers for power machines
- Refrigeration and air-conditioning units

We look at the heat exchanger application in case of combined gas-steam power cycles (see Fig. 4.3), where the objective of this cycle is to combine a Brayton cycle

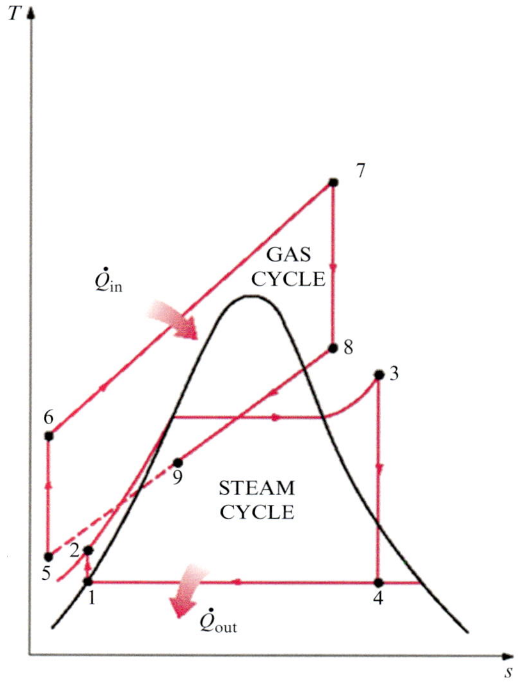

Fig. 4.4 *T–s* diagram of a combined gas (Brayton)-steam (Rankine) cycle

with a Rankine cycle. This combination improves significantly the thermal efficiency of the power plant.

The idea behind a combined gas-steam power cycle is to use the exhaust high temperature gases from the gas turbine to heat the steam within the boiler of the steam turbine.

If we consider such combined cycle, the *T–s* diagram is presented by Fig. 4.4.

Heat exchanger components are defined and the required designs are summarized as follows:

- Heat exchanger designs have significant impacts on both the efficiency and cost of the PCS. For a given heat exchanger type, higher effectiveness must be balanced against the increased size or pressure drop implications. Using small passages increases heat exchanger surface area per unit volume, but those same small passages tend to reduce the heat transfer coefficient due to laminar flow. Higher pressures may be utilized to force those flows back into the turbulent region, but those higher pressures force the construction of a more robust pressure boundary and increase pumping power.
- The recuperator effectiveness and total heat exchanger (HX) pressure drop is a significant impact on the cycle efficiency and there is significant leverage in optimizing the recuperator design for both high heat transfer effectiveness and minimum pressure drop. For modular recuperators, careful attention must be

paid to the module configuration and duct design to obtain equal flow rates to each module.
- Material limitations may limit the operating temperatures for many components, including the reactor vessel and heat exchangers. But because of the large flexibility of the Brayton cycle, high efficiency systems can still be designed within these limitations. Fabrication techniques will probably differ between intermediate, pre and intercooler, and recuperator heat exchangers, because of their operating temperature ranges. It would appear that transients could be tolerated by most of these heat exchanger designs.

A heat exchanger is a device for transferring heat from one fluid to another. There are three main categories:

1. *Recuperative*: in which the two fluids are at all times separated by a solid wall
2. *Regenerative*: in which each fluid transfers heat to or from a matrix of material
3. *Evaporator*: in which the enthalpy of vaporization of the fluid is used to provide a cooling effect

The most commonly used type is the recuperative heat exchanger and it can be classified by flow arrangement, where numerous possibilities exist for flow arrangement in heat exchangers.

In these types of heat exchangers, the two fluids can flow [3–7]

1. In *parallel flow*
2. In *counterflow*
3. In *cross-flow*

Heat exchangers may be classified according to their flow arrangement.

1. *Parallel flow*: In parallel flow heat exchanger configuration, the hot and cold fluids enter at the same end of the heat exchanger, flow through in the same direction (i.e., in parallel to one another to the other side), and leave together at the other end, as illustrated in Fig. 4.5a [5].
2. *Counterflow*: In counterflow heat exchanger configuration, the hot and cold fluids enter in the opposite ends of the heat exchanger and flow through in opposite directions, as illustrated Fig. 4.5b [5].
3. *Cross-flow*: In the cross-flow heat exchanger configuration, the two fluids usually flow at right angles to each other, as illustrated in Fig. 4.5c [5]. In the cross-flow arrangement, the flow may be called *mixed* or *unmixed*, depending on the design. The countercurrent design is most efficient, in that it can transfer the most heat. In a cross-flow heat exchanger, the fluids travel roughly perpendicular to one another through the exchanger.

Figure 4.6a shows a simple arrangement in which both hot and cold fluids flow through individual channels formed by corrugation; therefore the fluids are not free to move in the transverse direction. Then each fluid stream is said to be unmixed. Figure 4.6b illustrates a typical temperature profile for the outlet temperatures when both fluids are unmixed, as shown in Fig. 4.6a. The inlet temperatures for both

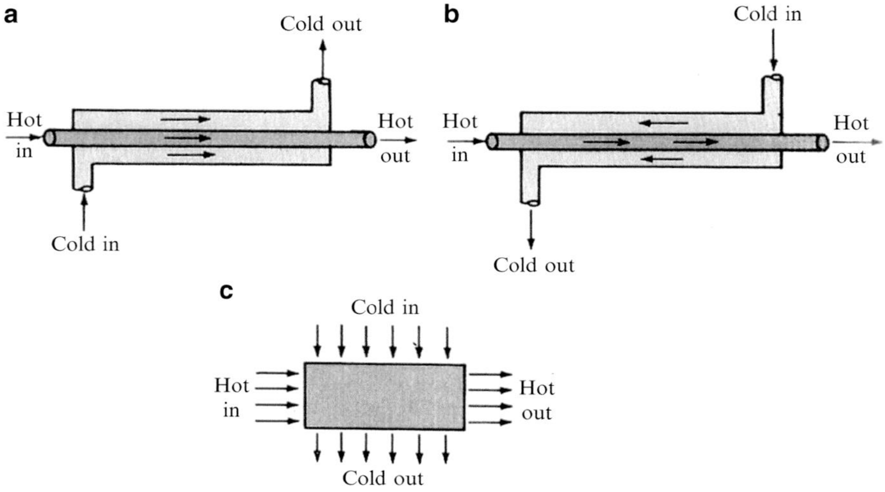

Fig. 4.5 (**a**) Parallel flow, (**b**) counterflow, and (**c**) cross-flow [4]

fluids are assumed to be uniform, but the outlet temperatures exhibit variation transverse to the flow.

In the flow, arrangement shown in Fig. 4.6c, the cold fluid flows inside the tubes and so is not free to move in the transverse direction. Therefore, the cold fluid is said to be unmixed. However, the hot fluid flows over the tubes and is free to move in the transverse direction. Therefore, the hot fluid stream is said to be mixed. The mixing tends to make the fluid temperature uniform in the transverse direction; therefore, the exit temperature of a mixed stream exhibits negligible variation in the crosswise direction.

In general, in a cross-flow exchanger, three idealized flow arrangements are possible:

1. Both fluids are unmixed
2. One fluid is mixed, and the other is unmixed
3. Both fluids are mixed

The last arrangement is not commonly used.

In a shell-and-tube exchanger, the presence of a large number of baffles serves to "mix" the shell-side fluid in a sense discussed above; that is, its temperature tends to be uniform at any cross-section.

The driving temperature across the heat transfer surface varies with position, but an appropriate means temperature can be defined. In most simple systems this is the Logarithmic Mean Temperature Difference (LMTD). Sometimes direct knowledge of the LMTD is not available and the Number of Transfer Units (NTU) method is used.

In summary, the NTU method is used to calculate the rate of heat transfer in heat exchangers (especially countercurrent exchangers) when there is insufficient

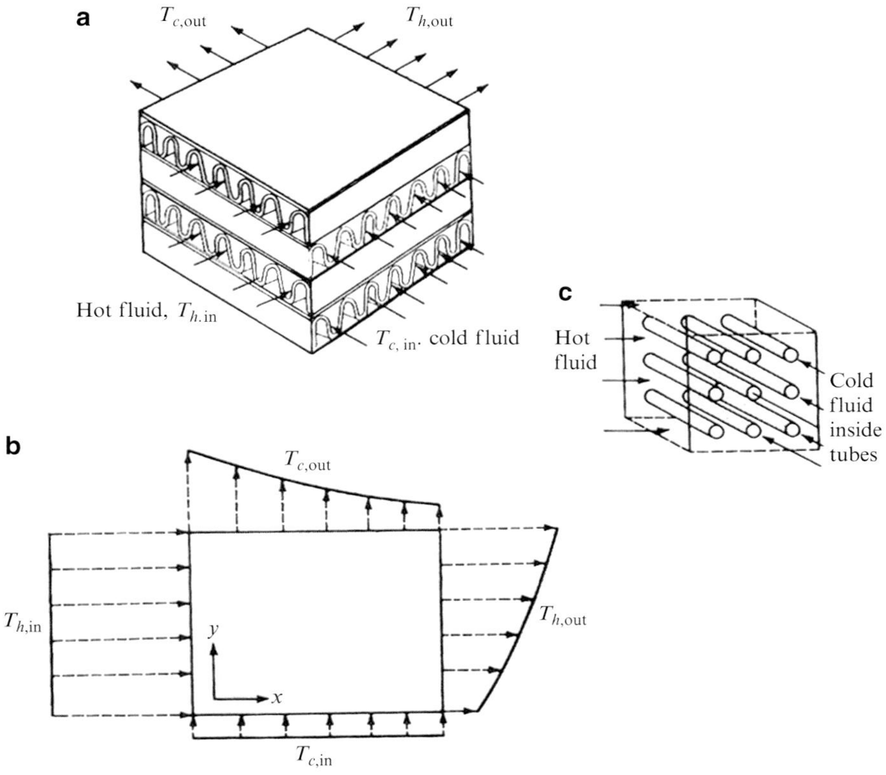

Fig. 4.6 Cross-flow arrangements: (**a**) both fluids unmixed; (**b**) temperature profile when both fluids are unmixed; (**c**) cold fluid unmixed, hot fluid mixed [4]

information to calculate the LMTD. In heat exchanger analysis, if the fluid inlet and outlet temperatures are specified or can be determined by simple energy balance, the LMTD method can be used; but when these temperatures are not available the NTU or the *effectiveness method* is used.

Now, we summarize here the principle ones as shown in Fig. 4.5a–c. The temperature variations of each type are shown in Fig. 4.5a–c. It can also be shown that for all three cases of Fig. 4.5 the true temperature difference is the LMTD given by [5–7]:

$$\text{LMTD} = \frac{(\Delta t_1 - \Delta t_2)}{\ln(\Delta t_1 / \Delta t_2)} \tag{4.1}$$

On the other hand, in case of NTU or effectiveness method analysis, in order to define the effectiveness of a heat exchanger we need to find the maximum possible heat transfer that can be hypothetically achieved in a counterflow heat exchanger of infinite length. Therefore, one fluid will experience the maximum possible temperature difference, which is the difference of $T_{\mathrm{hot,inlet}} - T_{\mathrm{cold,inlet}}$ (the temperature difference between the inlet temperature of the hot stream $T_{\mathrm{hot,inlet}} = T_{\mathrm{h,i}}$ and the inlet temperature of the cold stream $T_{\mathrm{cold,inlet}} = T_{\mathrm{c,i}}$). The method proceeds by calculating the heat capacity rates (i.e., mass flow rate multiplied by specific heat) C_{h} and C_{c} for the hot and cold fluids, respectively, and denoting the smaller one as $C_{\min}$. References such as Ozisik [5] or Incropera and DeWitt [6], and Incropera, DeWitt, Bergman, and Lavine [7] present a detailed analysis of the NTU method and we just write the result of finding the effectiveness that is given in the form of Eq. (4.2)

$$E = 1 - \exp[-\mathrm{NTU}] \tag{4.2}$$

As we said for a more detailed analysis of these two methods readers should refer themselves to the above-mentioned classical textbooks.

4.3.1.1 Recuperative Heat Exchanger

Industrial process facilities recover or "recuperate" otherwise wasted heat energy using heat exchangers, but on a much larger scale.

A recuperator is a special purpose counterflow energy recovery heat exchanger positioned within the supply and exhaust air streams of an air handling system, or in the exhaust gases of an industrial process, in order to recover the waste heat.

In many types of processes, combustion is used to generate heat, and the recuperator serves to recuperate, or reclaim this heat, in order to reuse or recycle it. The term recuperator refers as well to liquid–liquid counterflow heat exchangers used for heat recovery in the chemical and refinery industries and in closed processes such as ammonia–water or LiBr–water absorption refrigeration cycles.

Recuperators are often used in association with the burner portion of a heat engine, to increase the overall efficiency. For example, in a gas turbine engine, air is compressed, mixed with fuel, which is then burned and used to drive a turbine. The recuperator transfers some of the waste heat in the exhaust to the compressed air, thus preheating it before entering the fuel burner stage. Since the gases have been preheated, less fuel is needed to heat the gases up to the turbine inlet temperature. By recovering some of the energy usually lost as waste heat, the recuperator can make a heat engine or gas turbine significantly more efficient.

In a recuperator, heat exchange takes place between the flue gases and the air through metallic or ceramic walls. Ducts or tubes carry the air for combustion to be preheated; the other side contains the waste heat stream. A recuperator for recovering waste heat from flue gases is shown in Fig. 4.7.

There are different type manufactured recuperator are in existence which depends on their application and they are described in the following sections.

Figure 4.8 illustrates various recuperators that are manufactured by industries.

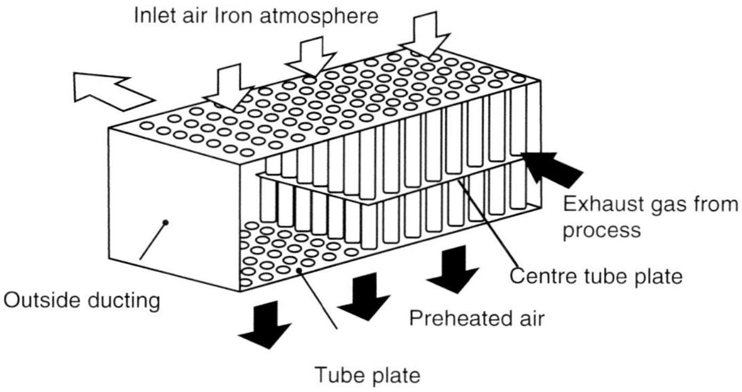

Fig. 4.7 Waste heat recovery using recuperator

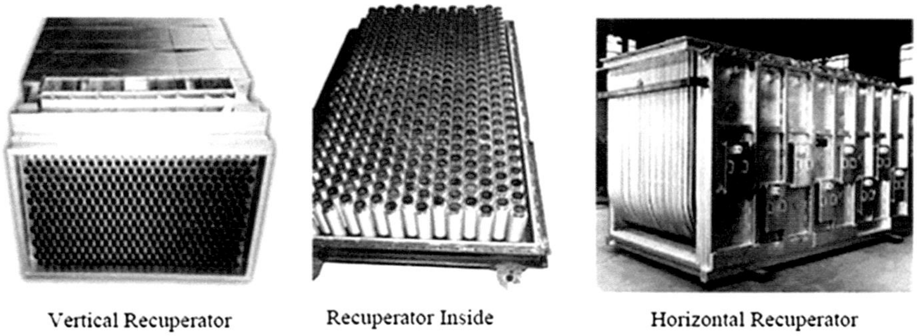

Fig. 4.8 Illustration of different types of recuperators

Metallic Radiation Recuperator

The simplest configuration for a recuperator is the metallic radiation recuperator, which consists of two concentric lengths of metal tubing as shown in Fig. 4.9.

The inner tube carries the hot exhaust gases while the external annulus carries the combustion air from the atmosphere to the air inlets of the furnace burners. The hot gases are cooled by the incoming combustion air, which now carries additional energy into the combustion chamber. This is the energy, which does not have to be supplied by the fuel; consequently, less fuel is burned for a given furnace loading. The saving in fuel also means a decrease in combustion air and therefore, stack losses are decreased not only by lowering the stack gas temperatures but also by discharging smaller quantities of exhaust gas.

The radiation recuperator gets its name from the fact that a substantial portion of the heat transfer from the hot gases to the surface of the inner tube takes place by radiative heat transfer. The cold air in the annuals, however, is almost transparent to infrared radiation so that only convection heat transfer takes place to the incoming

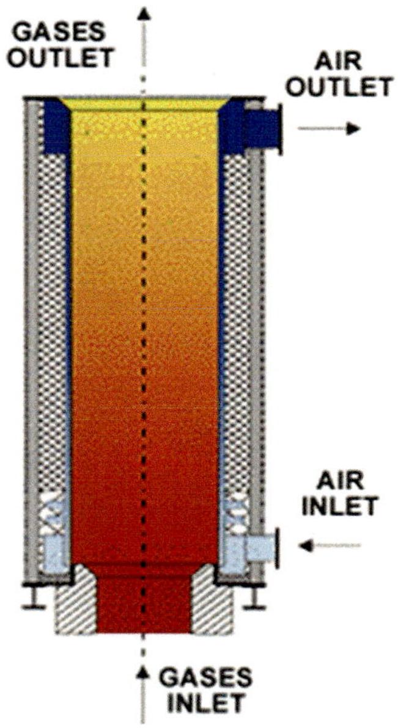

Fig. 4.9 Metallic radiation recuperator

air. As shown in the diagram, the two gas flows are usually parallel, although the configuration would be simpler and the heat transfer would be more efficient if the flows were opposed in direction (or counterflow). The reason for the use of parallel flow is that recuperators frequently serve the additional function of cooling the duct carrying away the exhaust gases and consequently extending its service life.

Convective Recuperator

A second common configuration for recuperator is called the tube type or convective recuperator. As seen in Fig. 4.10, the hot gases are carried through a number of parallel small diameter tubes, while the incoming air to be heated enters a shell surrounding the tubes and passes over the hot tubes one or more times in the direction normal to their axes.

If the tubes are baffled to allow the gas to pass over them twice, the heat exchanger is termed a two-pass recuperator; if two baffles are used, a three-pass recuperator, etc.

Although baffling increases both the cost of the exchanger and the pressure drop in the combustion air path, it increases the effectiveness of heat exchange. Shell and tube type recuperators are generally more compact and have a higher effectiveness than radiation recuperators, because of the larger heat transfer area made possible through the use of multiple tubes and multiple passes of the gases.

Using steel plain tubes, in these recuperators the heat transfer between the primary and secondary fluid is made by means of convection (see Fig. 4.11).

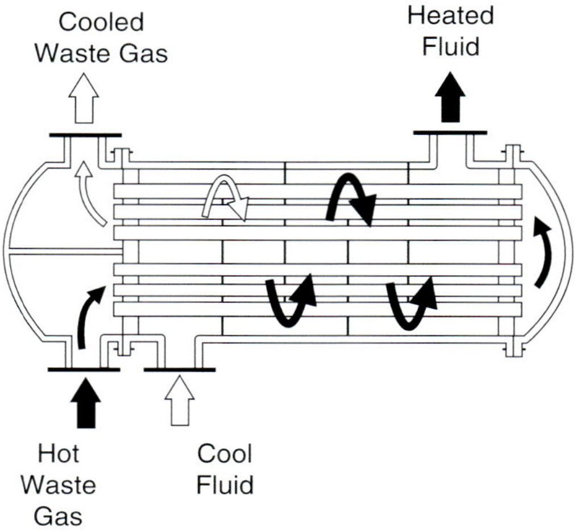

Fig. 4.10 Convective recuperator

Fig. 4.11 Typical commercial cross-flow convective recuperators

Hybrid Recuperator

For maximum effectiveness of heat transfer, hybrid recuperators are used. These are combinations of radiation and convective designs, with a high-temperature radiation section followed a by convective section (see Fig. 4.12).

These are more expensive than simple metallic radiation recuperators, but are less bulky.

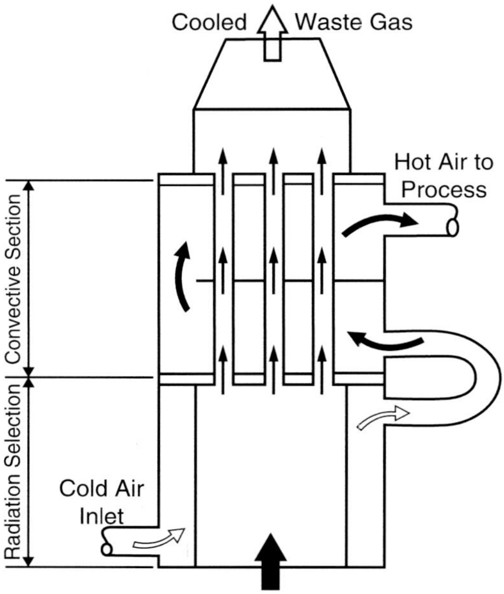

Fig. 4.12 Hybrid recuperator

Ceramic Recuperator

The principal limitation on the heat recovery of metal recuperators is the reduced life of the liner at inlet temperatures exceeding 1100 °C. In order to overcome the temperature limitations of metal recuperators, ceramic tube recuperators have been developed whose materials allow operation on the gas side to be at 1550 °C and on the preheated air side to be 815 °C on a more or less practical basis. Early ceramic recuperators were built of tile and joined with furnace cement, and thermal cycling caused cracking of joints and rapid deterioration of the tubes. Later developments introduced various kinds of short silicon carbide tubes, which can be joined by flexible seals located in the air headers (see Fig. 4.13).

Earlier designs had experienced leakage rates from 8 % to 60 %. The new designs are reported to last 2 years with air preheat temperatures as high as 700 °C, with much lower leakage rates.

4.3.1.2 Regenerative Heat Exchanger

A regenerative heat exchanger, or more commonly a regenerator, is a type of heat exchanger where heat from the hot fluid is intermittently stored in a thermal storage medium before it is transferred to the cold fluid. To accomplish this, the hot fluid is brought into contact with the heat storage medium; then the fluid is displaced with the cold fluid, which absorbs the heat.

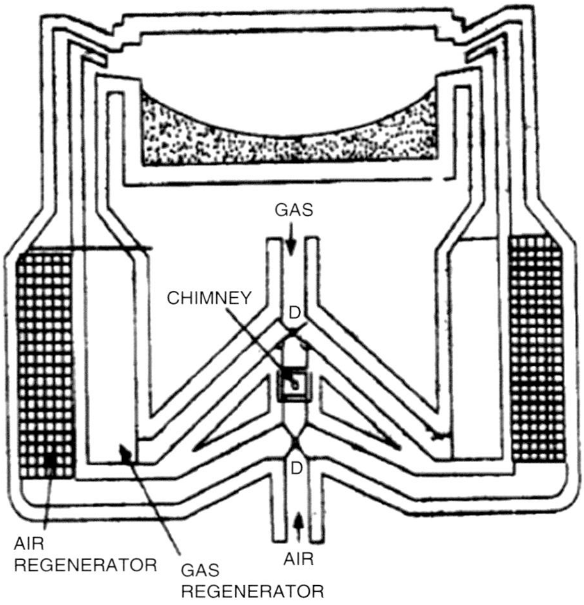

Fig. 4.13 Schematic of ceramic recuperator

In regenerative heat exchangers, the fluid on either side of the heat exchanger can be the same fluid. The fluid may go through an external processing step, and then it is flowed back through the heat exchanger in the opposite direction for further processing. Usually the application will use this process cyclically or repetitively.

The regenerator represents a class of heat exchangers in which heat is alternately stored and removed from a surface. This heat transfer surface is usually referred to as the matrix of the regenerator and for continuous operation; the matrix must be moved into and out of the fixed hot and cold fluid streams. In this case, the regenerator is called a rotary regenerator. If, on the other hand, the hot and cold fluid streams are switched into and out of the matrix, the regenerator is referred to as a fixed matrix regenerator.

Figure 4.14 illustrates typical regenerator heat exchangers, and Fig. 4.14a is a depiction of rotary, Fig. 4.14b fixed-matrix, and finally Fig. 4.14c shows a rotating hood of this type of heat exchanger.

Typical rotary regenerators or heat wheels are demonstrated in Fig. 4.15.

Any further detailed analysis, including advantages and disadvantages of this type of heat exchanger, is beyond the intended scope of this book and we refer readers to do their own researches and investigations.

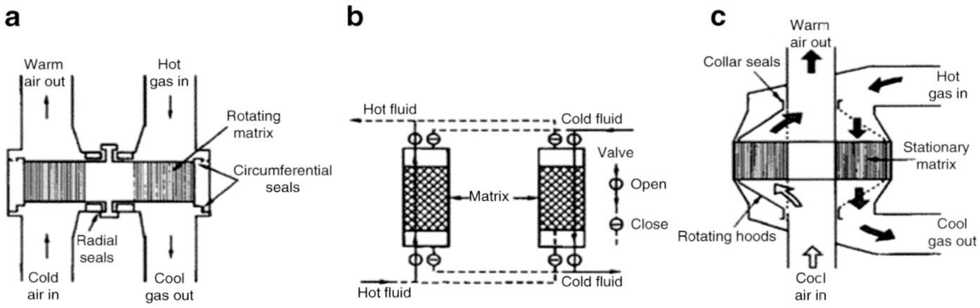

Fig. 4.14 (**a**) Rotary, (**b**) fixed-matrix, and (**c**) rotating hoods

Fig. 4.15 Typical rotary regenerators or heat wheels

4.3.1.3 Evaporative Heat Exchanger

Normally this type of heat exchanger can be found in any cooling tower of power plants all types and as its name of this heat exchanger sounds, it uses evaporation process to the heat transfer processes.

In an open circuit cooling tower, warm water from the heat source is evenly distributed via a gravity or pressurized nozzle system directly over a heat transfer surface called "fill" or "wet deck," while air is simultaneously forced or drawn through the tower, causing a small percentage of the water to evaporate. The evaporation process removes heat and cools the remaining water, which is collected in the tower's cold water basin and returned to the heat source (typically a water-cooled condenser or other heat exchanger). Figure 4.16 illustrates a cooling tower along with evaporative heat exchanger.

Maintenance frequency will depend largely upon the condition of the circulating water, the cleanliness of the ambient air used by the tower, and the environment in which the tower is operating.

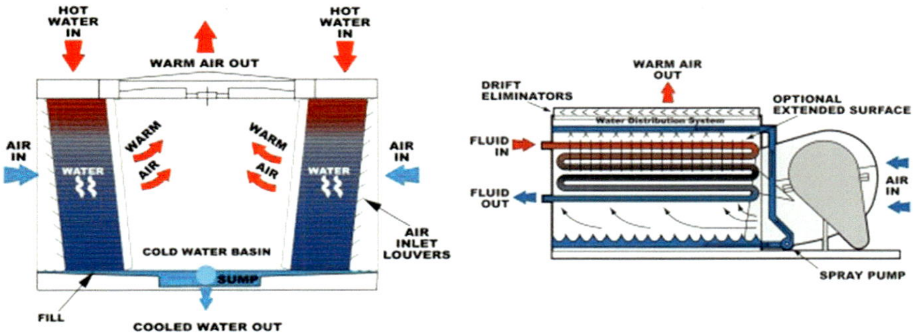

Fig. 4.16 Illustration of a cooling tower along with evaporative heat exchanger

4.3.2 *Compact Heat Exchangers*

A type of heat exchanger which is specifically designed to realize the larger heat transfer area per unit volume is designated or known as Compact Heat Exchanger. Compact heat exchangers are commonly used in gas-to-gas and gas-to-liquid (or liquid-to-gas) heat exchangers to counteract the low heat transfer coefficient associated with gas flow with increased surface area.

Figure 4.17 illustrates a typical configuration and structure of a gas-to-liquid compact heat exchanger for a residential air conditioning system.

Compact heat exchangers are a class of heat exchangers that incorporate a large amount of heat transfer surface area per unit volume, where the area density β parameter comes to play.

The ratio of the heat transfer surface area of a heat exchanger to its volume is called the area density β. A heat exchanger with $\beta = 700\ \mathrm{m^2/m^3}$ (or 200 $\mathrm{ft^2/ft^3}$) is classified as being compact. Examples of compact heat exchangers are car radiators (1000 $\mathrm{m^2/m^3}$), glass ceramic gas turbine heat exchangers (6000 $\mathrm{m^2/m^3}$), the regenerator of a Stirling engine (15,000 $\mathrm{m^2/m^3}$), and the human lung (20,000 $\mathrm{m^2/m^3}$).

A heat exchanger with $\beta > 700\ \mathrm{m^2/m^3}$ is classified as being a compact heat exchanger, for example,

- Car radiator ($\beta \approx 1000\ \mathrm{m^2/m^3}$)
- Gas ceramic in gas turbine heat exchanger ($\beta \sim 6000\ \mathrm{m^2/m^3}$)
- Human lung ($\beta \approx\ 20{,}000\ \mathrm{m^2/m^3}$)

Most automotive heat exchangers would come into the compact heat exchanger category since space is an extreme constraint for automotive applications. By the same talking recent research and study open air combined cycles at University of New Mexico by Zohuri et al. [8–11] and CO_2 closed combined cycle by a team of engineers at Sandia National Laboratory [12–14] and other folks cross universities and industry, utilizing the Brayton and Rankine cycles as a means of driving the efficiencies on next-generation (GEN IV) nuclear power plant with an Small

Fig. 4.17 A residential gas-to-liquid compact heat exchanger

Modular Reactors (SMR) Approaches, shows a promising application of this types of heat exchangers.

Types of compact heat exchangers are categorized as follows [15, 16]:

- Tubular heat exchanger
- Fin-plate heat exchanger
- Tube-fin heat exchanger
- Plate-frame heat exchanger
- Regenerative heat exchanger

Designing this type of heat exchangers is an easy task and the reasoning to argue why this is not that simple can be summarized below:

- Perform the required heat transfer and
  - Minimize size and weight
  - Minimize pressure drop
  - Meet required life
  - Be resistant to fouling and contamination
  - Minimize cost

Bear in mind, another important factor for designing these types of heat exchangers (Compact) in particular in their application for high temperature combined cycle for the next-generation nuclear power plants is pinch point and imposed golden rules driven by pinch technology, which is mentioned in Chap. 5 of the this text here, which makes them very expensive to manufacture.

Comparing the compact heat exchangers with traditional shell and tube heat exchangers can be summarized as follows:

- They occupy 80 % less space as compared to shell and tube heat exchangers.

- Overall heat transfer coefficient is three or four times higher as compared to shell and tube heat exchangers. These are very important advantages that are required by combined cycles in SMR to drive the higher thermal efficiency out from the cost of ownership to make them more desirable.
- Low pressure and low temperature devices.

Limitations of the compact heat exchangers that can be worth mentioning are:

- Maximum design pressure: 25 bar.
- Maximum design temperature: 200 °C.
- It is only used for clean fluids.

For further detailed information and related tables as well as correlated experimental data, we encourage readers to refer to the classical and well-known book by Kays and London [15] as well as a book edited by Shah [16].

4.4 Development of Gas Turbine

The gas turbine has experienced phenomenal progress and growth since its first successful development in the 1930s. The early gas turbines built in the 1940s and even 1950s had simple-cycle efficiencies of about 17 % because of the low compressor and turbine efficiencies and low turbine inlet temperatures due to metallurgical limitations of those times. Therefore, gas turbines found only limited use despite their versatility and their ability to burn a variety of fuels. The efforts to improve the cycle efficiency concentrated in three areas:

1. Increasing the turbine inlet (or firing) temperatures.
2. Increasing the efficiencies of turbo-machinery components.
3. Add modifications to the basic cycle. The simple-cycle efficiencies of early gas turbines were practically doubled by incorporating intercooling, regeneration (or recuperation), and reheating. The back work ratio of a gas-turbine cycle improves as a result of intercooling and reheating. However, this does not mean that the thermal efficiency will also improve. Intercooling and reheating will always decrease the thermal efficiency unless they are accompanied by regeneration. This is because intercooling decreases the average temperature at which heat is added, and reheating increases the average temperature at which heat is rejected. Therefore, in gas-turbine power plants, intercooling and reheating are always used in conjunction with regeneration. These improvements, of course, come at the expense of increased initial and operation costs, and they cannot be justified unless the decrease in fuel costs offsets the increase in other costs. In the past, the relatively low fuel prices, the general desire in the industry to minimize installation costs, and the tremendous increase in the simple-cycle efficiency due to the first two figures (Fig. 4.18 *T–s* diagram) increased efficiency options to approximately 40 % and left little desire for incorporating these modifications. With continued expected rise in demand and cost of producing electricity, these

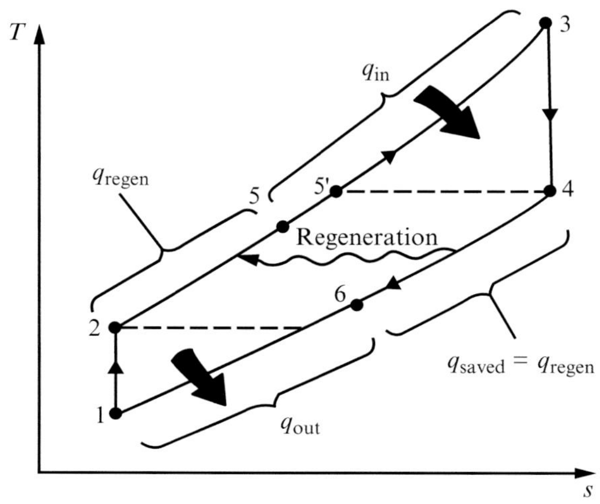

Fig. 4.18 *T–s* diagram of a Brayton cycle with regeneration

options will play an important role in the future of gas-turbine power plants. The purpose of this book is to explore this third option of increasing cycle efficiency via intercooling, regeneration, and reheating.

Gas turbines installed until the mid-1970s suffered from low efficiency and poor reliability. In the past, large coal and nuclear power plants dominated the base-load electric power generation (Point 1 in Fig. 4.19). Base load units are on line at full capacity or near full capacity almost all of the time. They are not easily nor quickly adjusted for varying large amounts of load because of their characteristics of operation [17]. However, there has been a historic shift toward natural gas-fired turbines because of their higher efficiencies, lower capital costs, shorter installation times, better emission characteristics, the abundance of natural gas supplies, and shorter start up times (Point 1 in Fig. 4.19). Now electric utilities are using gas turbines for base-load power production as well as for peaking, making capacity at maximum load periods, and in emergency situations because they are easily brought on line or off line (2 in Fig. 4.19). The construction costs for gas-turbine power plants are roughly half that of comparable conventional fossil fuel steam power plants, which were the primary base-load power plants until the early 1980s, but peaking units are much higher in energy output costs. A recent gas turbine manufactured by General Electric uses a turbine inlet temperature of 1425 °C (2600 °F) and produces up to 282 MW while achieving a thermal efficiency of 39.5 % in the simple-cycle mode. Over half of all power plants to be installed in the foreseeable future are forecast to be gas turbine or combined gas-steam turbine types (Fig. 4.19).

An overall combined gas-steam power cycle was touched upon in Sect. 4.3.1 and Fig. 4.20 demonstrates the difference between gas and steam turbine blades.

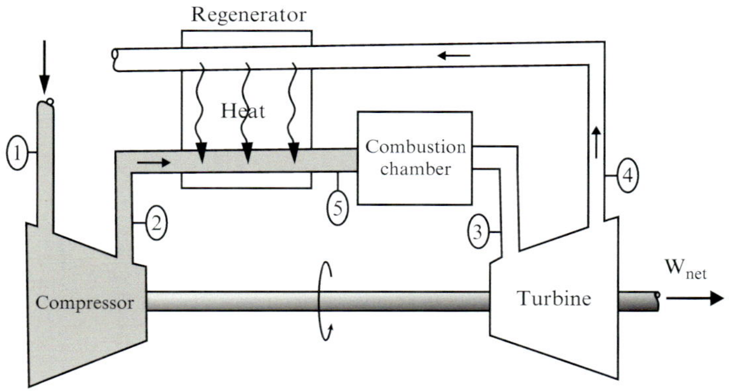

Fig. 4.19 A gas-turbine engine with recuperator

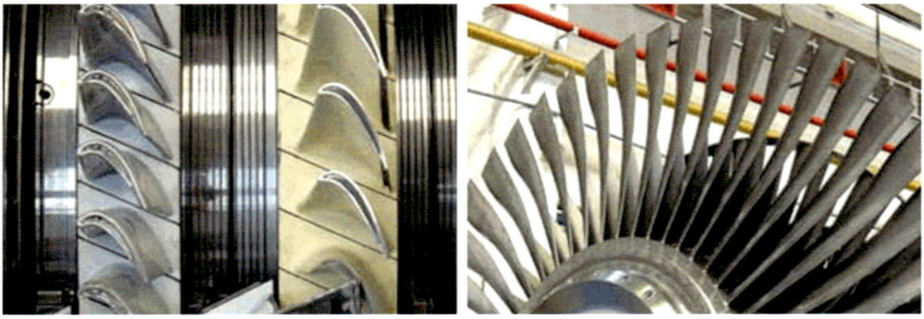

Fig. 4.20 Difference between gas turbine (*left*) and steam turbine (*right*) blades

Typical numerical examples for Brayton, Rankine, and combined cycles are presented below and they are supported by computer code that was developed by Zohuri and McDaniel [1] at University of New Mexico, Nuclear Engineering Department. The flow chart and overall capability of this code based on static approach is explained in Chaps. 8–10 of this book.

Numerical example:

- Simple Brayton efficiency: 36.8 %
- Simple Rankine efficiency: 35 %
- Combined Brayton and Rankine cycles: 56.4 %

4.5 Turbomachinery

We have used thermal engines widely since they were invented in the seventeenth century. There are many kinds of engines, and they are used in our life. The two turbomachinery that we face in nuclear plants are:

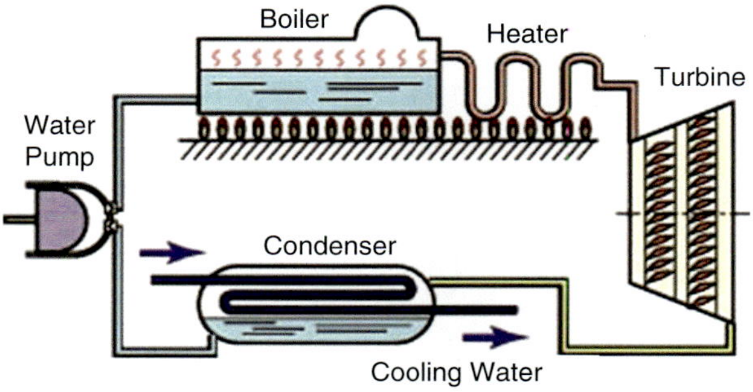

Fig. 4.21 Schematic of steam turbine

1. Gas turbine
2. Steam turbine

Our main focus will be on gas turbine where the Brayton cycle plays a big role and we extensively touch upon it in Chap. 5 and it is worth briefly talking about steam turbine here.

Figure 4.21 depicts a steam turbine layout in a simple form. The steam turbine has rotating blades instead of the piston and the cylinder of the reciprocating steam engine. This engine is used as the power source in the thermal and nuclear power plants. The steam turbine utilizes dynamic pressure of the steam, and converts a thermal energy to a mechanical energy, though the reciprocating steam engine utilizes the static pressure of the steam. Both engines use the energy that is obtained at the expansion of the steam.

Turbo machinery that is used in the new generation of nuclear power systems (GEN-IV) plays a significant role in commercial applications in order to produce the electricity of the future.

It is also good to know that any time we talk about steam, we will be dealing with Rankine cycle as rule of thumb.

- First-order estimates of key turbine and compressor design and performance characteristics can be made with low-level analysis. For the reference systems, key turbo machinery design parameters, (speed, stages, stage diameters, blade heights, blade clearances) will be similar to current commercial gas turbine engines.
- At lower reactor thermal powers He compressors will require greater than 3600 rpm operation to achieve efficiency goals (800 MW(t) allows 3600 rpm operation).
- Maximum system temperatures in the reference designs are near the limit for uncooled turbines.
- For both direct and indirect designs, the seals, housing, and bearing components will be fundamentally different than current gas turbines, requiring extensive development with the associated cost and risk.

These observations illustrate the complex interactions of the many design choices that will be considered in the NGNP PCS. It is clear that detailed and integrated design efforts must be performed on candidate designs before quantitative evaluations are possible. The assessment described in that study helped illuminate those critical design choices and the resulting implications for the cost and performance of the future NGNP PCS design.

4.6 Heat Transfer Analysis

Further analysis of heat transfer for any recuperative heat exchanger shows that the following governing equation stands by:

$$Q = \dot{m}_{H}c_{H}(t_{H1} - t_{H2}) = \dot{m}_{C}c_{C}(t_{C1} - t_{C2}) = UA_{0}(\text{LMTD})K \tag{4.3}$$

where $\dot{m}_H$ and $\dot{m}_C$ are the mass flow rates of the hot and cold fluids; c_H and c_C are the specific heats of the hot and cold fluids; A_0 is the outside area of the wall separating the two fluids; U is the overall heat transfer coefficient based on the outside area; and K is a multiplying factor for cross-flow and mixed-flow types ($K = 1$ for counterflow and parallel flow).

The temperature variations as shown in Fig. 4.22a–c can be shown for all three cases, where Fig. 4.22a shows temperature variations in counterflow with $\dot{m}_C c_C > \dot{m}_H c_H$ and Fig. 4.22b indicates temperature variations in counterflow with $\dot{m}_C c_C < \dot{m}_H c_H$ while Fig. 4.22c illustrates the temperature variations in parallel flow.

Using the concept of thermal resistance in series we have:

$$\frac{1}{UA_o} = \frac{1}{h_i A_i} + R_w + \frac{1}{h_o A_o} + F_i + F_o \tag{4.4}$$

where subscription o and i refer to the outside and inside surfaces of the separating wall; where F represents fouling factor that for most heat exchangers a deposit of salts, oil, or other contaminants will gradually build up on the heat transfer surfaces. This is allowed for in the design by using a *fouling factor*, F, in the form of an

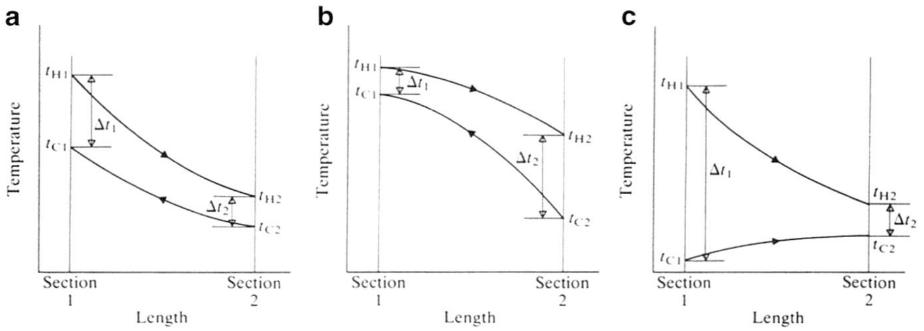

Fig. 4.22 Illustration of temperature variations

additional thermal resistance as in Eq. (4.4). Cleaning of the heat exchanger takes place when the fouling has reached the design value. As part of fouling and treatment, for convenience in general, we can classify them under one of the following six headings, depending on the mechanism causing the deposition and they are:

- Crystallization or precipitation fouling
- Particulate fouling (silting)
- Biological fouling
- Corrosion fouling
- Chemical reaction fouling
- Freezing or solidification fouling

4.7 Combined Cycle Power Plant

A combined cycle gas turbine power plant is essentially an electrical power plant in which a gas turbine and a steam turbine are used in combination to achieve greater efficiency than would be possible independently. The gas turbine drives an electrical generator while the gas turbine exhaust is used to produce steam in a heat exchanger, called a Heat Recovery Steam Generator (HRSG), to supply a steam turbine whose output provides the means to generate more electricity. If the steam were used for heat then the plant would be referred to as a cogeneration plant.

It is important first to distinguish between a closed cycle power plant (or heat engine) and an open cycle power plant. In a closed cycle, fluid passes continuously round a closed circuit, through a thermodynamic cycle in which heat is received from a source at higher temperature, and heat rejected to a sink at low temperature and work output is delivered usually to drive an electric generator.

A gas turbine power plant may simply operate on a closed circuit as shown in Fig. 4.23.

Most gas turbine plants operate in "open circuit," with an internal combustion system as shown in Fig. 4.24. Air fuel passes cross the single control surface into the compressor and combustion chamber, respectively, and combustion products leave the control surface after expansion through the turbine.

The classical combined cycle for power production in a gas turbine and steam plant is normally associated with the names of Brayton and Rankine, respectively.

Figure 4.25 is a simple representation of Combined Cycle Gas Turbine (CCGT) system. It demonstrates the fact that a CCGT system is two heat engines in series. The upper engine is the gas turbine. The gas turbine exhaust is the input to the lower engine (a steam turbine). The steam turbine exhausts heat to a circulating water system that cools the steam condenser.

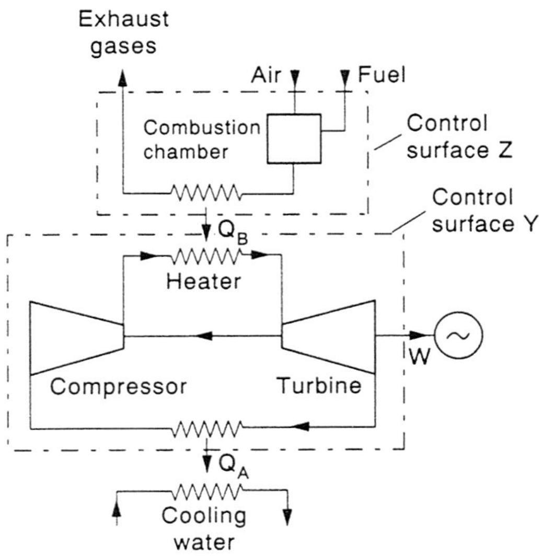

Fig. 4.23 Closed circuit gas turbine plant

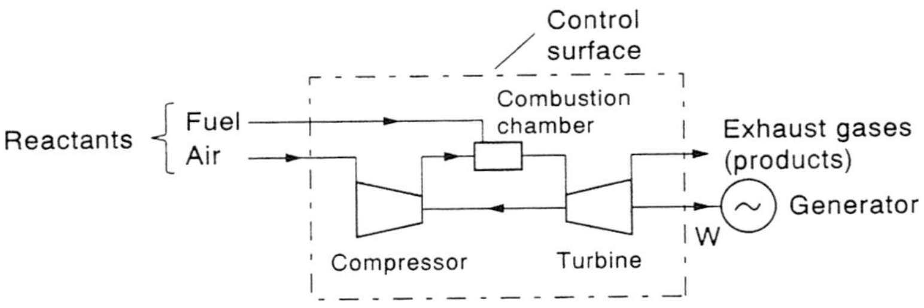

Fig. 4.24 Open circuit gas turbine plant

An approximate combined cycle efficiency (η_{CC}) is given by the equation

$$\eta_{CC} = \eta_B + \eta_R - (\eta_B \times \eta_R) \tag{4.5}$$

Equation (4.5) states that the sum of the individual efficiencies minus the product of the individual efficiencies equals the combined cycle efficiency. This simple equation gives significant insight to why combined cycle systems are successful.

For example, suppose the gas turbines efficiency (Brayton) η_B is 40 % (a reasonable value for a today's gas turbines) and that the steam turbine efficiency (Rankine) η_R is 30 % (a reasonable value for a Rankine cycle steam turbine).

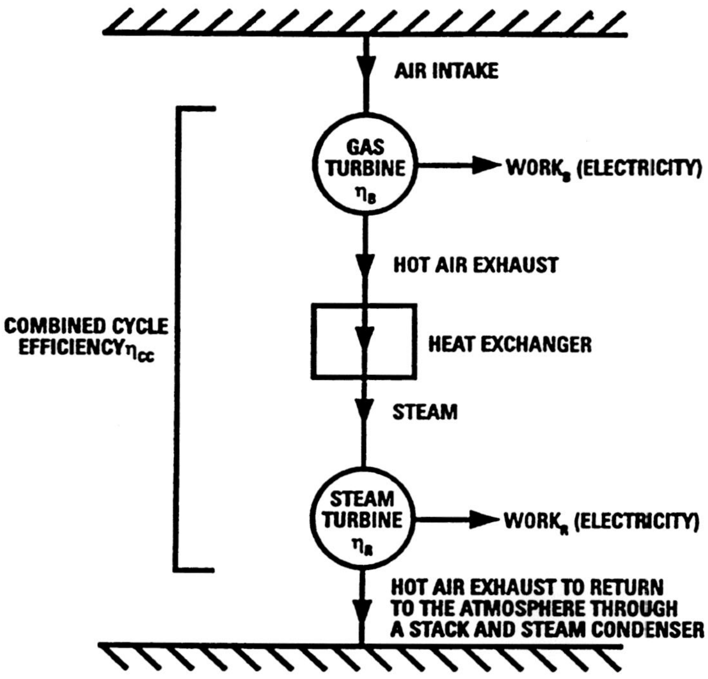

Fig. 4.25 Schematic of combined cycle (CCGT) plant

Utilizing Eq. (4.5) would lead to the following conclusion:

$$\eta_{CC} = 0.4 + 0.3 - (0.4 \times 0.3)$$
$$\eta_{CC} = 0.58$$
$$\eta_{CC} = 58\,\%$$

The combined cycle efficiency of 58 % is much greater than either the gas turbine or the steam turbine efficiencies separately. The 58 % value is slightly misleading in that system losses were ignored. However, efficiency values in the 60 % range have been recorded for CCGT systems in the past few years [7].

CCGT power plants come in many different configurations. Some companies choose to treat the gas turbine exhaust bypass stack as a commodity; others choose to incorporate a diverter damper into the turbine exhaust gas path. The diverter damper allows for the rapid configuration of the power plant as a combined cycle or simple cycle system. The initial cost of the diverter damper is much higher than the cost of treating the gas turbine exhaust stack as a commodity. However, the diverter damper allows for the gas turbines to be operated in simple cycle when HRSG or steam turbine repair or maintenance is required.

4.8 Advanced Computational Materials Proposed for Gen-IV Systems

A renewed interest in nuclear reactor technology has developed in recent years, in part as a result of international interest in sources of energy that do not produce CO_2 as a by-product. One result of this interest was the establishment of the Generation IV International Forum, which is a group of international governmental entities whose goal is facilitating bilateral and multilateral cooperation related to the development of new nuclear energy systems.

Historically, both the fusion and fission reactor programs have taken advantage of and built on research carried out by the other program. This leveraging can be expected to continue over the next 10 years as both experimental and modeling activities in support of the Gen-IV program grow substantially. The Gen-IV research will augment the fusion studies (and vice versa) in areas where similar materials and exposure conditions are of interest. However, in addition to the concerns that are common to both fusion and advanced fission reactor programs, designers of a future DT fusion reactor have the unique problem of anticipating the effects of the 14 MeV neutron source term. For example, advances in computing hardware and software should permit improved (and in some cases the first) descriptions of relevant properties in alloys based on ab initio calculations. Such calculations could provide the basis for realistic inter-atomic potentials for alloys, including alloy-He potentials that can be applied in classical molecular dynamics simulations. These potentials must have a more detailed description of many-body interactions than accounted for in the current generations which are generally based on a simple embedding function. In addition, the potentials used under fusion reactor conditions (very high pK_a energies) should account for the effects of local electronic excitation and electronic energy loss. The computational cost of using more complex potentials also requires the next generation of massively parallel computers. New results of ab initio and atomistic calculations can be coupled with ongoing advances in kinetic and phase field models to dramatically improve predictions of the non-equilibrium, radiation-induced evolution in alloys with unstable microstructures. This includes phase stability and the effects of helium on each microstructural component.

However, for all its promise, computational materials science is still a house under construction. As such, the current reach of the science is limited. Theory and modeling can be used to develop understanding of known critical physical phenomena, and computer experiments can, and have been used to, identify new phenomena and mechanisms, and to aid in alloy design. However, it is questionable whether the science will be sufficiently mature in the foreseeable future to provide a rigorous scientific basis for predicting critical materials' properties, or for extrapolating well beyond the available validation database.

Two other issues remain even if the scientific questions appear to have been adequately answered. These are licensing and capital investment. Even a high degree of scientific confidence that a given alloy will perform as needed in a particular Gen-IV or fusion environment is not necessarily transferable to the reactor licensing or capital market regimes. The philosophy, codes, and standards employed for reactor licensing are properly conservative with respect to design data requirements.

Experience with the US Nuclear Regulatory Commission suggests that only modeling results that are strongly supported by relevant, prototypical data will have an impact on the licensing process. In a similar way, it is expected that investment on the scale required to build a fusion power plant (several billion dollars) could only be obtained if a very high level of confidence existed that the plant would operate long and safely enough to return the investment.

These latter two concerns appear to dictate that an experimental facility capable of generating a sufficient, if limited, body of design data under essentially prototypic conditions (i.e., with ~14 MeV neutrons) will ultimately be required for the commercialization of fusion power. An aggressive theory and modeling effort will reduce the time and experimental investment required to develop the advanced materials that can perform in a DT fusion reactor environment. For example, the quantity of design data may be reduced to that required to confirm model predictions for key materials at critical exposure conditions. This will include some data at a substantial fraction of the anticipated end-of-life dose, which raises the issue of when such an experimental facility is required. Long lead times for construction of complex facilities, coupled with several years irradiation to reach the highest doses, imply that the decision to build any fusion-relevant irradiation facility must be made on the order of 10 years before the design data is needed.

Two related areas of research can be used as reference points for the expressed need to obtain experimental validation of model predictions. Among the lessons learned from ASCI, the importance of code validation and verification has been emphasized at the workshops among the courtiers involved with such research.

Because of the significant challenges associated with structural materials applications in these advanced nuclear energy systems, the *Workshop on Advanced Computational Materials Science*: *Application to Fusion and Generation IV Fission Reactors* was convened by the US Department of Energy's Office of Science and the Office of Nuclear Energy, Science and Technology to ensure that research funded by these programs takes full advantage of ongoing advancements in computational science and the department's investment in computational facilities. In particular, participants in the workshop were asked to:

1. Examine the role of high-end computing in the prediction of materials behavior under the full spectrum of radiation, temperature, and mechanical loading conditions anticipated for advanced structural materials that are required for future Generation IV fission and fusion reactor environments.
2. Evaluate the potential for experimentally validated computational modeling and simulation to bridge the gap between data that is needed to support the design of these advanced nuclear technologies and both the available database and data that can be reasonably obtained in currently available irradiation facilities.

Like the requirements for advanced fusion reactors, the need to develop materials capable of performing in the severe operating environments expected in Generation IV reactors represents a significant challenge in materials science. There is a range of potential Gen-IV fission reactor design concepts and each concept has its own unique demands. Improved economic performance is a major goal of the Gen-IV designs. As a result, most designs call for significantly higher operating temperatures than the current generation of LWRs to obtain higher thermal efficiency. In many cases, the desired operating temperatures rule out the use of the structural alloys employed today. The very high operating temperature (up to 1000 °C) associated with the Next Generation Nuclear Plant (NGNP) is a prime example of an attractive new system that will require the development of new structural materials.

The operating temperatures, neutron exposure levels, and thermomechanical stresses for proposed Gen-IV fission reactors are huge technological challenges among material scientists and engineers. In addition, the transmutation products created in the structural materials by the high energy neutrons produced in this generation of nuclear power reactors can profoundly influence the microstructural evolution and mechanical behavior of these materials.

4.9 Material Classes Proposed for Gen-IV Systems

The types of materials that were proposed in a DOE workshop in March of 2004 are tabulated in Table 4.3.

4.10 Generation IV Materials Challenges

A summary of these challenges for the next-generation nuclear power plants are presented here. They are:

- Higher temperature/larger temperature ranges
  - Examples

    VHTR coolant outlet temperature near 1000 °C
    GFR transient temps to 1600–1800 °C, gradient across core of ~400 °C
    LFR to 800 °C steady-state outlet
  - Issues

    Creep
    Fatigue
    Toughness
    Corrosion/SCC

Table 4.3 Structural materials

System	Structural materials						
	Ferritic martensitic stainless steel alloys	Austenitic stainless steel alloys	Oxide dispersion strengthened steels	Ni-based alloys	Graphite	Refractory alloys	Ceramics
GFR	P	P	P	P	P		P
LFR	P	P	S			S	S
MSR				P	P	S	S
SFR	P	P	P				
SCWR-thermal spectrum	P	P	S	S			
SCWR-fast spectrum	P	P	S	S			
VHTR	S			P	P	S	P

P primary, *S* secondary

- *Must drive modeling toward a predictive capability of materials properties in complex alloys across a wide temperature range.*
- High flounce dose
  - Examples

    LFR, SFR cladding
    SCWR core barrel
    GFR matrix
  - Issues

    Swelling
    Creep, stress relaxation
- *Must drive modeling toward a predictive capability of materials properties in complex alloys to large radiation dose.*
- Unique chemical environments
  - Examples

    Pb and Pb-Bi eutectic
    Supercritical water
    High temperature oxidation in gas-cooled systems
    Molten salts
  - Issues

    Corrosion
    SCC/IASCC
    Liquid metal embrittlement
- *Must drive modeling toward a predictive capability of chemical interactions in complex alloys to large radiation dose.*

4.11 Generation IV Materials Fundamental Issues

The coevolution of all components of the microstructure, and their roles in the macroscopic response in terms of swelling, anisotropic growth, irradiation creep, and radiation-induced phase transformations should be studied within the science of complex systems. See Fig. 4.26.

In summary, we can conclude that

- Six concepts have been identified with the potential to meet the Generation IV goals.
- Concepts operate in more challenging environments than current LWRs and significant material development challenges must be met for any of the Generation IV systems to be viable.

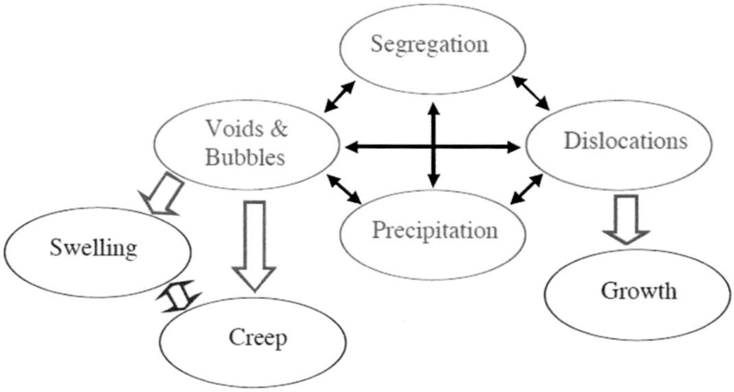

Fig. 4.26 Flow chart of materials fundamental issues

- Experimental programs cannot cover the breadth of materials and irradiation conditions for the proposed Gen-IV reactor designs.
- Modeling and microstructural analysis can provide the basis for a material selection that is performed based on an incomplete experimental database and that requires considerable judgment to carry out the necessary interpolation and extrapolation.

4.12 Capital Cost of Proposed Gen IV Reactors

Different Power Conversion System (PCS) design trades may have substantial effects on the system capital cost. System optimization is typically complex, because, for example, increased PCS cost can increase cycle efficiency, reducing the reactor capital cost. The Generation IV Economic Modeling Working Group [EMWG, 2004] recommends two methodologies for modeling economics costs, a top-down method based on scaling and detailed information about similar systems, and a bottom-up method based on detailed accounting for all construction commodities, plant equipment, and labor-hours. For top-down methods, the EMWG recommends [EMWG, 2004]:

> The first task is to develop a reference design to which cost estimating techniques can be applied. The cost estimating part of this task generally is accomplished by considering the costs of equipment used for similar type projects and then scaling the equipment upwards or downwards. As an example, one might start cost estimating work on the Very High Temperature Reactor (VHTR) by scaling reactor plant equipment from a project for which detailed estimates are available, such as the General Atomics HTGR.

For the purpose of system comparison, the top-down method was adopted to estimate PCS parameters that are important in scaling relative capital costs.

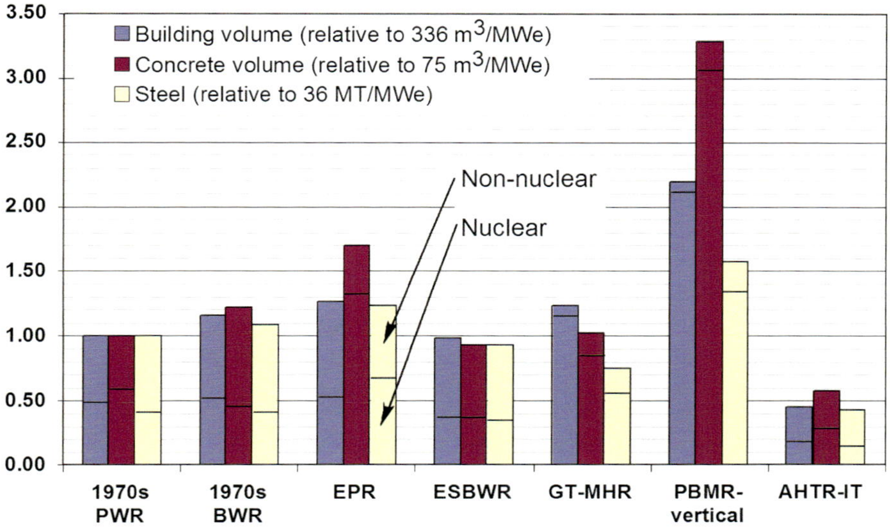

Fig. 4.27 Comparison of the total building volumes, and total plant steel and concrete inputs, for the reference HTR and LWR systems considered

The measures selected were those typically calculated to provide input for system cost estimates, and thus provide a basis for rough comparisons of system options. To provide an approximate baseline for comparison, where possible, comparisons were made with Gen II and Gen III+ light water reactor values. Fig. 4.26 shows such a comparison, quantifying steel and concrete inputs for the reference systems considered in the study. Several insights can be drawn from Fig. 4.26. For example, the 1500-MWe passive ESBWR light-water reactor has slightly smaller inputs than the 1970s light water reactors, as well as the evolutionary Environmental Program Requirements (EPR).

But Fig. 4.27 also shows that it is possible to build high-temperature gas-cooled reactors, e.g., the 286 MWe GT-MHR, with smaller construction material inputs than for light water reactors, due to the higher thermodynamic efficiency and power density. This shows that it is possible, with high-temperature gas power cycle technology, to break the economic scaling of the large light water reactors. This study also suggests that high-temperature, high efficiency gas-cycle power conversion can be adapted to other advanced reactor systems. For example, the even smaller inputs for the high-temperature, liquid-cooled, 1235-MWe AHTR-IT show that scaling economies may exist for high-temperature reactors. However, the material inputs for high-temperature reactors can be sensitive to equipment design choices and configurations, as shown by the differences in Fig. 4.26 between the GT-MHR and the PBMR. Thus, careful attention to design trade-offs is clearly important in the design of power conversion systems.

The selected capital costs which have been calculated for the reference systems in the study and are presented by the report from the UC Berkeley team [18] in Chap. 2 in more detail are based on the volumes of materials used:

- Structures costs:
  - Building volume ($m^3/MW(e)_{ave}$) (nuclear/non-nuclear)
  - Concrete volume ($m^3/MW(e)_{ave}$) (nuclear/non-nuclear)
- Reactor and PCS cost:
  - Reactor power density ($m^3/MW(e)_{ave}$)
  - PCS power density ($m^3/MW(e)_{ave}$) (nuclear/non-nuclear)
  - System-specific steel ($MT/MW(e)_{ave}$) (nuclear/non-nuclear)
  - Turbo machinery specific volume ($m^3/MW(e)_{ave}$)
  - System specific helium ($kg/MW(e)_{ave}$) (nuclear/non-nuclear) (nonrenewable resource, correlates with building volume (blow-down))

For each of these figures of merit, the values for the nuclear and non-nuclear portions of the plant were estimated. This division recognizes the difference in costs for procuring and installing nuclear-grade materials. For example, for concrete and reinforcing steel, material costs are estimated to be 65 % greater for nuclear-grade materials, and installation costs 30 % greater.

4.12.1 Economic and Technical of Combined Cycle Performance

The output and efficiency of combined cycle plants can be increased during the design phase by selecting the following features [19]:

- Higher steam pressure and temperature
- Multiple steam pressure levels
- Reheat cycles

Additional factors are considered if there is a need for peak power production. They include gas turbine power augmentation by water or steam injection or a supplementary-fired Heat Recovery Steam Generator (HRSG). If peak power demands occur on hot summer days, gas turbine inlet evaporative cooling and chilling should be considered. Fuel heating is another technique that has been used to increase the efficiency of combined cycle plants.

The ability of combined cycle plants to generate additional power beyond their base capacity during peak periods has become an important design consideration. During the last decade, premiums were paid for power generated during the summer peak periods. The cost of electricity during the peak periods can be 70 times more expensive than off-peak periods. Since the cost during the peak periods is much higher, most of the plant's profitability could be driven by the amount of power generated during these peak periods. Thus, plants that can generate large quantities of power during the peak periods can achieve the highest profits

4.12.2 Economic Evaluation Technique

Plant output and efficiency are carefully considered during the initial plant design because they impact the cost of electricity in combination with fuel costs, plant capital cost, cost of capital, and electricity sales. These factors will drive the gas turbine selection as well as the bottoming cycle design in combined-cycle operation.

As fuel costs increase, cycle selections typically include higher steam pressures, multiple steam pressure levels, reheat cycles, and higher steam temperatures. Once these selections have been made, other factors are addressed. Is there a need for peak power production with premiums paid for the resulting power? If so, gas turbine power augmentation by way of water or steam injection or a supplementary fired heat recovery steam generator (HRSG) may be the solution. Do peak power demands occur on a hot day (summer peaking)? This may suggest a potential benefit from some form of gas turbine inlet evaporative cooling or chilling [20].

For existing plants, some performance enhancement options can also be economically retrofitted to boost power output and efficiency. Although this research's primary focus is on options that enhance output, a brief discussion of fuel gas heating, which is a technique used to enhance combined-cycle plant efficiency, is provided.

The ability of utilities and Independent Power Producers (IPPs) to generate additional power beyond a plant's base capacity during summer peak power demand periods has become an important consideration in the design of combined-cycle plant configurations. In recent years, utilities and IPPs within the USA have received premiums for power generation capacity during summer peak power demand periods. The price of electricity varies greatly as a function of annual operating hours. The variation is also highly region dependent. With price-duration curves that are sharply peaked, implying a few hours annually with very high rates, the majority of a plant's profitability could be driven by the high peak energy rates that can be achieved over a relatively short period of time. Thus, a plant that can economically dispatch a large quantity of additional power could realize the largest profits.

While current market trends should be considered during the design and development phase of a combined-cycle facility, forecasts of future market trends and expectations are equally important and warrant design considerations.

One of the primary challenges facing developers of new combined-cycle plants, as well as owner/operators of existing plants, is the optimization of plant revenue streams. As a result of escalating peak energy rates and peak demand duration, significant emphasis has been placed on developing plant designs that maximize peak power generation capacity while allowing for cost-effective, efficient operation of the plant during non-peak power demand periods.

In addition to maximizing plant profitability in the face of today's marketplace, expectations of future market trends must be considered. Therefore, the goal is to determine which performance-enhancement options or combination of options can be applied to a new or existing combined-cycle plant to maximize total plant profits on a plant life-cycle basis.

With very few exceptions, the addition of power-enhancement techniques to a base plant configuration will impact base load performance negatively and, hence, affect a plant's net revenue generating capability adversely during nonpeak periods [21].

In general, efficiency is the predominating economic driver during non-peak generating periods, while capacity dominates the economic evaluation during peak power demand periods. Thus, it is extremely important to develop an economic model that considers both the Cost of Electricity (COE) during non-peak periods while taking into consideration expectations of peak energy rates.

After having established baseline peak and nonpeak period performance levels for the various power-enhancement alternatives, a COE analysis technique is applied to determine alternatives that would afford the best overall life-cycle benefit. In addition to including both peak and non-peak performance levels, the COE model includes the split between annual peak and non-peak operating hours, the premium paid for peak power generation capacity, the cost of fuel, plant capital cost, the incremental capital cost of the enhancements, and the cost to operate and maintain the plant. This COE model then can be used to determine the sensitivity of a given power-enhancement alternative with respect to the economic parameters included within it [21].

Most peak power enhancement opportunities exist in the topping cycle (gas turbine) as opposed to the bottoming cycle (HRSG/steam turbine). In general, with the exception of duct firing within the HRSG, there are few independent design enhancements that can be made to a bottoming cycle that has already been fully optimized to achieve maximum plant performance. However, in general, performance enhancements to the gas turbines will carry with them an increase in bottoming cycle performance due to an associated increase in gas turbine exhaust energy [21].

4.12.3 Output Enhancement

The two major categories of plant output enhancements are:

1. Gas turbine inlet air cooling
2. Power augmentation

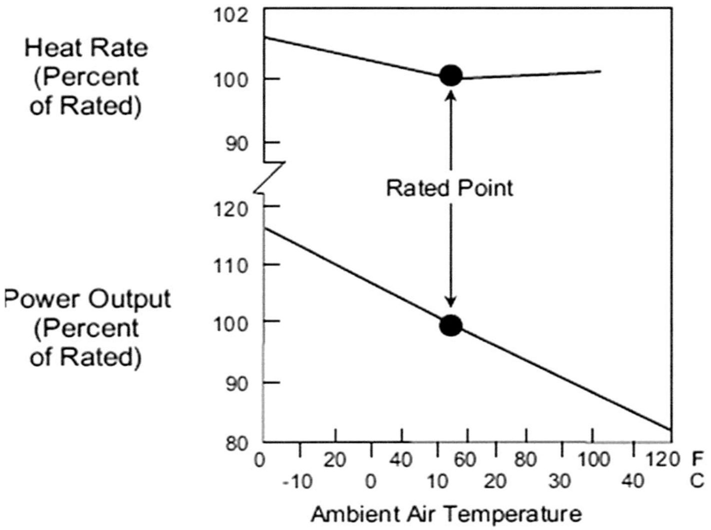

Fig. 4.28 Combined-cycle system performance variation with ambient air temperature [21]

4.12.3.1 Gas Turbine Inlet Air Cooling

Industrial gas turbines operating at constant speed have a constant volumetric flow rate. Since the specific volume of air is directly proportional to temperature, cooler air has a higher mass flow rate. It generates more power in the turbine. Cooler air also requires less energy to be compressed to the same pressure as warmer air. Thus, gas turbines generate higher power output when the incoming air is cooler [22].

A gas turbine inlet air cooling system is a good option for applications where electricity prices increase during the warm months. It increases the power output by decreasing the temperature of the incoming air. In combined cycle applications, it also results in improvement in thermal efficiency. A decrease in the inlet dry-bulb temperature by 10 °F (5.6 °C) will normally result in around a 2.7 % power increase of a combined cycle using heavy duty gas turbines. The output of simple-cycle gas turbines is also increased by the same amount.

Figure 4.28 shows that a 10 °F (5.6 °C) reduction in gas turbine inlet dry-bulb temperature for heavy-duty gas turbines improves combined cycle output by about 2.7 %. The actual change is somewhat dependent on the method of steam turbine condenser cooling being used. Simple cycle output is improved by a similar percentage.

Several methods are available for reducing gas turbine inlet temperature. There are two basic systems currently available for inlet cooling. The first and perhaps the most widely accepted system is evaporative cooling. Evaporative coolers make use of the evaporation of water to reduce the gas turbine's inlet air temperature.

The second system employs various ways to chill the inlet air. In this system, the cooling medium (usually chilled water) flows through a heat exchanger located in the inlet duct to remove heat from the inlet air.

Evaporative cooling is limited by wet-bulb temperature. Chilling, however, can cool the inlet air to temperatures that are lower than the wet bulb temperature, thus providing additional output albeit at a significantly higher cost.

Depending on the combustion and control system, evaporative cooling may reduce NOx emissions; however, there is very little benefit to be gained from current dry low NOx technology. This is another avenue that requires further analysis and investigation as well as collaboration between scientific communities, national laboratories, and industries.

4.12.3.2 Power Augmentation

Three basic methods are available for power augmentation: water or steam injection, HRSG supplementary firing, and peak firing.

1. Gas turbine steam/water injection
2. Supplementary fired HRSG
3. Peak firing

These are the three methods that general electric is suggesting and they need to be investigated further by nuclear power manufacturers and the community involved with enhancing Nuclear Power Energy Efficiency [19] using combined cycle technology.

Others aspects of the cost of producing electricity are generally expressed in US $/MWh or US cts/kWh, depending on following parameters [22]:

- Capital cost of the project
- Fuel cost
- Operation and maintenance cost

The capital cost per unit of electricity for a given power plant depends on the following elements:

- Investment cost
- Financing structure
- Interest rate and return on equity
- Load factor of the plant (or equivalent utilization time)

The investment costs are the sum of the following positions:

- Power plant contract prices(s)
- Interest during construction (depending upon the construction time)
- Owner's cost for the realization of the project (project manager, owner's engineer, land cost, etc.)

The financing structure is defined by the debt-to-equity ratio of the financing and the return on equity is the return expected by the investors on their capital. Both are linked to the risks of the project.

The load factor results from the type of application the plant is intended for: base, intermediate or peak load operation, and the availability and reliability of the power station.

Fuel costs per unit of electricity are proportional to the specific price of the fuel, and inversely proportional to the average electrical efficiency of the installation. This average electrical efficiency must not be mixed up with the electrical efficiency at rated load. It is defined as follows:

$$\overline{\eta} = \eta \cdot \eta_{\text{Oper}} \tag{4.6}$$

where:

η is the electrical net efficiency at rated load. (This is the % of the fuel that is converted into electricity at rated load for a new and clean condition.)

η_{Oper} is the operating efficiency, which takes into account the following losses:

- Start-up and shutdown losses
- Higher fuel consumption for part load operation
- Aging and fouling of the plant

References

1. Zohuri B. (2014). *Innovative combined Brayton open cycle systems for the next generation nuclear power plants.* Ph.D. Dissertation, Nuclear Engineering Department, University of New Mexico.
2. Eastop, T. D., & Croft, D. R. (1990). *Energy efficiency*. London: Longman.
3. Zohuri, B., & McDaniel, P. J. (2014). *Thermodynamics in nuclear power plant systems.* New York: Springer.
4. Zohuri, B., & Fathi, N. (2015). *Thermal-hydraulic analysis of nuclear reactors.* New York: Springer.
5. Necati Ozisik, M. (1985). *Heat transfer: A basic approach.* New York: McGraw-Hill.
6. Incropera, F. P., & DeWitt, D. P. (1990). *Fundamentals of heat and mass transfer* (3rd ed., pp. 658–660). New York: Wiley.
7. Incropera, F. P., DeWitt, D. P., Bergman, T. L., & Lavine, A. S. (2006). *Fundamentals of heat and mass transfer* (6th ed., pp. 686–688). New York: Wiley.
8. Zohuri, B., McDaniel, P. J., & de Olivera, C. R. (2015). Advanced Nuclear Open Air-Brayton Cycles for Highly Efficient Power Conversion. *Nuclear Technology*, *192*(1), 48–60.
9. Zohuri, B., McDaniel, P. J., & de Olivera, C. (2014, June). A comparison of a recuperated open cycle (Air) Brayton power conversion system with the traditional steam Rankine cycle for the next generation nuclear power plant. *ANS Transactions*, Reno, NV.
10. McDaniel, P. J., Zohuri, B., & de Olivera, C. (2014, November). A combined cycle power conversion system for small modular LMFBRs. *ANS Transactions*, Anaheim, CA.
11. McDaniel, P. J., Zohuri, B., de Oliveira, C., & Cole, J. (2012, November). A combined cycle power conversion system for the next generation nuclear power plant. *ANS Transactions*, San Diego, CA.

12. http://energyfromthorium.com/2014/04/04/closed-loop-brayton-cycle-sandia-national-laboratory/.
13. Pasch, J., Conboy, T., Fleming, D., & Rochau, G. (2012). Supercritical CO_2 recompression Brayton cycle: Completed assembly description. Sandia Report. SAND2012-9546. Unlimited release. Printed Oct 2012.
14. http://www.netl.doe.gov/publications/proceedings/11/utsr/pdf/wed/Wright%20SCO2%20Power%20Cycle%20Summary%20UTSR%202011%20v2a.pdf.
15. Kays, W. M., & London, A. L. (1984). *Compact heat exchangers* (3rd ed.). New York: McGraw-Hill.
16. Shah, R. K. (Ed.). (1997). *Compact heat exchangers for the process industries*. New York: Begell House, Inc.
17. Pansini, A. J., Smalling, K. D., & Lilburn, G. A. (1991). *Guide to electric power generation*. Lilburn, GA: The Fairmont Press, Inc.
18. Peterson, P., Zhao, H., Ballinger, R., Fuller, R., Forsha, M., Oh, C., et al. (2004). *Next generation nuclear plant power conversion study: Technology options assessment*. Washington, DC: US Department of Energy.
19. Jones, C., & Jacob III, J. Economic and technical considerations for combined-cycle.
20. Langston, L. S., & Opdyke, G. (1997). *Introduction to gas turbine for non-engineers*. Global Gas Turbine News (Vol. 37, No. 2).
21. Jones, C., & Jacobs, J. A., III. (2000). *Economic and technical considerations for combined-cycle performance-enhancement options*. Schenectady, NY: GE Power Systems. GER-4200.

Chapter 5
Air Brayton Cycles for Nuclear Power Plants

Power generation is an important issue today, especially on the West Coast. Demand is outweighing supply because of lack of incentives for the utilities industry to build additional power plants over the past 10–20 years.

5.1 Background

The major growth in the electricity production industry in the last 30 years has centered on the expansion of natural gas power plants based on gas turbine cycles. The most popular extension of the simple Brayton gas turbine has been the combined cycle power plant with the Air-Brayton cycle serving as the topping cycle and the Steam-Rankine cycle serving as the bottoming cycle. The Air-Brayton cycle is an open air cycle and the Steam-Rankine cycle is a closed cycle. The Air-Brayton cycle for a natural gas driven power plant must be an open cycle, where the air is drawn in from the environment and exhausted with the products of combustion to the environment. The hot exhaust from the Air-Brayton cycle passes through a Heat Recovery Steam Generator (HRSG) prior to exhausting to the environment in a combined cycle. The HRSG serves the same purpose as a boiler for the conventional Steam-Rankine cycle.

In 2007, gas turbine combined cycle plants had a total capacity of 800 GW and represented 20 % of the installed capacity worldwide. They have far exceeded the installed capacity of nuclear plants, though in the late 1990s they had less than 5 % of the installed capacity worldwide [1]. There are a number of reasons for this. First natural gas is abundant and cheap. Second combined cycle plants achieve the greatest efficiency of any thermal plant. And third, they require the least amount of waste heat cooling water of any thermal plant.

A typical gas turbine plant consists of a compressor, combustion chamber, turbine, and an electrical generator. A combined cycle plant takes the exhaust

B. Zohuri, *Application of Compact Heat Exchangers For Combined Cycle Driven Efficiency In Next Generation Nuclear Power Plants*,
DOI 10.1007/978-3-319-23537-0_5

from the turbine and runs it through a Heat Recovery Steam Generator (HRSG) before exhausting to the local environment. The HRSG serves the function of the boiler for a typical closed cycle steam plant. The steam plant consists of a steam turbine, a condenser, a water pump, an evaporator (boiler), and an electrical generator. In a combined cycle plant the gas turbine and steam turbine can be on the same shaft to eliminate the need for one of the electrical generators. However the two shafts, two generator systems provide a great deal more flexibility at a slightly higher cost. In addition to the closed loop for the steam, an open loop circulating water system is required to extract the waste heat from the condenser. The waste heat extracted by this "circulating" water system is significantly less per megawatt for a combined cycle system as the open Brayton cycle exhausts its waste heat directly to the air.

The layout for the components of a typical combined cycle power plant is given in Fig. 5.1 as a typical example of such cycle.

GE currently markets a system that will produce 61 % efficiency at design power and better than 60 % efficiency down to 87 % of design power [2] for gas turbine combined cycle plants.

An approximate efficiency can be calculated for a combined cycle power plant by the following simple argument [3].

Brayton cycle efficiency $= \frac{W_B}{Q_{in}} = \eta_B$

Heat to Rankine cycle $= Q_R = (1 - \eta_B)Q_{in}$

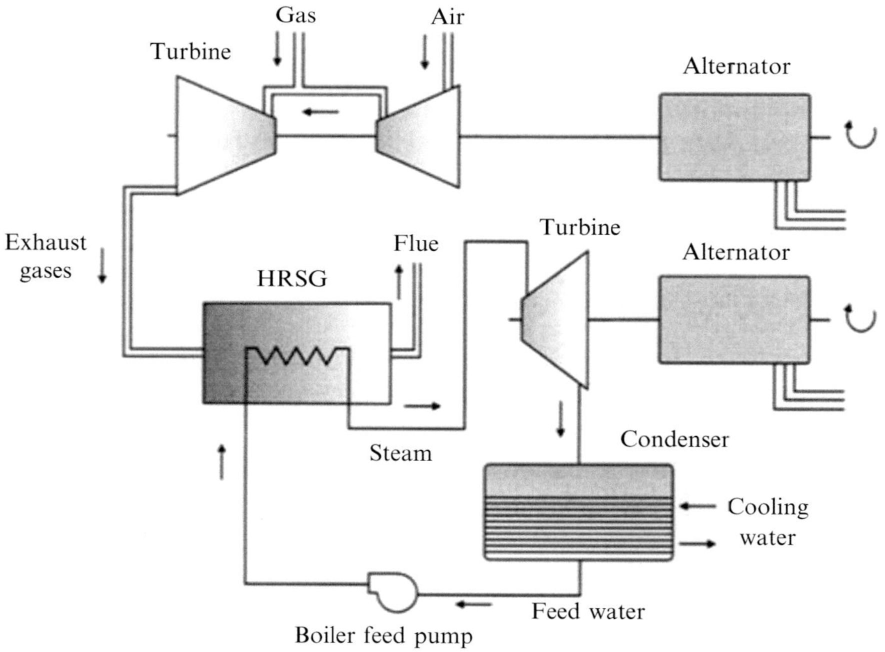

Fig. 5.1 Typical gas turbine combined cycle power plant

Rankine cycle efficiency $= \frac{W_R}{Q_R} = \eta_R$

Overall efficiency $= \frac{W_B + W_R}{Q_{in}} = \eta_T = \frac{\eta_B Q_{in} + \eta_R Q_R}{Q_{in}} = \frac{\eta_B Q_{in} + \eta_R (1-\eta_B) Q_{in}}{Q_{in}} = \eta_B + \eta_R - \eta_B \eta_R$

$$\eta_T = \eta_B + \eta_R - \eta_B \eta_R$$

This efficiency has to be corrected for pressure losses and assumes that all of the heat in the Brayton exhaust is used in the HSRG. For a combustion gas turbine this is not usually possible if condensation of the water in the exhaust products is to be avoided. The detailed models developed in this effort give a more accurate answer.

For a nuclear system to take advantage of combined cycle technology, there are a number of changes to the plant components that have to be made. The most significant of course is that the combustion chamber has to be replaced by a *Heat Exchanger* in which the working fluid from the nuclear reactor secondary loop is used to heat the air. The normal Brayton cycle is an internal combustion one where the working fluid is heated by the combustion of the fuel with the air in the combustion chamber. The walls of the combustion chamber can be cooled and peak temperatures in the working fluid can be significantly above the temperature that the walls of the chamber can tolerate for any length of time.

For the nuclear reactor system the heat transfer is in the opposite direction. All reactor components and fluids in the primary and secondary loops must be at a higher temperature than the peak temperature of the gas exiting the heat exchanger. This severely restricts the peak temperature that can be achieved for the air entering the turbine. However, all is not lost.

In a typical combustion system, there are pressure losses approaching 5 % of the total pressure to complete the combustion process [4]. Heat exchangers can be built with significantly lower pressure drops than 5 % approaching 1 % [5]. So the most straightforward method to overcome this severe temperature limitation is to borrow a technique from steam power plants and implement multiple reheat cycles. That is the first heat exchanger heats the air to its peak temperature. Then the air is expanded through the first turbine. The air is then reheated to the same peak temperature and expanded through the second turbine. Based on the relative pressure losses that appear possible, up to five turbines might be considered. All five turbines will be driving the same compressor. Multiple compressors on concentric shafts [4] driven by different sets of turbines might be possible, but that has not been considered here.

Multiple reheat cycles allow more heat to be put into the working fluid at a higher temperature. This improves the efficiency of the overall cycle. Interestingly enough it does not improve the efficiency of the Brayton cycle, but because the exit temperature from the last turbine is higher, it does improve the efficiency of the Rankine part of the cycle with a net gain for the overall cycle. For this to work, the reactor coolant temperature must reach temperatures significantly higher than current Light Water Reactor temperatures. Even sodium cooled reactor exit temperatures in the 550 °C range are not quite high enough to get a nuclear Air Brayton combined cycle to compete with the efficiency of a pure Steam Rankine cycle. But when the coolant exit temperatures reach the 650–700 °C ranges, the combined cycle systems with multiple turbines surpass the performance of Steam Rankine systems. So the analysis that follows is targeted at a molten salt reactor, or a lead coolant reactor. A pressurized sodium reactor that could reach these temperatures is another possibility but no one is proposing such a system at this time. It could also apply to a High Temperature Gas Reactor but the heat exchangers would be quite different. Gas to gas heat exchangers have not been considered as the primary heat exchangers at this point, but will be addressed in the recuperated systems.

Liquid metal and molten salt heat exchangers were developed and tested successfully during the Aircraft Nuclear Propulsion program in the late 1950s [6–8]. They were conventional tube and plate exchangers and were tested for over 1000 h at temperatures up to 1100 K. The largest size tested transferred 55 MW of heat in a package of approximately 1.2 m^3. The heat transfer area on the air side had a surface area per unit volume of 1180 m^2/m^3. Certainly some development will be needed to bring this technology up to modern standards and get NRC approval for a power producing reactor. But the tasks involved do not appear insurmountable. A number of additional heat exchangers were designed in this work in an attempt to estimate sizes of components and validate that pressure drop criteria could be met. The heat from the exhaust of the Brayton cycle transfers heat to vaporize the steam in the Rankine cycle in an HRSG of fairly conventional design. This includes air to steam superheaters as well as an economizer and evaporator section. A condenser of conventional design is included. For this work all heat exchangers were considered to be counterflow designs.

The heat exchanger design procedures and experimental data were taken from the text by Kays and London [9]. All of the data presented in this text was developed from steam to air heat exchangers and should be particularly applicable to the types of heat exchangers developed here. The only ones not using these two fluids are the molten salt or liquid metal to air heat exchangers referenced above, thus there is a reality basis for all of the design calculations performed here to estimate power conversion system performance and sizing.

One of the significant advantages of the combined cycle power system over current LWR power systems is its reduced requirement for circulating water in the waste heat rejection loop for the Rankine cycle. The typical combined cycle plant considered here produces approximately 50 % of its power from the Rankine cycle and 50 % of its power from the Brayton cycle. This automatically reduces the cooling water requirement by half. In addition, the combined cycle plant achieves 45 % efficiency so that only 55 % of the heat generated has to be released as waste. Thus, a typical 25 MW system will only need to get rid of 6.9 MW of heat via a circulating water system. A current LWR plant generating 25 MW at an efficiency of 33 % would need to dump 16.8 MW of heat. This represents a major savings in circulating water requirements.

Since the combined cycle reduced the circulating water requirements so significantly, the natural question arises as to whether they can be eliminated completely. With a recuperated Air Brayton cycle they can be. All of the waste heat can be rejected directly to the atmosphere. At first it was thought that the efficiency of a multi-turbine recuperated cycle could not compete with a combined cycle plant. However, after performing the detailed analysis, the efficiencies of a recuperated cycle come within 1 or 2 % of predicted combined cycle efficiencies [10]. This would seem to be a minor penalty to pay for being free of a circulating water requirement. However, achieving these high efficiencies requires a very effective recuperator which can become quite large.

5.2 Typical Cycles

Figure 5.2a and b provide a schematic of a four turbine combined cycle system and its thermodynamic cycles on a temperature-entropy plot along with pinch point between Rankine and Brayton cycle. This turns out to be the near optimum combined cycle system.

For a nuclear system to take advantage of combined cycle technology, there are a number of changes to the plant components that have to be made. The most significant of course is that the combustion chamber has to be replaced by a heat exchanger in which the working fluid from the nuclear reactor secondary loop is used to heat the air. The normal Brayton cycle is an internal combustion one where the working fluid is heated by the combustion of the fuel with the air in the combustion chamber. The walls of the combustion chamber can be cooled and peak temperatures in the working fluid can be significantly above the temperature that the walls of the chamber can tolerate for any length of time [11].

Figure 5.3a and b provides a schematic layout and a temperature-entropy diagram for a three-turbine recuperated system. This turns out to be the optimum recuperated system.

5.3 Methodology

The approach taken in the Combined Cycle (CC) code developed for this effort is to model the thermodynamics of the components making up the power conversion systems as real components with nonideal efficiencies [10, 11]. Pressure drops are included for every component except the connected piping. The compressor design is modeled with a small stage polytropic efficiency to take into account state of the art designs [4]. The gas turbines are likewise modeled with a polytropic efficiency. The steam turbines are modeled with a simple overall thermal efficiency. Pressure drops in each of the heat exchangers are included. The input files specify the pressure drops and the heat exchangers are designed to meet these specifications if possible.

The code begins with an estimated compressor pressure ratio and then calculates the state points for both the Brayton and Rankine cycles. Then the code iterates on the compressor ratio to deliver an exit air pressure slightly above atmospheric. In a sub-iteration, it calculates the ratio of mass flows in the Rankine and Brayton cycles. Once the cycle state points have been identified, the output from the cycle is normalized to the desired power level. This sets the total mass flows of air and water. Once the mass flow rates have been calculated, it is possible to size all of the components. The compressor and turbines are sized based on a correlation

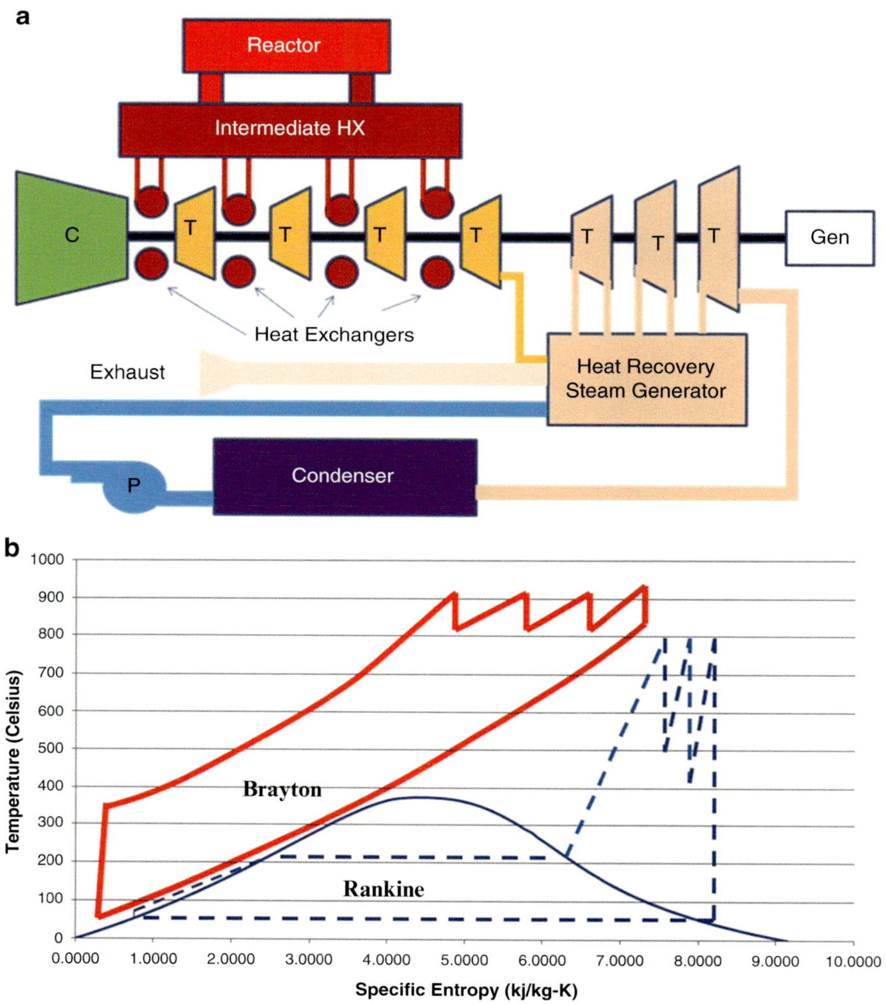

Fig. 5.2 (**a**) Layout for four turbine combined cycle. (**b**) Temperature entropy diagram for combined cycle. (**c**) *T–s* diagram of temperature entropy for combined cycles showing pinch point

with state of the art components and simple scaling rules. Each of the heat exchangers is designed based on the configuration chosen from the Kays and London text. Finally the volumes of all of the components are summed to get an estimate of system size.

In order to optimize the efficiency of the combined cycle, there are two main parameters that must be varied. These are the gas turbines' exit temperatures and the steam cycle peak pressure. The peak gas turbine inlet temperature is set as an input parameter. It can be varied, but it is obvious that the higher the gas turbine inlet temperature, the better the efficiency will be and the same can be said for most of the

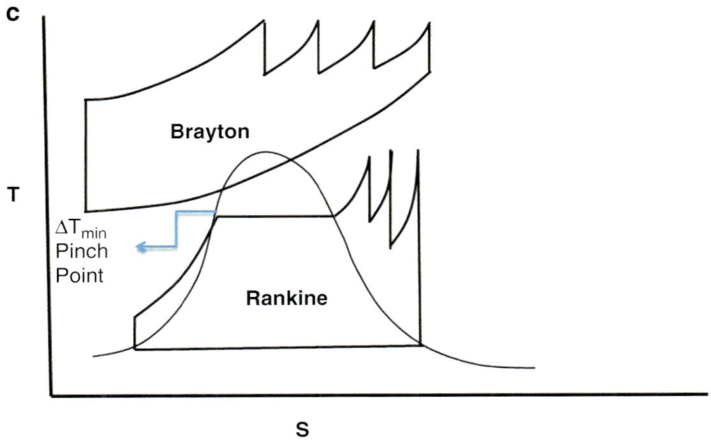

Fig. 5.2 (continued)

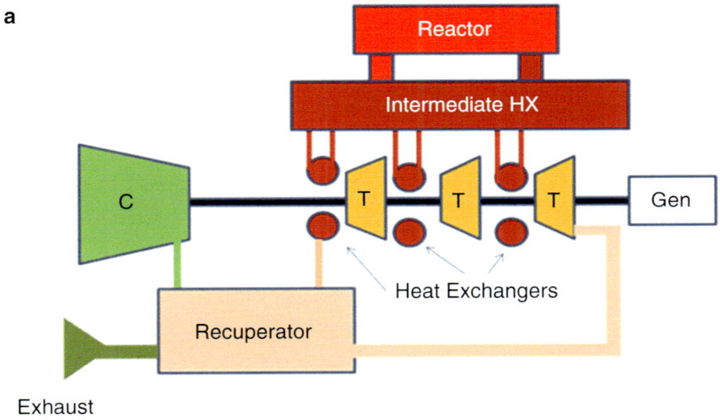

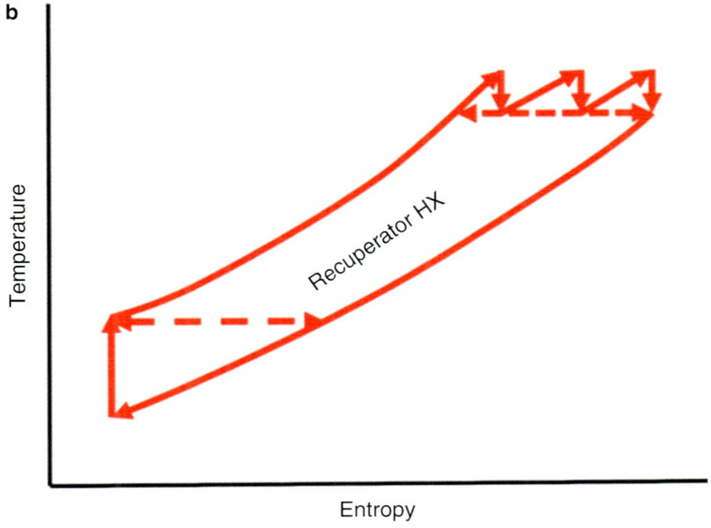

Fig. 5.3 (**b**) Recuperated System Temperature Entropy Diagram

components. If their efficiency is higher, the efficiency of the cycle will be higher. The same is true of atmospheric conditions. The colder the input air and the circulating water, the better the efficiency will be. Therefore, most of the input parameters are chosen based on nominal values. It is not obvious though what the values of gas turbine exit temperatures and steam pressure should be to get an optimum efficiency. These must be varied to identify the peak efficiency achievable [12–14].

The coding for the recuperated system is much simpler than the combined cycle coding because the steam cycle does not have to be modeled. The compressor pressure ratio iteration is much simpler, but still sets the pressure ratio so as to meet an exit pressure slightly above atmospheric. The new calculation in the recuperated system is the air to air recuperator itself. Since the recuperator will be the largest component in the system it does not make sense to prescribe anything other than a *counterflow heat exchanger*. In this case the pressure drops for the hot and cold fluids cannot be set independently for simple heat transfer correlations. So the pressure drop on the hot air side was chosen as the flow path setting parameter. In this case, also, most of the parameters in the code models are set to nominal values. The only parameter that must be varied to optimize the efficiency of the system is the gas turbines' exit temperatures. The choice of the exit temperature can be made to achieve peak efficiency.

5.4 Combined Cycle Code Equations

The combined cycle code tracks the conversion of thermal energy to mechanical energy as the working fluid moves through the system and is developed based on steady-state assumption (transient analysis mode needs to be investigated for better and accurate approach). It does this by calculating the temperatures, or enthalpies, and pressures at a sequence of state points between the major system components. The following are the major state points.

Air Path

State 0: Ambient air
State 1: Inlet to compressor
State 2: Exit from compressor (Inlet to recuperator)
State 3: Inlet to first heater (Exit from recuperator)
State 4: Exit from heater, Turbine inlet
State 5: Turbine exit, Recuperator inlet
State 6: Recuperator exit, Superheater inlet
State 7: Superheater exit, Evaporator inlet
State 8: Evaporator exit, Economizer inlet
State 9: Economizer exit, Exhaust to atmosphere

Note: The Heat Recovery Steam Generator contains the superheaters, evaporator, and economizer. State 5 and 6 are the same for the combined cycle and State 6 and 9 are the same for the recuperated cycle.

Steam Path

State 20: Pump inlet, Condenser exit (saturated water)
State 21: Pump exit, Economizer inlet (liquid water)
State 23: Economizer exit, Pinch Point, Evaporator inlet (saturated water)
State 24: Evaporator exit, First Superheater inlet (saturated steam)
State 25: First Superheater exit, High Pressure Turbine inlet (superheated steam)
State 26: High Pressure Turbine exit, Second Superheater inlet (superheated steam)
State 27: Second Superheater exit, Medium Pressure Turbine inlet (superheated steam)
State 28: Medium Pressure Turbine exit, Third Superheater inlet (superheated steam)
State 29: Third Superheater Exit, Low Pressure Turbine inlet (superheated steam)
State 30: Low Pressure Turbine exit, Condenser inlet (superheated steam)

The major equations are:

$p_2 = p_1 \times \mathrm{CPR}$ — CPR = Compressor pressure ratio

$T_2 = T_1 \times \mathrm{CPR}^{\frac{\gamma-1}{\eta_{\mathrm{pc}}\gamma}}$ — γ = Ratio of specific heats, η_{pc} = Compressor polytropic efficiency

$W_{\mathrm{c}} = \dot{m}_{\mathrm{air}} C_{\mathrm{p}} (T_2 - T_1)$ — W_{c} = Compressor work, $\dot{m}_{\mathrm{air}}$ = Mass flow rate of air, C_{p} = Constant pressure-specific heat for air

$\varepsilon = \frac{T_3 - T_2}{T_5 - T_2} = \frac{T_5 - T_9}{T_5 - T_2}$ — ε = Recuperator effectiveness

$Q_1 = \dot{m}_{\mathrm{air}} C_{\mathrm{p}} (T_4 - T_3)$ — Q_1 = Heat input in first heater

$W_{\mathrm{t1}} = \dot{m}_{\mathrm{air}} C_{\mathrm{p}} (T_4 - T_5)$ — W_{t1} = Work from first turbine

$p_4 = \mathrm{PR}_{\mathrm{h1}} \times p_3$ — $\mathrm{PR}_{\mathrm{h1}}$ = Pressure ratio across first heater

$p_5 = p_4 \times \left(\frac{T_5}{T_4}\right)^{\frac{\eta_{\mathrm{pt}}\gamma}{(\gamma-1)}}$ — η_{pt} = Turbine polytropic efficiency

$p_9 = p_5 \times \mathrm{PR}_{\mathrm{rec}}$ — $\mathrm{PR}_{\mathrm{rec}}$ = Pressure ratio across the recuperator

$T_{23} = T_8 - \mathrm{PPDT}$ — PPDT = Pinch point delta temperature

$T_{25} = T_5 - \mathrm{STDT}$ — STDT = Superheater terminal temperature difference

$\dot{m}_{\mathrm{w}} (h_{25} - h_{22} + h_{27} - h_{26} + h_{29} - h_{28}) = \dot{m}_{\mathrm{air}} C_{\mathrm{p}} (T_5 - T_8)$ — $\dot{m}_{\mathrm{w}}$ = Mass flow rate of water

$\dot{m}_{\mathrm{w}} (h_{30} - h_{29}) = \dot{m}_{\mathrm{air}} C_{\mathrm{p}} (T_8 - T_9)$

$W_{\mathrm{p}} = \dot{m}_{\mathrm{w}} \frac{(p_{21} - p_{22})}{\rho_{\mathrm{w}}}$ — ρ_{w} = Water density, W_{p} = Pump work

$W_{\mathrm{hp}} = \dot{m}_{\mathrm{w}} (h_{25} - h_{26})$ — W_{hp} = Work from high pressure turbine

$W_{\mathrm{mp}} = \dot{m}_{\mathrm{w}} (h_{27} - h_{28})$ — W_{mp} = Work from medium pressure turbine

$W_{\mathrm{lp}} = \dot{m}_{\mathrm{w}} (h_{29} - h_{30})$ — W_{lp} = Work from low pressure turbine

$\eta_{\mathrm{B}} = \frac{\sum_i W_{\mathrm{t}i} - W_{\mathrm{c}}}{\sum_i Q_{\mathrm{h}i}}$ — η_{B} = Brayton cycle efficiency

$\eta_{\mathrm{R}} = \frac{W_{\mathrm{hp}} + W_{\mathrm{mp}} + W_{\mathrm{lp}} - W_{\mathrm{p}}}{(h_{25} - h_{24}) + (h_{27} - h_{26}) + (h_{29} - h_{28})}$ — η_{R} = Rankine cycle efficiency

$\eta_{\mathrm{overall}} = \frac{\sum_i W_{\mathrm{t}i} - W_{\mathrm{c}} + \frac{\dot{m}_{\mathrm{w}}}{\dot{m}_{\mathrm{air}}} \left[W_{\mathrm{hp}} + W_{\mathrm{mp}} + W_{\mathrm{lp}} - W_{\mathrm{p}}\right]}{\sum_i Q_{\mathrm{h}i}}$ — η_{overall} = Overall efficiency

5.5 Computer Code Flowchart

The following presentation is the top level flowchart of computer code that was developed for the purpose of analysis for Air Brayton Cycles for Nuclear Power Plant Systems.

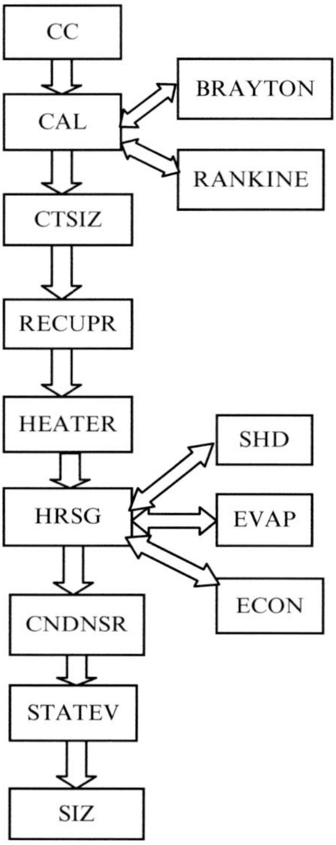

CC-reads input and controls the start and stopping process

CAL - Manages the calculation

BRAYTON- Calculates the Brayton cycle

RANKINE-Calculates the Rankine cycle

CTSIZ - Sizes the cycles to meet the desired power

RECUPR - Designs the recuperator

HEATER - Designs the primary heaters

HRSG-Designs the HRSG

SHD - Designs a superheater

EVAP - Designs the evaporator

ECON -Designs the economizer

CNDNSR - Designs the condenser

STATEV - Summarizes all state variables

SIZ - Summarizes the volumes of all components

There is an additional set of small routines that calculates material properties as a function temperature. The property tables have been taken from Incropera [15]. In addition a new set of steam properties were calculated as needed based on the IAPWS 1997 formulation [16].

5.6 Validation of Methodology

Before proceeding to estimate the performance of advanced systems, it is useful to validate the methodology by estimating the performance of a currently deployed system. By the year 200 General Electric (GE) had over 893 combined cycle and cogeneration systems installed worldwide producing more than

Table 5.1 Turbine temperatures for several GE engines

GE engines	CPR	T5		T4	
GE10-1	15.5	482	C	1449	K
PGT16	20.2	491	C	1562	K
PGT20	15.7	475	C	1440	K
PGT25	17.9	525	C	1585	K
PGT25+	21.5	500	C	1604	K
PGT+G4	23.2	510	C	1654	K
LM6000	28	455	C	1608	K
LMS100	40	417	C	1659	K
MS5002E	17	511	C	1538	K
MS6001B	12.2	548	C	1488	K
MS7001EA	12.6	537	C	1480	K
MS9001E	12.6	543	C	1491	K
	Average =			1546	K
	Range =			1440–1659 K	

67,397 MW(e) [17], so one of their systems was chosen to model. The system chosen was an S107FA [18] 60 Hz, 250 MW Gas Turbine with a three pressure reheat steam cycle. It is a single shaft system and most of the nominal performance parameters are available in the GE literature. The only performance parameter not readily available was the gas turbine inlet temperature. The listed efficiency for the combined cycle plant is 56.5 % at standard sea level conditions.

In order to estimate the turbine inlet temperature (T4) a set of given turbine exit temperatures (T5) and pressure ratios were available for 12 GE engines. The T4s were then estimated assuming a polytropic efficiency of 0.95. The results are presented in Table 5.1.

The average temperature of 1546 K was chosen for the baseline comparison. With this temperature and a single turbine model, the CC code gives an estimated efficiency of 56.55 %. If instead of the average temperature, the two extremes of 1440 K and 1659 K are considered, the code estimates efficiencies of 53.90 % and 58.97 %, so the calculation is within ~2 % at its worst and within 0.05 % at nominal conditions. It should be pointed out that the simple model for the efficiency,

$$\eta_T = \eta_B + \eta_R - \eta_B\eta_R = 0.426 + 0.421 - 0.426 \times 0.421 = 0.668$$

Or it overestimates the combined cycle efficiency by 10.3 %. The detailed model developed here does significantly better.

5.7 Nominal Analysis Parameters

Having demonstrated that the CC code calculation predicts a reasonable efficiency for current gas turbine combined cycle systems, it can be used to predict the performance of Nuclear Air Brayton Combined Cycle Systems and Nuclear

Recuperated Air Brayton Cycle Systems. A nominal set of conditions has been chosen as a best estimate for environmental conditions and component performance. A peak turbine inlet temperature of 933 K was chosen as the baseline condition. It is anticipated that this will be achievable by both the molten salt reactors and the liquid lead, or lead-bismuth reactors. The High Temperature Gas Reactors will do better, but helium to air heat exchangers have not been considered here. For the combined cycle system, the number of turbines, the turbine exit temperature, and the steam pressure in the Rankine bottoming cycle were varied to achieve the maximum thermodynamic efficiency. After the optimum efficiency was determined, the sensitivity of this result to important parameters was estimated. For the recuperated cycle the number of turbines and the turbine exit temperatures were varied to achieve the maximum efficiency. The nominal input parameters for the two systems follow.

5.8 Combined Cycle System Baseline

Here is the list of Combine Cycle baseline list below:

Number of turbines—varied
Turbine inlet temperature—933 K
Turbine exit temperature—varied
Turbine polytropic efficiency—0.90
Compressor pressure ratio—calculated
Compressor polytropic efficiency—0.90
Main heater pressure ratios—0.99
Atmospheric pressure—1 atm
Atmospheric temperature—288 K (15 °C)
Circulating water input temperature—288 K (15 °C)
Ratio of exhaust pressure to atmospheric—0.98
Air pressure ratios across each superheater—0.99
Air pressure ratio across the evaporator—0.99
Air pressure ratio across the economizer—0.99
Pinch point temperature difference—10 K
Terminal temperature difference at steam exit to superheaters—15 K
Peak Rankine cycle pressure—varied
Intermediate Rankine cycle pressure—varied, 1/4 of peak
Low Rankine cycle pressure—varied, 1/16 of peak
Condenser pressure—9 kPa
Steam turbines thermal efficiency—0.90
Power level—25 MW

5.9 Recuperated Cycle System Baseline

Here is the list of recuperated cycle baseline listed below:

Number of turbines—varied
Turbine inlet temperature—933 K
Turbine exit temperature—varied
Turbine polytropic efficiency—0.90
Compressor pressure ratio—calculated
Compressor polytropic efficiency—0.90
Main heater pressure ratios—0.99
Recuperator effectiveness—0.95
Recuperator pressure ratio on hot side—0.99
Atmospheric pressure—1 atm
Atmospheric temperature—288 K (15 °C)
Circulating water input temperature—288 K (15 °C)
Ratio of exhaust pressure to atmospheric—0.98
Power level—25 MW

5.10 Nominal Results for Combined Cycle Model

The turbine exit temperatures and the peak pressure in the steam bottoming cycle were varied for systems using one to five turbines. The best efficiency achievable in each case is plotted in Fig. 5.4. The efficiency is a monotonic function of the number of turbines, with the five-turbine system only slightly better than the four-turbine system. So the four-turbine case was chosen as the baseline representative combined cycle system. The peak efficiency for the four-turbine case is 45.88 % and for the five-turbine case is 46.52 %.

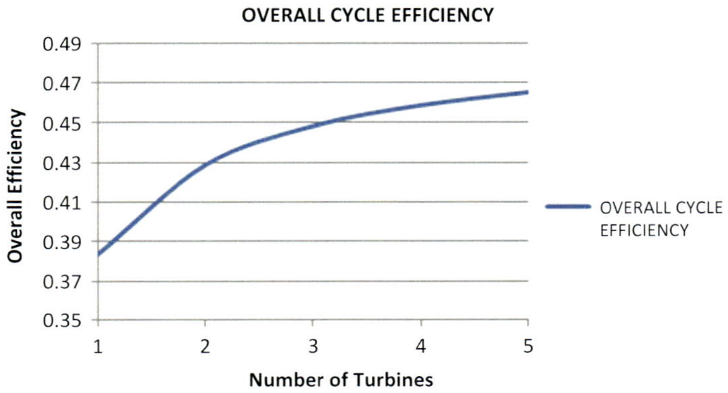

Fig. 5.4 Peak overall thermal efficiency versus number of turbines

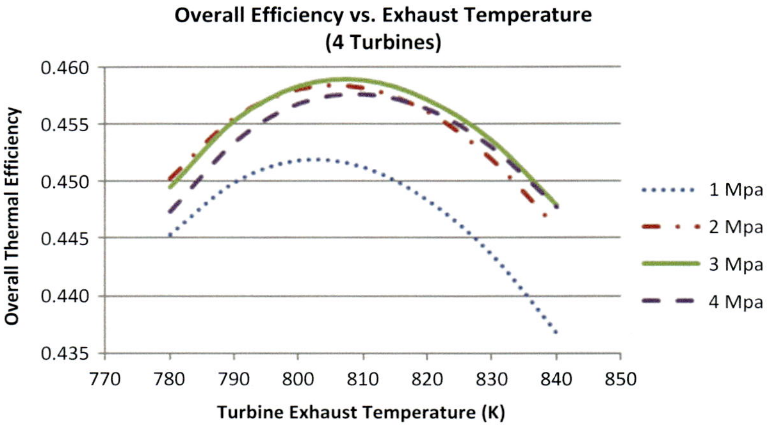

Fig. 5.5 Efficiency versus exhaust temperature for various steam pressures

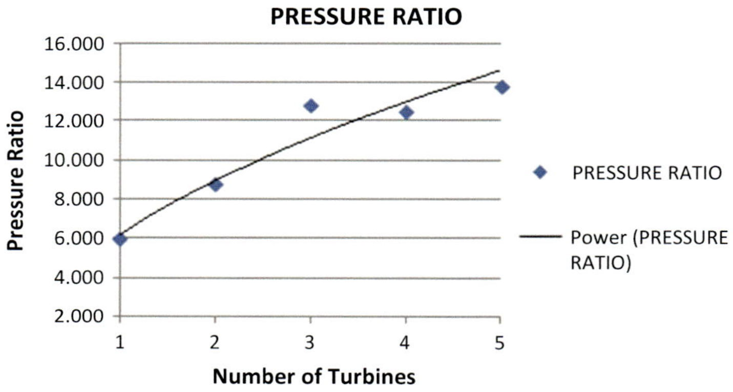

Fig. 5.6 Optimum pressure ratio versus number of turbines

The optimum turbine exit temperature for the four-turbine case is 810 K with a best steam pressure in the bottoming cycle of 3 MPa. The detailed results are presented in Fig. 5.5. The system underperforms with a steam pressure in the bottoming cycle of only 1 MPa, but from 2 to 4 MPa, the results are very similar. The optimum exhaust temperature shifts slightly from 800 to 810 K.

The optimum compressor ratio to achieve peak efficiency and keep the exhaust pressure above the environmental pressure is plotted in Fig. 5.6 for each of the turbine models.

Since the exhaust temperature and the steam pressure in the bottoming cycle are varying in a discrete manner in fairly large steps, the calculated points do not follow a smooth curve, so a power fit to the data has been included.

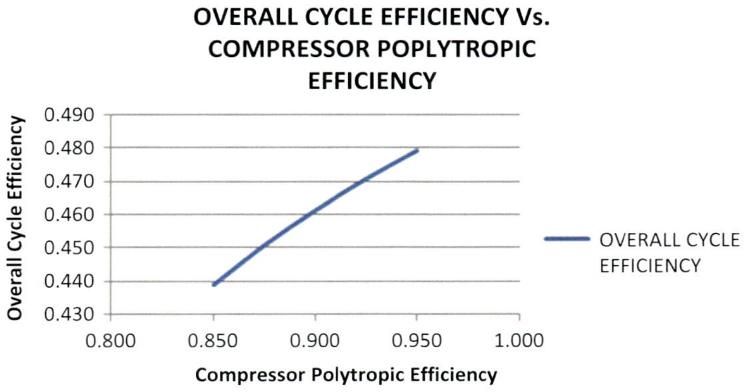

Fig. 5.7 Overall cycle efficiency versus compressor polytropic efficiency

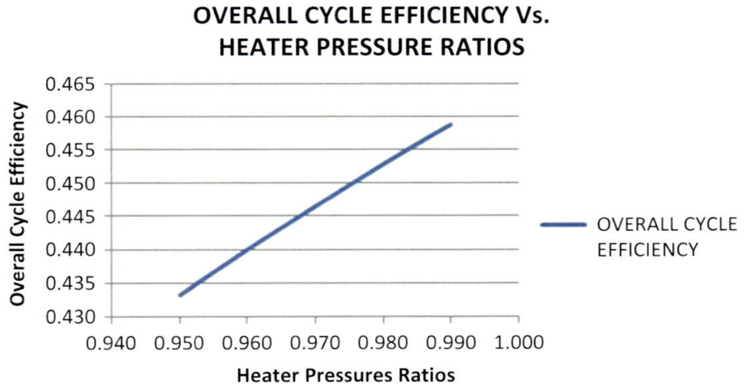

Fig. 5.8 Overall cycle efficiency versus heater pressure ratio (four turbines)

Three major sensitivities were considered—compressor polytropic efficiency, pressure drops (or ratios) in the main heaters, and the environmental temperature. Figure 5.7 presents the dependence of the overall thermal efficiency on the gas compressor polytropic efficiency.

The overall thermal efficiency increases about 0.4 % for every 1 % increase in the compressor polytropic efficiency.

The pressure drops in the main heaters are analogous to the pressure drop in a combustion chamber.

Nominally the pressure drop in each heater was set to 1 %, but the effect of these pressure drops on the overall efficiency was estimated (see Fig. 5.8). The overall efficiency drops approximately 0.5 % for every 1 % increase in the pressure drop in the main heater for the four-turbine system.

The standard day conditions of 15 °C and 1 atm pressure are not likely to be met very often during the operation of a typical power plant, so the variation of

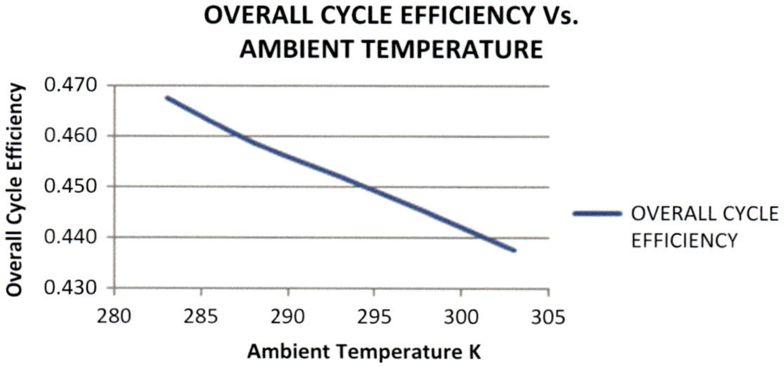

Fig. 5.9 Overall efficiency versus ambient temperature (four turbines)

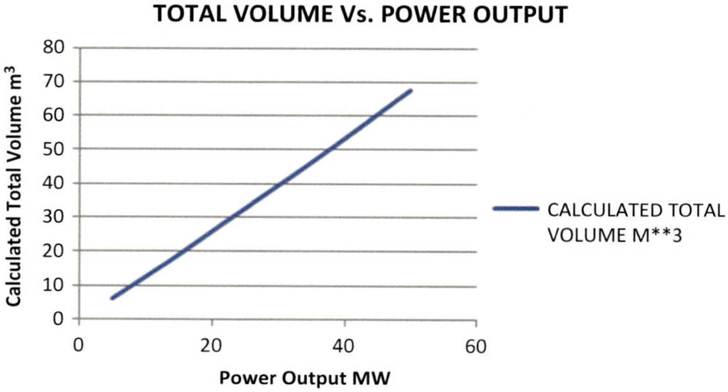

Fig. 5.10 System volume versus power output

efficiency with the ambient temperature was estimated. The results are presented in Fig. 5.9.

The overall efficiency drops about 1.4 % for every 10 K increase in temperature.

The main purpose of designing the heat exchangers and estimating sizes for pump, compressor, and turbine components was to get an overall estimate of the size for the complete power conversion system. This estimate is provided in Fig. 5.10.

5.11 Nominal Results for Recuperated Cycle Model

For this case, one to five turbines were considered also. However, the only parameter of interest for optimizing the overall cycle efficiency is the turbine exhaust temperature. The results for all five turbine models are presented in Fig. 5.11.

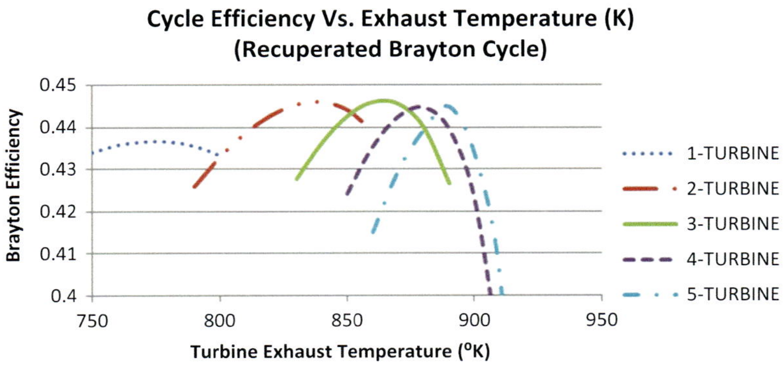

Fig. 5.11 Recuperated Brayton cycles versus turbine exhaust temperatures

Table 5.2 Peak cycle efficiencies for recuperated Brayton system

Turbines	Overall efficiency
1	0.4366
2	0.4458
3	0.4461
4	0.4447
5	0.4447

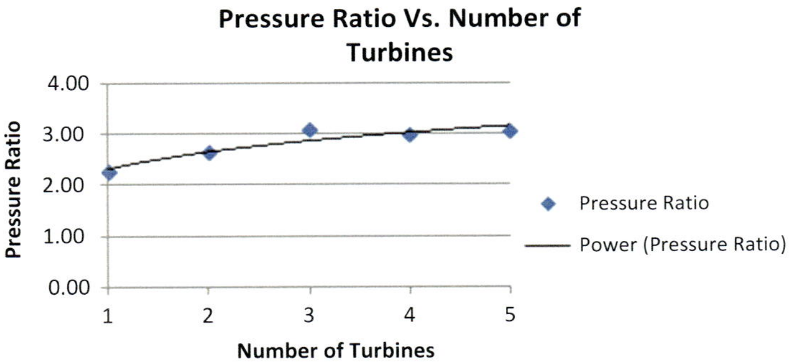

Fig. 5.12 Compressor pressure ratios for peak efficiency versus number of turbines

For the recuperated Brayton cycles, the efficiency peaks with three turbines under the nominal conditions, but the peak efficiencies for two and four turbines are very close. The actual numbers are given in Table 5.2.

Therefore, all sensitivity studies were done on the three-turbine model. It is interesting to note that the peak efficiency achieved for the recuperated system of 44.61 % is only 1.27 % less than the peak efficiency obtained for the combined cycle system.

The compressor pressure ratios for the peak efficiencies are plotted in Fig. 5.12.

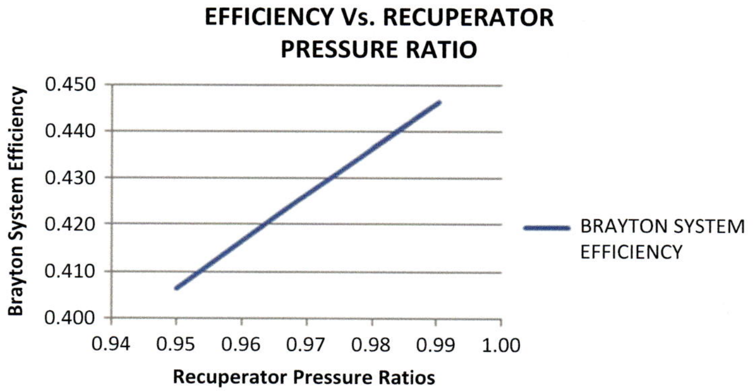

Fig. 5.13 System efficiency versus recuperator pressure ratio

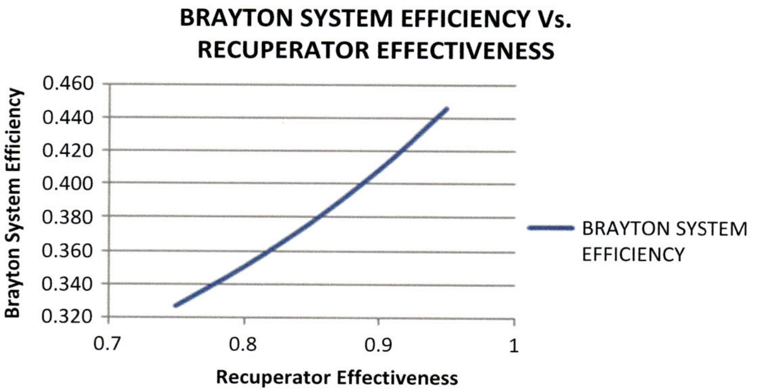

Fig. 5.14 Efficiency versus recuperator effectiveness (three turbines)

As expected for a recuperated system, the compressor pressure ratios are significantly less than those for an unrecuperated system.

The sensitivity of the results to recuperator pressure drops and recuperator effectiveness were estimated. The sensitivity to the recuperator pressure drops is presented in Fig. 5.13.

The cycle efficiency drops 0.6 % for every 1 % increase in the recuperator pressure drop. This is one of the more sensitive parameters because the pressure drops occur on both the cold leg and the hot leg, and the compressor pressure ratio has to compensate for both of them.

The recuperator effectiveness of 0.95 for the baseline cases produces a very large recuperator, so the effectiveness of the recuperator was varied to determine its effect on the overall cycle efficiency and the volume of the resulting system. The effectiveness of the recuperator's impact on the overall efficiency is plotted in Fig. 5.14.

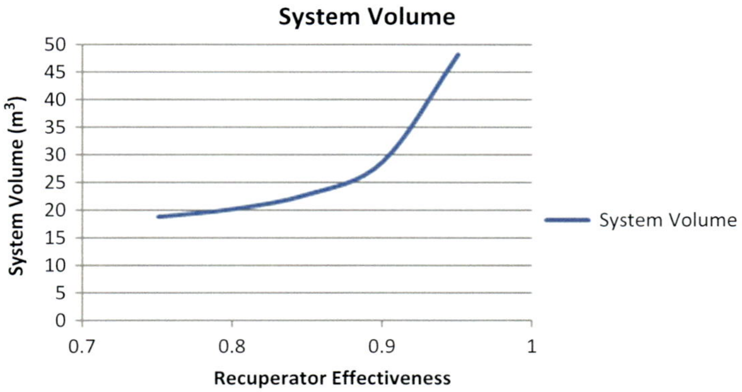

Fig. 5.15 Recuperated system volume versus recuperator effectiveness

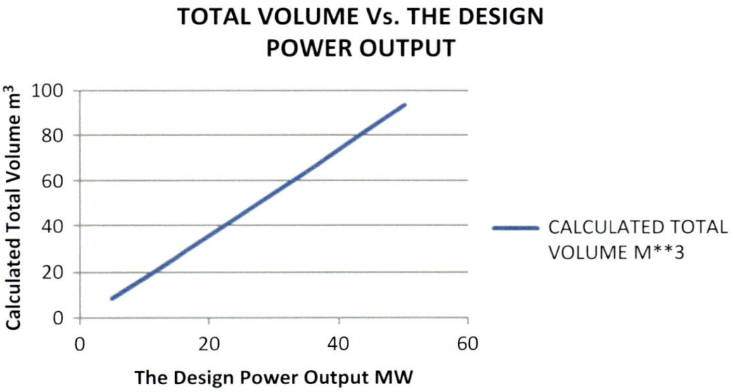

Fig. 5.16 System volume versus system power for the recuperator system

The overall system efficiency also drops approximately 0.6 % for every 1 % decrease in the recuperator's effectiveness. The recuperator volume estimates for the baseline 25 MW system are plotted in Fig. 5.15.

It appears that asking for an effectiveness of 0.95 causes a very significant size increase compared to an effectiveness of 0.9. Giving up 5 % in effectiveness causes a 4 % loss in system efficiency, but reduces the size of the system by 40 %.

Figure 5.16 gives the recuperator system sizes as a function of total system power for the range from 5 to 50 MW.

Clearly the recuperated system requires a larger volume than the combined cycle system, but it has fewer components and is free of the circulating water requirements.

5.12 Extension of Results Versus Peak Turbine Temperatures

The turbine inlet temperature of 933 K chosen for this study is aggressive but within the range projected for the molten salt reactor and the lead or lead-bismuth cooled reactor. Should it be possible to achieve even higher temperatures in the future, the CC code was used to estimate the efficiencies that might be achieved. Figure 5.17 gives the anticipated efficiencies that can be achieved by a nuclear combined cycle system up to about a 1100 K turbine inlet temperature.

Figure 5.18 does the same extrapolation for the recuperated system.

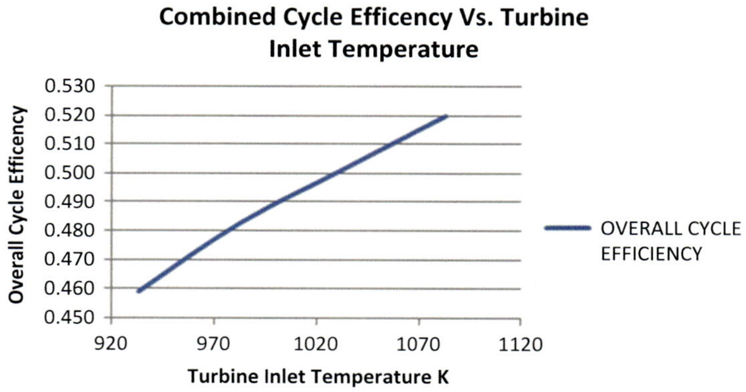

Fig. 5.17 Combined cycle efficiency versus increases in turbine inlet temperature

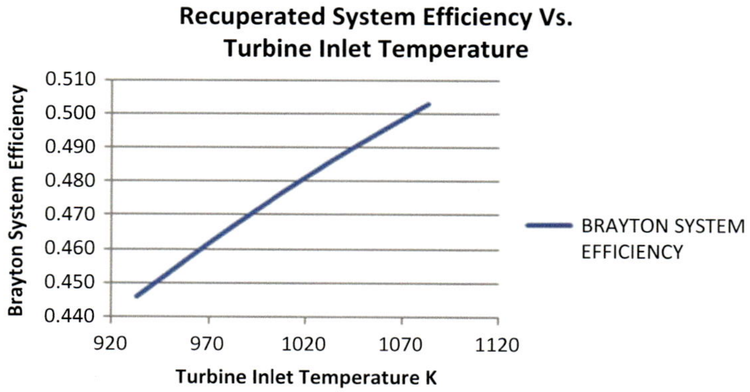

Fig. 5.18 Recuperated system efficiency versus turbine inlet temperature

5.13 Conclusions

The combined cycle code methodology does an excellent job of modeling combined cycle plants and allows prediction of performance based on a number of plant design features. It predicts plant efficiencies above 45 % for reactor temperatures predicted to be in the 700 °C range. It also predicts reduction in circulating water requirements by over 50 % compared to current generation reactors. This will be a major saving in water resources and should reduce the capital investment in large cooling towers. In fact the circulating water requirements can be eliminated completely if a recuperator is used instead of the bottoming steam cycle. The efficiency penalty for going to a recuperated system appears to be only about 1–2 %. The size of the system increases by about 50 % however.

No attempt to analyze a gas-cooled reactor was undertaken in this study. Given that efficiencies will continue to increase with even higher reactor outlet temperatures, it could prove fruitful to consider this possibility in the future. However, the primary heaters will be large, but they will operate at higher pressure so they will not grow like the recuperator does. This brings up another possibility for improving the recuperated system's performance. It might be possible to exhaust at higher pressure and reduce the size of a 0.95 effective recuperator. Little has been done in this study to optimize the recuperated design. This study started as a combined cycle effort and the recuperator showed up as an added on component. Future work may well be able to reduce the size of the recuperator and keep a high efficiency, which is a driving factor for writing this book.

Though this study focused on a 25 MW(e) power conversion system, it only considered one loop. Typically an LWR system has three or four loops and so the results are easily extended to a 100 MW(e) plant. This should cover the range for most of the Generation IV Small Modular Reactors that are being considered. There is also talk of replicating Small Modular Reactors to build up to a significant power station. Therefore, the 25 MW(e) system components appear to be a reasonable point to start hardware development.

References

1. Kehlhofer, R., et al. (2009). *Combined cycle gas & steam turbine power plants* (3rd ed.). Tulsa, OK: PennWell.
2. GE Energy. (2012). *FlexEfficiency 50 combined cycle power plant*, e-brochure.
3. Horlock, J. H. (1997). *Cogeneration-combined heat and power (CHP)*. Malabar, FL: Krieger Publishing Company.
4. Mattingly, D. (1996). *Elements of gas turbine propulsion*. New York: McGraw-Hill.
5. Fraas, A. P. (1989). *Heat exchanger design* (2nd ed.). New York: Wiley.
6. MacPherson, R. E., Amos, J. C., & Savage, H. W. (1960). Development testing of liquid metal and molten salt heat exchangers. *Nuclear Science and Engineering, 8*, 14–20.
7. Fraas, A. P. (1960). Design precepts for high temperature heat exchangers. *Nuclear Science and Engineering, 8*, 21–31.

8. Yarosh, M. (1960). Evaluation of the performance of liquid metal and molten salt heat exchangers. *Nuclear Science and Engineering, 8*, 32–43.
9. Kays, W. M., & London, A. L. (1964). *Compact heat exchangers* (2nd ed.). New York: McGraw-Hill.
10. Zohuri, B. (2015). *Combined cycle driven efficiency for next generation nuclear power plants: An innovative design approach*. New York: Springer.
11. Zohuri, B. (2014). *Innovative open Air Brayton combined cycle systems for the next generation nuclear power plants*. Albuquerque, NM: University of New Mexico Publications.
12. McDaniel, P. J., Zohuri, B., & de Oliveira, C. R. E. (2014, September). A combined cycle power conversion system for small modular LMFBRs. *ANS Transactions*.
13. Zohuri, B., McDaniel, P., & de Oliveira, C. R. E. (2014, June). A comparison of a recuperated open cycle (Air) Brayton power conversion system with the traditional steam Rankine cycle for the next generation nuclear power plant. *ANS Transactions*.
14. McDaniel, P. J., de Oliveira, C. R. E., Zohuri, B., & Cole, J. (2012, November). A combined cycle power conversion system for the next generation nuclear power plant. *ANS Transactions*.
15. Incropera, F. P. (2007). *Introduction to heat transfer* (5th ed.). Danvers, MA: Wiley.
16. International Association for the Properties of Water and Steam. (1997). *IAWPS Industrial Formulation 1997 for the thermodynamic properties of water and steam*. Buenos Aires, Argentina: IAPWS.
17. Maslak, C. E., & Tomlinson, L. O. (2000). *GE combined cycle experience, GER-3651D*. Schenectady, NY: GE Industrial & Power Systems.
18. Chase, D. L. (2001). *Combined-cycle development evolution and future, GER-4206*. Schenectady, NY: GE Power Systems.

Chapter 6
Heat Exchangers

A number of technologies are being investigated for the Next Generation Nuclear Plant (NGNP) that will produce heated fluids at significantly higher temperatures than current generation power plants. The higher temperatures offer the opportunity to significantly improve the thermodynamic efficiency of the energy conversion cycle. One of the concepts currently under study is the Molten Salt Reactor. The coolant from the Molten Salt Reactor may be available at temperatures as high as 800–1000 °C. At these temperatures, an open Brayton cycle combined with a Rankine bottoming cycle appears to have some strong advantages. Thermodynamic efficiencies approaching 50 % appear possible. Requirements for circulating cooling water will be significantly reduced. However, to realistically estimate the efficiencies achievable it is essential to have good models for the heat exchangers involved as well as the appropriate turbo-machinery. This study has concentrated on modeling all power conversion equipment from the fluid exiting the reactor to the energy releases to the environment.

6.1 Heat Exchanger Types

A heat exchanger is a heat transfer device that exchanges heat between two or more process fluids. Heat exchangers have widespread industrial and domestic applications. Many types of heat exchangers have been developed for use in steam power plants, chemical processing plants, building heat and air conditioning systems, transportation power systems, and refrigeration units.

The actual design of heat exchangers is a complicated problem. It involves more than heat-transfer analysis alone. Cost of fabrication and installation, weight, and size play important roles in the selection of the final design from a total cost of ownership point of view. In many cases, although cost is an important consideration, size and footprint often tend to be the dominant factors in choosing a design.

B. Zohuri, *Application of Compact Heat Exchangers For Combined Cycle Driven Efficiency In Next Generation Nuclear Power Plants*,
DOI 10.1007/978-3-319-23537-0_6

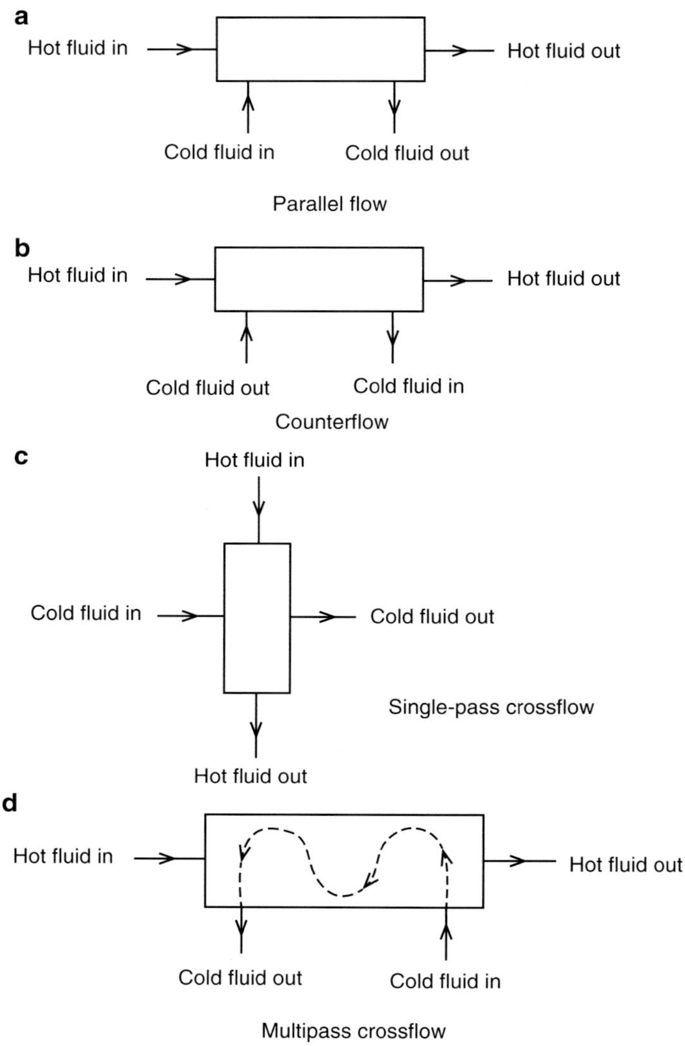

Fig. 6.1 Types of flow path configuration through heat exchanger

Most heat exchangers may be classified as one of several basic types. The four most common types, based on flow path configuration, are illustrated in Fig. 6.1 [1].

1. In *concurrent,* or *parallel-flow*, units the two fluid streams enter together at one end, flow through in the same direction, and leave together at the other end.
2. In *countercurrent*, or *counterflow*, units the two streams move in opposite directions.

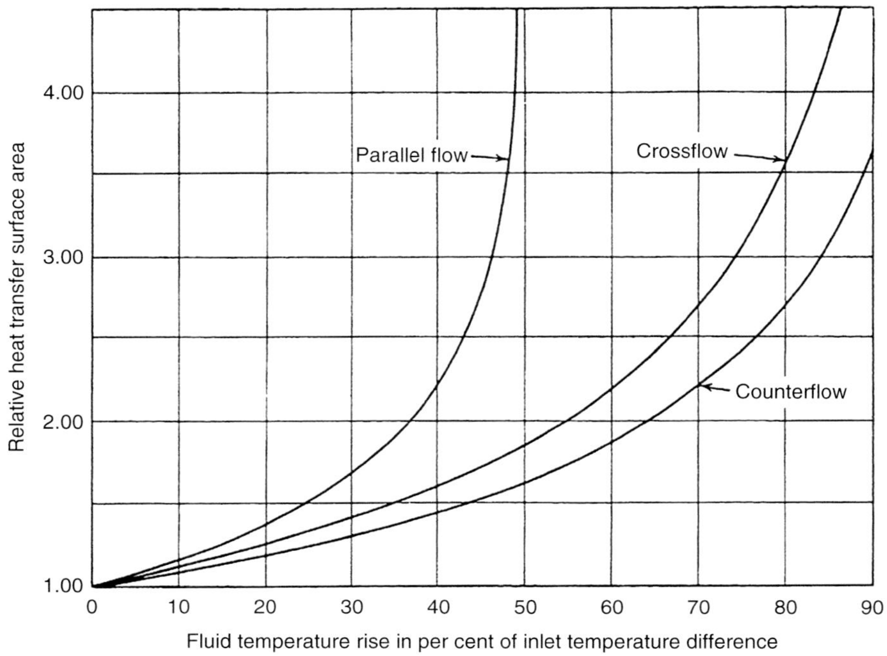

Fig. 6.2 The required relative heat transfer surface area as a function of the ratio of the temperature rise (or drop) in the fluid stream having the greater change in temperature to the difference in temperature between the inlet streams

3. In *single-pass cross-flow* units one fluid moves through the heat transfer matrix at right angles to the flow path of other fluid.
4. In *multipass cross-flow* units one fluid stream shuttles back and forth across the flow path of the other fluid stream, usually giving a cross-flow approximation to counterflow.

The most important difference between these four basic types lies in the relative amounts of heat transfer surface area required to transfer the desired amount of heat between the two fluids.

Figure 6.2 shows the relative area required for each type as a function of the change in temperature of the fluid with the largest temperature change requirement for a typical set of conditions. In the region in which the fluid temperature change across the heat exchanger is a small percentage of the difference in temperature between the two entering fluid streams, all the units require roughly the same area. The parallel-flow heat exchanger is of interest primarily for applications in this region. Cross-flow units have a somewhat broader range of application and are peculiarly suited to some types of heat exchanger construction that have special advantages. The counterflow heat exchanger requires the least area. Furthermore, it is the only type that can be employed in the region in which the temperature change in one or both of the fluid streams closely approaches the temperature difference between the entering fluid streams.

In addition, heat exchangers may be classified as direct contact or indirect contact. In the direct-contact type, heat transfer takes place between two immiscible fluids, such as a gas and a liquid, coming into direct contact. For example, cooling towers, jet condensers for water vapor, and other vapors utilizing water spray are typical examples of direct-contact exchangers.

An *Immiscible Fluids* are incapable of is being mixed or blended together. Immiscible liquids that are shaken together eventually separate into layers. Oil and Water are typical immiscible fluids.

In the indirect-contact type of heat exchangers, such as automobile radiators, the hot and cold fluids are separated by an impervious surface, and they are referred to as *surface heat exchangers*. There is no mixing of the two fluids.

6.2 Classification According to Transfer Processes

Heat exchangers are classified according to transfer processes into indirect and direct contact types.

6.2.1 Indirect Contact Type Heat Exchangers

In an indirect-contact heat exchanger, the fluid streams remain separate and the heat transfers continuously through an impervious dividing wall or into and out of a wall in a transient manner. Thus, ideally, there is no direct contact between thermally interacting fluids. This type of heat exchanger also referred to as a surface heat exchanger can be further classified into direct-transfer type, storage type, and fluidized-bed exchangers.

6.2.2 Direct Contact Type Heat Exchangers

In this type, heat transfers continuously from the hot fluid to the cold fluid through a dividing wall. Although a simultaneous flow of two (or more) fluids is required in the exchanger, there is no direct mixing of the two (or more) fluids because each fluid flows in separate fluid passages. In general, there are no moving parts in most such heat exchangers. This type of exchanger is designated as a recuperative heat exchanger or simply as a recuperator. Some examples of direct transfer type heat exchangers are tubular, plate-type, and extended surface exchangers.

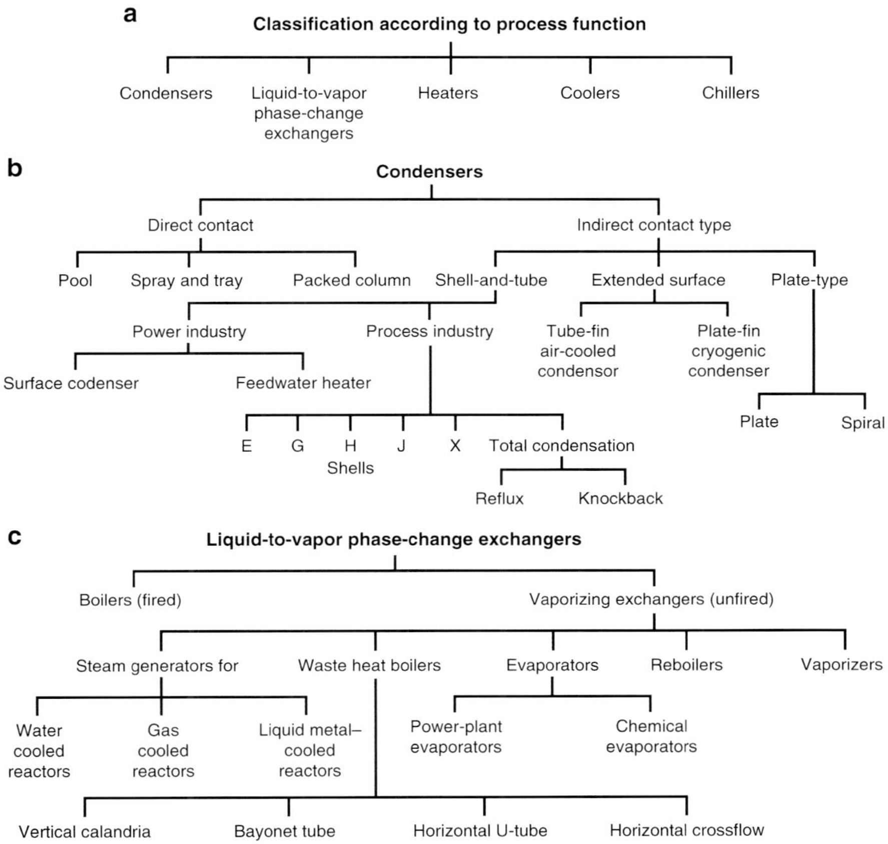

Fig. 6.3 (**a**) Classification according to process function; (**b**) classification of condensers; (**c**) classification of liquid-to-vapor phase-change exchangers

Note that the term recuperator is not commonly used in the process industry for shell-and-tube and plate heat exchangers, although they are also considered as recuperators. Recuperators are further subclassified as prime surface exchangers and extended-surface exchangers. Prime surface exchangers do not employ fins or extended surfaces on any fluid side. Plain tubular exchangers, shell-and-tube exchangers with plain tubes, and plate exchangers are good examples of prime surface exchangers. Recuperators constitute a vast majority of all heat exchangers (Fig. 6.3).

6.3 Classification of Heat Exchanger by Construction Type

Heat exchangers also can be classified according to their construction features. For example, there are tubular, plate, plate–fin, tube–fin, and regenerative exchangers. An important performance factor for all heat exchangers is the amount of heat transfer surface area within the volume of the heat exchanger. This is called its *compactness factor* and is measured in square meters per cubic meter.

6.3.1 *Tubular Heat Exchangers*

Tubular exchangers are widely used, and they are manufactured in many sizes, flow arrangements, and types. They can accommodate a wide range of operating pressures and temperatures. The ease of manufacturing and their relatively low cost have been the principal reason for their widespread use in engineering applications. A commonly used design, called the *shell-and-tube* exchanger, consists of round tubes mounted on a cylindrical shell with their axes parallel to that of the shell.

Figure 6.4 illustrates the main features of a shell-and-tube exchanger having one fluid flowing inside the tubes and the other flowing outside the tubes. The principle components of this type of heat exchanger are the tube bundle, shell, front and rear end headers, and baffles. The baffles are used to support the tubes, to direct the fluid flow approximately normal to the tubes, and to increase the turbulence of the shell fluid. There are various types of baffles, and the choice of baffle type, spacing, and geometry depends on the flow rate allowable shell-side pressure drop, tube support requirement, and the flow-induced vibrations. Many variations of shell-and-tube exchanger are available; the differences lie in the arrangement of flow configurations and in the details of construction.

Baffled heat exchangers with one shell pass and two tubes passes and with two shell passes and four tube passes are shown in Fig. 6.5a and b, respectively [3].

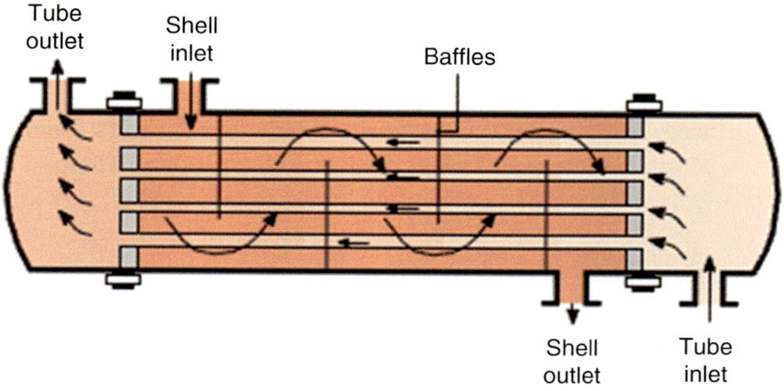

Fig. 6.4 A shell-and-tube heat exchanger; one shell pass and one tube pass [2]

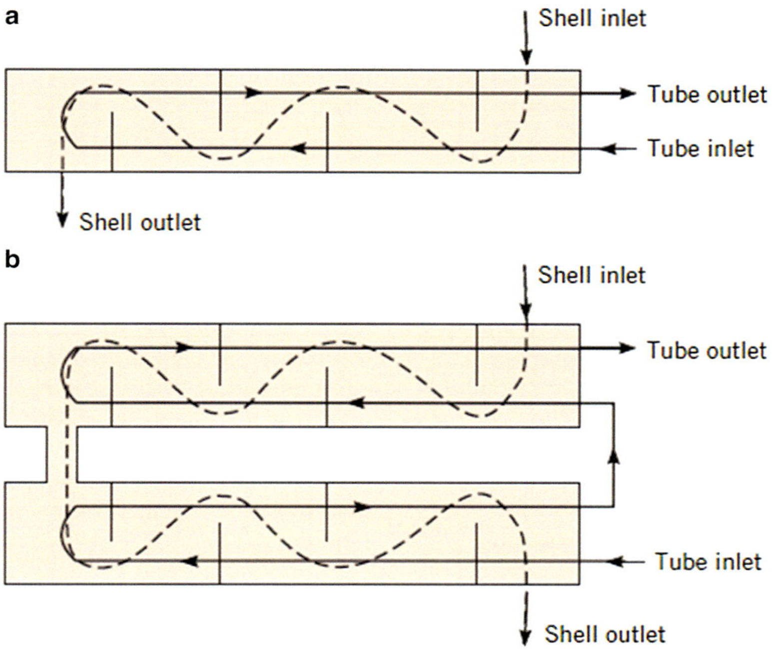

Fig. 6.5 Shell-and-tube heat exchangers. (**a**) One shell pass and two tube passes. (**b**) Two shell passes and four tube passes [3]

The character of the fluids may be *liquid-to-liquid*, *liquid-to-gas*, or *gas-to-gas*. Liquid-to-liquid exchangers have the most common applications. Both fluids are pumped through the exchangers; hence, the heat transfer on both the tube side and the shell side is by forced convection. Since the heat transfer coefficient is high with the liquid flow, generally there is no need to use fins [2].

The liquid-to-gas arrangement is also commonly used; in such cases, the fins usually are added on the gas side of the tubes, where the heat transfer coefficient is low.

Gas-to-gas exchangers are used in the exhaust-gas and air preheating recuperators for gas-turbine systems, cryogenic gas-liquefaction systems, and steel furnaces. Internal and external fins generally are used in the tubes to enhance heat transfer.

6.3.2 Plate Heat Exchangers

As the name implies, plate heat exchangers usually are constructed of thin plates. The plates may be smooth or may have some form of corrugation. Since the plate geometry cannot accommodate as high pressure and/or temperature differentials as a circular tube, it is generally designated for moderate temperature and/or pressure differentials of the compactness factor for plate exchanger ranges from about 120 to 230 m^2/m^3.

6.3.3 Plate Fin Heat Exchangers

The compactness factor can be significantly improved (i.e., up to about 6000 m^2/m^3) by using the plate–fin type of heat exchanger. Figure 6.6 illustrates typical plate–fin configurations. Flat plates separate louvered or corrugated fins. Cross-flow, counterflow, or parallel-flow arrangements can be obtained readily by properly arranging the fins on each side of the plate. Plate–fin exchangers are generally used for gas-to-gas applications, but they are used for low-pressure applications not exceeding about 10 atm (i.e., 1000 kPa). The maximum operating temperatures are limited to about 800 °C. Plate–fin heat exchangers have also been used for cryogenic applications.

6.3.4 Tube Fin Heat Exchangers

When a high operating pressure or an extended surface is needed on one side, tube–fin exchangers are used. Figure 6.7 illustrates two typical configurations, one with round tubes and the other with flat tubes. Tube–fin exchangers can be used for a wide range of tube fluid operating pressures not exceeding about 30 atm and

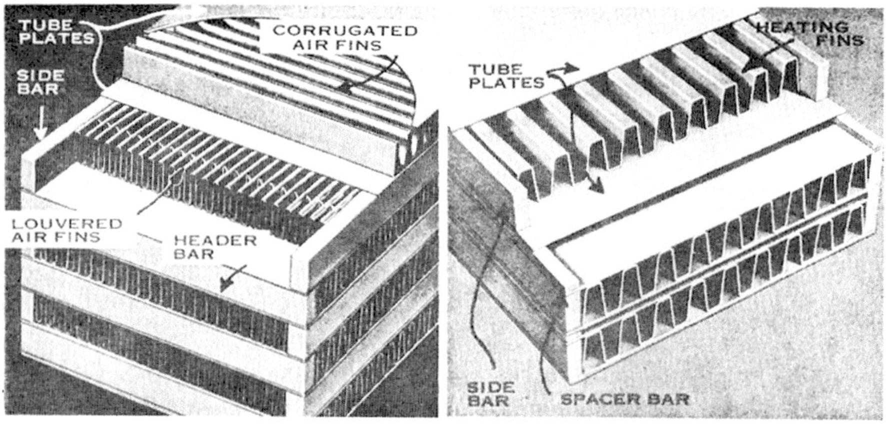

Fig. 6.6 Plate–fin heat exchangers (Courtesy of Harrison Radiator Division of General Motors Corporation)

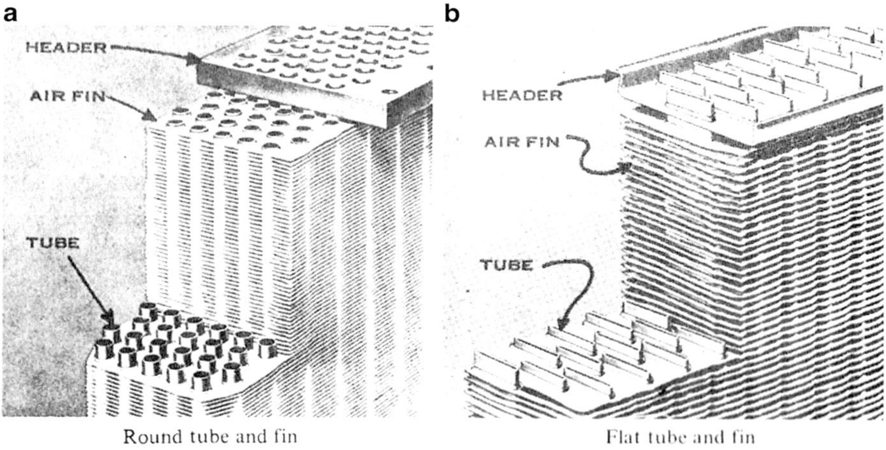

Fig. 6.7 Tube fin heat exchangers (Courtesy of Harrison Radiator Division of General Motors Corporation)

operating temperatures from low cryogenic applications to about 870 °C. The maximum compactness ratio is somewhat less than that obtainable with plate–fin exchangers.

The tube–fin heat exchangers are used in gas turbine, nuclear, fuel cell, automobile, airplane, heat pump, refrigeration, electronics, cryogenics, air conditioning, and many other applications.

6.3.5 Regenerative Heat Exchangers

Regenerative heat exchangers use a heat transfer matrix that is heated by one fluid and then cooled by the second fluid. The flow over the matrix is switched as a function of time with both fluids flowing over the same surfaces of the matrix. They have seen little use in fixed station power plants and will not be emphasized here.

6.4 Condensers

Condensers are used for such varied applications as steam power plants, chemical processing plants, and nuclear electric plants for space vehicles. The major types include the *surface condensers, jet condensers,* and *evaporative condensers*. The most common type is the surface condenser, which has the feed-water system [5]. Figure 6.8 shows a section through a typical two-pass surface condenser for a large steam turbine in a power plant. Since the steam pressure at the turbine exit is

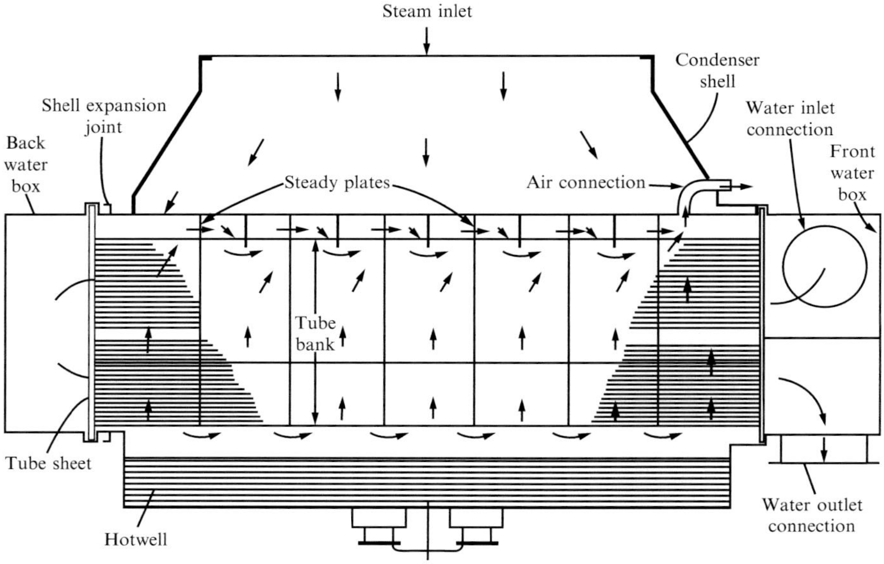

Fig. 6.8 Section through a typical two-pass surface condenser for a large plant (Courtesy of Allis-Chalmers Manufacturing Company)

only 1.0–2.0 in Hg absolute, the steam density is very low and the volume rate of flow is extremely large. To minimize the pressure loss in transferring steam from the turbine to the condenser, the condenser is normally mounted beneath and attached to the turbine. Cooling water flows horizontally inside the tubes, while the steam flows vertically downward from the large opening at the top and passes transversely over the tubes. Note that provision is made to aspirate cool air from the regions just above the center of the hot well. This is important because the presence of noncondensable gas in the steam reduces the heat transfer coefficient for condensation.

6.5 Boilers

Steam boilers are one of the earliest applications of heat exchangers. The term *steam generator* is often applied to boiler in which the heat source is a hot fluid stream rather than the products of combustion.

An enormous variety of boiler types exist, ranging from small units for house heating applications to huge, complex, expensive units for modern power stations.

6.6 Classification According to Compactness

The ratio of the heat transfer, surface area on one side of the heat exchanger to the volume can be used as a measure of the compactness of heat exchangers. A heat exchanger having a surface area density on any one side greater than about 700 m^2/m^3 quite arbitrarily is referred to as a compact heat exchanger regardless of its structural design. For example, automobile radiators having an area density approximately 1100 m^2/m^3 and the glass ceramic heat exchangers for some vehicular gas-turbine engines having an area density approximately 6600 m^2/m^3 are compact heat exchangers. The human lungs, with an area density of about 20,000 m^2/m^3 are the most compact heat-and-mass exchanger. The very fine matrix regenerator for the Stirling engine has an area density approaching that of the human lung.

On the other hand extreme of the compactness scale, plane tubular and shell-and-tube type exchangers, having an area density in the range of 70–500 m^2/m^3, are not considered compact [2].

The incentive for using compact heat exchangers lies in the fact that a high value of compactness reduces the volume for a specified heat exchanger performance.

When heat exchangers are to be employed for automobiles, marine uses, aircraft, aerospace vehicles, cryogenic systems, and refrigeration and air conditioning, the weight and size—hence the compactness—become important. To increase the effectiveness or the compactness of heat exchangers, fins are used. In a gas-to-liquid heat exchanger, for example, the heat transfer coefficient on the gas side is an order of magnitude lower than for the liquid side. Therefore, fins are used on the gas side to obtain a balanced design; the heat transfer surface on the gas side becomes much more compact.

6.7 Types of Applications

Heat exchangers are often classified based on the application for which they are intended, and special terms are employed for major types. These terms include *boiler*, *steam generator*, *condenser*, *radiator*, *evaporator*, *cooling tower*, *regenerator*, *recuperator*, *heater*, and *cooler*. The specialized requirements of the various applications have led to the development of many types of construction, some of which are unique to particular applications [4].

6.8 Cooling Towers

In locations where the supply of water is limited, heat may be rejected to the atmosphere very effectively by means of cooling towers such as those in Fig. 6.9a, b, and c here. A fraction of the water sprayed into these towers

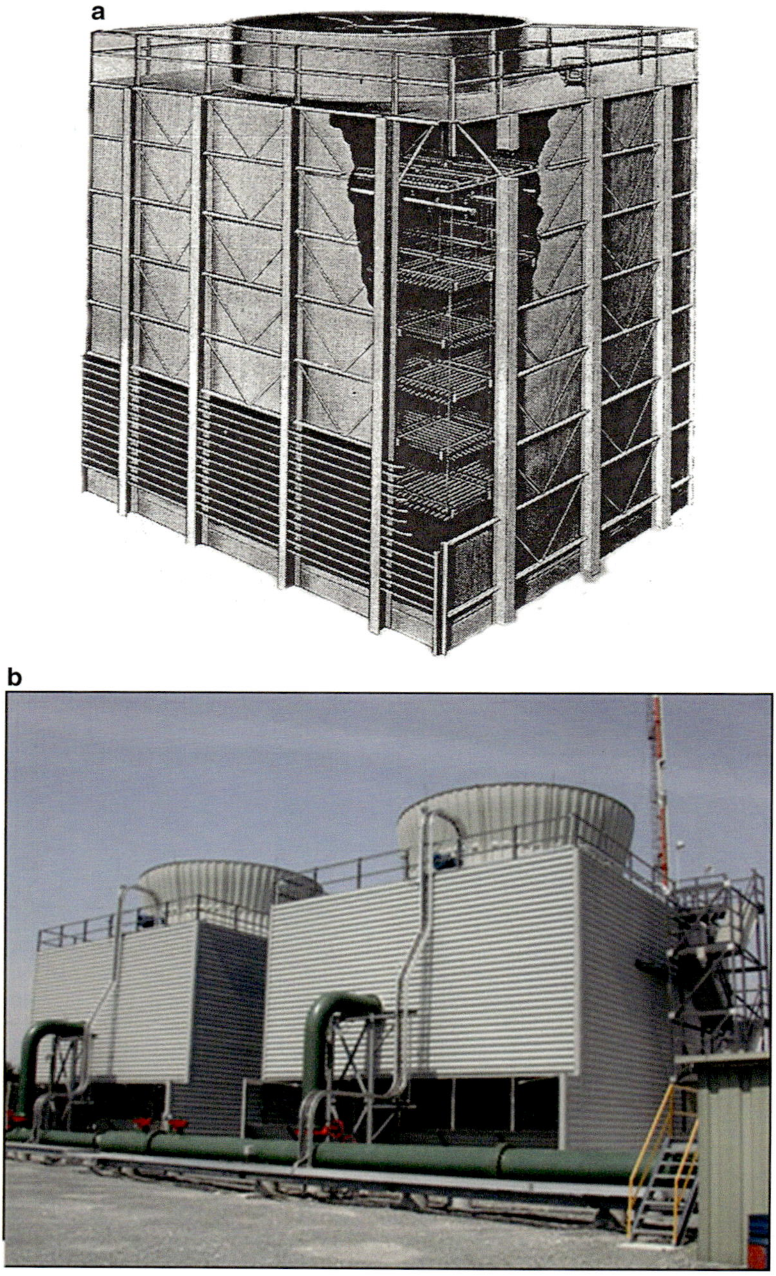

Fig. 6.9 (a) Vertical induced draft-cooling tower (Courtesy Foster Wheeler Corp.) [1]. (b) Schematic of cooling tower. (c) Forced convection cooling tower with draft induced by a fan [1]

evaporates, thus cooling the balance. Because of the high heat of vaporization of water, the water consumption is only about 1 % as much as would be the case if water were taken from a lake or a stream and heated at 10 or 20 °F.

Cooling towers may be designed so that the air moves through them by thermal convection, or fans may be employed to provide forced air circulation. To avoid contamination of the process water, shell-and-tube heat exchangers are sometimes employed to transmit heat from the process water to the water recirculated through the cooling tower.

6.9 Regenerators and Recuperators

The thermal efficiency of both gas-turbine power plants can be greatly increased if heat can be extracted from the hot gases that are leaving the gas turbine and added to the air being supplied to the combustion chamber. For a major gain in thermal efficiency, it is necessary to employ a very large amount of heat transfer, surface area. This is particularly noticeable in gas-turbine plants, where even with counterflow the size of the heat exchanger required for good performance is inclined to be large compared to the size of the turbine and compressor. This characteristic can be observed even in the small, portable gas turbine (about 3 ft in diameter) shown in Fig. 6.10. Note that in this device the hot combination gases leave the radial in-flow turbine wheel at the right end of the shaft and enter a set of heat exchanger cores arranged in parallel around the central axis.

Figure 6.10 shows a close-up view of one of these cores. In each core, the hot gases from the turbine flow roughly radially outward through one set of gas passages. Air from the centrifugal compressor wheel, which is at the center of the shaft, flows to the right through the space just inside of the outer casing and axially into the other set of gas passages through the core. The air being heated makes two passes, flowing first to the right in the outer portion of the core and then back to the

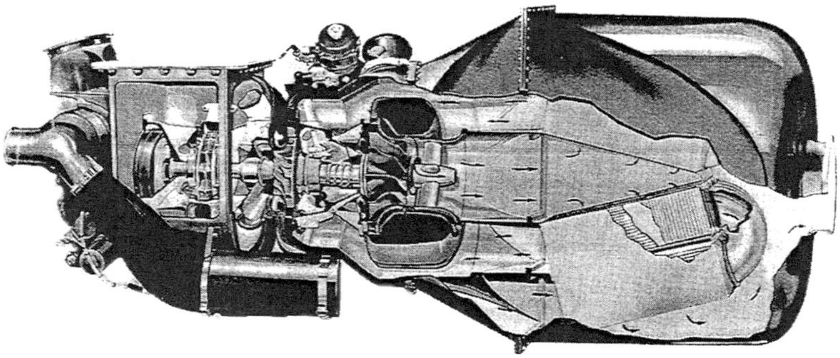

Fig. 6.10 A small gas-turbine power plant fitted with a recuperator to improve the fuel economy (Courtesy of AiResearch Manufacturing Company) [4]

left through the inner portion, thus giving a two-pass cross-flow approximation to counterflow. (The flow passages through the combustion chamber are not shown in this view.)

As can be seen in Fig. 6.11, the heat exchanger core is constructed of alternate layers of flat and corrugated sheets. The flat sheets separate the hot and cold fluid streams, while the corrugated sheets act as fins that roughly triple the heat transfer surface area per unit of volume. Note also that the axis of the corrugations is at right angles in alternate layers to provide a cross-flow pattern for the two fluid streams.

One of several recuperator units to be mounted in parallel in a much larger gas turbine plant is shown in Fig. 6.12. The hot exhaust gas from the turbine enters vertically at the bottom, flows upward through the heat transfer matrix, and discharges vertically from the top. The air from the compressor enters a large circular port at the top at the right end, flows vertically downward in pure counterflow, and leaves a second circular port at the bottom to flow to the combustion chamber. The hot exhaust gas passages are formed by corrugated sheets sandwiched between flat plates that extend all the way from the bottom to the top of the unit. The air to be heated flows horizontally from the long plenum at the top into the spaces between the walls of the exhaust gas passages. Curved space strips guide the air through a 90° bend and then downward between the heated walls. A similar header arrangement is used at the bottom. Note that both the flow passage area and the heat transfer surface area for the hot exhaust gas are about three times as great as the corresponding values for the air being heated. This comes about because the two fluid streams differ in density by a factor of about four.

The air pre-heaters in steam power plants are usually quite different from the units just described for gas turbines. Rotary regenerators are often used.

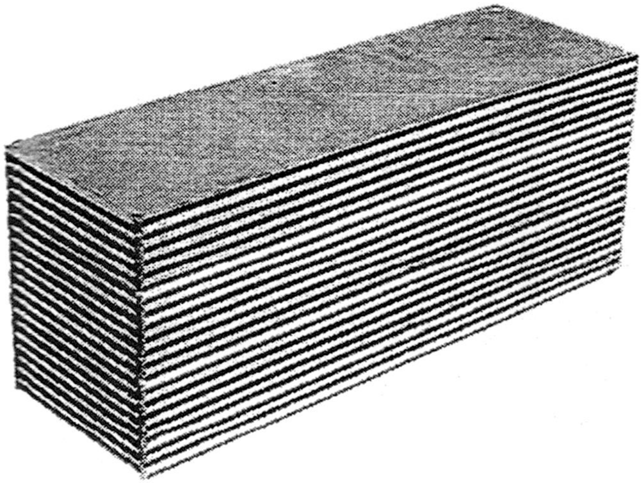

Fig. 6.11 A brazed plate–fin recuperator core for the gas turbine of Fig. 6.9 (Courtesy of AiResearch Manufacturing) [4]

Fig. 6.12 A welded steel recuperator for a large gas-turbine power plant (Courtesy Harrison Radiator Division, General Motors Corp.) [4]

These consist of a cylindrical drum filled with a heat transfer matrix made of alternately flat and corrugated sheets. The drum is mounted so that the hot gas heats a portion of the matrix as it passes from the furnace to the stack. The balance of the matrix gives up its stored heat to the fresh air in route from the forced draft fans to the furnace. The ducts are arranged so that the two gas streams move through the drum in counterflow fashion while it is rotated, so that the temperature of any given element of the metal matrix fluctuates relatively little as it is cycled from the hot to the cold gas streams.

In the steam- and gas-turbine power plant fields a distinction is sometimes made between air pre-heaters that involve a conventional heat transfer matrix with continuous flow on both sides of a stationary heat transfer surface and those through which the fluids flow periodically. The hot fluid is heating one section of the matrix, while the cold fluid is removing heat from another section. Where this distinction is made, the term regenerator is applied to the periodic-flow type of heat exchanger, since this term has long been applied to units of this type employed for blast furnaces and steel furnaces, whereas the term recuperator is applied to units through which the flow is continuous.

Recuperators are used for gas turbine, but the gas turbines installed until the mid-1970s suffered from low efficiency and poor reliability. In the past, large coal and nuclear power plants dominated the base-load electric power generation (Point 1 in Fig. 6.13). Base load units are on line at full capacity or near full capacity almost all of the time. They are not easily nor quickly adjusted for varying large amounts of load because of their characteristics of operation [7]. However, there has been a historic shift toward natural gas-fired turbines because of their higher efficiencies, lower capital costs, shorter installation times, better emission

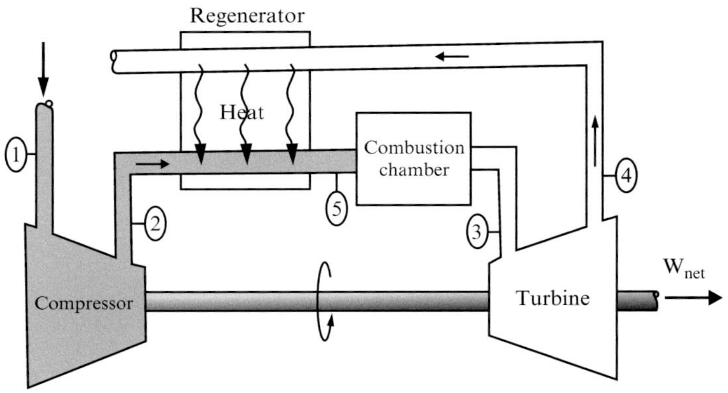

Fig. 6.13 A gas-turbine engine with recuperator

characteristics, the abundance of natural gas supplies, and shorter start up times (Point 1 in Fig. 6.13). Now electric utilities are using gas turbines for base-load power production as well as for peaking, making capacity at maximum load periods and in emergency, situations because they are easily brought online or off-line (Point 2 in Fig. 6.13). The construction costs for gas-turbine power plants are roughly half that of comparable conventional fossil fuel steam power plants, which were the primary base-load power plants until the early 1980s, but peaking units are much higher in energy output costs. A recent gas turbine manufactured by General Electric uses a turbine inlet temperature of 1425 °C (2600 °F) and produces up to 282 MW while achieving a thermal efficiency of 39.5 % in the simple-cycle mode. Over half of all power plants to be installed in the near future are forecast to be gas turbine or combined gas-steam turbine types (Fig. 6.13).

In gas turbine engines with the Brayton cycle that includes recuperator, the temperature of the exhaust gas leaving the turbine is often considerably higher than the temperature of the air leaving the compressor. Therefore, the high-pressure air leaving the compressor can be heated by transferring heat to it from the hot exhaust gases in a counterflow heat exchanger, which is also known as a regenerator or recuperator (see Fig. 6.14 on Point 1). Gas turbine regenerators are usually constructed as shell-and-tube type heat exchangers using very small diameter tubes, with the high-pressure air inside the tubes and low-pressure exhaust gas in multiple passes outside the tubes [8]. The thermal efficiency of the Brayton cycle increases as a result of regeneration since the portion of energy of the exhaust gases that is normally rejected to the surroundings is now used to preheat the air entering the combustion chamber. This, in turn, decreases the heat input (thus fuel) requirements for the same net work output. Note, however, that the use of a regenerator is recommended only when the turbine exhaust temperature is higher than the compressor exit temperature. Otherwise, heat will flow in the reverse direction (to the exhaust gases), decreasing the efficiency. This situation is encountered in gas turbines operating at very high-pressure ratios (Point 1 on Fig. 6.14).

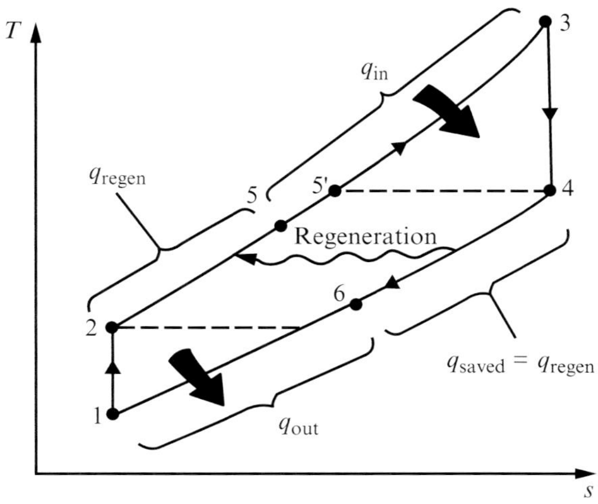

Fig. 6.14 *T–s* diagram of a Brayton cycle with regeneration

The highest temperature occurring within the regenerator is the temperature of the exhaust gases leaving the turbine and entering the regenerator (Point 1 on Fig. 6.14). The gas turbine recuperator receives air from the turbine compressor at pressures ranging from 73.5 to 117 psia and temperatures from 350 to 450 °F (Point 3 on Fig. 6.14). Under no conditions can the air be preheated in the regenerator to a temperature above this value. In the limiting (ideal) case, the air will exit the regenerator at the inlet temperature of the exhaust gases. Air normally leaves the regenerator at a lower temperature (Point 1 on Fig. 6.14). Gas turbine exhaust gas passes over the other side of the recuperator at exhaust temperatures ranging from 750 to 1000 °F. Compressor air temperatures are now raised to higher temperatures up to about 750–900 °F as it enters the combustor. Turbine exhaust gases are then reduced to between 500 and 650 °F from the original 750 to 1000 °F. This heat recovery contributes appreciably to the turbine fuel rate reduction and increase in efficiency (Point 3 on Fig. 6.14). The regenerator is well insulated and any changes in kinetic and potential energies are neglected.

A regenerator with a higher effectiveness will save a greater amount of fuel since it will preheat the air to a higher temperature prior to combustion (Point 1 on Fig. 6.14). However, achieving a higher effectiveness requires the use of a larger regenerator, which carries a higher price tag and causes a larger pressure drop because shaft horsepower is reduced. Pressure drop through the regenerator or recuperator is important and should be kept as low as practical on both sides. Generally, the air pressure drop on the high-pressure side should be held below 2 % of the compressor total discharge pressure. The gas pressure drop on the exhaust side (hot side) should be held below 4 in. of water. Therefore, the use of a regenerator with a very high effectiveness cannot be justified economically unless the savings from the fuel costs exceed the additional expenses involved. The effectiveness of most regenerators used in practice is below 0.85. The thermal efficiency of an ideal Brayton cycle with regeneration depends on the ratio of the minimum to maximum temperatures as well as the pressure ratio. Regeneration is most effective at lower pressure ratios and low minimum-to-maximum temperature ratios.

Gas-to-air recuperators (or regenerators) are also used on marine type industrial and utility open-cycle gas turbine applications. In this application, recuperator receives air from the turbine compressor at pressure and temperature ranging as above, where gas turbine exhaust gas passes over the other side of the recuperator at exhaust temperature, depending on the turbine. The air side (high pressure side) of the recuperator is in the system between the compressor and the combustor and compressor air is raised to a higher temperature than what is mentioned above as it enters the combustor. Obviously, pressure drop through the regenerator or recuperator is important and should be kept as low as practical on both sides.

6.10 Heat Exchanger Analysis: Use of the LMTD

Utilizing the Logarithmic Mean Temperature Difference (LMTD) method is one way to design or to predict the performance of a heat exchanger; it is essential to relate the total heat transfer rate to measurable quantities such as the inlet and outlet fluid temperatures, the overall heat transfer coefficient, and the total surface area for heat transfer. Two such relations may readily be obtained by applying overall energy balances to the hot and cold fluids, as shown in Fig. 6.15. In particular, if q is the total rate of heat transfer between the hot and cold fluids and there is negligible heat transfer between the exchanger and its surroundings, as well as negligible potential and kinetic energy changes, application of the steady flow energy equation gives

$$q_{\text{total}} = \dot{m}_{\text{h}}(h_{\text{h,i}} - h_{\text{h,o}}) \tag{6.1a}$$

and

$$q_{\text{total}} = \dot{m}_{\text{c}}(h_{\text{c,o}} - h_{\text{c,i}}) \tag{6.1b}$$

where h is the fluid enthalpy. The subscripts h and c refer to the hot and cold fluids, whereas the subscripts i and o designate the fluid inlet and outlet conditions. If the

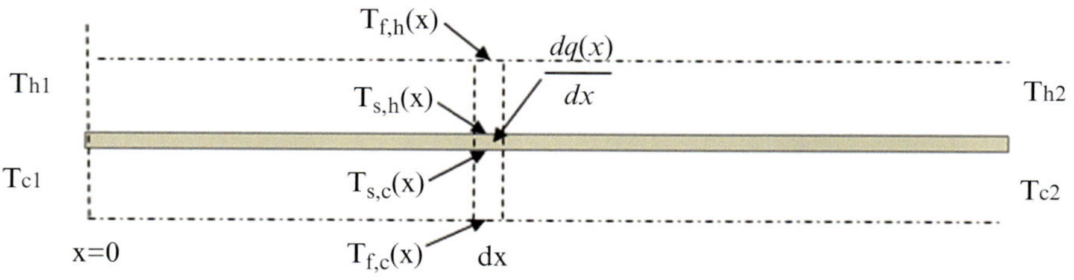

Fig. 6.15 Heat transfer between two moving fluids separated by a solid boundary

fluids are not undergoing a phase change and constant specific heats are assumed, these expressions reduce to

$$q_{\text{total}} = \dot{m}_{\text{h}} c_{\text{p,h}} (T_{\text{h,i}} - T_{\text{h,o}}) \quad (6.2a)$$

and

$$q_{\text{total}} = \dot{m}_{\text{c}} c_{\text{p,c}} (T_{\text{c,o}} - T_{\text{c,i}}) \quad (6.2b)$$

where the temperatures appearing in the expressions refer to the *mean* fluid temperatures at the designated locations. Note that Eqs. (6.1a, 6.1b) and (6.2a, 6.2b) are independent of the flow arrangement and heat exchanger type.

Now consider the heat transfer at a particular point, x, on the heat transfer surface. At x there will be a bulk hot fluid temperature given by $T_{\text{f,h}}(x)$, a wall surface temperature on the hot fluid side given by $T_{\text{s,h}}(x)$, a wall surface temperature on the cold fluid side given by $T_{\text{s,c}}(x)$, and a cold fluid bulk temperature given by $T_{\text{f,c}}(x)$. The total temperature drop from the hot fluid at x to the cold fluid at x is given by

$$\begin{aligned} \Delta T &= T_{\text{f,h}}(x) - T_{\text{f,c}}(x) \\ &= T_{\text{f,h}}(x) - T_{\text{s,h}}(x) + T_{\text{s,h}}(x) - T_{\text{s,c}}(x) + T_{\text{s,c}}(x) - T_{\text{f,c}}(x) \\ &= \Delta T_{\text{f,h}} + \Delta T_{\text{s}} + \Delta T_{\text{f,c}} \end{aligned}$$

Then the heat flux leaving the hot fluid is given by

$$\frac{\text{d}q(x)}{\text{d}x} = h_{\text{f,h}} A_{\text{f,h}} (T_{\text{f,h}}(x) - T_{\text{s,h}}(x)) = h_{\text{f,h}} \text{d}A_{\text{f,h}} \Delta T_{\text{f,h}}(x) \quad \Delta T_{\text{f,h}}(x) = \frac{\frac{\text{d}q(x)}{\text{d}x}}{h_{\text{f,h}} \text{d}A_{\text{f,h}}}$$

The heat flux crossing the wall between the two fluids is given by,

$$\frac{\text{d}q(x)}{\text{d}x} = \frac{k_{\text{s}}}{\delta_{\text{s}}} \text{d}A_{\text{s}} (T_{\text{s,h}}(x) - T_{\text{s,c}}(x)) = \frac{k_{\text{s}}}{\delta_{\text{s}}} \text{d}A_{\text{s}} \Delta T_{\text{s}}(x) \quad \Delta T_{\text{s}}(x) = \frac{\frac{\text{d}q(x)}{\text{d}x}}{\frac{k_{\text{s}}}{\delta_{\text{s}}} \text{d}A_{\text{s}}}$$

And the heat flux into the cold fluid is given by,

$$\frac{\text{d}q(x)}{\text{d}x} = h_{\text{f,c}} \text{d}A_{\text{f,c}} (T_{\text{s,c}}(x) - T_{\text{f,c}}(x)) = h_{\text{f,c}} \text{d}A_{\text{f,c}} \Delta T_{\text{f,c}}(x) \quad \Delta T_{\text{f,c}}(x) = \frac{\frac{\text{d}q(x)}{\text{d}x}}{h_{\text{f,c}} \text{d}A_{\text{f,c}}}$$

Then the difference in the bulk temperatures of the two fluids can be written as

$$\begin{aligned} T_{\mathrm{f,h}}(x) - T_{\mathrm{f,c}}(x) &= \frac{\frac{\mathrm{d}q(x)}{\mathrm{d}x}}{h_{\mathrm{f,h}}\mathrm{d}A_{\mathrm{f,h}}} + \frac{\frac{\mathrm{d}q(x)}{\mathrm{d}x}}{\frac{k_{\mathrm{s}}}{\delta_{\mathrm{s}}}\mathrm{d}A_{\mathrm{s}}} + \frac{\frac{\mathrm{d}q(x)}{\mathrm{d}x}}{h_{\mathrm{f,c}}\mathrm{d}A_{\mathrm{f,c}}} \\ &= \frac{\mathrm{d}q(x)}{\mathrm{d}x}\left[\frac{1}{h_{\mathrm{f,h}}\mathrm{d}A_{\mathrm{f,h}}} + \frac{\delta_{\mathrm{s}}}{k_{\mathrm{s}}\mathrm{d}A_{\mathrm{s}}} + \frac{1}{h_{\mathrm{f,c}}\mathrm{d}A_{\mathrm{f,c}}}\right] \end{aligned} \tag{6.3}$$

Note that δ_{s} will depend on the geometry. For slab or plate geometry

$$\delta_{\mathrm{s}} = \Delta t \quad \text{the wall thickness}$$

For cylindrical geometry typical of tubes,

$$\delta_{\mathrm{s}} = r_{\mathrm{in}}\ln\left(\frac{r_{\mathrm{out}}}{r_{\mathrm{in}}}\right) \qquad r_{\mathrm{out}} - r_{\mathrm{in}} = \text{the tube wall thickness}$$

Also note that the differential areas do not all have to be equal. There will be a slight difference if the bounding surface is a tube, but the addition of fins to either the hot or the cold side could change the effective area significantly and that is the area that must be used in Eq. (6.3). Also, note that the areas are areas per unit length. That is why they have been written as dA:

$$T_{\mathrm{f,h}}(x) - T_{\mathrm{f,c}}(x) = \frac{\mathrm{d}q(x)}{\mathrm{d}x}\left(\frac{1}{U\frac{\mathrm{d}A}{\mathrm{d}x}}\right) \quad \frac{1}{U\frac{\mathrm{d}A}{\mathrm{d}x}} = \frac{1}{h_{\mathrm{f,h}}\mathrm{d}A_{\mathrm{f,h}}} + \frac{\delta_{\mathrm{s}}}{k_{\mathrm{s}}\mathrm{d}A_{\mathrm{s}}} + \frac{1}{h_{\mathrm{f,c}}\mathrm{d}A_{\mathrm{f,c}}}$$

$$\frac{\mathrm{d}q(x)}{\mathrm{d}x} = U\frac{\mathrm{d}A}{\mathrm{d}x}(T_{\mathrm{f,h}} - T_{\mathrm{f,c}})$$

Then the heat lost by the hot fluid is given by

$$\frac{\mathrm{d}q(x)}{\mathrm{d}x} = -\dot{m}_{\mathrm{f,h}}C_{\mathrm{p,h}}\frac{\mathrm{d}T_{\mathrm{f,h}}(x)}{\mathrm{d}x} \tag{6.4}$$

And the heat gained by the cold fluid is given by

$$\frac{\mathrm{d}q(x)}{\mathrm{d}x} = \dot{m}_{\mathrm{f,c}}C_{\mathrm{p,c}}\frac{\mathrm{d}T_{\mathrm{f,c}}(x)}{\mathrm{d}x} \tag{6.5}$$

Combining these two equations gives

$$\frac{\mathrm{d}T_{\mathrm{f,h}}(x)}{\mathrm{d}x}-\frac{\mathrm{d}T_{\mathrm{f,c}}(x)}{\mathrm{d}x}=-\frac{\mathrm{d}q(x)}{\mathrm{d}x}\left(\frac{1}{\dot{m}_{\mathrm{h}}C_{\mathrm{p,h}}}+\frac{1}{\dot{m}_{\mathrm{c}}C_{\mathrm{p,c}}}\right)=-U\frac{\mathrm{d}A}{\mathrm{d}x}(T_{\mathrm{f,h}}-T_{\mathrm{f,c}})$$

$$\Delta T(x)=T_{\mathrm{f,h}}(x)-T_{\mathrm{f,c}}(x)$$

$$\frac{\mathrm{d}\Delta T(x)}{\mathrm{d}x}=-U\frac{\mathrm{d}A}{\mathrm{d}x}\Delta T\left(\frac{1}{\dot{m}_{\mathrm{h}}C_{\mathrm{p,h}}}+\frac{1}{\dot{m}_{\mathrm{c}}C_{\mathrm{p,c}}}\right)$$

$$\frac{\mathrm{d}\Delta T(x)}{\Delta T(x)}=-U\left(\frac{1}{\dot{m}_{\mathrm{h}}C_{\mathrm{p,h}}}+\frac{1}{\dot{m}_{\mathrm{c}}C_{\mathrm{p,c}}}\right)\mathrm{d}A \quad \frac{\mathrm{d}A}{\mathrm{d}x}\mathrm{d}x=\mathrm{d}A$$

Integrating gives

$$\ln\left(\frac{\Delta T_2}{\Delta T_1}\right)=-UA\left(\frac{1}{\dot{m}_{\mathrm{h}}C_{\mathrm{p,h}}}+\frac{1}{\dot{m}_{\mathrm{c}}C_{\mathrm{p,c}}}\right) \tag{6.6}$$

Now for the hot fluid flowing from left to right Eq. (6.2a) becomes

$$q_{\mathrm{total}}=\dot{m}_{\mathrm{h}}C_{\mathrm{p,h}}(T_{\mathrm{f,h,1}}-T_{\mathrm{f,h,2}}) \quad \frac{1}{\dot{m}_{\mathrm{h}}C_{\mathrm{p,h}}}=\frac{(T_{\mathrm{f,h,1}}-T_{\mathrm{f,h,2}})}{q_{\mathrm{total}}}$$

In addition, for the cold fluid also flowing from left to right (parallel flow) Eq. (6.2b) becomes

$$q_{\mathrm{total}}=\dot{m}_{\mathrm{c}}C_{\mathrm{p,c}}(T_{\mathrm{f,c,2}}-T_{\mathrm{f,c,1}}) \quad \frac{1}{\dot{m}_{\mathrm{c}}C_{\mathrm{p,c}}}=\frac{(T_{\mathrm{f,c,2}}-T_{\mathrm{f,c,1}})}{q_{\mathrm{total}}}$$

Plugging these into Eq. (6.3) gives

$$\ln\left(\frac{\Delta T_2}{\Delta T_1}\right)=-UA\left(\frac{T_{\mathrm{f,h,1}}-T_{\mathrm{f,h,2}}}{q_{\mathrm{total}}}+\frac{T_{\mathrm{f,c,2}}-T_{\mathrm{f,c,1}}}{q_{\mathrm{total}}}\right)=\frac{UA}{q_{\mathrm{total}}}(T_{\mathrm{f,h,2}}-T_{\mathrm{f,c,2}}-T_{\mathrm{f,h,1}}+T_{\mathrm{f,c,1}})$$

$$\ln\left(\frac{\Delta T_2}{\Delta T_1}\right)=\frac{UA}{q_{\mathrm{total}}}(\Delta T_2-\Delta T_1) \quad q_{\mathrm{total}}=UA\frac{(\Delta T_2-\Delta T_1)}{\ln\left(\frac{\Delta T_2}{\Delta T_1}\right)}$$

$$q_{\mathrm{total}}=UA\Delta T_{\mathrm{lm}} \quad \Delta T_{\mathrm{lm}}=\frac{(\Delta T_2-\Delta T_1)}{\ln\left(\frac{\Delta T_2}{\Delta T_1}\right)} \tag{6.7}$$

This looks a lot like Newton's law of cooling with ΔT_{lm} playing the role of the standard ΔT. ΔT_{lm} is called the log-mean temperature difference.

Now consider the counterflow arrangement. In this case Eq. (6.5) becomes,

$$\frac{\mathrm{d}q(x)}{\mathrm{d}x}=-\dot{m}_{\mathrm{f,c}}C_{\mathrm{p,c}}\frac{\mathrm{d}T_{\mathrm{f,c}}(x)}{\mathrm{d}x}$$

Moreover, Eq. (6.6) becomes,

$$\ln\left(\frac{\Delta T_2}{\Delta T_1}\right) = -UA\left(\frac{1}{\dot{m}_\mathrm{h} C_{\mathrm{p,h}}} - \frac{1}{\dot{m}_\mathrm{c} C_{\mathrm{p,c}}}\right)$$

Then Eq. (6.2b) becomes

$$q_\mathrm{total} = \dot{m}_\mathrm{c} c_{\mathrm{p,c}} (T_{\mathrm{f,c,1}} - T_{\mathrm{f,c,2}})$$

This gives

$$\ln\left(\frac{\Delta T_2}{\Delta T_1}\right) = -UA\left(\frac{T_{\mathrm{f,h,1}} - T_{\mathrm{f,h,2}}}{q_\mathrm{total}} - \frac{T_{\mathrm{f,c,1}} - T_{\mathrm{f,c,2}}}{q_\mathrm{total}}\right) = \frac{UA}{q_\mathrm{total}}(T_{\mathrm{f,h,2}} - T_{\mathrm{f,c,2}} - T_{\mathrm{f,h,1}} + T_{\mathrm{f,c,1}})$$

$$\ln\left(\frac{\Delta T_2}{\Delta T_1}\right) = \frac{UA}{q_\mathrm{total}}(\Delta T_2 - \Delta T_1) \quad q_\mathrm{total} = UA\frac{(\Delta T_2 - \Delta T_1)}{\ln\left(\frac{\Delta T_2}{\Delta T_1}\right)}$$

which is the identical equation for the parallel flow heat exchanger. It is important to remember how the ΔTs are defined.
Parallel Flow:

$$\Delta T_1 = T_{\mathrm{f,h.in}} - T_{\mathrm{f,c,in}} \quad \Delta T_2 = T_{\mathrm{f,h,out}} - T_{\mathrm{f,c,out}}$$

Counterflow: $\Delta T_1 = T_{\mathrm{f,h,in}} - T_{\mathrm{f,c,out}}$ $\quad \Delta T_2 = T_{\mathrm{f,h,out}} - T_{\mathrm{f,c,in}}$

Example 6.1 A counterflow, concentric tube heat exchanger is used to cool the lubricating oil for a large industrial gas turbine engine. The flow rate of cooling water through the inner tube ($D_\mathrm{i} = 25$ mm) is 0.2 kg/s, while the flow rate of oil through the outer annulus ($D_0 = 45$ mm) is 0.1 kg/s. The oil and water enter at temperatures of 100 and 30 °C, respectively. How long must the tube be made if the outlet temperature of the oil is to be 60 °C? (The steel tube that separates the two flows is so thin that the temperature drop across it may be neglected.). See Fig. 6.16.

Solution:
Known: Fluid flow rates and inlet temperature for a counterflow, concentric tube heat exchanger of prescribed inner and outer diameter.

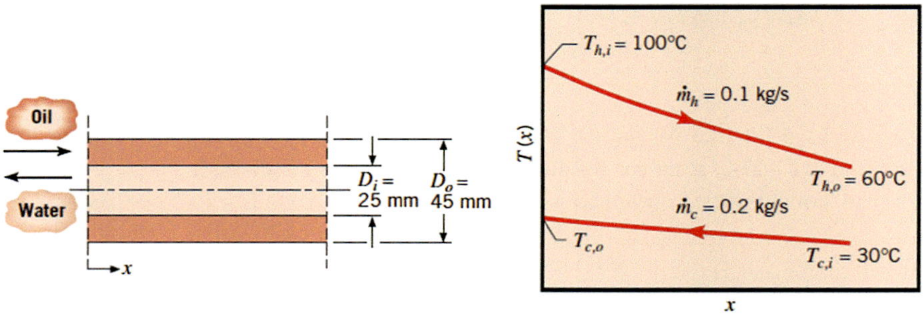

Fig. 6.16 Schematic of Example 6.1 [4]

Find: Tube length to achieve a desired hot fluid outlet temperature.
Assumptions:

1. Negligible heat loss to the surroundings.
2. Negligible kinetic and potential energy changes.
3. Constant properties.
4. Negligible tube wall thermal resistance and fouling factors.
5. Fully developed conditions for the water and oil (U independent of x).

Properties:
Table A.5, unused engine oil $\left(\overline{T}_{\mathrm{h}} = 80\,^{\circ}C = 353\,\mathrm{K}\right)$:

$$c_{\mathrm{p}} = 2131\,\mathrm{J/kg \cdot K}, \quad \mu = 3.25 \times 10^{-2}\mathrm{N \cdot s/m^2}, \quad k = 0.138\,\mathrm{W/m \cdot K}.$$

Table A.6, water $\left(\overline{T}_{\mathrm{c}} \approx 35\,^{\circ}C = 308\,\mathrm{K}\right)$:

$$c_{\mathrm{p}} = 4178\,\mathrm{J/kg \cdot K}, \quad \mu = 725 \times 10^{-6}\mathrm{N \cdot s/m^2}, \quad k = 0.625\,\mathrm{W/m \cdot K}, \quad \mathrm{Pr} = 4.85\,.$$

Analysis:
The required heat transfer rate may be obtained from the overall energy balance for the hot fluid, Eq. (6.1a)

$$q = \dot{m}_{\mathrm{h}} c_{\mathrm{p,h}} (T_{\mathrm{h,i}} - T_{\mathrm{h,o}})$$
$$q = 0.1\,\mathrm{kg/s} \times 2131\,\mathrm{J/kg \cdot K}(100 - 60)\,^{\circ}\mathrm{C} = 8524\,\mathrm{W}$$

Applying Eq. (6.2b), the water outlet temperature is

$$T_{\mathrm{c,o}} = \frac{q}{\dot{m}_{\mathrm{c}} c_{\mathrm{p,c}}} + T_{\mathrm{c,i}}$$
$$T_{\mathrm{c,o}} = \frac{8524\,\mathrm{W}}{0.2\,\mathrm{kg/s} \times 2131\,\mathrm{J/kg \cdot K}} + 30\,^{\circ}\mathrm{C} = 40.2\,^{\circ}\mathrm{C}$$

Accordingly, use of $\overline{T}_{\mathrm{c}} = 35\,^{\circ}C$ to evaluate the water properties was a good choice. The required heat exchanger length may now be obtained from Eq. (6.7),

$$q = UA\Delta T_{\mathrm{lm}}$$

where

$$A = \pi D_{\mathrm{i}} L$$

$$\Delta T_{\mathrm{lm}} = \frac{(T_{\mathrm{h,i}} - T_{\mathrm{c,o}}) - (T_{\mathrm{h,o}} - T_{\mathrm{c,i}})}{\ln[(T_{\mathrm{h,i}} - T_{\mathrm{c,o}}) - (T_{\mathrm{h,o}} - T_{\mathrm{c,i}})]} = \frac{59.8 - 30}{\ln(59.8/30)} = 43.2\,^{\circ}\mathrm{C}$$

The overall heat transfer coefficient is

$$U = \frac{1}{(1/h_\mathrm{i}) + (1/h_\mathrm{o})}$$

For water flow through the tube,

$$\dot{m} = \rho AV \quad \frac{\dot{m}}{A} = \rho V \ \mathrm{Re_D} = \frac{\dot{m}_\mathrm{c} D_\mathrm{i}}{A\mu} = \frac{4\dot{m}}{\pi D_\mathrm{i}\mu}$$

$$= \frac{4 \times 0.2\,\mathrm{kg/s}}{\pi (0.025\,\mathrm{m}) 725 \times 10^{-6}\,\mathrm{N \cdot s/m^2}} = 14{,}050$$

Accordingly, the flow is turbulent and the convection coefficient may be computed from the following equation:

$$\mathrm{Nu_D} = 0.023 \mathrm{Re_D^{4/5}} \mathrm{Pr}^{0.4}$$

$$\mathrm{Nu_D} = 0.023(14{,}050)^{4/5}(4.85)^{0.4} = 90$$

Hence

$$h_\mathrm{i} = \mathrm{Nu_D} \frac{k}{D_\mathrm{i}} = \frac{90 \times 0.625\,\mathrm{W/m \cdot K}}{0.25\,\mathrm{m}} = 2250\,\mathrm{W/m^2 \cdot K}$$

For the flow of oil through the annulus, the hydraulic diameter is

$$D_\mathrm{h} = \frac{4A}{P} = \frac{4\left(\pi r_\mathrm{o}^2 - \pi r_\mathrm{i}^2\right)}{2\pi (r_\mathrm{o} - r_\mathrm{i})} = 2(r_\mathrm{o} - r_\mathrm{i}) = D_\mathrm{o} - D_\mathrm{i}$$

and the Reynolds number is

$$\mathrm{Re_D} = \frac{\rho u_\mathrm{m} D_\mathrm{h}}{\mu} = \frac{\dot{m}(D_\mathrm{o} - D_\mathrm{i})}{A\mu} = \frac{\dot{m}(D_\mathrm{o} - D_\mathrm{i})}{\mu\pi\left(D_\mathrm{o}^2 - D_\mathrm{i}^2\right)/4}$$

$$\mathrm{Re_D} = \frac{4\dot{m}}{\pi (D_\mathrm{o} + D_\mathrm{i})\mu} = \frac{4(0.1\,\mathrm{kg/s})}{\pi (0.045 + 0.025) 3.25 \times 10^{-2}\,\mathrm{kg/s \cdot m}} = 56.0$$

The annular flow is therefore laminar. For a constant heat flux the laminar correlation is given by the following equation:

$$\mathrm{Nu_i} = \frac{h_\mathrm{o} D_\mathrm{h}}{k} = 4.36$$

and

$$h_\mathrm{o} = 4.36 \frac{0.138\,\mathrm{W/m \cdot K}}{0.020\,\mathrm{m}} = 30.1\,\mathrm{W/m^2 \cdot K}$$

The overall convection coefficient is then

$$U = \frac{1}{(1/2250\ \mathrm{W/m^2 \cdot K}) + (1/30.1\ \mathrm{W/m^2 \cdot K})} = 29.7\ \mathrm{W/m^2 \cdot K}$$

and from the rate equation it follows that

$$L = \frac{q}{U \pi D_\mathrm{i} \Delta T_\mathrm{lm}} = \frac{8524\,\mathrm{W}}{29.7\,\mathrm{W/m^2 \cdot K}\pi(0.025\,\mathrm{m})(43.2\,^\circ\mathrm{C})} = 84.6\,\mathrm{m}$$

Comments:

1. The hot side convection coefficient controls the rate of heat transfer between the two fluids, and the low value of h_o is responsible for the large value of L.
2. Because $h_\mathrm{i} \gg h_\mathrm{o}$, the tube wall temperature will follow closely that of the coolant water.

Many heat exchangers have been designed based on a log-mean temperature difference. However, there are some problems with proceeding based on the log-mean temperature difference equation. First, it says nothing about cross-flow heat exchangers, which are very common due to the ease of construction of this type of exchanger. Secondly, it often requires iterative calculations for a design if all of the inlet and exit temperatures are not known a priori. In the example above, we had the outlet temperature of the oil not being specified, an iterative solution would have been required. Iterative solutions can certainly be accurate, but they often require more work. Thirdly, the log-mean temperature difference method does not provide a feel for the maximum heat transfer possible given the entering conditions of the fluids. Sometimes this is an important parameter to understand, if the design is to be optimized.

6.11 Effectiveness-NTU Method for Heat Exchanger Design

A better method has been developed for heat exchanger design that uses some of the preceding analysis. This method is called the effectiveness-NTU (Number of Transfer Unit) method [9]. It starts by considering the fluid heat transfer capacity rates defined as

$$\begin{array}{ll} \text{Cold Fluid Capacity Rate} & C_\mathrm{c} = \dot{m}_\mathrm{c} C_{\mathrm{p,c}}\,\mathrm{W/K} \\ \text{Hot Fluid Capacity Rate} & C_\mathrm{h} = \dot{m}_\mathrm{h} C_{\mathrm{p,h}}\,\mathrm{W/K} \end{array} \tag{6.8}$$

Then the maximum amount of heat that can be transferred between the two fluids is the Minimum Fluid Capacity Rate times the difference in temperature of the hot fluid entering the exchanger and the cold fluid entering the exchanger. Or

$$C_{\min} = \min(C_{\mathrm{c}}, C_{\mathrm{h}})$$
$$q_{\max} = C_{\min}(T_{\mathrm{h,in}} - T_{\mathrm{c,in}}) \tag{6.9}$$

Then the heat exchanger effectiveness is defined as

$$\varepsilon = \frac{q_{\mathrm{act}}}{q_{\max}} = \frac{C_{\mathrm{h}}(T_{\mathrm{h,in}} - T_{\mathrm{h,out}})}{C_{\min}(T_{\mathrm{h,in}} - T_{\mathrm{c,in}})} = \frac{C_{\mathrm{c}}(T_{\mathrm{c,in}} - T_{\mathrm{c,out}})}{C_{\min}(T_{\mathrm{h,in}} - T_{\mathrm{c,in}})} \tag{6.10}$$

The number of heat exchanger transfer units is then defined as

$$\mathrm{NTU} = \frac{UA}{C_{\min}} \tag{6.11}$$

where U and A are defined as above. The heat Capacity-Rate Ratio is defined as,

$$C_{\mathrm{r}} = \frac{C_{\min}}{C_{\max}} \tag{6.12}$$

Then in general it is possible to express the effectiveness as

$$\varepsilon = \varepsilon(\mathrm{NTU}, C_{\mathrm{r}}, \mathrm{Flow\ \ Arrangement})$$

Different functions for ε have been developed for many flow arrangements [9]. The three main flow arrangements of interest are parallel flow, counterflow, and cross-flow.

Parallel Flow:
For parallel flow the expression is

$$\varepsilon = \frac{1 - \mathrm{e}^{-\mathrm{NTU}(1+C_{\mathrm{r}})}}{1 + C_{\mathrm{r}}} \tag{6.13}$$

There are two interesting limits,

$$\begin{aligned} C_{\mathrm{r}} \to 0 \quad & \varepsilon = 1 - \mathrm{e}^{-\mathrm{NTU}} \\ C_{\mathrm{r}} \to 1 \quad & \varepsilon = \frac{1}{2} \end{aligned} \tag{6.14}$$

The $C_{\mathrm{r}} = 0$ limit corresponds to one fluid vaporizing or condensing and the heat capacity rate for this fluid becomes immense. The other limit $C_{\mathrm{r}} = 1.0$ corresponds to both fluids having the same Heat Capacity Rate.

Counterflow:

$$\varepsilon = \frac{1 - e^{-NTU(1-C_r)}}{1 - C_r e^{-NTU(1-C_r)}} \tag{6.15}$$

Addressing the same two limits

$$\begin{aligned} C_r \to 0 \quad & \varepsilon = 1 - e^{-NTU} \\ C_r \to 1 \quad & \varepsilon = \frac{NTU}{1 + NTU} \end{aligned} \tag{6.16}$$

The $C_r = 0$ limit is the same but the $C_r = 1$ limit is twice as effective for large values of NTU. This essentially is the known performance advantage of counterflow heat exchangers.

Cross-Flow: Cross-flow has to be broken down into three different types. The performance is different depending on whether or not the fluids are allowed to mix with themselves as they move through the exchanger. A typical tube and shell exchanger would have the fluid moving through the tubes described as unmixed and the fluid moving through the shell would be mixed.

Cross-Flow—Both Fluids Unmixed
This case requires a series numerical solution and the curves for values of various C_r values are given in Kays and London. For the case of $C_r = 0$, the solution is the same as for counterflow and parallel flow:

$$C_r = 0 \quad \varepsilon = 1 - e^{-NTU}$$

However, all of the curves for any value of C_r asymptotically approach 1.0 like the counterflow exchanger. For all $C_r > 0$ the effectiveness is less than for a counter-flow exchanger with the same C_r.

Cross-Flow—One Fluid Mixed
For the case of

$$\begin{aligned} C_{max} = C_{unmixed} \quad & C_{min} = C_{mixed} \\ \varepsilon = 1 - e^{-\Gamma/C_r} \quad & \Gamma = 1 - e^{-NTU C_r} \end{aligned} \tag{6.17}$$

And for the case of

$$\begin{aligned} C_{max} = C_{mixed} \quad & C_{min} = C_{unmixed} \\ \varepsilon = \frac{1 - e^{-\Gamma' C_r}}{C_r} \quad & \Gamma = 1 - e^{-NTU} \end{aligned} \tag{6.18}$$

Once again for $C_r = 0$, this gives the same behavior as the counterflow heat exchanger. For $C_r = 1.0$, it gets complicated but it is important to note that if a choice is possible it is better to have the fluid with the smaller heat capacity rate to be the mixed fluid (Eq. 6.17).

Cross-Flow—Both Fluids Mixed
The closed form solution is

$$\varepsilon = \frac{\text{NTU}}{\frac{\text{NTU}}{(1-e^{-\text{NTU}})} + \frac{C_r\text{NTU}}{1-e^{-\text{NTU}C_r}} - 1} \tag{6.19}$$

As always, for the case of $C_r = 0$, the results are the same as the counterflow exchanger.

For $C_r = 1.0$ as NTU becomes large, the effectiveness goes to 1/2. However, this is the only case that a better effectiveness can be obtained at a lower NTU. The effectiveness actually decreases after an NTU of about 3–5.

There are many other configurations reported by Kays and London, but these three are the most important. The availability of solutions for the common cross-flow case of one fluid mixed makes this technique very useful.

Example 6.2 Consider a gas-to-gas recuperator of the shell-and-tube design. The tubes are 2 cm diameter tubes spaced on 4 cm centers with a 2 mm thickness made of aluminum. The flow cross section is a 2 m by 2 m square. The pressure ratio for the compressor is 20. Both fluids are air and the cold fluid is in the tubes. The hot fluid enters at 783 K and exits at 670 K and is atmospheric pressure. The cold fluid enters at 655 K and exits at 768 K and is at 20 atm. The flow rate is 2.5E+5 kg/h, for a 10 MW power plant.

Solution:
Start with the hot fluid—calculate Re No.

$$\text{N tubes} = 2401, \quad A_{\text{flow}} = 2^2 - 2401 \times \pi(0.012)^2 = 2.9138\,\text{m}^2$$

$$\dot{m} = \rho AV \quad \rho V = \frac{\dot{m}}{A} = \frac{69.44\,\text{kg/s}}{2.9138} = 23.8\,\text{kg/s/m}^2$$

$$\mu = 3.65\text{E} - 5 \quad D = \frac{4\left(0.04^2 - \pi 0.012^2\right)}{2\pi 0.012} = 0.0609\,\text{m}$$

$$\text{Re} = \frac{23.8 \times 0.0609}{3.7\text{E} - 05} = 39,789.2$$

This is clearly in the turbulent range.

Using the same equations, the cold fluid Re = 56,731.1

$$\begin{aligned}
\text{Pr}_{\text{hot}} &= C_p\text{m/k} = 1078.8 \times 3.65\text{E} - 5/0.0564 = 0.697 \\
\text{Pr}_{\text{cold}} &= C_p\text{m/k} = 1076.7 \times 3.25\text{E} - 5/0.050 = 0.700 \\
\text{Nu}_{\text{hot}} &= 0.023 \times 39,789.2^{0.8} \times 0.697^{0.3} = 98.7 \\
h_{\text{hot}} &= 98.7 \times 0.0564/0.0609 = 91.5\,\text{w/m}^2 \cdot \text{K} \\
\text{Nu}_{\text{cold}} &= 0.023 \times 56,731.1^{0.8} \times 0.700^{0.4} = 127.0 \\
h_{\text{cold}} &= 127.0 \times 0.0500/0.02 = 316.0\,\text{w/m}^2 \cdot \text{K}
\end{aligned}$$

This allows us to calculate UA as a function of L

$$A_{\text{hot}} = 2\pi(0.012) \times 2401 \times L = 724.8 \times L\,\text{m}^2$$
$$A_{\text{cold}} = 2\pi(0.01) \times 2401 \times L = 603.4 \times L\,\text{m}^2$$
$$A_{\text{tube}} = 2\pi(0.011) \times 2401 \times L = 663.74 \times L\,\text{m}^2$$

$$\begin{aligned}\frac{1}{UA} &= \left(\frac{1}{h_{\text{hot}}A} + \frac{t}{kA} + \frac{1}{h_{\text{cold}}A}\right)\\ &= \frac{1}{91.5(724.8)L} + \frac{0.002}{218(663.74)L} + \frac{1}{316(603.4)L}\\ &= (1.508\text{E}-5 + 1.38\text{E}-8 + 5.25\text{E}-6)/L = \frac{2.03\text{E}-5}{L}\ \text{w/K}\end{aligned}$$

$$\dot{Q} = 8.46E+6\,\text{w} \quad \Delta T_{\text{in}} = 15\,\text{K} \quad \Delta T_{\text{out}} = 15\,\text{K}$$
$$\Delta T_{\text{lmn}} = 15\,\text{K} \quad \dot{Q} = \left(\frac{UA}{L}\right)L\Delta T_{\text{lmn}} \quad L = \frac{\dot{Q}}{(UA/L)\Delta T_{\text{lmn}}}$$
$$L = \frac{8.46\text{E}+6}{4.92\text{E}+4 \times 15} = 11.5\,\text{m}$$

Now try the NTU-effectiveness method

$$C_{\text{hot}} = 2.5\text{E}+5/3600 \times 1078.8 = 7.49\text{E}+04$$
$$C_{\text{cold}} = 2.5\text{E}+5/3600 \times 1076.7 = 7.48\text{E}+04 = C_{\min}$$
$$C_r = 0.998 \sim 1.0 \quad \varepsilon = \text{NTU}/(\text{NTU}+1) \quad \text{NTU} = \varepsilon/(1-\varepsilon)$$
$$\varepsilon = \frac{C_{\text{hot}}(T_{\text{h,in}} - T_{\text{h,out}})}{C_{\min}(T_{\text{h,in}} - T_{\text{c,in}})} = \frac{7.49\text{E}+4 \times (783-670)}{7.48\text{E}+4 \times (783-655)} = 0.8846$$
$$\text{NTU} = 7.67 \quad \frac{\text{NTU}}{L} = \frac{UA/L}{C_{\min}} = 0.6576 \quad L = \frac{7.67}{0.6576} = 11.66\,\text{m}$$

Note that the largest resistance to heat transfer was in the hot side convection and the resistance of the tube wall was negligible.

So add fins to the hot side channel by putting a 2 mm thick web between the tubes. See Fig. 6.17.

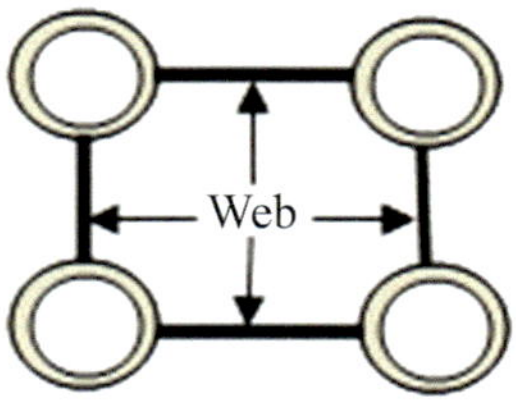

Fig. 6.17 Web channel

First, recalculate the hydraulic diameter for the hot side.
Treating the webs as wetted perimeter gives a new $D_h = 0.0329$ m
This gives a new

$$\begin{aligned} \text{Re} &= 21,521.3 \\ \text{Nu} &= 60.4 \\ h &= 104.0. \end{aligned}$$

Dividing the web in half for a fin for each tube,

$$w = 0.01, \quad t = 0.002, \quad P = 0.02, \quad A = 0.00002, \quad m = 21.79, \quad L = 0.008$$
$$mL = 0.1743, \quad h_{\text{fin}} = 0.9899, \quad h_{\text{o}} = 0.9915, \quad A_{\text{tot}} = 0.1394$$
$$h_{\text{o}} \times A_{\text{tot}} = 0.1382, \quad A_{\text{hot}}/A_{\text{cold}} = 2.1997, \quad A_{\text{hot}}/L = 1327.3$$

The cold side and tube resistances do not change, so

$$\frac{1}{UA} = \frac{1}{1.25\text{E}-5 \times L} \qquad L = \frac{\dot{Q}}{\frac{UA}{L}\Delta T_{\text{lmn}}} = \frac{8.47\text{E}+6}{7.98\text{E}+4 \times 15} = 7.1\,\text{m}$$

Now to get the pressure drops

$$\begin{aligned} C_{\text{f,hot}} &= 0.046\text{Re}^{-0.2} = 0.046 \times 21521.3^{-0.2} = 6.25\text{E}-3 \\ C_{\text{f,cold}} &= 0.046\text{Re}^{-0.2} = 0.046 \times 56731.1^{-0.2} = 5.15\text{E}-3 \end{aligned}$$

This gives

$$\tau_{\text{hot}} = 6.25\text{E}-3 \times \frac{1}{2}\rho V^2 = \frac{3.13\text{E}-3}{\rho_{\text{hot}}}\left(\frac{\dot{m}}{A}\right)^2_{\text{hot}}$$
$$\tau_{\text{cold}} = 5.15\text{E}-3 \times \frac{1}{2}\rho V^2 = \frac{2.58\text{E}-3}{\rho_{\text{cold}}}\left(\frac{\dot{m}}{A}\right)^2_{\text{cold}}$$
$$\rho_{\text{hot}} = \frac{101,325 \times 28.9669}{8314.4 \times 783} = 0.451\,\text{kg/m}^3 \quad \rho_{\text{cold}} = \frac{20 \times 101,325 \times 28.9669}{768 \times 8314.4} = 9.193\,\text{kg/m}^3$$
$$\tau_{\text{hot}} = \frac{3.13\text{E}-3}{0.451}(23.83)^2_{\text{hot}} = 3.941\,\text{Pa}$$
$$\tau_{\text{cold}} = \frac{2.58\text{E}-3}{9.193}(92.1)^2_{\text{cold}} = 2.38\,\text{Pa}$$

Clearly these pressure drops are negligible compared to atmospheric pressure. This says that the heat exchanger could be made a lot more compact by adding more surface area per unit volume—m^2/m^3.

6.12 Special Operating Conditions

It is useful to note certain special conditions under which heat exchangers may be operated. Figure 6.18a shows temperature distributions for a heat exchanger in which the hot fluid has a heat capacity rate, $C_h \equiv \dot{m}_h C_{p,h}$, which is much larger than that of the cold fluid, $C_c \equiv \dot{m}_c C_{p,c}$.

For this case the temperature of the hot fluid remains approximately constant throughout the heat exchanger, while the temperature of the cold fluid increases. The same condition is achieved if the hot fluid is a condensing vapor. Condensation occurs at constant temperature, and, for all practical purposes, $C_h \to \infty$. Conversely, in an evaporator or a boiler (Fig. 6.18b), it is the cold fluid that experiences a change in phase and remains at a nearly uniform temperature ($C_c \to \infty$). The same effect is achieved without phase change if $C_h \ll C_c$. Note that, with condensation or evaporation, the heat rate is given by Eq. (6.1a) or (6.1b), respectively. Conditions illustrated in Fig. 6.18a or b also characterize an internal tube flow (or *single stream heat exchanger*) exchanging heat with a surface at constant temperature or an external fluid at constant temperature.

The third special case (Fig. 6.18c) involves a counterflow heat exchanger for which the heat capacity rates are equal ($C_h = C_c$). The temperature difference ΔT must then be constant throughout the exchanger, in which case $\Delta T_1 = \Delta T_2 = \Delta T_{lm}$.

6.13 Compact Heat Exchangers and Their Classifications

A heat exchanger is quite arbitrarily referred to as a *compact heat exchanger*, providing that it has a surface area density greater than 700 m^2/m^3, which is characterized for the ratio of their surface area per unit volume of the exchanger.

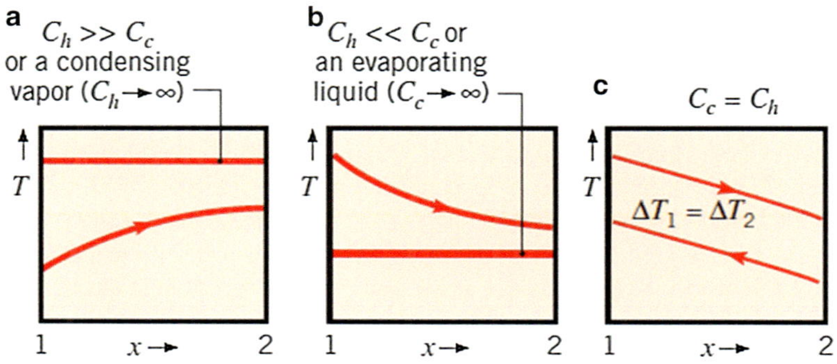

Fig. 6.18 Special heat exchanger conditions. (**a**) $C_h \gg C_c$ or a condensing vapor. (**b**) An evaporating liquid or $C_h \ll C_c$. (**c**) A counterflow heat exchanger with equivalent fluid heat capacities $C_h = C_c$ [4]

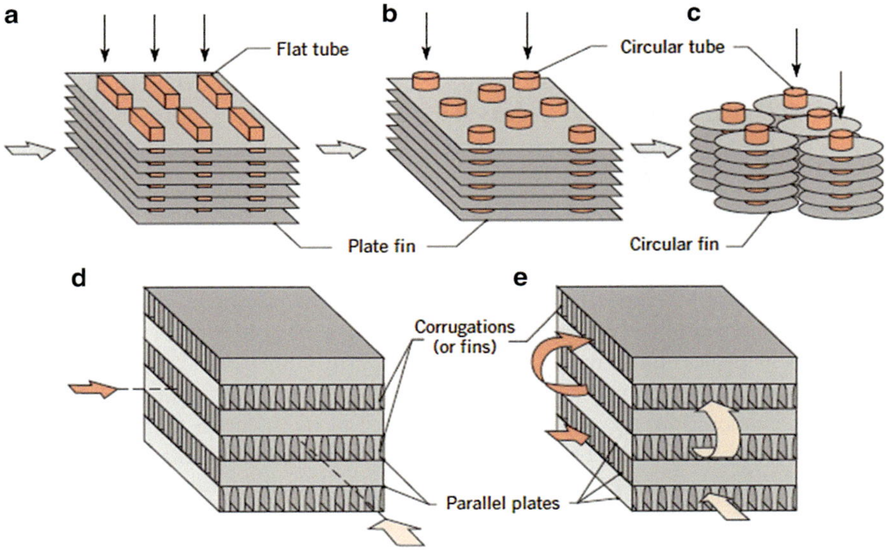

Fig. 6.19 Compact heat exchanger cores. (**a**) Fin–tube (flat tubes, continuous plate fins). (**b**) Fin–tube (circular tubes, continuous plate fins). (**c**) Fin–tube (circular tubes, circular fins). (**d**) Plate–fin (single pass). (**e**) Plate–fin (multipass) [4]

Such ratio is quite arbitrarily that referred to as a compact heat exchanger. By a large heat transfer surface area per unit volume of the exchanger, resulting in reduced space, weight, support structure and footprint, energy requirements and cost, as well as improved process design and plant layout and processing conditions, together with low fluid inventory. This is done by virtue of increasing the heat transfer surface area via fins per unit volume (m^2/m^3) and there many variations. See Fig. 6.19a–c.

A special and important class of heat exchangers is used to achieve a very large ($\geq 400\,m^2/m^3$ for liquids such as liquid-to-gas type exchangers and $\geq 700\,m^2/m^3$ for gases such as gas-to-gas type one) heat transfer surface area per unit volume. Termed compact heat exchangers, these devices have dense arrays of finned tubes or plates and are typically used when at least one of the fluids is a gas, and hence is characterized by a small convection coefficient. The tubes may be flat or circular, as in Fig. 6.19a, b and c, respectively, and the fins may be plate or circular, as in Fig. 6.19a, b and c, respectively. Parallel-plate heat exchangers may be finned or corrugated and may be used in single-pass (Fig. 6.19d) or multipass (Fig. 6.19e) modes of operation. Flow passages associated with compact heat exchangers are typically small ($D_h \leq 5\,mm$), D_h is the magnitude of the *hydraulic diameter*, and the flow is often laminar. Many of the geometries are far too complicated to apply deterministic methods to predict their performance. So many for these compact heat exchangers have had their performance determined experimentally.

Kays and London [8] have studied a wide variety of configurations for heat transfer matrices and catalogued their heat transfer and pressure drop characteristics. Figure 6.20 shows typical heat transfer materials for compact heat exchangers [4]. Figure 6.20a shows a *circular finned-tube array* with fins on individual tubes; Fig. 6.20b shows a *plain plate–fin matrix* formed by corrugation, and Fig. 6.20c shows a *finned flat-tube* matrix [4].

The heat transfer and pressure drop characteristics of such configurations for use as compact heat exchangers have been determined experimentally as explained above. Figures 6.21, 6.22, and 6.23 show typical heat transfer and friction factor data for three different configurations.

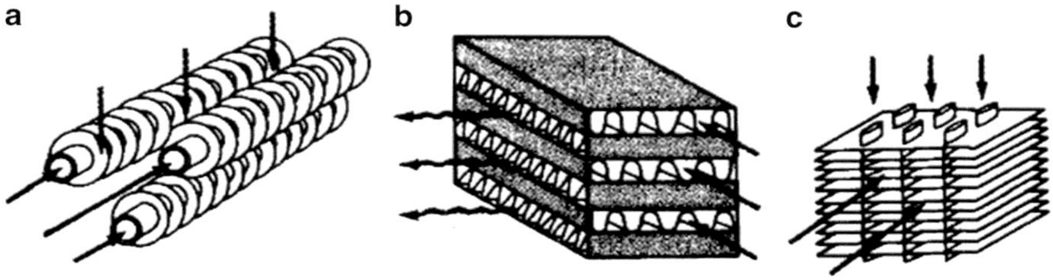

Fig. 6.20 Typical heat transfer matrices for compact heat exchangers: (**a**) circular finned-tube matrix; (**b**) plain plate–fin matrix; (**c**) finned flat-tube matrix [4]

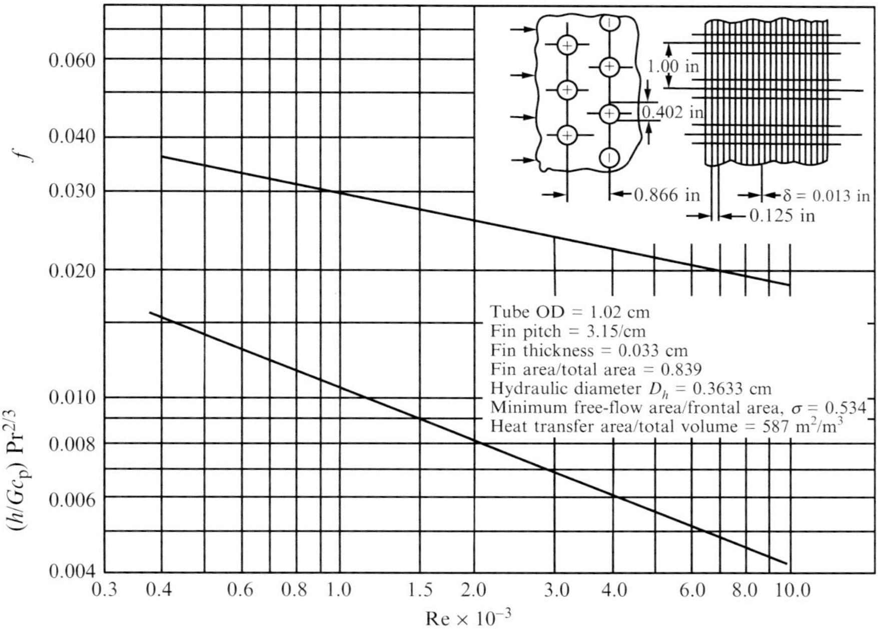

Fig. 6.21 Heat transfer and friction factor for flow across plate-finned circular tube matrix

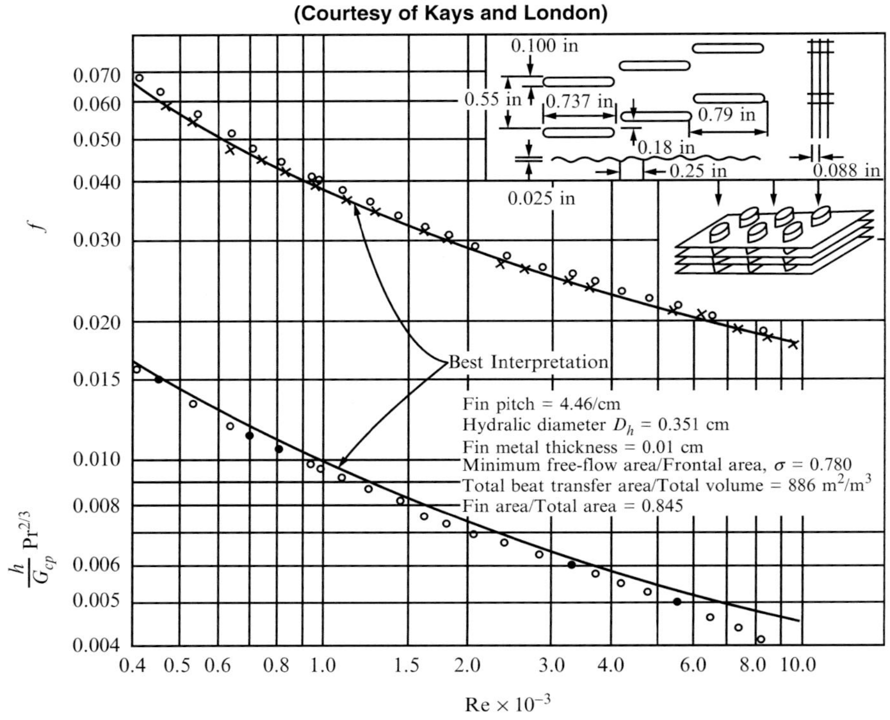

Fig. 6.22 Heat transfer and friction factor for flow across finned flat-tube matrix (Courtesy of Kays and London)

Note that the principal dimensionless groups governing these correlations are the Stanton, Prandtl, and Reynolds numbers [4]:

$$\mathrm{St} = \frac{h}{Gc_\mathrm{p}} \quad \mathrm{Pr} = \frac{c_\mathrm{p}\mu}{k} \quad \mathrm{Re} = \frac{GD_\mathrm{h}}{\mu} \tag{6.20}$$

Here G is the *mass velocity* defined as

$$G = \frac{m}{A_\mathrm{min}}\,\mathrm{kg}/\left(\mathrm{m}^2 \cdot \mathrm{s}\right) \tag{6.21}$$

where m = total mass flow rate of fluid (kg/s) and A_min minimum free-flow cross-sectional area (m^2) regardless of where this minimum occurs.

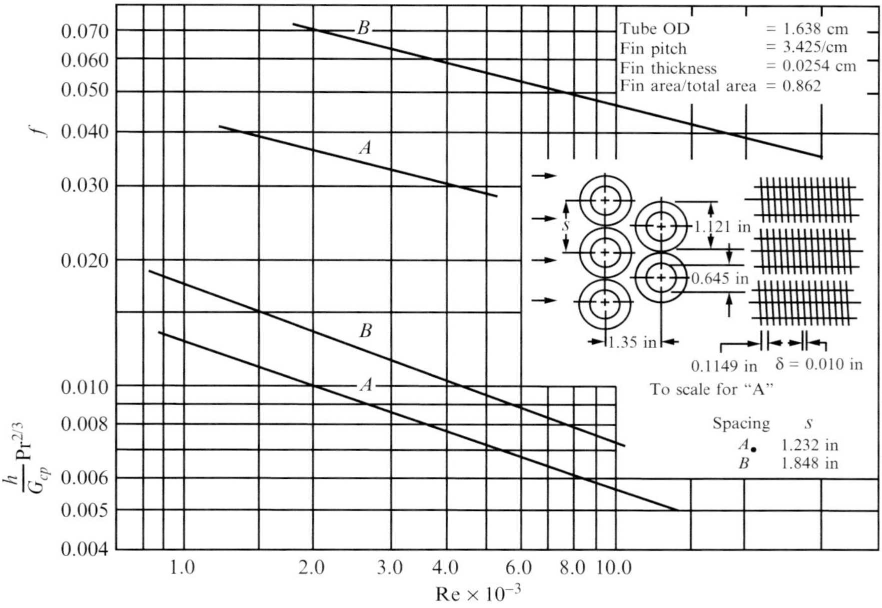

	A	B
Hydraulic diameter D_h	= 0.5477	1.1674 cm
Minimum free-flow area/frontal area, σ	= 0.443	0.628
Heat transfer area/total volume	= 323.8	215.6 m^2/m^3

Fig. 6.23 Heat transfer and friction factor for flow across circular finned-tube matrix (Courtesy of Kays and London)

The magnitude of the *hydraulic diameter* D_h for each configuration is specified in Figs. 6.21, 6.22, and 6.23. The hydraulic D_h is defined as:

$$D_h = 4\frac{LA_{min}}{A} \tag{6.22}$$

where A is the total heat transfer area and the quantity LA_{min} can be regarded as the minimum free-flow passage volume, since L is the flow length of the heat exchanger matrix.

Thus, once the heat transfer and the friction factor charts such as those shown in Figs. 6.20, 6.21, 6.22, and 6.23 are available for a specified matrix and the Reynolds number Re for the flow is given, the heat transfer coefficient h and the friction f for flow across the matrix can be evaluated. Then the rating and sizing problem associated with the heat exchanger matrix can be performed by utilizing either the LMTD or the effectiveness-NTU method of analysis.

References

1. Fraas, A. P. (1989). *Heat exchangers design* (2nd ed.). New York: Wiley Interscience.
2. Necati Ozisik, M. (1985). *Heat transfer: A basic approach*. New York: McGraw-Hill.
3. Incropera, F., Dewitt, D., Bergman, T., & Lavine, A. (2007). *Introduction to heat transfer* (5th ed.). John Wiley, October 30, 2006
4. Incropera, F., Dewitt, D., Bergman, T., & Lavine, A. (2011). Fundamentals of heat and mass transfer (7th ed.).
5. Kays, W. M., & London, A. L. (1964). *Compact heat exchangers* (2nd ed.). New York: McGraw-Hill.
6. Bowman, R. A., Mueller, A. C., & Nagle, W. M. (1940). Mean Temperature Differences in Design. *Transactions of the ASME, 62*, 283.
7. Pansini, A. J., & Smalling, K. D. (1991). *Guide to electric power generation*. Liburn, GA: The Fairmont Press, Inc.
8. Boyen, J. L. (1975). *Practical heat recovery*. New York: Wiley.
9. Kays, W. M., & London, A. L. (1998). *Compact heat exchangers* (3rd ed.). Malabar, FL: Krieger Publishing Company.

Chapter 7
Effective Design of Compact Heat Exchangers for NGNP

A number of technologies are being investigated for the Next Generation Nuclear Plant (NGNP) that will produce heated fluids at significantly higher temperatures than current generation power plants. The higher temperatures offer the opportunity to significantly improve the thermodynamic efficiency of the energy conversion cycle. Selection of the technology and design configuration for the NGNP must consider both the cost and risk profiles to ensure that the demonstration plant establishes a sound foundation for future commercial deployments. The NGNP challenge is to achieve a significant advancement in nuclear technology while at the same time setting the stage for an economically viable deployment of the new technology in the commercial sector soon after 2020. Energy is the elixir of life for the world's economy and for individual prosperity. Efforts being made for greater energy efficiency—especially in industrial countries—have indeed proved effective: as evidenced by the fact that energy consumption throughout the world is growing slower than gross domestic products. At the same time, however, the hunger for energy in quickly growing economies is leading to shifts in energy mix—which drives undesired CO_2 emissions upward. In conversion of primary energy to final and useful energy, technology from universities and national laboratories plus industry toward design of NGP has made key contributions to efficient handling of the resources of our planet.

7.1 Introduction

As we see in Chap. 5, for a nuclear system to take advantage of combined cycle technology, there are a number of changes to the plant components that have to be made. The most significant of course is that the combustion chamber has to be replaced by a *Heat Exchanger* in which the working fluid from the nuclear reactor secondary loop is used to heat the air (Fig. 7.1). The normal Brayton cycle is an

B. Zohuri, *Application of Compact Heat Exchangers For Combined Cycle Driven Efficiency In Next Generation Nuclear Power Plants*,
DOI 10.1007/978-3-319-23537-0_7

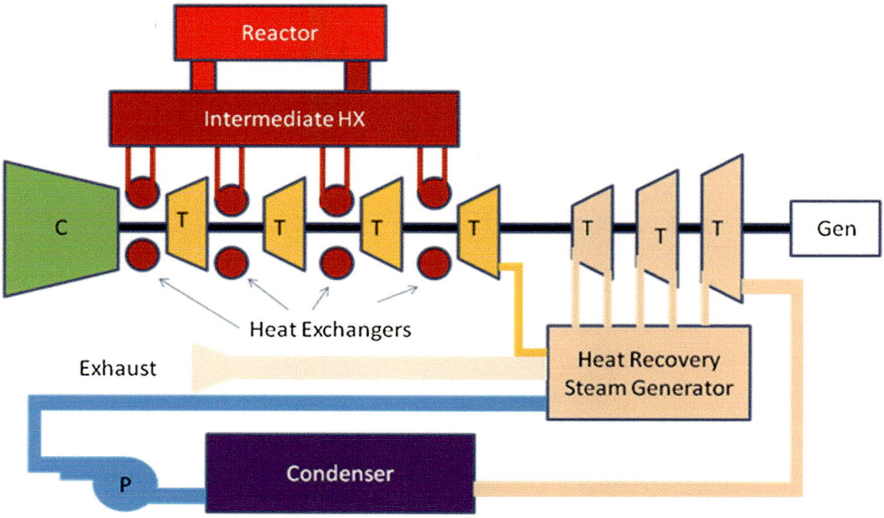

Fig. 7.1 Layout for four-turbine combined cycle with intermediate heat exchanger

internal combustion one where the working fluid is heated by the combustion of the fuel with the air in the combustion chamber. The walls of the combustion chamber can be cooled and peak temperatures in the working fluid can be significantly above the temperature that the walls of the chamber can tolerate for any length of time.

For the nuclear reactor system the heat transfer is in the opposite direction. All reactor components and fluids in the primary and secondary loops must be at a higher temperature than the peak temperature of the gas exiting the heat exchanger. This severely restricts the peak temperature that can be achieved for the air entering the turbine.

Out of six GEN-IV configurations, the Department of Energy (DOE) has selected the High Temperature Gas-cooled Reactor (HTGR) design for the Next Generation Nuclear Plant (NGNP) Project. The NGNP will demonstrate the use of nuclear power for electricity and hydrogen production. It will have an outlet gas temperature in the range of 850–950 °C and a plant design service life of 60 years [1]. As also we have showed as part of air Brayton Combined Cycle (BCC) power system [2–8] and Nuclear Air-Brayton Combined Cycle (NACC) power systems [9], the higher the temperature for outlet gas, the better the opportunity to increase our overall thermal efficiency of power plant in order to make it cost-effective and be more compatible with natural gas plant cost wise from Total Cost of Ownership (TCO) and Return On Investment (ROI) for the owner of such nuclear plants. The reactor design will be a graphite moderated, helium-cooled, and prismatic or pebble-bed reactor, and use low-enriched uranium, TRISO-coated fuel. The plant size, reactor thermal power, and core configuration will ensure passive decay heat removal without fuel damage or radioactive material releases during accidents.

The NGNP Materials Research and Development (R&D) Program is responsible for performing R&D on likely NGNP materials in support of the NGNP design, licensing, and construction activities.

As part of Intermediate Heat Exchanger (IHX) dosing and its application in NGNP, the compact designs are attractive to minimize the capital investment in materials by driving overall capital savings in footprint, volume, and structural support costs; however, they represent a significant technical risk at this stage of their development. Qualification of diffusion bonding methods and development of in service inspection methods represent significant schedule risk.

By looking at the compact heat exchanger as alternative solution will drive overall capital savings in footprint, volume, and structural support costs. A reduction in overall unit footprint, volume, and weight by up to 85 % when compared to traditional technologies such as shell and tube generates significant cost savings in the following areas:

- Reduced deck and skid volume required for implementation and installation
- Reduced structure support requirements
- Reduced pipe-work runs and sizing
- Improved access for maintenance
- Reduced relief and flare systems
- Topside craneage for installation and maintenance

Compact heat exchangers would allow the exceptionally high heat transfer efficiency-to-volume ration using a method such as printed circuit heat exchangers. Also, the plate heat exchangers are improving thermal management and economy in both fossil-fired and nuclear stations. Compact size, trusted-quality products, and ease of maintenance make them an attractive choice for new construction, retrofits, or capacity expansions. The compact heat exchangers are a perfect fit in heating/condensing, liquid, gas or two-phase operations where reliability, safety, and uptime are important considerations.

The definition of compactness is quite an arbitrary matter [10]. The ratio of the heat transfer surface area on one side of the heat exchanger to the volume can be used as a measure of the compactness of heat exchangers. A heat exchanger having a surface area density on any one side greater than about 700 m^2/m^3 quite arbitrarily is referred to as compact heat exchanger regardless of its structural design. For example, automobile radiators having an area density on the order of 1100 m^2/m^3 and the glass ceramic heat exchangers for some vehicular gas-turbine engines having an area density on the order of 6600 m^2/m^3 are compact heat exchangers. The human lungs, with an area density of about 20,000 m^2/m^3, are the most compact heat-and-mass exchanger. The very fine matrix regenerator for the Stirling engine has an area density approaching that of the human lung.

On the other hand extreme of the compactness scale, plane tubular and shell-and-tube type exchangers, having and area density in the range of 70–500 m^2/m^3, are not considered compact [10]. The incentive for using compact heat exchangers lies in the fact that a high value of compactness reduces the volume for a specified heat exchanger performance.

When heat exchangers are to be employed for automobiles, marine uses, aircraft, aerospace vehicles, nuclear power plants, or any other generating elasticity power plants such as gas system or fossil fueled one, cryogenic systems, refrigeration and air conditioning, the weight and size—hence the compactness—become important. To increase the effectiveness or the compactness of heat exchangers, fins are used. In a gas-to-liquid heat exchanger, for example, the heat transfer coefficient on the gas side is an order of magnitude lower than for the liquid side. Therefore, fins are used on the gas side to obtain a balance design; the heat transfer surface on the gas side becomes much more compact.

In order to obtain such a balance design for heat transfer between two sides, we have to be concerned about pinch point. The pinch point is defined as the point where the temperature difference is a minimum. The temperature difference at the pinch depends on the decision of heat exchanger; in general, the smaller the temperature difference the more expensive the heat exchanger.

If the simple Brayton cycle is modified to include a recuperator (which will transfer heat from the turbine exhaust to preheat compressed high pressure air before going to the combustion chamber), it will require less fuel to obtain the desired turbine inlet temperature of compressed air and also the optimum pressure ratio (either for compressor or turbine) is reduced to typically 3–4. This improves the thermal efficiency of the cycle. Alternatively, a regenerator can also be used replacing a recuperator. A number of regenerative cycles are presented by McDonald and Wilson [11].

7.2 Classification of Heat Exchangers

As we have learned so far heat exchangers are devices that are designed for exchanging heat which is a very broad category of devices, so first we need to restrict them to type of heat exchangers, which are exchanging heat between two fluids. These fluids can be gases as liquids. Even with this constraint, it is still difficult to have an overview, and a classification needs to be made. It is possible to classify heat exchangers in a number of ways.

1. The first classification of heat exchangers depends on the basic of the fluid paths through the heat exchanger. Thus, the differences that are made based on these fluid paths per our previous chapter are identified as:

    - Parallel flow
    - Counterflow
    - Cross-flow

    Each of these categories is described in Chap. 6 of the book and is briefly described here to remind you again.

    Parallel flow are those devices in which the warmed and cooled fluids flow past each other in the same direction, in contrast with the counterflow where these two flow in the opposite direction. In case of a cross-flow, fluids flows pass at right angles to each other.

2. The second classification made depends on the state of the media in the heat exchanger.

   - Liquid-to-liquid exchangers are those in which two liquids interact. Also gas-to-gas heat exchangers like air preheaters in steam plants and helium-cooled reactor gas turbine plants have to be mentioned. These devices operate with heat transfer coefficients that are between 10 and 100 times lower than the coefficients of liquid-to-liquid exchangers.
   - Gas-to-gas exchangers are general much larger and heavier if a same amount of transferred heat is demanded.
   - A third type is the liquid-to-gas heat exchanger (or vice versa), usually water and air are used, for instance in automotive radiators. Because of the lower heat transfer coefficients on the gas-side there are usually fines placed on the exchanging surfaces.

3. A third classification method is based the purpose of heat exchanger. In difference with the other classifications, this is not a designer's choice but a direct demand to fulfill the need for let us say an evaporator. So any demand based on this classification is generally a starting point, from which the designer needs to make decisions about the other classifications, like the choice between counter-flows or cross-flow.

   Some other examples of purpose classification are briefly the cooler, which cools liquids or gases by means of water, the chiller, which cools a fluid with a refrigerant such as Freon, to below a temperature that would be obtainable if water was used, and condensers, that condenses a vapor, often in the presence of a non-condensable gas (only shell tube condensers; classification on where condensation occurs: horizontal in-shell, vertical in-shell, horizontal in-tube, and vertical in-tube).

4. The last classification is actually the most important choice of the designer of a heat exchanging system. This is the choice of what kind of construction he is going to use. Below the two most common options are discussed: double pipe heat exchangers and shell-and-tube heat exchangers.

In summary all these calcifications of heat exchangers are depicted in the following Figs. 7.2, 7.3, and 7.4.

Notice that in Fig. 7.2 the parameter β is a designation of compactness or surface area density for heat exchanger. Quantitatively this parameter arbitrarily defines a compact heat exchange surface as one that has an area density greater than 700 m^2/m^3 (213 ft^2/ft^3). The range of surface area density and hydraulic diameter for various types of compact heat exchanger and heat exchange surfaces can be found in Kay and London book [12].

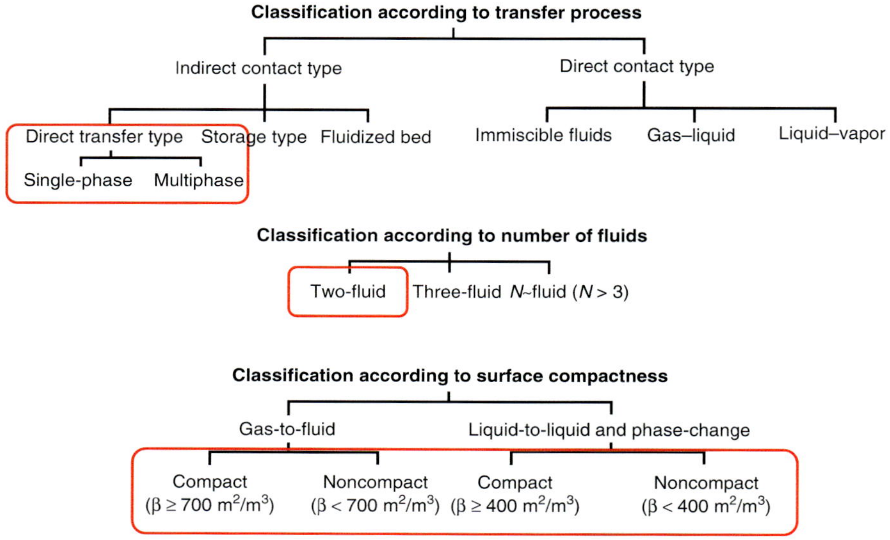

Fig. 7.2 Classification of heat exchangers according to transfer process

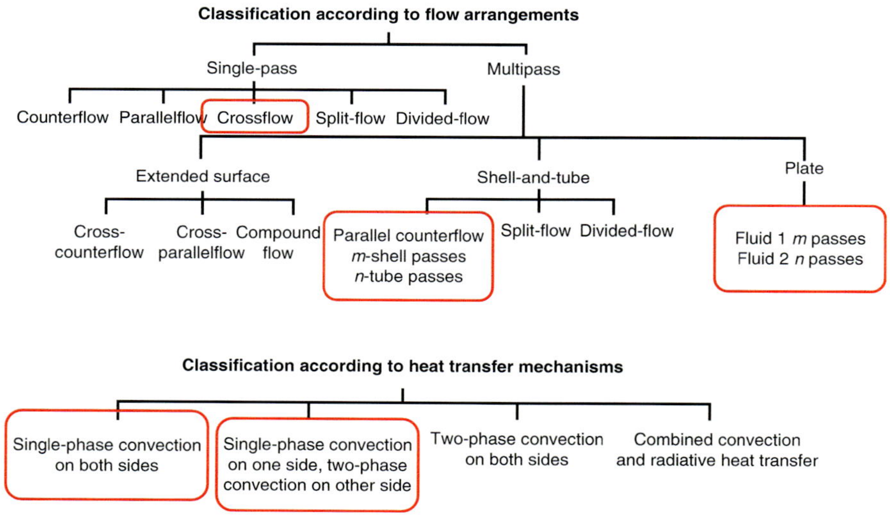

Fig. 7.3 Classification of heat exchangers according to flow arrangement

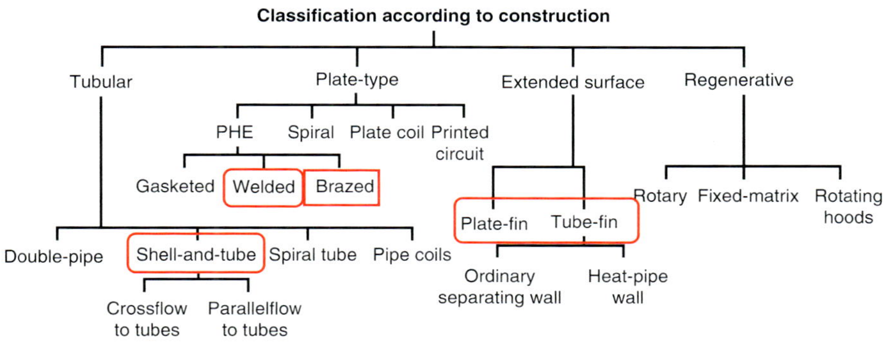

Fig. 7.4 Classification of heat exchangers according to construction

7.3 Compact Heat Exchanger-Driven Efficiencies in Brayton Cycle

As we analyze and demonstrate the modeling of a computer code in Chap. 5 via steady-state approach (i.e., transient-analysis should be considered for better accuracy sooner than later), if the simple Brayton cycle is modified to include a recuperator (which will transfer heat from the turbine exhaust to preheat compressed high pressure air before going to the combustion chamber), it will require less fuel to obtain the desired turbine inlet temperature of compressed air and also the optimum pressure ratio (either for compressor or turbine) is reduced to typically 3–4. This improves the thermal efficiency of the cycle. Alternatively, a regenerator can also be used replacing a recuperator.

With distributed power generation market, and renewing attention to nuclear power plant in particular new generation and small modular reactors (SMRs), one of the most economical solutions today is to generate power through small gas turbine systems in the form of Brayton cycle combined with these reactors. These gas turbines arbitrarily can be categorized as microturbines with output of (5–200 kW) and mini-turbines with output of (200–500 kW). The thermal efficiency of such microturbines is about 20 % or less if no recuperator is used in the system. Using a recuperator (regenerator can also be considered but has a number of problems) operating at 87 % effectiveness, the efficiency of the gas turbine system increases to about 30 %, a substantial performance improvement. However, the cost of the recuperator is about 25–30 % of the total power plant, and therefore the total cost of ownership and return on investments are not very well justified.

This means that the heat exchanger (recuperator) must be designed to get high performance with minimum cost. While the offset strip fins geometry is one of the highest performing surfaces it is also quite expensive to manufacture. This necessitates the use of all prime surface heat exchangers with no brazing.

As we present in Chap. 6, compact heat exchangers transfer more energy with a cost-effective manner than other heat exchangers and save more energy when compared with standard technology. Compact heat exchangers are used to transfer heat from one fluid either gas or liquid to another. Compact Heat Exchangers (CHEs) have a significantly greater surface area per unit volume and are more conventional types of heat exchangers. For the purposes of the Energy Efficiency, Performance, and Capacity (ECA) Scheme, a CHE is defined as a heat exchanger with a surface to volume ratio of >200 m^2/m^3.

Energy Efficiency, Performance, and Capacity (ECA) Scheme

The Scheme calls for manufacturers and importers of refrigeration equipment to establish that their products conform to certain requirements on energy efficiency, performance, and capacity. Such equipment is then "ECA Registered" and appears on what is known as the "Energy Technology List." Anyone purchasing equipment from the list will be entitled to 100 % tax relief on their investment in the first year.

The Scheme has been developed by the Inland Revenue and DEFRA (the Department for Environment, Food & Rural Affairs) and is managed by the Carbon Trust.

In order to meet the requirements of the new ECA Scheme, manufacturers must have their refrigeration equipment tested to prove that it conforms to various requirements on energy efficiency, performance, and capacity.

Compact heat exchangers are characterized by a high surface area per unit volume, which can result in a higher efficiency than conventional heat exchangers; in a significantly smaller volume (typically compact heat exchangers can achieve efficiencies of over 95 % CF (Cubic Foot). Eighty percent for non-compact heat exchangers).

Compact heat exchangers transfer more energy in a cost-effective manner than other heat exchangers and save more energy when compared to standard technology. Investments in CHEs can only qualify for Enhanced Capital Allowances if the specific product is named on the Energy Technology Product List.

Three types of compact heat exchanger are covered by the ECA Scheme. These are:

1. Plate heat exchangers
2. Plate-finned heat exchangers
3. Printed circuit heat exchangers
4. Compact heat exchangers with precision formed surfaces

Each of these above compact heat exchangers is briefly described below:

1. Plate Heat Exchangers

   A plate type heat exchanger is a type of heat exchanger that uses metal plates to transfer heat between two fluids. This has a major advantage over a conventional heat exchanger in that the fluids are exposed to a much larger surface area because the fluids spread out over the plates. This facilitates the transfer of

Fig. 7.5 Typical schematic of plate heat exchangers (courtesy of United Heat Exchanger Company, India) (http://www.heatexchanger.co.in/plate-heat-exchanger.html)

heat and greatly increases the speed of the temperature change. It is not as common to see plate heat exchangers because they need well-sealed gaskets to prevent the fluids from escaping, although modern manufacturing processes have made them feasible. See Fig. 7.5.

The concept behind a heat exchanger is the use of pipes or other containment vessels to heat or cool one fluid by transferring heat between it and another fluid. In most cases, the exchanger consists of a coiled pipe containing one fluid that passes through a chamber containing another fluid. The walls of the pipe are usually made of metal or another substance with a high thermal conductivity, to facilitate the interchange, whereas the outer casing of the larger chamber is made of a plastic or coated with thermal insulation, to discourage heat from escaping from the exchanger. The plate heat exchanger (PHE) was invented by Dr. Richard Seligman in 1923 and revolutionized methods of indirect heating and cooling of fluids.

Plate type heat exchanger one is composed of multiple, thin, slightly separated plates that have very large surface areas and fluid flow passages for heat transfer. This stacked-plate arrangement can be more effective, in a given space, than the shell and tube heat exchanger. Advances in gasket and brazing technology have made the plate-type heat exchanger increasingly practical. In HVAC applications, large heat exchangers of this type are called plate-and-frame; when used in open loops, these heat exchangers are normally of the gasket type to allow periodic disassembly, cleaning, and inspection. There are many types of permanently bonded plate heat exchangers, such as dip-brazed and vacuum-brazed plate varieties, and they are often specified for closed-loop applications such as refrigeration. Plate heat exchangers also differ in the types of plates that are used, and in the configurations of those plates. Some plates may be stamped with "chevron" or other patterns, where others may have machined fins and/or grooves.

As part of the specifications by the manufacturer, one can identify that:

- Liquid foods such as milk, fruit juices, beers, wines, and liquid eggs are pasteurized using plate-type heat exchangers.
- Wine and fruit juices are normally desecrated prior to pasteurization in order to remove oxygen and minimize oxidative deterioration of the products.

- Plate-type heat exchangers consist of a large number of thin, vertical steel plates that are clamped together in a frame.

As part of their applications the following could be presented:

- The plates produce an extremely large surface area, which allows for the fastest possible transfer.
- Making each chamber thin ensures that the majority of the volume of the liquid contacts the plate, again aiding exchange.
- The troughs also create and maintain a turbulent flow in the liquid to maximize heat transfer in the exchanger.
- A high degree of turbulence can be obtained at low flow rates and high heat transfer coefficient can then be achieved.

Few advantages worth mentioning for this type of CHE are as follows:

- Easy maintenance and suitable for Clean-In-Place (CIP), plate pack easily accessible
- High heat transfer coefficients
- Flexibility to change plate arrangement and to add or remove plate
- No mixing of product
- Compact constructions
- Optimized heat recovery

Disadvantages of the plate type heat exchanger are:

- A bonding material between plates limits the operating temperature of the cooler.
- Overtightening of the clamping bolts results in increased pressure drop across the cooler.
- Initial cost is high since titanium plates are expensive and titanium is a noble metal, and other parts of the cooling system are susceptible to corrosion.
- It requires, therefore, careful dismantling and assembling.

2. Plate Finned Heat Exchangers
   A plate–fin heat exchanger is made of layers of corrugated sheets separated by flat metal plates, typically aluminum, to create a series of finned chambers. Separate hot and cold fluid streams flow through alternating layers of the heat exchanger and are enclosed at the edges by side bars. Heat is transferred from one stream through the fin interface to the separator plate and through the next set of fins into the adjacent fluid.

   The fins also serve to increase the structural integrity of the heat exchanger and allow it to withstand high pressures while providing an extended surface area for heat transfer. See Fig. 7.6.

   A high degree of flexibility is present in plate–fin heat exchanger design as they can operate with any combination of gas, liquid, and two-phase fluids. Heat transfer between multiple process streams is also accommodated, with a variety

Fig. 7.6 Typical schematic of plate finned heat exchangers (courtesy of United Heat Exchanger Company, India) (http://www.heatexchanger.co.in/plate-heat-exchanger.html)

of fin heights and types as different entry and exit points available for each stream.

This is a type of heat exchanger design that uses plates and finned chambers to transfer heat between fluids. It is often categorized as a compact heat exchanger to emphasize its relatively high heat transfer surface area to volume ratio. The plate–fin heat exchanger is widely used in many industries, including the aerospace industry for its compact size and lightweight properties, as well as in cryogenics where its ability to facilitate heat transfer with small temperature differences is utilized. Plate–fin heat exchangers are generally applied in industries where the fluids have little chances of fouling.

As part of the specifications by the manufacturer, one can identify that:

- The main four types of fins are plain, which refers to simple straight-finned triangular or rectangular designs; herringbone where the fins are placed sideways to provide a zig-zag path; and serrated and perforated, which refer to cuts and perforations in the fins to augment flow distribution and improve heat transfer.

As part of their applications the following could be presented:

- Natural gas liquefaction
- Cryogenic air separation
- Ammonia production
- Offshore processing
- Nuclear engineering
- Syngas production

Few advantages that can be expressed for this type of CHE are as follows:

- High thermal effectiveness and close temperature approach (temperature approach as low as 3 K between single phase fluid streams and 1 K between boiling and condensing fluids is fairly common)
- Large heat transfer surface area per unit volume (typically 1000 m^2/m^3)
- Low weight
- Multi-stream operation (up to ten process streams can exchange heat in a single heat exchanger)

a b

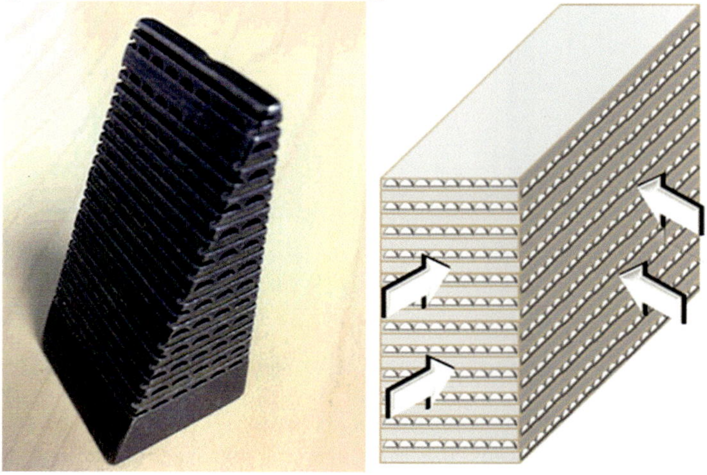

Fig. 7.7 A typical example of PCHE block

- True counterflow operation (unlike the shell and tube heat exchanger, where the shell side flow is usually a mixture of cross-flow and counterflow)

3. Printed Circuit Heat Exchangers (PCHE)
   This new type of exchanger has been developed by Heatric Pty Ltd over the last 7 years [13–15]. Flat plates are photo-chemically etched with heat-transfer passages and then diffusion bonded together to form a solid block. The exchanger is illustrated in Fig. 7.7a and b and a schematic cross section is shown in the figure. It is clear that the heat-transfer surface is effectively all primary. Entry and exit ports may be formed within the block, or alternatively headers may be welded onto the edges as with plate–fin exchangers.

   Printed Circuit Heat Exchanger (PCHE) is one of the compact types of heat exchangers available as alternatives to shell and tube heat exchangers. Its name is derived from the procedure used to manufacture the flat metal plates that form the core of the heat exchanger which is done by chemical milling. These plates are then stacked and diffusion bonded, converting the plates into a solid metal block containing precisely engineered fluid flow passages. These channels are typically semicircular in cross section with a depth of 1.5–3 mm. PCHEs are typically built from stainless steels and can operate at temperatures from cryogenic to 800 °C (1500 °F).

Few advantages PCHE, which are worth mentioning are given below:

- Channels are optimized for countercurrent flow.
- High heat transfer surface area per unit volume of the exchanger, resulting in reduced weight, space, and supporting structure.
- Used for temperature ranges of −200 to 900 °C.
- High pressure drops in excess of 600 bar.
- Reduced energy requirement and cost.
- Improved process design, plant layout, and processing conditions.
- Low fluid inventory compared to conventional designs such as shell-and-tube exchanger.
- Four to six times smaller and lighter than conventional designs such as shell-and-tube exchanger.
- Extremely high heat transfer coefficients are achievable with small-hydraulic diameter flow passages.
- Used for gases, liquids, and two-phase flows.

As part of disadvantages that we can mention and are listed below:

- Expensive when compared to shell and tube units.
- Fluid needs to be extremely clean.
- Blockages can occur easily due to the fine channels (0.5–2 mm).
- Blockages can be avoided by fine filtration (typically 300 μm), but will be additional cost.
- Filters need to be cleaned regularly.
- Blockages require chemical cleaning, which can be difficult in some installation and the system needs to be designed for this.
- Galvanic compatibility with the piping material can be an issue. Insulation kit or coated spool piece may be needed.

4. Compact Heat Exchangers with Precision Formed Surfaces
As part of being Compact, Efficient and Economical Heat Exchangers, companies like Tranter[1] heat exchangers share one thing in common—heat transfer through plates instead of tubes. As part of precision formed surface, their Spiral Heat Exchanger transfers heat through two spirally wound plates. Turbulent flow at low velocity produces high heat transfer efficiency and low fouling. The result is compact units with small heat transfer areas compared to conventional shell and tube exchangers. Plate exchangers conserve both materials and labor resources, reducing up-front costs and future operational costs. Their Prime Surface Heat Exchangers with configuration is known as PLATECOIL™ and looks like tank, vessel, or reactor unit fabrications, and is called "IPF"—Infinite PLATECOIL Flexibility™. Among their shapes are tangent bends, pancakes,

[1] http://www.tranter.com/Pages/products/prime-surface/configurations.aspx.

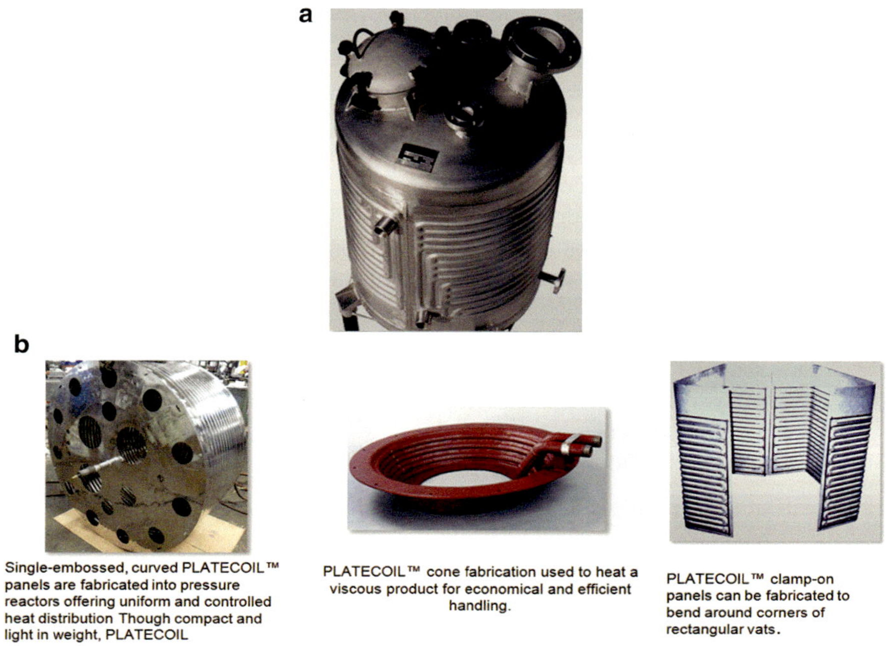

Fig. 7.8 (**a**) Single-embossed, curved PLATECOIL™ panels are fabricated into pressure reactors offering uniform and controlled heat distribution. (**b**) Different View Angle of PLATECOIL™ (courtesy of Tranter Company)

dished heads, cones, and bends around curves all with precision and dimensional stability unequaled by dimple sheet or pipe fabrications. See Fig. 7.8a and b.

Though compact and light in weight, PLATECOIL™ panels can attain surprisingly high jacket operating pressure ratings. And because they deliver higher flow velocities than other technologies, heat transfer is improved, fouling reduced.

Fouling Factor

Fouling is a generic term for the deposition of foreign matter on a heat transfer surface. Deposits accumulating in the small channels of a compact heat exchanger affect both heat transfer and fluid flow. Fouling deposits constricting passages in a compact heat exchanger are likely to increase the pressure drop and therefore reduce the flow rate. Reduced flow rate may be a process constraint; it reduces efficiency and increases the associated energy use and running costs. Maintenance costs will also increase. Fouling remains the area of greatest concern for those considering the installation of compact heat exchangers.

(continued)

Fouling problems cannot be avoided in many heat exchanger operations, and it is necessary to introduce defensive measures to minimize fouling and the cost of cleaning. The fouling control measures used during either design or operation must be subjected to a thorough economic analysis, taking into consideration all the costs of the fouling control measures and their projected benefits in reducing costs due to fouling. Under some conditions, nearly asymptotic fouling resistances can be obtained, and this suggests a somewhat different approach to the economics.

The widespread installation of compact heat exchangers has been hindered by the perception that the small passages are more strongly affected by the formation of deposits.

For further information refer to a paper by Asadi and Khoshkhoo [16].

Vessel sides can be easily designed with two or more zones to efficiently satisfy diverse process requirements. Also, panels can be configured as internal baffles or mixers with heat exchanger surfaces for reduced cycle time. Heavy gauge materials and special reinforcing features effectively withstand agitation forces.

Specially configured curved PLATECOIL™ panels are an economical means of converting existing unjacketed vessels to heated reactors, or of upgrading the thermal capacity of existing reactors. Standard units are available in 7 widths and 12 lengths, or in customized variations.

- Jacketed tanks and vessels
- Clamp-on upgrades
- Cryogenic shrouds
- Drum warmers
- Pipe coolers
- Gas cylinder heaters
- Heavy wall vessels and platens

Compact heat exchangers are employed in many different applications because of their high surface area density. Plate–fin heat exchangers in particular are well suited for gas-to-gas and air-to-air recuperators and heat recovery units, among many other applications. In this thesis, constant property, fully or periodically developed laminar flows of air ($Pr = 0.72$) inside a variety of different inter-fin channels of plate–fin heat exchangers are studied computationally, with the goal of achieving better understanding of plate–fin heat exchangers and providing new designs with superior performance to the existing ones.

The majority of plate–fin channels have rectangular, trapezoidal, or triangular cross-sectional shapes. Their convective behavior for air flows is investigated and solutions and polynomial equations to predict the Nusselt number are provided. Besides the limiting cases of a perfectly conducting and insulated fin, the actual conduction in the fin is also considered by applying a conjugate conduction–convection boundary condition at the fin surface between partition plates. For the latter, new sets of solutions and charts to determine the heat transfer coefficient based on the fin materials, channel aspect ratio, and fin density are presented.

Furthermore, while large fin density increases the heat transfer surface area, the convection coefficient can be increased by geometrical modification of the fins. To this end, two different novel plate–fin configurations are proposed by Huzayyin [17] and their convective behavior investigated in this thesis. These include:

1. Slotted plate-fins with trapezoidal converging–diverging corrugations
2. Offset-strip fins with in-phase sinusoidal corrugations

The enhanced heat transfer performance of the plate–fin compact core with perforated fin-walls of symmetric, trapezoidally profiled, converging–diverging corrugations is modeled computationally. Air flow rates in the range $10 \leq \mathrm{Re} \leq 1000$ are considered in a two-dimensional duct geometry described by the trapezoid inclination angle, the convergent–divergent amplitude ratio, the dimensionless corrugation pitch, and a surface porosity β of 10 %. The fin-wall flow transpiration is seen to promote enhanced heat transfer by inducing cross stream mixing, and periodic disruption and restarting of boundary layers. With uniform heat flux H_1 at the fin walls, an unusual performance is obtained where a higher Nusselt number is accompanied with reduction in the corresponding friction factor, relative to a non-slotted geometry of the same dimensions.

In the case of sinusoidal wavy offset-strip channels, he showed that the performance enhancement can be evaluated for air flows in the range of $10 \leq \mathrm{Re} \leq 1000$, with fins at constant wall temperature T, the effect of the wavy-fin amplitude, inter-fin spacing, and fin offset position on the thermal-hydraulic performance is reported. It is generally seen that S-shaped offset channels perform better than C-shaped ones. An average of 400 % reduction in volume of a plate–fin heat exchanger can be achieved with S-shaped offset fins when compared to that with plain parallel fins [17].

The use of compact heat exchangers for both single- and two-phase applications in the process industries is being actively encouraged. In this chapter, the benefits of compact heat exchangers, as well as their limitations, are briefly reviewed, followed by a description of a number of types of compact heat exchangers, some well-established and others relative newcomers to the market.

7.4 Thermal Energy Transfer for Process Heat Application in Enhanced Mode

Recent technological developments in next-generation nuclear reactors have created renewed interest in nuclear process heat for industrial applications. The Next Generation Nuclear Plant (NGNP) will most likely produce electricity and process heat for hydrogen production. Process heat is not restricted to hydrogen production, but is also envisioned for various other technologies such as the extraction of iron ore, coal gasification, and enhanced oil recovery. To utilize process heat, a thermal device is needed to transfer the thermal energy from NGNP to the hydrogen plant in the most efficient way possible. There are several options to transferring multi-megawatt thermal power over such a distance. One option is simply to produce only electricity, transfer it by wire to the hydrogen plant, and then reconvert the electric energy to heat via Joule or induction heating. Electrical transport, however, suffers energy losses of 60–70 % because of the thermal-to-electric conversion inherent in the Brayton cycle. A second option is to transport thermal energy via a single phase forced convection loop where a fluid is mechanically pumped between heat exchangers at the nuclear and hydrogen plants. High temperatures, however, present unique challenges for materials and pumping. Single phase, low pressure helium is an attractive option for NGNP, but is not suitable for a single-purpose facility dictated to hydrogen production because low pressure helium requires higher pumping power and makes the process very inefficient. A third option is two-phase heat transfer utilizing a high temperature *Thermosyphon*. Heat transport occurs via evaporation and condensation, and the heat transport fluid is recirculated by gravitational force. Thermosyphons have the ability to transport heat at high rates over appreciable distances, virtually isothermally, and without any requirement for external pumping devices.[2]

Heat pipes and thermosyphons have the ability to transport very large quantities of heat over relatively long distances with small temperature loss [18]. The applications of heat pipes and thermosyphons require heat sources for heating and heat sinks for cooling. The development of the heat pipe and thermosyphon was originally directed toward space applications. However, the recent emphasis on energy conservation has promoted the use of heat pipes and thermosyphons as components in terrestrial heat recovery units and solar energy systems. Thermosyphons have less thermal resistance, wider operating limits (the integrity of the wick material might not hold in heat pipes at very high temperatures), and lower fabrication costs than capillary heat pipes, which makes a thermosyphon a better heat recovery thermal device.

Perhaps the most important aspect of thermosyphon technology is that it can easily be turned off when required, whereas a heat pipe cannot be turned off. This safety feature makes the licensing of NGNP process heat transfer systems comparatively easier. This section describes the thermosyphon system and the potential benefits of using it in order to transfer process heat from the nuclear plant to the hydrogen production plant.

For hot fluid inlet temperature of less than about 675 °C, stainless steel material can be used for the heat exchanger, which has reasonable cost. However, for higher inlet temperatures in heat exchangers associated with higher turbine inlet temperatures, super alloys are essential which increase the material cost of the exchanger alone by a factor of 4–5.

---

[2] www.intechopen.com.

The power generation is generally in Megawatts. There is a need for small power generation for remote area, not enough grid power availability, emergency power, uninterrupted power requirement, and other reasons. With the decontrol on centralized power generation monopoly, more and more use of distributed power generation is taking place. The common mode is to generate the power by a Diesel engine. This is a costly power generation. The alternative way is to generate electricity using a gas turbine in a simple Brayton cycle. The gas turbine technology has advanced considerably over the last 60 years and the power generation on a large scale (in Megawatts) is common particularly in hydro and thermal power plants. While the gas turbine technology with smaller power range (to produce power in 5–500 kW range) has been developed, it is very costly [19].

The gas turbines developing power in the 5–200 kW range are referred to as microturbines and those in the 200–500 kW range as miniturbines [20]. We now briefly summarize the microturbine technology.

Definition of Microturbine

A "microturbine" implies a small compact gas turbine based power system and includes a turbo-compressor (a turbine and compressor on a single shaft), a combustion chamber and a generator, with recuperator as an optional component. However, almost all microturbines require recuperators to achieve desirable system thermodynamic efficiency.

As we saw in Chap. 5, if the simple Brayton cycle is employed and modified to include a recuperator (which will transfer heat from the turbine exhaust to preheat compressed high pressure air before going to the combustion chamber), it will require less fuel to obtain the desired turbine inlet temperature of compressed air and also the optimum pressure ratio (either for compressor or turbine) is reduced to typically 3–4. This improves the thermal efficiency of the cycle [19].

Alternatively, a regenerator can also be used replacing a recuperator. A number of regenerative cycles are presented by McDonald and Wilson [11]. However, the durability and air-to-gas leakage problems are serious enough that the recuperator is not being considered after over 50 years of development. The regenerator development also started after the Second World War. Very high performance brazed plate–fin type recuperators have been developed and are being used in large systems today. With cost pressures, the modern recuperator designs for microturbine systems use prime surfaces on both fluid sides with no brazing, just stacking, and welded at the side edges to form air flow passage, to prevent the leaks and mixing of the fluids. This allows high heat transfer performance with low pressure drop, an essential design requirement today. Since both fluids are gases (compressed air and turbine exhaust gas) in the heat exchanger, the design of inlet and outlet manifolds is challenging to ensure good flow distribution through the core on both fluid sides [19].

Brazing Process
Brazing is a metal-joining process whereby a filler metal is heated above melting point and distributed between two or more close-fitting parts by capillary action. The filler metal is brought slightly above its melting (liquids) temperature while protected by a suitable atmosphere, usually a flux. It then flows over the base metal (known as wetting) and is then cooled to join the workpieces together. It is similar to soldering, except the temperatures used to melt the filler metal are higher for brazing.

When using a recuperator in a microturbine, the recuperator cost is about 25–30 % of the microturbine system. When a significant cost reduction is necessary, the brazed plate–fin type costly recuperator is not acceptable. The alternative is to use a high performance prime surface recuperator without any brazing. This avoids the costly fin manufacturing and brazing thus reducing the cost of the recuperator without performance reduction [19].

As part of design criteria for such recuperator one should follow the major steps toward the design and development of a gas-to-gas recuperator of compact heat exchanger in this case [19] as follows:

- Find out approximate core size using prior empirical data and finite difference tools.
- Manufacture heat transfer surface and test to determine j (Colburn factor $j = St\ Pr^{2/3}$), and f (Fanning friction factor, some textbooks show that as f_c) nodes vs. Reynolds (Re) number design data. The basic performance data for an enhanced surface are shown as curves of the Colburn factor j, and the Fanning friction factor f, plotted vs. Reynolds number Re. Kays and London [12] present j and f vs. Re for a large number of compact surfaces, in one of the first comprehensive collections of data on enhanced surfaces for compact heat exchangers.
- Determine core size and tool sample plates for manufacturing development and test cores.
- Analyze flow and temperature in the core using Computational Fluid Dynamics (CFD) to predict flow and temperature distribution, as well as verify performance.
- Compute thermal stresses using transient temperature distribution models input into finite element analysis program. If thermal stresses are not acceptable, modify appropriately the heat transfer surface design.
- Build cores, instrument, and test to verify thermal models.
- Refine the design to mitigate risks brought to light by analysis and test results.

Some of the materials used for the compact heat exchangers (recuperator) are: 300 series stainless steel (AISI 347 SS) for temperatures below about 675 °C, Inconel 625, Inconel 803, Haynes 120, Haynes 214, and PM2000 materials up to about 900 °C. For a 50 kW microturbine, the recuperator would weigh about 40 kg and the thin foil stainless steel would cost about \$12/kg. Thus the rounded off recuperator material cost would be about \$500 (refer to McDonald [21, 22]).

While the plate–fin recuperator technology and manufacturing processes are known, there is good design flexibility and the recuperator would be light weight. There are some important limitations for the plate–fin design: high material and capital cost, long braze cycle, potential for high repair rate, limited material flexibility, complicated assembly, and difficult automated manufacturing. Thus, the current emphasis is on the development of a recuperator using primary surface only with the following attributes [19]:

- Basic core construction consists of a Laser welded stack of stamped plates (one or two parts).
- Simple construction leads to highly robust design.
- With the fully automated laser welding process it is possible to seal side edges and form flow passages on one fluid side. Laser welding eliminates high cost of nickel braze materials that are traditionally used in high temperature heat exchangers.

For combined cycle in the form of Brayton cycle configuration to be economically viable, the recuperator or heat exchanger implemented in it as one of the components of the system must be cost-effective and compact for applications that require compact packaging. One such application is hybrid engine with a microturbine generating electricity at the maximum efficiency and the hybrid vehicle then running on the electricity or direct-drive vehicles. For an automotive application with engine power rating of 65–100 kW and cost \$25/kW, the recuperator should be manufactured for about \$150. Such an exchanger should be operating at high temperatures, have low-cost manufacturing methods, and be easy to replace or maintain [11].

In summary, when one is looking for a special surface geometry to enhance heat transfer in an industrial heat exchanger application, in particular for the next generation nuclear plants (NGNP), we have a large number of options to choose from. Now the question is:

How can one compare the performance improvement given by various enhanced surfaces?

Certainly, one can judge the relative heat transfer enhancement for selected geometries by comparing the heat transfer coefficients or dimensionless heat transfer parameters (i.e., Nusselt number, Stanton number, etc.) yielded by each enhanced surface. But this will only give a partial indication of performance [23, 24].

Enhanced surfaces do provide a greater heat transfer coefficient, but they also lead to increased fluid flow friction and pressure drop. Sometimes, the benefits gained from heat transfer enhancement are not great enough to offset the increased friction losses. Clearly, then, the performance goal is to gain maximum enhancement of heat transfer with minimum penalty on pumping power. However, this balance is difficult to quantify in a manner that allows straightforward comparisons between various enhanced surface geometries.

Generally speaking, heat transfer surfaces can be used for three purposes:

1. To make heat exchangers more compact in order to reduce their overall volume, and possibly their cost
2. To reduce the pumping power required for a given heat transfer process
3. To increase the overall *UA* value of the heat exchanger

Heat Exchanger Value of *UA*

Nearly all heat transfer equipment can be a analyzed from the following basic design equation

$$q = UA\Delta T \tag{1}$$

which relates the heat flow rate or duty (q) to the driving force or temperature difference (ΔT) by a proportionality constant (UA), where U is a coefficient and A is the heat transfer area. Heat exchanger analyses usually involve the process of determining the *UA* required for the process to operate and comparing that to the *UA* available.

The *UA* required is normally determined based only on thermodynamics. The *UA* available is calculated from the heat exchanger geometry and the transport properties of the fluid.

A higher *UA* value can be exploited in either of two ways:

1. To obtain an increased heat exchange rate for fixed fluid inlet temperatures.
2. To reduce the mean temperature difference for the heat exchange; this increases the thermodynamic process efficiency, which can result in a saving of operating costs.

Enhancement techniques can be separated into two categories [25]:

1. *Passive*:
   Passive methods require no direct application of external power. Instead, passive techniques employ special surface geometries or fluid additives which cause heat transfer enhancement
2. *Active*:
   On the other hand, active schemes such as electromagnetic fields and surface vibration do require external power for operation.

The majority of commercially interesting enhancement techniques are passive ones. Active techniques have attracted little commercial interest because of the costs involved, and the problems that are associated with vibration or acoustic noise [26].

Special surface geometries provide enhancement by establishing a higher hA per unit base surface area. Where h is the heat transfer coefficient and A is the heat transfer surface area on same fluid side for which the value of h is mentioned.

Clearly, there are three basic ways of accomplishing this [26]:

1. Increase the effective heat transfer surface area (A) per unit volume without appreciably changing the heat transfer coefficient (h). Plain fin surfaces enhance heat transfer in this manner.
2. Increase h without appreciably changing A. This is accomplished by using a special channel shape, such as a wavy or corrugated channel, which provides mixing due to secondary flows and boundary-layer separation within the channel. Vortex generators also increase h without a significant area increase by creating longitudinally spiraling vortices exchange fluid between the wall and core regions of the flow, resulting in increased heat transfer.
3. Increase both h and A. Interrupted fins (i.e., offset strip and louvered fins) act in this way. These surfaces increase the effective surface area and enhance heat transfer through repeated growth and destruction of the boundary layers.

In forced-convection heat transfer between a gas and a liquid, the heat transfer coefficient of the gas may be 10–50 times smaller than that of the liquid. The use of specially configured surfaces can be used to reduce the gas-side thermal resistance. For heat transfer between two gases, the difficulty in inducing the desired heat exchange is even more pronounced. In this case especially, the use of enhanced surfaces can substantially reduce heat exchanger size. This is the motivation behind the design of a category of heat exchangers with reduced size and greatly enhanced gas-side heat transfer, which are referred to as "compact."

A compact heat exchanger is generally defined as one which incorporates a heat transfer surface having a high "area density." In other words, it possesses a high ratio of heat transfer surface area to volume. This does not necessarily mean that a compact heat exchanger is of small mass or volume.

Figure 7.9 shows a spectrum of surface area density for heat exchangers. The range of surface area density and hydraulic diameter is given for various types of heat exchange surfaces, with the dividing line for compactness clearly marked.

Compact surfaces are utilized to yield a specified heat exchanger performance $q/\Delta T_{\text{mean}}$, within acceptable mass and volume constraints, which is calculated by using Eq. (7.1a, 7.1b):

$$\frac{q}{\Delta T_{\text{mean}}} = U\beta V \tag{7.1a}$$

or

$$\frac{q}{\Delta T_{\text{mean}}\beta} = UV \tag{7.1b}$$

Analyzing Eq. (7.1b), it is obvious that a high β decreases volume. Furthermore, compact surfaces generally result in higher overall conductance, U. And since compact surfaces can achieve structural strength and stability with thinner sections,

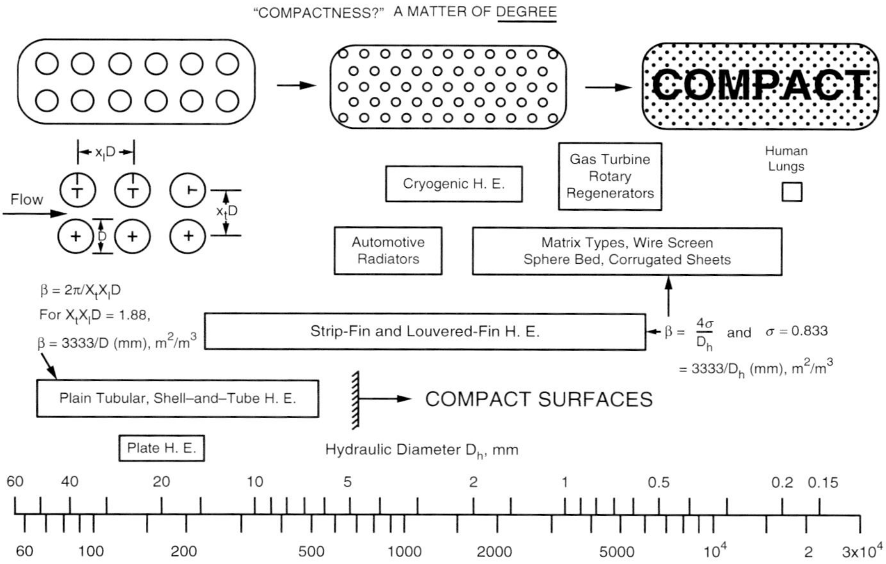

Fig. 7.9 Heat transfer surface area density β, m^2/m^3

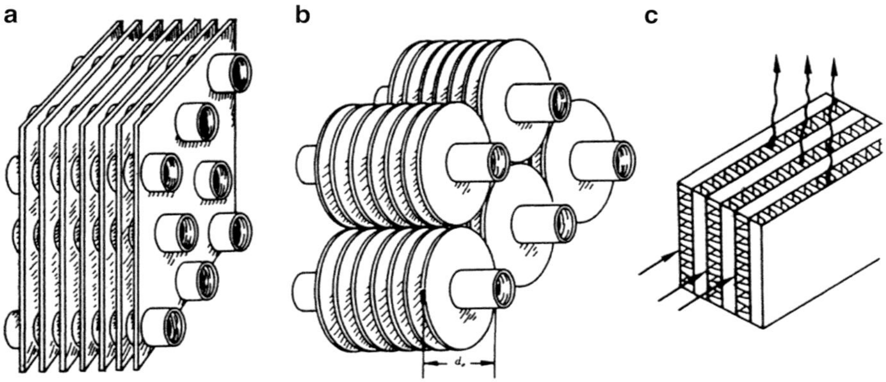

Fig. 7.10 (**a**) Finned-tube heat exchanger with flat fins, (**b**) individually finned tubes, (**c**) plate–fin heat exchanger [12]

the reduction in heat exchanger mass is even more pronounced than the reduction in volume [27].

Various techniques can be used to make heat exchangers more compact. Figure 7.10 shows three general types of extended surface geometries which can be used to increase gas-side heat transfer coefficients.

These include:

(a) A finned-tube heat exchanger with flat fins
(b) A finned-tube heat exchanger with individually finned tubes
(c) A plate–fin heat exchanger

This chapter focuses on various types of plate–fin geometries.

A typical element of a plate–fin heat exchanger consists of a die-formed **fin** (type of material) plate sandwiched between flat metal separator plates as shown in Fig. 7.11.

These heat exchangers are widely used in cryogenic application because of their low cost, small size, low weight, high thermal capacity, and effectiveness relative to other types of heat exchangers. The result of the improved effectiveness is the achievement of true countercurrent flow where there is an increase in the temperature spread and a closer approach to ideality. This means that the refrigerant cooling curve is closer to the natural gas cooling curve [24].

The exchanger is made up of manifolds or headers which consist of elements. A manifold and an element are shown below, Figs. 7.6 and 7.12, respectively.

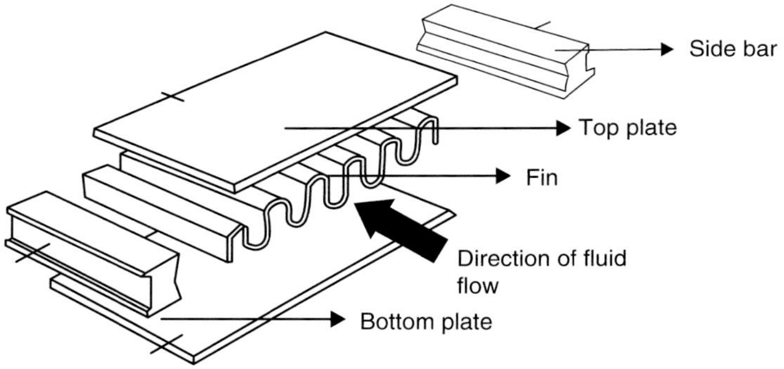

Fig. 7.11 Construction details of a typical single-element plate–fin heat exchanger

Fig. 7.12 Manifold or header (courtesy xchanger.com)

An element is made up of a corrugated die-formed fin plate placed between flat metal separator plates. There are side bars along the outside of the fin sections. A stack of the elements is welded to form a rigid matrix and can be designed to meet any configuration and size. The stacks are welded onto the manifolds. Depending on the application, a number of manifolds can be assembled to form the heat exchanger.

The wavy configuration of the fin promotes (Fig. 7.10) turbulence and therefore improves heat transfer. This increase in heat transfer is accompanied by an increase in pressure drop. This is a problem with low density fluids like gases because of the extra work required to surmount the pressure drop.

Figure 7.13a and b shows the side bars are located along the outer edges of the **fin** sections, while Fig. 7.14 shows the installed stack of these manifold headers.

This work is often much higher than the increase in heat transfer acquired from the fins. For applications where any fin configuration other than the simplest is proposed, a thorough analysis of the effect on the system should be conducted.

In designing the plate fin heat exchanger, it is possible to have different heights of the alternating fin plates. There is no requirement to have the same height or spacing of separator plates. This is a useful freedom to have in situations where the difference in density of the hot and cold fluids is large. In cryogenic systems, the refrigerant stream entering the expander has a higher density than the stream coming out of the expander. In such a case as this, it is necessary to use a larger

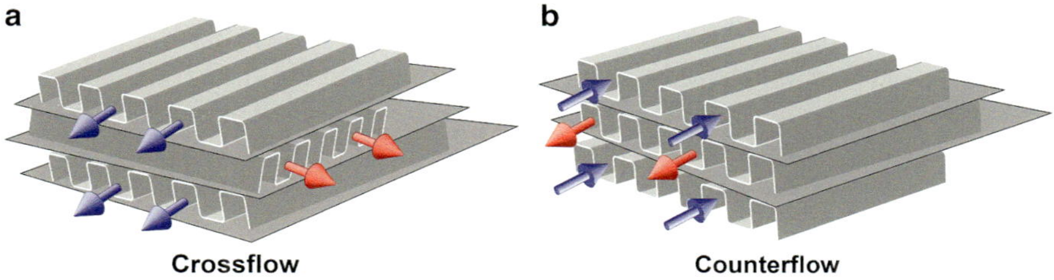

Fig. 7.13 (**a**) Cross-flow and (**b**) counterflow arrangements of plate–fin heat exchangers

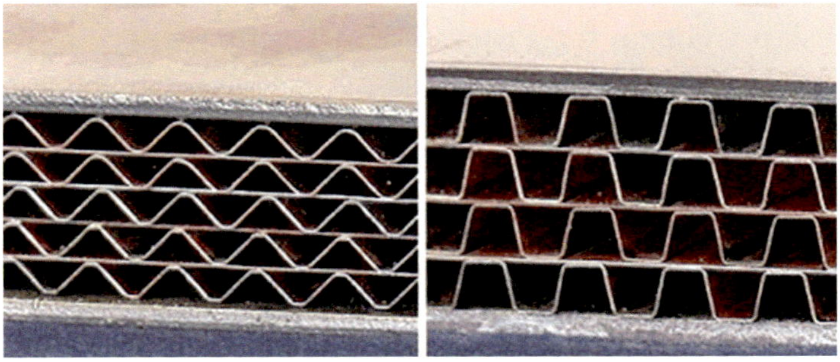

Fig. 7.14 Fin configurations

Fig. 7.15 Large matrix of plate–fin heat exchanger elements [24]

height for the lower density stream so that a common Reynolds number and therefore heat transfer coefficient, U, can be attained.

Aluminum is the usual material of construction of plate fin heat exchangers for cryogenic applications.

Stacks of such elements are then welded or dip-brazed to form large heat exchange devices like those in Fig. 7.15. A wide variety of plate–fin geometries have been used to obtain enhanced heat transfer, and engineers are constantly working to develop new and more effective enhanced surfaces [24].

Six commonly used plate–fin geometries are shown in Fig. 7.16.

Typical fin spacings are 300–800 fins/m. Due to their small hydraulic diameter and the low density of gases, these surfaces are usually operated in the Reynolds number range $500 < \text{Re} < 1500$. As a result, plated-fin enhancement geometries must be effective in the low Reynolds number regime. For example, surface roughness has been shown to promote heat transfer in the turbulent regime, but it does not provide appreciable enhancement in the lower Reynolds number range [24].

7.5 Design Criteria for Process Heat Exchangers

There are some criteria that a process heat exchanger must satisfy which are easily enough stated if we confine ourselves to a certain process. The criteria include the following

The heat exchanger must meet the process requirements. This means that it must effect the desired change in thermal condition of the process stream within the allowable pressure drops. At the same time, it must continue doing this until the next scheduled shut down for maintenance.

The heat exchanger must withstand the service conditions of the environment of the plant which includes the mechanical stresses of installation, startup, shutdown,

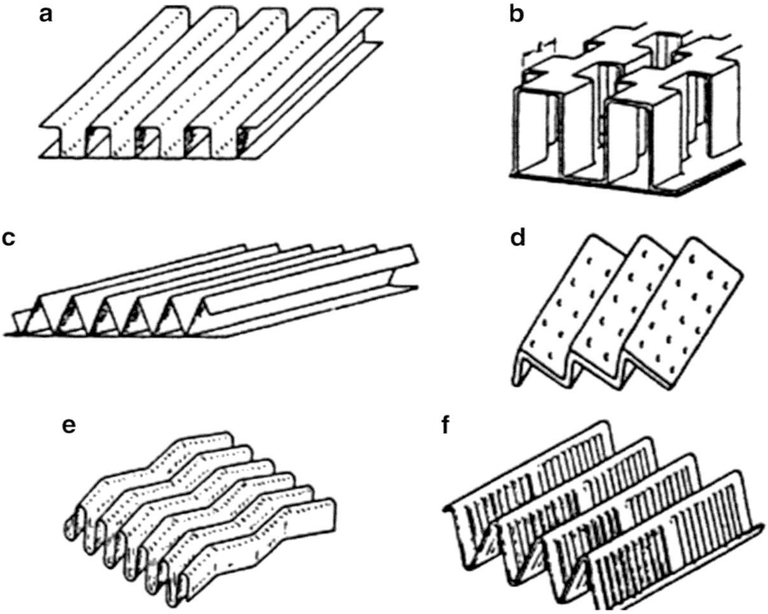

Fig. 7.16 (**a**) Plain rectangular fins, (**b**) plain triangular fins, (**c**) wavy fins, (**d**) offset strip fins, (**e**) perforated fins, and (**f**) louvered fins [24]

normal operation, emergencies, and maintenance. Besides, the heat exchanger must also resist corrosion by the environment, processes, and streams. This is mainly a matter of choosing materials of construction, but mechanical design does have some effect.

The heat exchanger must be maintainable, which usually implies choosing a configuration that permits cleaning and replacement. In order to do this, the limitations are positioning the exchanger and providing clear space around it. Replacement usually involves tubes and other components that may be especially vulnerable to corrosion, erosion, or vibration.

The cost of the heat exchanger should be consistent with requirements. The cost here implies the cost of installation. Operation cost and cost of lost production due to exchanger malfunction or unavailable should be considered earlier in the design.

The limitations of the heat exchanger: Limitations are on length, diameter, weight, and tube specifications due to plant requirements and process flow.

As part of design methodology, we should look at design in term an activity that is amid at providing complete descriptions of an engineering system such as our case of combined cycle of open air Brayton cycle or part of a system, or just of a single system component. By defining a well-defined design methodology, we can very clear path for specification of the system/components and their structure, size, and performance, as well as other characteristics that are important for manufacturing and utilization purpose such as next generation nuclear plants (NGNP).

This very important departure point for any combined cycle drive efficiency upward either for open or closed cycle.[3]

To look at the scoop and formulizing of these activities, one must understand that the design methodology faces a very complex structure and furthermore a design methodology for a heat exchanger as a subcomponent of a system must be consistent with the *life cycle design* of a system, which includes the following stages [28]:

- Problem formulation including interaction with end user
- Concept development such as selection of preliminary and workable designs
- Detailed exchanger design that includes design calculations and other important pertinent considerations
- Manufacturing and cost-effectiveness
- Utilization consideration such as operation, life cycle, phase-out, and final disposal

These various quantitative and qualitative design aspects and their interaction and interdependence would allow us to arrive at an optimum heat exchanger design. Figure 7.17 demonstrates a methodology for designing a new single heat exchanger from a holistic point of view and may be characterized as a *onetime case study* method which includes [28]:

- Process and design specifications
- Thermal and hydraulic design
- Mechanical design
- Manufacturing considerations and cost both for total cost of ownership and return on investment
- Trade-off factors and system-based optimization

These design considerations are usually not sequential; there could be strong interactions and feedback among the aforementioned considerations, as indicated by double-sided arrows in Fig. 7.16, and may require a number of iterations before the design is finalized.

As part of augmentation of above design methodology, based on the problem of specifications and experience, the heat exchangers construction type and flow arrangement are first front line of selection criteria for such subsystem of bigger system. For example, in the case of Open Air Brayton Cycle as part of enhancing its overall output efficiency as presented in Chap. 5, cross-flow or counterflow pattern for consideration of a compact heat exchanger as subsystem was most promising as per computer code out and related analyses and calculation. As we said these analyses were done based on the steady-state case and the code development was found as per this infrastructure.

---

[3] While the Department of Nuclear Engineering of University of New Mexico under leadership of Dr. McDaniel along with this author are involved in open Air Barton Cycle, Sandia National Laboratory is in pursue of CO_2 closed loop.

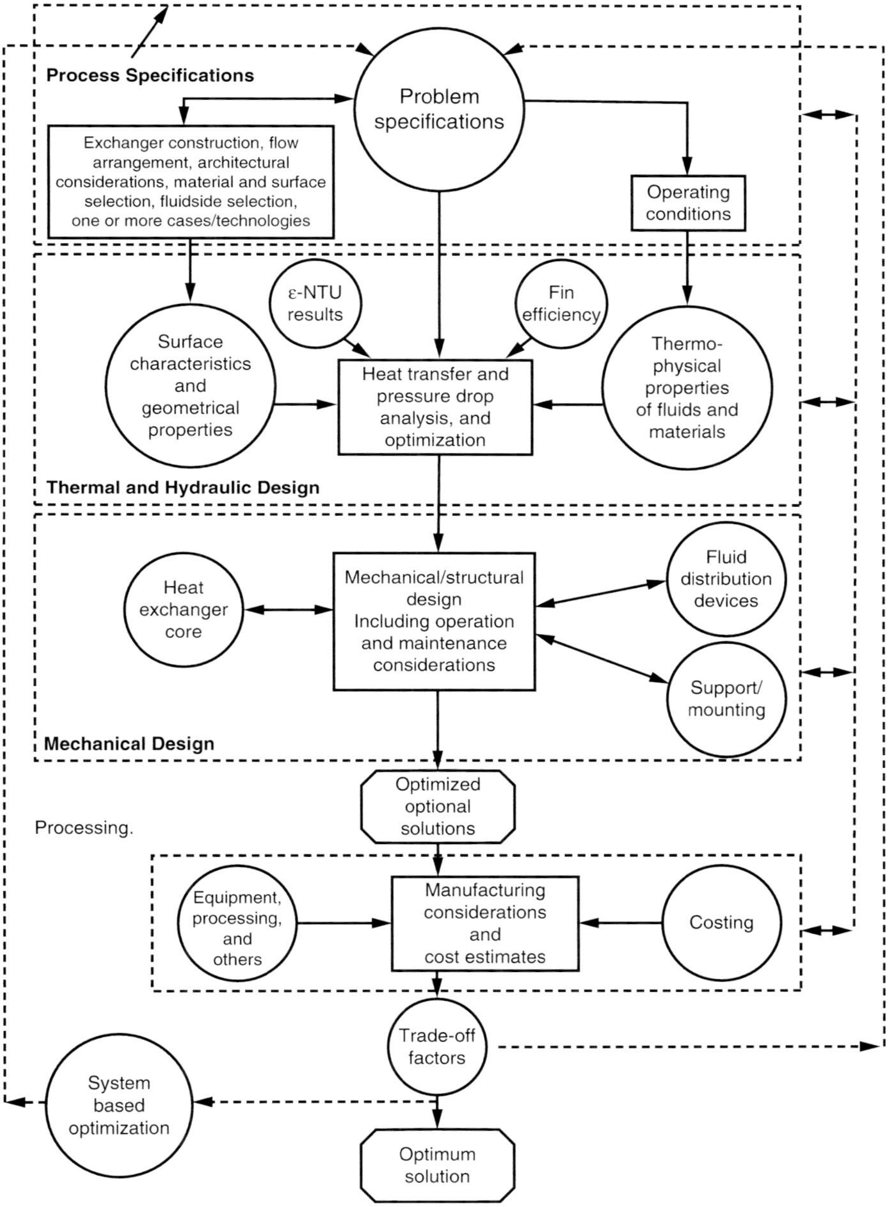

Fig. 7.17 Heat exchanger design methodology (modified from Shah and Sekulic [28]; Taborek, 1988; and Kays and London [12, 13]

In summary the selection criteria and construction type depends upon the following parameters as well:

1. Fluids (gas, liquids, or condensing/evaporating) used on each side of a two-fluid heat exchanger.
2. Operating pressure and temperature.
3. Fouling parameter.
4. Surface selection for purpose of enhanced mode of heat transfer in an efficient way.
5. Leakage or contamination of one fluid to the other is permitted or not.
6. Cost and available heat exchanger manufacture technology for production and commercial purpose not an R&D prototype.
7. Preventive Maintenance (PM) and intended life cycle.

As part of design methodology, one can also use the following chart in particular for the design of a compact heat exchanger, so one can go through processes faster. See Fig. 7.18.

Factors that influence the surface selection include the operating pressure, fouling, maintenance requirements, erosion, fabrication, cost, etc.

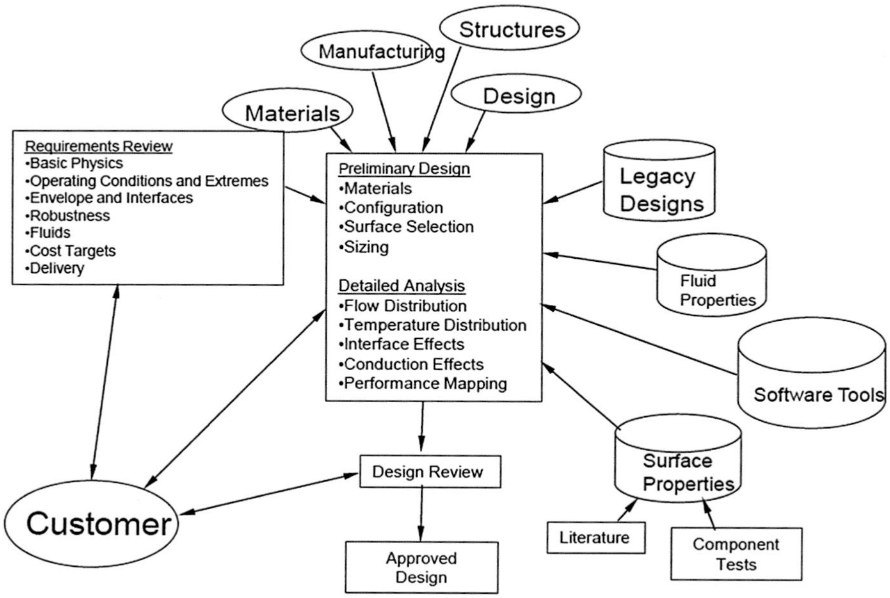

Fig. 7.18 Compact heat exchanger design methodology steps and process

7.6 Thermal and Hydraulic Design

Heat exchanger thermal and hydraulic design and analysis involves a complex analysis and quantitative heat transfer and pressure drop (in case of transient analysis rather than steady-state) or exchanger size. So in today's modern computer capabilities, most of thermodynamics design method and iteration of processes to find the most efficient gain and optimization are done via Computational Fluid Dynamics (CFD) and design methods and inputs to this analysis are summarized as follows, which depends on type of exchangers:

- *Basic thermo-hydraulic design methods*:
  P-NTU, LMTD, ε-NTU, or ψ-P methods that can be mentioned
- *Thermophysical properties*:
  For any analysis regarding pressure-drop and heat-transfer, the following thermophysical fluid properties (gas-to-gas or gas-to-fluid as media for heat exchange) are needed and they are defined as:

  - μ = dynamic viscosity
  - c_p = specific heat at constant pressure
  - k = thermal conductivity
  - γ = surface tension

  Note that for the conduction wall, thermal conductivity is needed.
- *Surface geometrical properties*:
  For purpose of pressure drop and heat transfer analysis, at minimum the following surface geometrical properties are required on each side of a two-fluid heat exchanger, and they are identified as:

  - A = heat transfer area, which includes both primary and secondary surface area if any.
  - A_{fr} = core frontal area
  - A_a = minimum free flow area
  - D_h = hydraulic diameter
  - L = flow length
  - V = core volume
  - T, W = fine thickness and fin conduction length
  - L_1, L_2, L_3 = core dimensions

  These quantities are obtained from computation of the basic dimensions layout of the core and heat-transfer surface. In case of shell-and-tube type heat exchangers, one should take under consideration various leakage and bypass flow areas as well.
- *Surface characteristics*:
  For the purpose of surface characteristics, the heat transfer j and flow friction f are key inputs for the exchanger heat transfer and pressure drop analysis, respectively [23]. Experimental data results for a variety of compact heat

	A	B	C	D	E	F	G	H	I
1	**Preliminary Heat Exchanger Design (U.S. units)**								
2	**Estimation of Heat Transfer Area Needed**								
3									
4	**Inputs**					**Calculations**			
5									
6	$Fluid_1$ mass flow					Heat Transfer Rate, Q =		925,000	Btu/hr
7	rate, m_1 =		25,000	lb/hr					
8						Log Mean Temp		79.58	°F
9	$Fluid_1$ temp. in, T_{1in} =		190	°F		Diff, ΔT_{lm} =			
10									
11	$Fluid_1$ temp. out, T_{1out} =		140	°F		Heat Transfer Area, A =		96.86	ft^2
12									
13	$Fluid_1$ sp. heat, C_{p1} =		0.74	Btu/lb-°F		$Fluid_2$ mass flow			
14						rate, m_2 =		55,000	lb/hr
15	$Fluid_2$ temp. in, T_{2in} =		50	°F					
16									
17	$Fluid_2$ temp. out, T_{2out} =		120	°F					
18									
19	$Fluid_2$ sp. heat, C_{p2} =		1.0	Btu/lb-°F					
20									
21	Overall heat transf.								
22	coeff. estim., U =		120.0	Btu/hr-ft²-°F					
23									
24	**Equations used for calculations:**								
25									
26		$Q = \pm (m_1)(C_{p1})(T_{1in} - T_{1out})$							
27									
28		$Q = \pm (m_2)(C_{p2})(T_{2in} - T_{2out})$							
29									
30		$\Delta T_{lm} = [(T_{1in} - T_{2out}) - (T_{1out} - T_{2in})]/\ln[(T_{1in} - T_{2out})/(T_{1out} - T_{2in})]$							
31									
32		$Q = U A \Delta T_{lm}$							

Fig. 7.19 Heat exchanger specification sample sheet

exchangers are presented in Kays and London [12]. More references on *j* and *f* factor are furnished in Shah and Sekulic [28], Chap. 4.

- *Heat exchanger specification sheet*:
  Heat exchanger specification sheet as an important fact sheet to help you as part of design methodology of your choice and type of the heat exchanger that is picked for intended application. So many Microsoft Excel Sheet Templates can be found on the Web and they are downloadable free of charge.[4] A simple sample of such sheet is shown here as Fig. 7.19.

7.6.1 Equations and Parameters

The heat exchanger design equation can be used to calculate the required heat transfer surface area for a variety of specified fluids, inlet and outlet temperatures, and types and configurations of heat exchangers, including counterflow or parallel flow. A value is needed for the overall heat transfer coefficient for the given heat exchanger, fluids, and temperatures. Heat exchanger calculations could be made for

[4] http://www.brighthubengineering.com/hvac/89972-preliminary-heat-exchanger-design-calculations-involved/#imgn_0.

the required heat transfer area or the rate of heat transfer for a heat exchanger of area that is given.

Heat exchanger theory leads to the basic heat exchanger design equation as:

$$q = UA\Delta T_{\mathrm{m}} \tag{7.2}$$

where q is the rate of heat transfer between the two fluids in the heat exchanger in Btu/h.

U is the overall heat transfer coefficient in Btu/h-ft^2-°F, A is the heat transfer surface area in ft^2, and ΔT_{m} is the log mean temperature difference (LMTD) in °F, calculated from the inlet and outlet temperatures of both fluids. Note that the product of UA is overall heat transfer coefficient and reference surface area.[5] This product is often called the heat transfer conductance.

For the design of heat exchangers, the basic heat exchanger design equation can be used to calculate the required heat exchanger area for known or estimated values of the three other parameters, q, U, and ΔT_{m}. Each of those parameters is now discussed in the following sections, but before we jump into these parameters we need to have some understanding of how to assess the basic concept and initial sizing of intended heat exchanger based on its application within the system, and where this exchanger will be used. Thus, the next section describes basic concepts and initial size assessment [29].

7.6.1.1 Basic Concept and Initial Size Assessment as Part of Design Process

As part of design methodology and design process, we have to have a basic concept initial sizing of exchanger per its application, and for that when pressure drop and thermal (i.e., heat transfer) specification have been fixed it is possible to do a quick analysis to determine the most appropriate heat exchanger type based on flow configuration which will be either cross-flow or counterflow, or often multipass overall counterflow arrangements (i.e., compact shell and tube exchanger). For this we need firstly to entertain the effectiveness design or method approach, and then the Logarithmic Mean Temperature Difference (LMTD) Methods as described in Chap. 6.

In Chap. 4 of this book, we touch upon the effectiveness method also known as NTU, as part of the design parameter for heat exchangers, and now we evaluate this parameter further on as part of our design methodology of a compact heat exchanger as part of our subsystem for next generation nuclear plant (NGNP).

[5] Note that any such value of overall coefficient U has to be uniquely associated with its reference surface area A, but the product UA is independent of reference area, having dimensions W/K or kW/K in metric or KMS.

Heat exchangers are usually analyzed using either the Logarithmic Mean Temperature Difference (LMTD) or the Effectiveness-Number of Transfer Units (ε-NTU) methods. The LMTD method is convenient for determining the overall heat transfer coefficient based on the measured inlet and outlet fluid temperatures. The ε-NTU method is more convenient for the prediction of the outlet fluid temperatures if the heat transfer coefficient and the inlet temperatures are known.

The analysis presented below assumes certain constraints which are:

1. There is no energy loss to the environment.
2. Heat exchanger is at a steady-state.
3. There are no phase changes in the fluids.
4. Heat capacities of the fluids are independent of temperature.
5. Overall heat transfer coefficient is independent of the fluid temperature and position within the heat exchanger.

Since all heat exchangers considered in this experiment have a single pass for both the hot and cold fluids, the discussion below is limited to single-pass heat exchangers. Qualitative dependence of the fluid temperature on position inside a single-pass heat exchanger is shown in Fig. 7.20. Although Fig. 7.20b shows a situation for parallel flow, our main target for intended compact heat exchanger for driven combined cycle efficiency for the next generation nuclear plant (NGNP) is focusing either on cross-flow or counterflow type of classification of these exchangers. This has been backed up by computer code and analyses that are presented in Chap. 5 of this book. Presentation of a parallel flow is a matter of compression for purpose effectiveness analyses matter only.

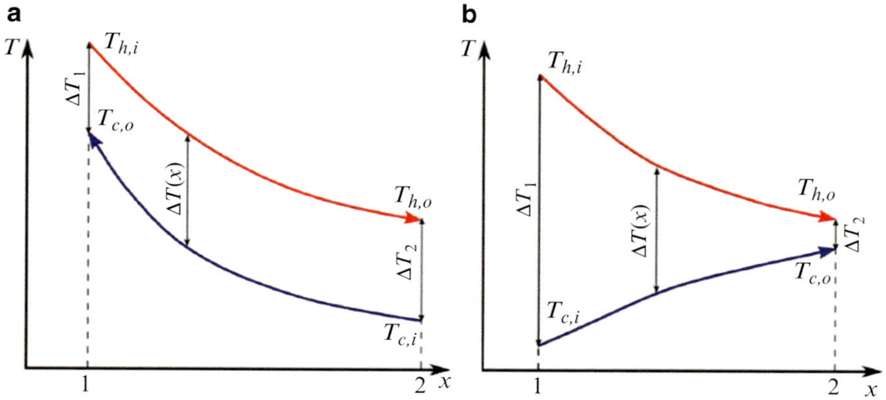

Fig. 7.20 Temperature profiles in (**a**) counterflow and (**b**) parallel flow single pass heat exchangers. Note that in a counterflow heat exchanger the outlet temperature of the cold fluid can exceed the outlet temperature of the hot fluid but this cannot happen in a parallel flow system

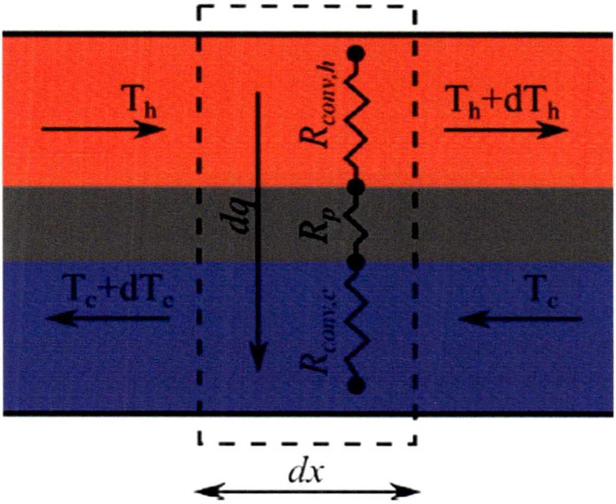

Fig. 7.21 Energy balance in a differential element of a single pass heat exchanger operated in the counterflow regime

We can analyze the total heat transfer rate q as follows:

$$q = \int_1^2 \mathrm{d}q' \tag{7.3}$$

where $\mathrm{d}q'$ is some aspect of differential form of Eq. (7.2) for a segment of a single-pass flow heat exchanger from point 1 to point 2 which is shown schematically in Fig. 7.21, in order to present overall heat transfer coefficient U. Equation (7.5) presents overall heat transfer coefficient U in this segment:

$$\mathrm{d}q'(x) = U\Delta T(x)\mathrm{d}A(x) \tag{7.4}$$

where U is the overall heat transfer coefficient, $\Delta T(x)$ is the local temperature difference between the hot and cold fluids within this segment, while $\mathrm{d}A(x)$ is the contact area in the differential segment.

Note that red, blue, and gray colors represent the hot fluid, cold fluid, and the partition between the fluids, respectively. The dashed rectangle shows a differential segment corresponding to the energy balance Eq. (7.4). Three resistances ($R_{\mathrm{conv,h}}$, R_{p}, and $R_{\mathrm{conv,c}}$) contributing to the total resistance to the heat transfer are indicated schematically. Therefore, for overall heat transfer coefficient U we can write:

$$U = \frac{1}{R_{\mathrm{Total}}} = \frac{1}{R_{\mathrm{conv,h}} + R_{\mathrm{p}} + R_{\mathrm{conv,c}}} \tag{7.5}$$

where:

$R_{\mathrm{conv,h}}$ = convective heat transfer, which is resistance to the convective heat transfer is inversely proportional to the convective heat transfer coefficient,

$h = 1/R_{\mathrm{conv,h}}$. The convective heat transfer coefficient depends on fluid properties, flow geometry, and the flow rate. It is convenient to describe this dependence using several dimensionless numbers, namely, the Reynolds number as:

$$\mathrm{Re} = \frac{L\nu\rho}{\mu} \tag{7.6}$$

The Prandtl number

$$\mathrm{Pr} = \frac{c_{\mathrm{p}}\mu}{k} \tag{7.7}$$

And finally the Nusselt number is given by

$$\mathrm{Nu} = \frac{hL}{k} \tag{7.8}$$

Here, ρ, μ, k, and c_{p} are the density, viscosity, thermal conductivity, and heat capacity of the fluid, ν is the flow velocity, and L is characteristic length. The choice of L depends on the system geometry. For example, for a flow in a circular pipe, L is the pipe diameter.

The relationship between Re, Pr, and Nu depends on the system geometry and whether the flow is laminar and turbulent. For example, for a turbulent flow inside a pipe with circular cross section of diameter D,

$$\mathrm{Nu} = 0.027\mathrm{Re}^{0.8}\mathrm{Pr}^{1/3} \tag{7.9}$$

R_{p} = conductive heat transfer, which is the resistance R_{p} to heat transfer through the partition depends on the system geometry. In the current experiment, you will need to consider heat conduction in the radial direction of a cylindrical tube and heat conduction across a thin plate. Resistances in both of these cases can be obtained analytically by solving the heat diffusion equation.

However, it is possible to obtain Q by combining Eq. (7.4) with energy balance in differential segments of the heat exchanger as:

$$\mathrm{d}q = -C_{\mathrm{p,h}}\mathrm{d}T_{\mathrm{h}} = C_{\mathrm{p,c}}\mathrm{d}T_{\mathrm{c}} = C_{\mathrm{p,k}}\mathrm{d}T_{\mathrm{k}} \tag{7.10}$$

Here, $\mathrm{d}T_{\mathrm{k}}$ is the temperature change of fluid k (k = c or h) in the interval under consideration, and $C_{\mathrm{p,k}}$ is the heat capacity rate of fluid k under constant pressure as:

$$C_{\mathrm{p,k}} = \dot{m}_{\mathrm{k}}c_{\mathrm{k}} \quad \text{for } \mathrm{k} = \mathrm{c} \text{ or } \mathrm{h} \tag{7.11}$$

where $\dot{m}_{\mathrm{k}}$ and c_{k} are the mass flow rate and heat capacity of fluid k, respectively. This analysis yields:

$$q = UA\Delta T_m \tag{7.12}$$

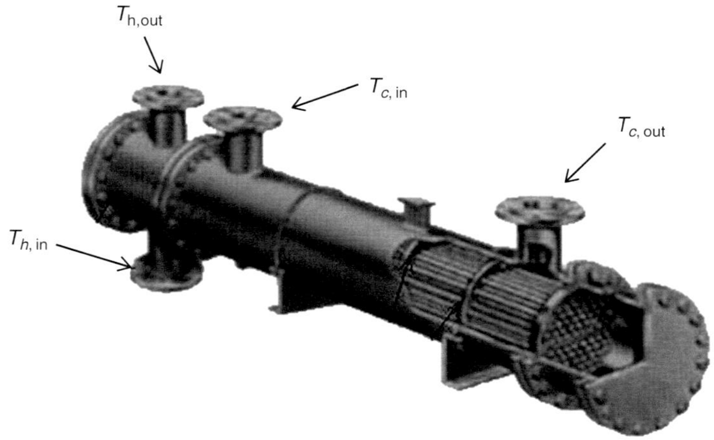

Fig. 7.22 Straight tube, two pass compact shell and tube heat exchanger

This is exactly the equation we introduced previously and is identified as Eq. (7.2), where A is the total contact area and ΔT_m is the logarithmic mean temperature difference (LMTD), and it is the subject of the next section.

7.6.1.2 Logarithmic Mean Temperature Difference (LMTD)

The driving force for any heat transfer process is a temperature difference. For heat exchangers, there are two fluids involved, with the temperatures of both changing as they pass through the heat exchanger, so some type of average temperature difference is needed. Many heat transfer textbooks have a derivation showing that the log mean temperature difference is the right average temperature to use for heat-exchanger calculations. In case of compact shell and tube heat exchanger (see Fig. 7.22), the mathematical relationship is as follows.

That log mean temperature is defined in terms of the temperature differences as shown in Eq. (7.13). $T_{\mathrm{h,in}}$ and $T_{\mathrm{h,out}}$ are the inlet and outlet temperatures of the hot fluid and $T_{\mathrm{c,in}}$ and $T_{\mathrm{c,out}}$ are the inlet and outlet temperatures of the cold fluid. Those four temperatures are shown in the diagram at the left for a straight tube, two pass shell and tube heat exchanger with the cold fluid as the shell side fluid and the hot fluid as the tube side fluid:

$$\Delta T_{\mathrm{m}} = \frac{\left(T_{\mathrm{h,in}} - T_{\mathrm{c,out}}\right) - \left(T_{\mathrm{h,out}} - T_{\mathrm{c,in}}\right)}{\ln \frac{\left(T_{\mathrm{h,in}} - T_{\mathrm{c,out}}\right)}{\left(T_{\mathrm{h,out}} - T_{\mathrm{c,in}}\right)}} \tag{7.13}$$

In general form Eq. (7.13) can be written as:

$$\Delta T_{\mathrm{m}} = \frac{\Delta T_1 - \Delta T_2}{\ln(\Delta T_1/\Delta T_2)} \tag{7.14}$$

Here, we can define a new term ΔT_{k} which refers to temperature difference the hot and cold fluids at point k (k = c or h), i.e.,

$$\Delta T_1 = T_{\mathrm{h,in}} - T_{\mathrm{c,out}} \quad \text{and} \quad \Delta T_2 = T_{\mathrm{h,out}} - T_{\mathrm{c,in}} \tag{7.15}$$

for the counterflow current and for parallel flow it yields to the following form:

$$\Delta T_1 = T_{\mathrm{h,in}} - T_{\mathrm{c,in}} \quad \text{and} \quad \Delta T_2 = T_{\mathrm{h,out}} - T_{\mathrm{c,out}} \tag{7.16}$$

Note that:

1. If heat capacity rates of the cold and hot fluids are the same and the heat exchanger is operated in the counterflow regime then ΔT is independent of position in the heat exchanger. In this case Eqs. (7.12) and (7.14) are not applicable and the total heat transfer rate Q should be obtained by direct integration of Eq. (7.4).
2. This result holds for single pass heat exchangers only. However, the LMTD method can be extended to more complex heat-exchanger designs (e.g., multi-pass and cross-flow systems) using a correction factor (see Ref. [30]).

Now that we have better understanding of LMTD methodology, we can see that this method is very useful for determining the overall heat transfer coefficient U based on experimental values of the inlet and outlet temperatures and the fluid flow rates. However, this method is not very convenient for the prediction of outlet temperatures if the inlet temperatures and U are known. In this case, one has to solve a nonlinear system of two equations (see Eq. (7.12) and the overall energy balance) for two unknowns ($T_{\mathrm{h,out}}$ and $T_{\mathrm{c,out}}$). This solution requires application of an iterative approach.

A more convenient method for predicting the outlet temperatures is the effectiveness method that is known as NTU method and sometimes written as ε-NTU also. This method can be derived from the LMTD method without introducing any additional assumptions. Therefore, the effectiveness-NTU and LMTD methods are equivalent.

An advantage of the effectiveness-NTU method is its ability to predict the outlet temperatures without resorting to a numerical iterative solution of a system of nonlinear equations.

7.6.1.3 Effectiveness-NTU Method

Now that we have better understanding of LMTD, we are going to define the heat-exchanger effectiveness ε as:

$$\varepsilon = \frac{q}{q_{\max}} = \frac{C_{\mathrm{h}}(T_{\mathrm{h,in}} - T_{\mathrm{h,out}})}{C_{\min}(T_{\mathrm{h,in}} - T_{\mathrm{c,in}})} = \frac{C_{\mathrm{c}}(T_{\mathrm{c,in}} - T_{\mathrm{c,in}})}{C_{\min}(T_{\mathrm{h,in}} - T_{\mathrm{c,in}})} \tag{7.17}$$

where Q is the actual rate of heat transfer from hot to cold fluid and $Q_{\max}$ is the maximum possible rate of heat transfer for given temperatures of the fluids as:

$$q_{\max} = C_{\min}(T_{\mathrm{h,in}} - T_{\mathrm{c,in}}) \tag{7.18}$$

In this equation, $C_{\min}$ represents the smaller of the two heat capacity rates between C_{c} and C_{h}.

where C_{c} and C_{h} are cold stream and hot stream heat capacity rates respectively, one of which will be $C_{\min}$ unless the exchanger is balanced. This allows the effectiveness of the exchanger to be designed and it can be determined directly from the terminal temperatures, if these temperatures are known, and the stream parameters as well. As we said beside this criterion the other major relationship for design that will be needed is given by Eq. (7.12).

If heat exchanger effectiveness is known, one can readily obtain Q from Eqs. (7.17) and (7.18) by solving these two equations simultaneously. After that, the outlet temperatures can be obtained from the energy balance.

The efficiency ε depends on the heat exchanger geometry, flow pattern (i.e., parallel flow, counterflow, cross-flow, etc.) and the Number of Transfer Units (NTU).

Now we define the Number of Transfer Units (NTU) for the exchanger as:

$$\mathrm{NTU} = \frac{UA}{C_{\min}} \tag{7.19}$$

Relationships between the effectiveness and NTU have been established for a large variety of heat exchanger configurations. Most of these relationships involve the *ratio of stream capacity rates* denoted as $C_{\mathrm{r}} = C_{\min}/C_{\max}$ of the smaller and the larger of the heat capacity rates C_{c} and C_{h}. For example, for a single pass heat exchanger in the parallel flow regime as:

$$\varepsilon = \frac{1 - \exp[-\mathrm{NTU}(1 + C_{\mathrm{r}})]}{1 + C_{\mathrm{r}}} \tag{7.20}$$

And for a single pass heat exchanger in the counterflow regime, the effectiveness is expressed as:

$$\begin{cases} \varepsilon = \dfrac{1 - \exp[-\mathrm{NTU}(1 + C_r)]}{1 + C_r \exp[-\mathrm{NTU}(1 - C_r)]} & \text{if } C_r < 1 \\ \varepsilon = \dfrac{\mathrm{NTU}}{1 + \mathrm{NTU}} & \text{if } C_r < 1 \end{cases} \tag{7.21}$$

Similarly for cross-flow regime, we can define the effectiveness as:

$$\varepsilon = \frac{1}{C_r}[1 - \exp\{-C_r[1 - \exp(-\mathrm{NTU})]\}] \tag{7.22}$$

Designing a heat exchanger today with the help of tools such as CAD (Computer Aided Design) and CAM (Computer Aided Manufacturing), some mathematical and algebraic relationships between all the above parameters and consequently selection of right heat exchanger for right application will be most helpful and make the design process easier than hand calculations. For most commonly used configuration Table 7.1 is proposed by Kays and London [12] and Hesselgreaves [29].

Considering typical temperature for common arrangements can be depicted as shown in Fig. 7.23.

Per Eq. (7.22), which is the representation of an idealized performance with infinite surface area, Fig. 7.24 shows the idealized temperature distribution diagram [29].

HHH = hot fluid has highest heat capacity rate (hot end pinch point)
CHC = cold fluid has highest heat capacity rate (cold end pinch point)

These acronyms allow us to remember the location of pinch point at either end of heat exchanger.

Analyzing Fig. 7.24 clearly shows that the maximum heat transferred is obtained when the stream of lowest heat capacity rate has an outlet temperature equal to the inlet temperature (for a counterflow configuration) of the other stream. For the parallel flow arrangement, the state is reached when both streams attain the same temperature at outlet and pinch point is the point of equal temperature. Note that per this idealized situation, the pinch point corresponds to zero temperature difference.

Further analysis of NTU is illustrated graphically in Fig. 7.25, which is reveling the product of required area and heat transfer coefficient UA based on Eq. (7.2) or (7.12), which relates these two parameters to the effectiveness, determined by the specified temperatures, and the stream capacity rates ratio C_r.

From figures in Fig. 7.24 for various heat exchanger types, it is obvious that the configuration makes some differences to the value of effectiveness for low C_r (i.e., $C_r < 0.25$), or NTU of less than 1 (i.e., NTU < 1). For this range of NTU < 1, we can see that effectiveness is low as well. Now analyzing the given mathematical relationships for the effectiveness of heat exchanges of different types that are

Table 7.1 Effectiveness-NTU relationships [29]

Counterflow $\varepsilon = \dfrac{1 - \exp[-\mathrm{NTU}(1 + C_\mathrm{r})]}{1 + C_\mathrm{r}\exp[-\mathrm{NTU}(1 - C_\mathrm{r})]}$

Asymptotic value = 1 as NTU $\rightarrow \infty$, for all C_r

Unmixed cross-flow (approximation from Kays and Crawford [30])

$$\varepsilon = 1 - \exp\left\{\frac{\left(\exp\left(-\mathrm{NTU}^{0.78}C_\mathrm{r}\right)-1\right)\mathrm{NTU}^{0.22}}{C_\mathrm{r}}\right\}$$

Asymptotic value = 1 as NTU $\rightarrow \infty$, for all C_r

Parallel, or co-current flow $\varepsilon = \dfrac{1 - \exp(-\mathrm{NTU}(1 + C_\mathrm{r}))}{1 + C_\mathrm{r}}$

Asymptotic value = $1/(1 + C_\mathrm{r})$, for all C_r

Cross-flow, C_{min} unmixed $\varepsilon = \frac{1}{C_\mathrm{r}}\{1 - \exp\left[-C_\mathrm{r}\left(1 - \exp(-\mathrm{NTU})\right]\right\}$

Asymptotic value = $[1 - \exp(-C_\mathrm{r})]/C_\mathrm{r}$ as NTU $\rightarrow \infty$

Cross-flow, C_{max} unmixed

$\varepsilon = 1 - \exp\left\{-\frac{1}{C_\mathrm{r}}[1 - \exp(-\mathrm{NTU}C_\mathrm{r})]\right\}$

Asymptotic value = $[1 - \exp(1/-C_\mathrm{r})]$ as NTU $\rightarrow \infty$

Cross-flow, both fluids mixed

$$\varepsilon = \frac{1}{\dfrac{\mathrm{NTU}}{1 - \exp(-\mathrm{NTU})} + \dfrac{\mathrm{NTU}C_\mathrm{r}}{1 - \exp(-\mathrm{NTU}C_\mathrm{r})} - 1}$$

Asymptotic vale = $1/(1 + C_\mathrm{r})$, for all C_r

Multipass overall counterflow, fluid mixed between passes

$$\varepsilon = \frac{\left(\dfrac{1 - \varepsilon_\mathrm{p}C_\mathrm{r}}{1 - \varepsilon_\mathrm{p}}\right)^n}{\left(\dfrac{1 - \varepsilon_\mathrm{p}C_\mathrm{r}}{1 - \varepsilon_\mathrm{p}}\right)^n - C_\mathrm{r}}$$

with ε_p = effectiveness of each pass [as a function of $\mathrm{NTU}_\mathrm{p} = (\mathrm{NTU}/n)$]

n = number of identical passes (i.e., each pass having the same ε_p)

and

$$\varepsilon_\mathrm{p} = \frac{\left(\dfrac{1 - \varepsilon C_\mathrm{r}}{1 - \varepsilon}\right)^{1/n} - 1}{\left(\dfrac{1 - \varepsilon C_\mathrm{r}}{1 - \varepsilon}\right)^{1/n} - C_\mathrm{r}}$$

Limiting value = $\varepsilon_{\mathrm{counterflow}}$, as $n \rightarrow \infty$

Multipass overall parallel flow, fluids mixed between passes:

$$\varepsilon = \frac{1}{1+C_\mathrm{r}}\left\{1 - \left[1 - (1 + C_\mathrm{r})\varepsilon_\mathrm{p}\right]^n\right\}$$

All configurations, $C_\mathrm{r} = 0$ for pure condensation and evaporation:

$$\varepsilon = 1 - \exp(-\mathrm{NTU})$$

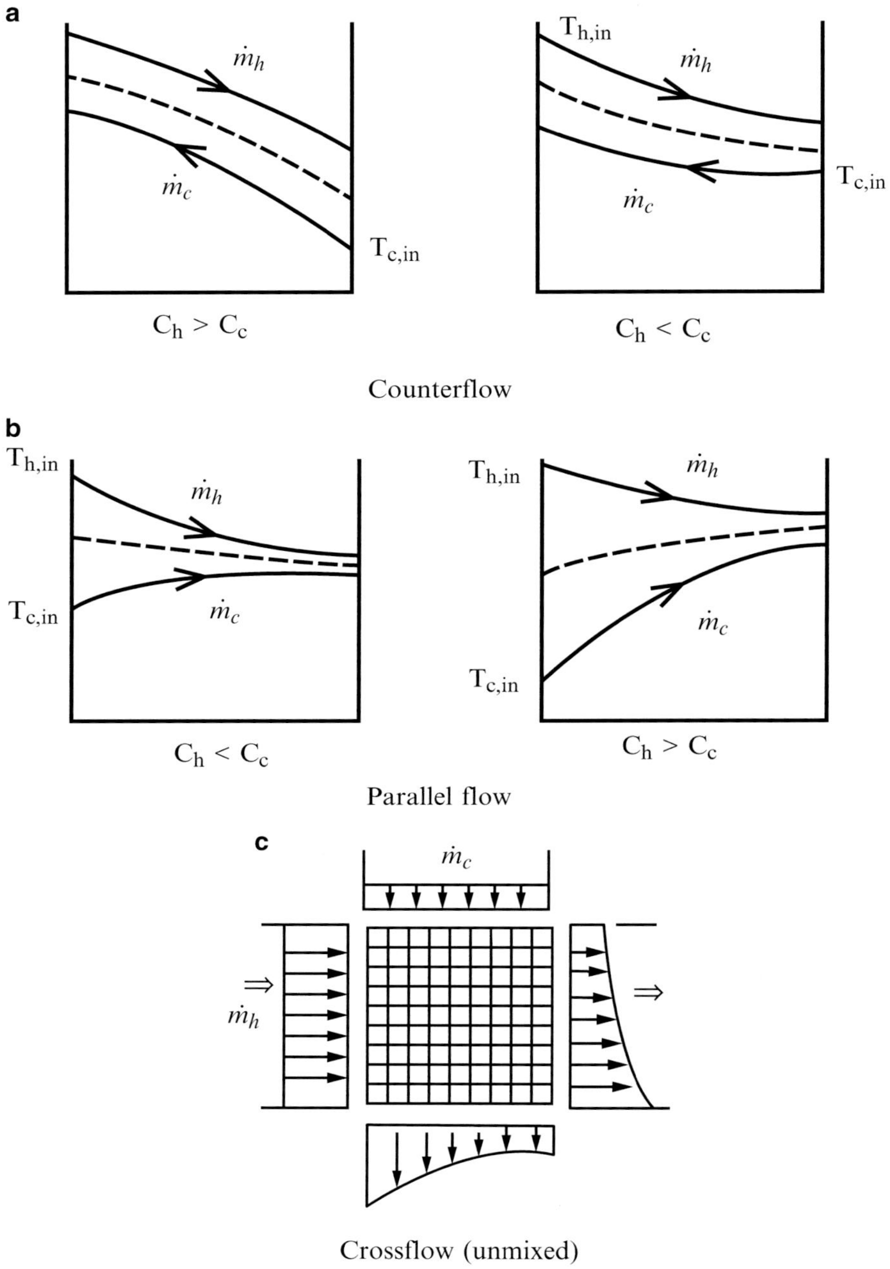

Fig. 7.23 Typical temperature distributions in heat exchangers [29]

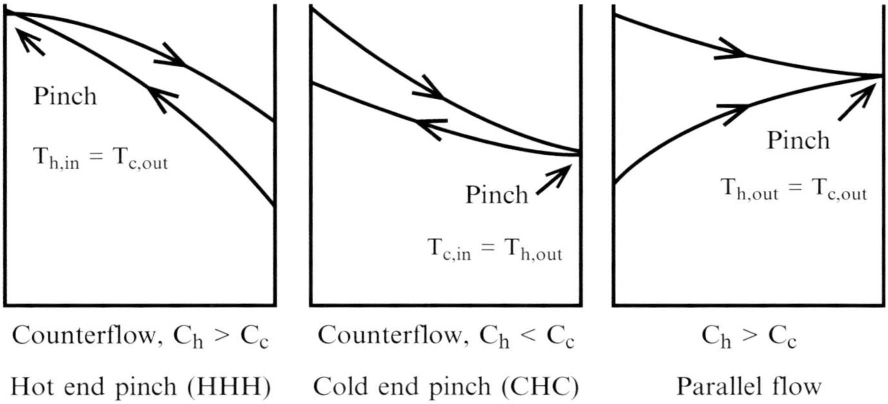

Fig. 7.24 Idealized temperature distributions showing the pinch point [29]

provided in Table 7.1, it is also clear for the purpose of CAD/CAM design, a choice of cross-flow type exchanger is most appropriate due to the simplicity of its relationship as shown in this table, which most often applies to liquid/gas compact heat exchangers and their application to enhance overall efficiency of next-generation nuclear power plants that are driven by combined cycle such as open air Brayton cycle. This choice in particular applies to the case of exchangers, for which the gas side heat transfer coefficient is low and dominates the overall conductance of *UA* product. On the other hand if C_r is higher than 0.25, and particularly if the required effectiveness is higher than 0.80 (i.e., $\varepsilon > 0.80$), a counterflow configuration will be a better choice and usually provide the most economical design approach. Figure 7.25c shows that for multipass cross-counterflow the pure counterflow value of effectiveness is closely approaching for three or more passes, which is preferred more often over the pure counterflow configuration because porting is an easier and cheaper design.

For better computer coding utilizing any CAD tool and the scope of calculations where the terminal temperature are often specified, it is better to have NTU as a function of effectiveness ε, where this parameter can be determined directly.

Therefore, it is better to create another table as Table 7.2 to explicitly define NTU in terms of ε. This then gives the overall conductance *UA*, and is directly equivalent to the LMTD method as discussed before.

A further aspect strongly affecting counterflow vs. cross-flow choice is that of pressure drop. These points are further explored in the discussion on mass velocity equation below, but before we continue with mass velocity equation we need to start our discussion from pressure drop phenomena first.

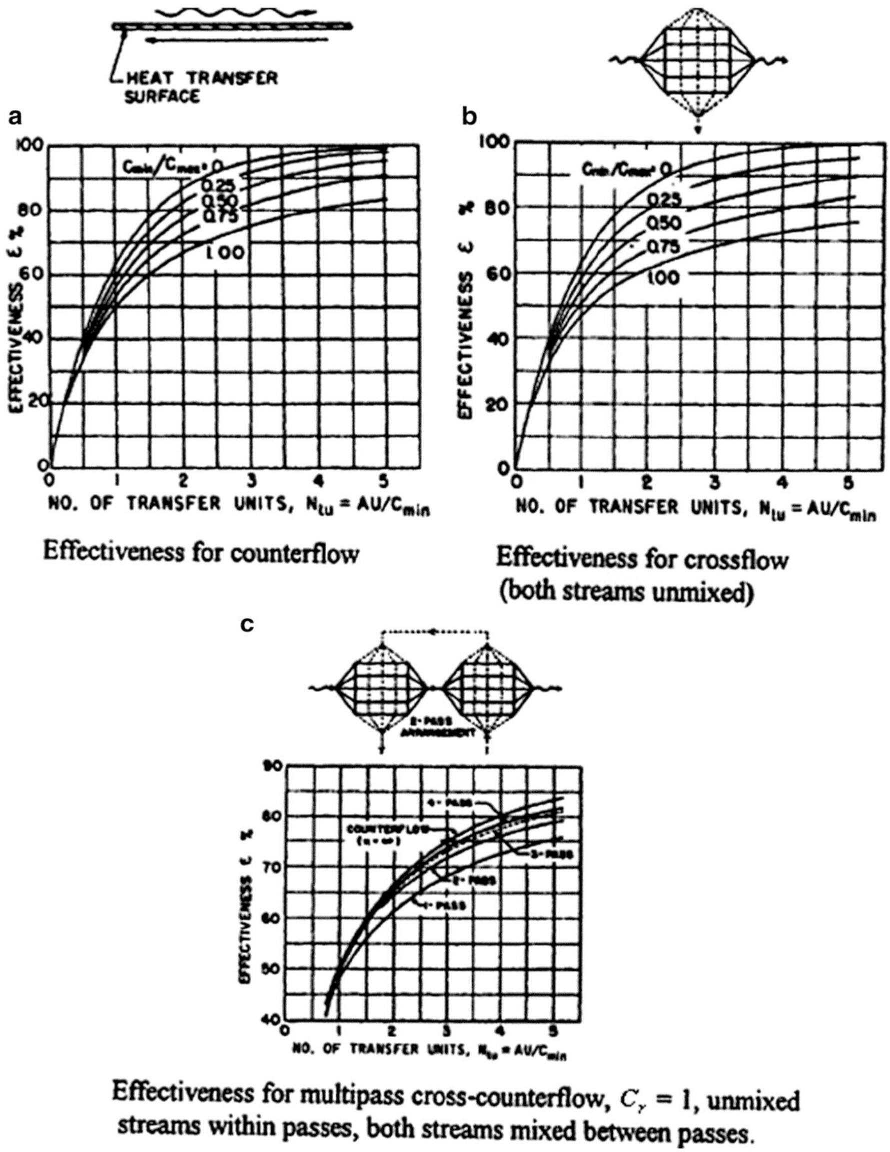

Fig. 7.25 Effectiveness vs. NTU curves for simple configurations [12]

7.6.1.4 Pressure Drop

For the application of compact heat exchanger in next generation nuclear plants (NGNP), it is necessary to have accurate design tools for predicting heat transfer and pressure drop. Until recently, this type of heat exchanger was not well studied,

Table 7.2 Inverse (NTU-ε) Relationships

Counterflow:	$\mathrm{NTU} = \frac{1}{C_r - 1}\ln\left(\frac{\varepsilon - 1}{C_r(\varepsilon - 1)}\right)$
Cross-flow ($C_r = 1$):	$\mathrm{NTU} = \frac{\varepsilon}{1 - \varepsilon}$
Parallel flow:	$\mathrm{NTU} = \frac{-\ln[1 - (1 + C_r)\varepsilon]}{1 + C_r}$
Cross-flow:	
C_{max} mixed, C_{min} unmixed	$\mathrm{NTU} = -\ln\left[1 + \frac{1}{C_r}\ln(1 - C_r\varepsilon)\right]$
C_{max} unmixed, C_{min} mixed	$\mathrm{NTU} = \frac{-1}{C_r}\ln[1 + \ln(1 - \varepsilon)]$
All configurations, $C_r = 0$	$\mathrm{NTU} = -\ln(1 - \varepsilon)$

and in the scientific literature there were large discrepancies between results reported by different investigators.

A fundamental requirement during heat exchanger design is the ability to predict heat transfer coefficients and pressure drop under the conditions of interest. Studies in the literature reporting on single and two phase fluid flow and heat transfer in compact heat exchangers are relatively few. However, extensive applications in process industries exist where the heat is transferred utilizing flows via confined spaces and which can provide more compact design and better performance.

Several features make compact heat exchangers attractive in industrial applications where the energy conservation, space and weight saving, and cost are important considerations. These features include high thermal effectiveness (ratio of actual heat transferred to the theoretical maximum heat which can be transferred), large heat transfer surface to volume ratio (surface area density), low weight per heat-transfer duty, opportunity for true counterflow operation, close temperature approach (as a result of the ability to design for true counterflow), design flexibility, and reduced fluid inventory. Flow maldistribution and the design of headers to minimize maldistribution remain inherent problems in the application of such heat exchangers, especially in the case of phase-change heat transfer. The potential for fouling of small flow passages represents a major disadvantage in the use of compact heat exchangers. However, fouling should not be a problem in applications that involve clean fluids.

The pressure drop Δp of fluid through a surface can be calculated as a function Darcy friction factor f (some textbooks show this as f_d or f_D so do not be confused by Fanning friction factor f_c or C_f) by Eq. (7.23), neglecting for practical purpose the relatively small contributions of entry and exit losses and flow acceleration:

$$\Delta p = \frac{1}{2}\rho u^2 \frac{4L}{d_h} f \tag{7.23}$$

where $u =$ is the flow velocity, and with mass flow rate $\dot{m}$, it is equal to

$$u = \frac{\dot{m}}{\rho A_c} \tag{7.24}$$

$\rho =$ flow density
$d_h =$ the hydraulic diameter
$L =$ the flow length
$A_c =$ the flow area that varies with flow length

Substituting (7.24) into (7.23) for velocity u, we have

$$\frac{2\rho \Delta p}{\dot{m}^2} = f \frac{4L}{d_h A_c} = \text{Constant for given conditions} \tag{7.25}$$

and we consider two surfaces denoted by 1 and 2, and with given pressure drop we can write:

$$\frac{\left(A_c^2\right)_1}{\left(A_c^2\right)_2} = \frac{f_1 L_1 (d_h)_2}{f_2 L_2 (d_h)_1} \tag{7.26a}$$

Therefore if the thermal performance is ignored, the flow areas for two surfaces 1 and 2 with similar friction factors are the same if the ratios L/d_h are the same.

And for given conditions the product $\text{Pr}^{2/3}N$ is fixed, therefore for two surfaces suffixed by 1 and 2 to be compared, we can write:

$$\frac{L_1}{L_2} = \frac{d_{h,1} j_2}{d_{h,2} j_1} \tag{7.26b}$$

Equation (7.26b) is showing the flow length is directly proportional to hydraulic diameter while inversely it is proportional to the j factor, which is Reynolds number dependent.

If we consider the dimensionless j, or Colburn factor in terms of Nusselt number $\text{Nu} = \frac{\alpha d_h}{\lambda}$, Reynolds number Re, and Prandtl number Pr as before (see Chap. 6) then we can write again as:

$$j = \frac{\text{Nu}}{\text{RePr}^{1/3}} = \text{StPr}^{2/3} \tag{7.27}$$

where $\text{St} = \frac{\alpha}{Gc_p}$ is Stanton number, where α is non-dimensionalized in terms of mass velocity G. Note that for a fixed G, j is proportional to α. Substituting the value of Nusselt number, Eq. (7.27) in terms of α reduces to:

$$\begin{cases} \dfrac{\alpha d_\mathrm{n}}{\lambda} = \dfrac{u d_\mathrm{n}}{v} j\mathrm{Pr}^{1/3} \\ \text{and} \\ \alpha = \dfrac{\dot{m}}{\rho v} \lambda \mathrm{Pr}^{1/3} \dfrac{j}{A_\mathrm{c}} \end{cases} \tag{7.28}$$

Assuming the specified load q is given by the heat transfer and rate equations for either side of exchanger analogous to Eq. (7.12) is given as:

$$q = \alpha A \overline{\Delta T} = \dot{m} c_\mathrm{p} (T_2 - T_1) \tag{7.29}$$

Manipulation of Eqs. (7.29) and (7.28) and given the value for Nusselt number, we obtain the following relationship:

$$j = \frac{A_\mathrm{c}}{A} \mathrm{Pr}^{2/3} N \tag{7.30}$$

where N = NTU (Number of Thermal Unit).

If we recall the relationship for hydraulic diameter d_h as:

$$d_\mathrm{h} = \frac{4 A_\mathrm{c} L}{A} \tag{7.31}$$

Then we have an alternative equation form for Eq. (7.30) as:

$$j = \frac{d_\mathrm{h}}{4L} \mathrm{Pr}^{2/3} N \tag{7.32}$$

Now combining Eqs. (7.25) and (7.32), we produce the *core mass velocity equation*, which can be written as:

$$\begin{cases} \dfrac{2\rho \Delta p}{\dot{m}^2} = \dfrac{f \mathrm{Pr}^{2/3} N}{j A_\mathrm{c}^2} \\ \text{and} \\ \dfrac{G^2}{2\rho \Delta p} = \dfrac{j/f}{\mathrm{Pr}^{2/3} N} \end{cases} \tag{7.33}$$

G being the mass velocity as per Eq. (6.13) $\dot{m}/A_\mathrm{c}$. As mentioned, G and therefore flow area A_c can be closely estimated from design specification methodology, and this equation, with the assumption of a typical value j/f (f is Fanning friction factor), is often used as a starting point for preliminary sizing as we said before [28].

Note that the basic elements of effect of the surface on thermal design, which is embodied in Eqs. (7.32) and (7.33) as above, are as follows:

- That flow length decreases as hydraulic diameter decreases
- That flow area is largely independent of hydraulic diameter

Thus if performance specification includes the pressure drop, as is normally the case, then increasing compactness only implies the reduction of flow length, with a change of shape, or aspect ratio, of the active block. In practice, this reduction in flow length can make longitudinal conduction a problem to be taken into account. In terms of comparing surfaces 1 and 2, based on Eqs. (7.25) and (7.26b), we can deduce the following equation:

$$\frac{(A_c)_1}{(A_c)_2} = \left[\frac{f_1}{f_2} \cdot \frac{j_2}{j_1}\right]^{1/2} = \left(\frac{j_2/f_2}{j_1/f_1}\right)^{1/2} \tag{7.34}$$

The form of Eq. (7.34) gives rise to the description of the ratio *j*/*f* as the *Flow Area Goodness Factor* (London) [31].

By now, you are getting a fair idea about one of design criteria in Sect. 7.4 of this chapter, as part of thermal energy transfer for process of heat application in enhanced mode, we need information provided by Eq. (7.30), to be able to manufacture heat transfer surface and testing to collected the correlation data in order to determine the basic performance of an enhanced surface for heat transfer in a compact heat exchanger.

7.6.1.5 The Core Mass Velocity Equation

In summary, the core mass velocity equation based on what we discussed earlier in the previous section can be summarized here, considering the sides are chosen and we drop the suffix h for hot side or c for cold side, and N's in the following form:

$$\frac{2\rho\Delta p}{\dot{m}^2} = \frac{f\mathrm{Pr}^{2/3}N}{jA_c^2} \quad \text{and} \quad \frac{G^2}{2\rho\Delta p} = \frac{j/f}{\mathrm{Pr}^{2/3}N} \tag{7.35}$$

All the parameters that are use in these equations are defined as before. For laminar flows, these equations yield to the following form:

$$\frac{2\rho\Delta p}{\dot{m}^2} = \frac{N\mathrm{Pr}}{A_c^2}\frac{k}{\mathrm{Nu}} \quad \text{and} \quad \frac{G^2}{2\rho\Delta p} = \frac{\mathrm{Nu}/k}{\mathrm{Pr}N} \tag{7.36}$$

where the friction factor f is give by $f = k/\mathrm{Re}$.

Note that, for given conditions of Pr, N, p and Δp, it is clear that G is only a function of j/f, or Nu/k, and most importantly is independent of the hydraulic diameter of the surface. The ratio j/f (f is fanning friction factor) is only a weak function of Reynolds number, being of the order of 0.2–0.3 for most compact surfaces. Thus G, and hence flow area A_c, can be closely estimated from the design specification [29].

The operating Reynolds number, with which *j*/*f* is a weak function and *j* is a strong function for most surfaces, in terms of prescribed side NTU and pressure drop:

$$\frac{\text{Re}}{d_{\text{h}}(j/f)^{1/2}} = \frac{1}{\eta}\left(\frac{2\rho\Delta p}{\text{Pr}^{2/3}N}\right)^{1/2} \tag{7.37}$$

and for laminar flow Eq. (7.37) results as:

$$\frac{\text{Re}}{d_{\text{h}}(\text{Nu}/k)^{1/2}} = \frac{1}{\eta}\left(\frac{2\rho\Delta p}{\text{Pr}N}\right)^{1/2} \tag{7.38}$$

If G has already been calculated and accounted for as above, we can simply write:

$$\text{Re} = \frac{Gd_{\text{h}}}{\mu} \tag{7.39}$$

where μ is dynamic viscosity of the fluid.

Note that this estimation of G effectively determines the through-flow velocity, which is also reflected in Eq. (7.38), since this velocity is proportional to Re/d_{h}. The velocity is, in addition, responsible for determining the controllable entropy generation rate, and features in economic optimization [32].

7.7 The Overall Heat Exchanger Design Process

Design of heat exchangers is an iterative (trial and error) process. Here is a set of steps for the process:

1. Calculate the required heat transfer rate, q, in Btu/h from specified information about fluid flow rates and temperatures.
2. Make an initial estimate of the overall heat transfer coefficient, U, based on the fluids involved.
3. Calculate the logarithmic mean temperature difference, ΔT_{m}, from the inlet and outlet temperatures of the two fluids.
4. Calculate the estimated heat transfer area required, using: $A = q/U\Delta T_{\text{m}}$.
5. Select a preliminary heat exchanger configuration.
6. Make a more detailed estimate of the overall heat transfer coefficient, U, based on the preliminary heat exchanger configuration.

7.7.1 Input Information Needed

In order to start the heat exchanger design process, several items of information are needed as follows:

1. The two fluids involved need to be identified.
2. The heat capacity of each fluid is needed.
3. The required initial and final temperatures for one of the fluids are needed.
4. The design value of the initial temperature for the other fluid is needed.
5. An initial estimate for the value of the Overall Heat Transfer Coefficient, U, is needed

Knowing the first four items allows determination of the required heat transfer rate, q, and the inlet and outlet temperatures of both fluids, thus allowing calculation of the log mean temperature difference, ΔT_{m}. With values now available for q, U, and ΔT_{m}, an initial estimate for the required heat transfer area can be calculated from the equation.

For the design of heat exchangers, the basic heat exchanger design equation can be used to calculate the required heat exchanger area for known or estimated values of the other three parameters, q, U, and ΔT_{m}. Each of those parameters is now discussed briefly.

1. Logarithmic Mean Temperature Difference, ΔT_{m}
   The driving force for any heat transfer process is a temperature difference. For heat exchangers, there are two fluids involved, with the temperatures of both changing as they pass through the heat exchanger, so some type of average temperature difference is needed. Many heat transfer textbooks have a derivation showing that the log mean temperature difference is the right average temperature to use for heat exchanger calculations. That log mean temperature is defined in terms of the temperature differences as shown in the equation at the right. $T_{\mathrm{h_{in}}}$ and $T_{\mathrm{h_{out}}}$ are the inlet and outlet temperatures of the hot fluid and $T_{\mathrm{c_{in}}}$ and $T_{\mathrm{c_{out}}}$ are the inlet and outlet temperatures of the cold fluid. Those four temperatures are shown in the diagram at the left for a straight tube, two pass shell and tube heat exchanger with the cold fluid as the shell side fluid and the hot fluid as the tube side fluid. See Fig. 7.22 as well.

$$\Delta T_{\mathrm{m}} = \frac{\left(T_{\mathrm{h_{in}}} - T_{\mathrm{c_{out}}}\right) - \left(T_{\mathrm{h_{out}}} - T_{\mathrm{c_{in}}}\right)}{\ln\left\{\frac{\left(T_{\mathrm{h_{in}}} - T_{\mathrm{c_{out}}}\right)}{\left(T_{\mathrm{h_{out}}} - T_{\mathrm{c_{in}}}\right)}\right\}} \tag{7.40}$$

2. Heat Transfer Rate q
   Heat Transfer Rate q can be calculated based on the above design rules. This information can be found by Eq. (7.12) and then an initial estimate for the required heat transfer area can be calculated from the relationship of $A = q/U\Delta T_{\mathrm{m}}$.

Heat exchanger calculations with the heat exchanger design equation require a value for the heat transfer rate, q, which can be calculated from the known flow rate of one of the fluids, its heat capacity, and the required temperature change. The following is the equation to be used:

$$q = \dot{m}_{\mathrm{h}} c_{\mathrm{p_h}} (T_{\mathrm{h_{in}}} - T_{\mathrm{h_{out}}}) = \dot{m}_{\mathrm{c}} c_{\mathrm{p_c}} (T_{\mathrm{c_{out}}} - T_{\mathrm{c_{in}}})$$

where (i.e., the units are given in British system and can easily be shown as M.K.S. System as well):

$\dot{m}_{\mathrm{h}}$ = mass flow rate of the hot fluid, slug/h
$c_{\mathrm{p_h}}$ = heat capacity of the hot fluid at constant pressure, BTU/slug °F
$\dot{m}_{\mathrm{c}}$ = mass flow rate of cold fluid, slug/h
$c_{\mathrm{p_c}}$ = heat capacity of the cold fluid at constant pressure, BTU/slug °F

and the temperatures are as defined in the previous section.

The required heat transfer rate can be determined from known flow rate, heat capacity and temperature change for either the hot fluid or the cold fluid. Then either the flow rate of the other fluid for a specified temperature change, or the outlet temperature for known flow rate and inlet temperature can be calculated.

3. Overall Heat Transfer Coefficient, U

The overall heat transfer coefficient, U, depends on the conductivity through the heat transfer wall separating the two fluids, and the convection coefficients on both sides of the heat transfer wall. For a shell and tube heat exchanger, for example, there would be an inside convective coefficient for the tube side fluid and an outside convective coefficient for the shell side fluid. The heat transfer coefficient for a given heat exchanger is often determined empirically by measuring all of the other parameters in the basic heat exchanger equation and calculating U. Typical ranges of U values for various heat exchanger/fluid combinations are available in textbooks, handbooks, and on websites for other types of heat exchangers, such as our intended compact heat exchanger [12]. A sampling is given in Table 7.3 below for shell and tube heat exchangers:

Table 7.3 Typical value of heat transfer coefficient

Typical ranges of values of U for shell and tube heat exchangers		
Hot fluid	Cold fluid	U, Btu/h-ft^2-°F
Water	Water	140–260
Steam	Water	260–700
Light oils	Water	62–159
Gases	Water	3–50
Water	Light organics	35–53

7.8 Design Summary

As part of design process and methodology for compact heat exchangers, they generally within scope of certain constraint which includes:

1. Implementation of basic heat exchange equation
2. Flow frontal entrance area
3. Volume, weight, and size
4. Pressure drop

With these constraints in hand, different heat transfer surfaces may be analyzed and plotted on graphs as part of specification excel sheet for comparison purpose and selection of final choice for indented application exchanger as candidate.

1. Basic Heat Transfer or Exchange Equation
   Preliminary heat exchanger design to estimate the required heat exchanger surface area can be done using the basic heat exchanger equation, $q = UA\Delta T_{\mathrm{m}}$, if values are known or can be estimated for q, A, and ΔT_{m}. Heat exchanger theory tells us that ΔT_{m} is the right average temperature difference to use. So we need the following basic heat transfer equation:

$$q = UA\Delta T_{\mathrm{m}} \tag{7.41}$$

   where estimated heat transfer area required using $A = q/U\Delta T_{\mathrm{m}}$ and LMTD is given as Eq. (7.40).
2. Flow Area
   Using the mass velocity $G \propto (\mathrm{St}/f)^{1/2}$ for a given pressure drop, we can write the following flow area relationship:

$$\mathrm{StPr}^{2/3}/f \ \text{ vs. } \ \mathrm{Re} \tag{7.42}$$

   Further investigation of Eq. (7.42) indicates that a higher $(\mathrm{St\,Pr}^{2/3}/f)$ surface characteristic implies a smaller flow frontal entrance area (A_{c}) for a give $\Delta p/p$. Again f is Fanning friction factor.
3. Exchanger Volume, Weight, and Size
   A useful comparison for purpose of exchanger volume, weight, and size is provided as:

$$h_{\mathrm{std}} \ \text{ vs. } \ E_{\mathrm{std}} \tag{7.43}$$

   where E_{std} is the friction power per unfit area in hp/ft^2. Here subscription std is an indication of an evaluation at reference fluid properties (i.e., air at 500 °F and 1 atm) [33].

Equation (7.43) implies that the surface with the higher h_{std} vs. E_{std} characteristic will tend to have the smallest core volume and result in better compactness. Hence, these surface goodness factors are not infallible and are usually indicative of the relative performance of the surfaces. For the heat exchanger surfaces that are operating at Lower Reynolds number near laminar flow, London [33] and Zohuri and Fathi [23] provide well worth classical analyses for laminar flow solutions for fully developed flow in cylindrical passages. These analyses could be used as a benchmark for several different type surfaces to be looked at as part of comparison for the final selection of exchangers for all intended purpose and application in mind (i.e., Open Air Brayton Cycle Driven Efficiency for NGNP) [2].

Table 7.4 provides such comparison for different cross section of these passages at near laminar flow. Some noteworthy points are [2, 33]:

1. The convective heat-transfer coefficient for constant flux conditions q (BTU/h ft) of flow length a constant is 9–28 % higher than for the constant or uniform wall temperature (i.e., that is known as *T-boundary condition*) boundary condition. The smallest difference is for the rectangular passage with a large aspect ratio and the largest for the triangular passage. See Table 7.4 here and for more detailed information refer to the book by Zohuri and Fathi [2], Chap. 5.
2. The ratio St $\mathrm{Pr}^{2/3}/f$ is constant for a given surface irrespective of Re.

Table 7.4 Laminar flow solutions. comparison of laminar-flow solutions for different cross sections

Cross section	$(\mathrm{Nu})_T$	$\frac{(u)_q}{(u)_T}$	$f\mathrm{Re}$	$\{\mathrm{StPr}^{2/3}\}/f = \frac{1.15\,(u)_T}{f\mathrm{Re}}$
	2.35	1.28	13.33	0.203
	3.66	1.19	16.0	0.264
	2.89	1.26	14.2	0.234
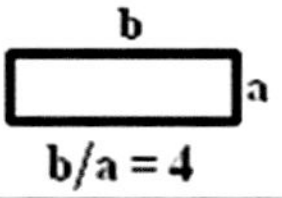	4.65	1.15	18.3	0.292
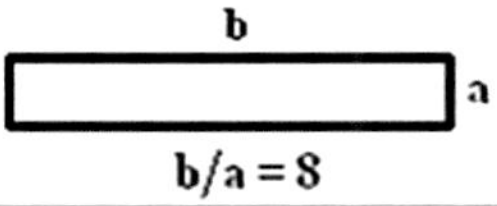	5.95	1.09	20.5	0.334
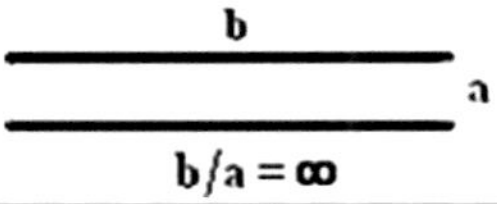	7.54	1.09	24.0	0.361

Note: (a) For $L/4r_h > 100$, (b) For $\mathrm{Pr} = 0.66$, where r_h is the hydraulic radius

3. Rectangular passages of large aspect ratio tend to have smaller frontal area requirements by about 1/3 relative to the triangular passage.

Note that the rectangular cross section with high aspect ratio (b/a) long and narrow offers the least frontage area.

If we use the given number from Table 7.4 to form h_{STD} vs. E_{STD}, we produce results for the surfaces all with a common magnitude of $4r_h = 2 \times 10^{-3}$ ft (for surface density β of $1200 < \beta < 1800$ ft^2/ft^3). Clearly, we can see that there exist large differences between surfaces, with the large-aspect-ratio rectangular passages again being superior to the triangular passages in a rather spectacular fashion as shown in Fig. 7.26 and that is a 3.2–1.0 higher heat transfer coefficient [31].

Figure 7.16 shows some of these passages that go with data from Table 7.4 and it is interesting to compare these surfaces with each other and to reference them to Fig. 7.26 and Table 7.4.

It is also important to make some preliminary points on the importance of obtaining accurate geometrical dimensions, both for test date reduction purposes and in comparison to different surface geometries as part of design methodology and these are a basis for measuring the actual surfaces for the exchanger that is

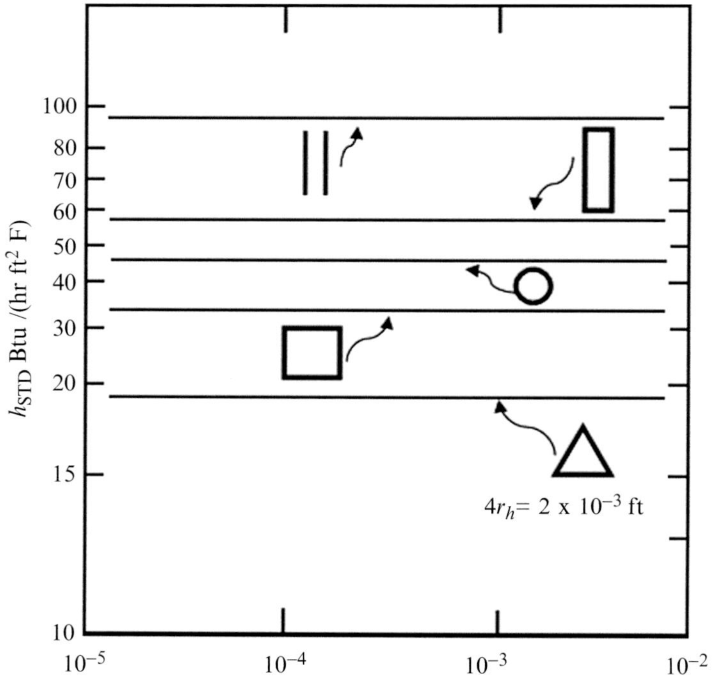

Fig. 7.26 Heat transfer power vs. friction power for fully developed lamina flow. Note again, the high-aspect ratio rectangular section shows superiority

finally selected for its intended application. Bear in your mind that for the surface geometrical factors of primary interest are three parameters and they are:

1. The porosity, p
2. The hydraulic radius r_h
3. The area density β

   Note that only two of these parameters are independent per Eq. (7.44) as shown below:

$$\beta = \frac{p}{r_\mathrm{h}} \tag{7.44}$$

The definitions of these parameters are mainly important as they specify the geometrical measurements that must be made as:

$$\begin{aligned} p &= \frac{A\ \text{(flow)}}{A_\mathrm{c}\ \text{(frontal)}} \\ p &= \frac{A\ \text{(flow)}\ \times L}{A\ \text{(heat transfer)}} \\ \beta &= \frac{A\ \text{(heat transfer)}}{A_\mathrm{c}\ \text{(frontal)}\ \times L} \end{aligned} \tag{7.45}$$

Out of any test measurement of different geometry (see Fig. 7.16) of flow passage, A_c (frontal) and L of the test core can be measured with good accuracy; however, A (flow) and A (heat transfer) are not easily established.

As part of data test reduction processes, it can be shown that:

$$\begin{aligned} &\mathrm{Re} \propto 1/\beta \\ &\mathrm{St} \propto p/\beta \\ &f \propto p^3/\beta \end{aligned} \tag{7.46}$$

For surface comparisons that are presented by Eqs. (7.42) and (7.43), it may be shown that

$$\begin{aligned} &\mathrm{StPr}^{2/3}/f \propto p^2 \\ &h_\mathrm{std} \propto p/\beta^2 \\ &E_\mathrm{std} \propto p^3/\beta^4 \end{aligned} \tag{7.47}$$

4. Pressure Drop for Fully Developed Flow

   Although we talk about pressure drop Δp as one of the quantities of interest as part of design methodology for a compact heat exchanger, we expand upon it from a fully developed laminar flow solution to a different cross section passage point of view. Analysis of pressure drop for flow within a pipe has direct impact on maintaining consistent fluid flow. In order to satisfy this requirement of consistency, we can easily note that $(\mathrm{d}p/\mathrm{d}x) = \text{constant}$, and integrating from $x = x_1$ where the pressure is p_1 to $x = x_1 + L$ where pressure is p_2 gives:

$$\frac{\mathrm{d}p}{\mathrm{d}x} = \frac{p_2 - p_1}{L} \tag{7.48}$$

Now taking another value known as average velocity V_{avg} into account for a laminar flow profile as shown Fig. 7.27, we can calculate this value at some stream-wise cross section determined from the requirement that the *conservation of mass* principle from fluid mechanics point of view must be satisfied and that is:

$$\dot{m} = \rho V_{\mathrm{avg}} A_{\mathrm{c}} = \int_{A_{\mathrm{c}}} \rho u(r) \mathrm{d}A_{\mathrm{c}} \tag{7.49}$$

where:

$\dot{m}$ = mass flow rate
ρ = fluid density
A_{c} = cross-sectional area
$u(r)$ = velocity profile

Hence, the average velocity for incompressible flow in a circular pipe of radius R can be expressed as:

$$V_{\mathrm{avg}} = \frac{\int_{A_{\mathrm{c}}} \rho u(r) \ \mathrm{d}A_{\mathrm{c}}}{\rho A_{\mathrm{c}}} = \frac{\int_{r=0}^{r=R} \rho u(r) 2\pi r \mathrm{d}r}{\rho \pi R^2} = \frac{2}{R^2} \int_{r=0}^{r=R} u(r) r \mathrm{d}r \tag{7.50}$$

Note that for fully developed laminar pipe flow V_{avg} is half of the maximum velocity.

Now considering the steady laminar flow of an incompressible fluid with constant properties in the fully developed region of a straight circular pipe, we can obtain the momentum equation by applying a momentum balance to a differential volume element, as shown in Fig. 7.28.

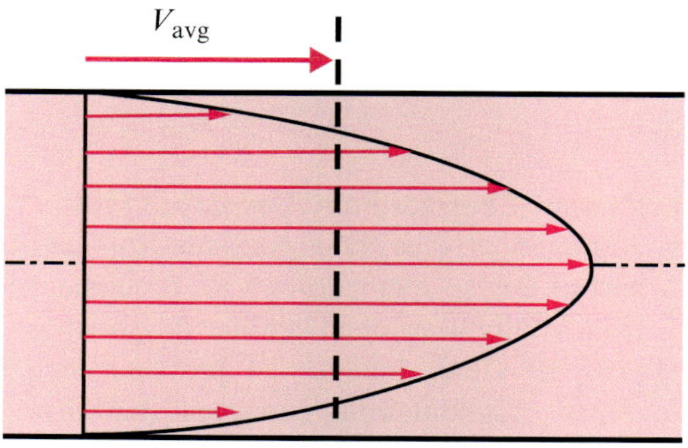

Fig. 7.27 Average velocity V_{avg} is defined as the average speed through a cross section

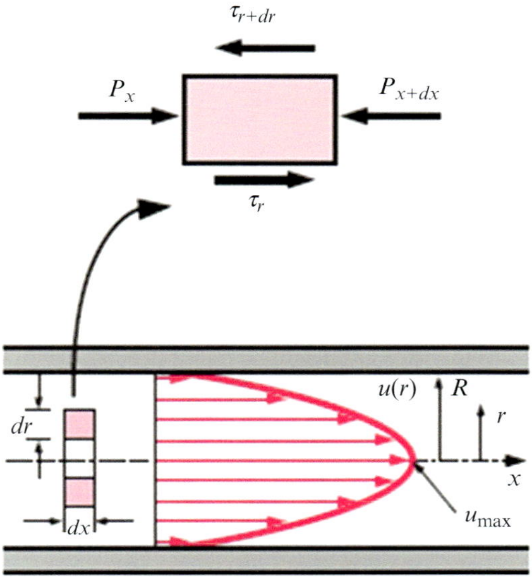

Fig. 7.28 Free-body diagram of a ring-shaped differential fluid element

Figure 7.28 is a depiction of a free-body diagram of a ring-shaped differential fluid element of radius r, thickness dr, and length dx oriented coaxially with a horizontal pipe in fully developed laminar flow. Under this demonstration, the volume element involves only pressure (p) and viscous effects (μ) and therefore the pressure and shear force τ must balance each other in given x and r directions (i.e., cylindrical coordinate and everything around azimuth angle θ is symmetrical; x is axial direction):

$$(2\pi r \mathrm{d}r\, p)_x - (2\pi r \mathrm{d}r\, p)_{x+\mathrm{d}x} + (2\pi r \mathrm{d}x \tau)_r - (2\pi r \mathrm{d}x)_{r+\mathrm{d}r} = 0 \qquad (7.51)$$

Equation (7.51) is established based on the fact that pressure force acting on a submerged plane surface is the product of the pressure at the centroid of the surface and surface area A and the force balance is given by this equation on the volume element in flow direction as shown in Fig. 7.28, which is an indication of fully developed flow in horizontal pipe.

Dividing both sides of Eq. (7.51) by $2\pi r \mathrm{d}r \mathrm{d}x$ and rearranging algebraically, this equation reduces to the following Eq. (7.52):

$$r\frac{p_{x+\mathrm{d}x} - p_x}{\mathrm{d}x} + \frac{(r\tau)_{r+\mathrm{d}r} - (r\tau)_r}{\mathrm{d}r} = 0 \qquad (7.52)$$

Taking the limit as dr, d$x \to 0$ yields:

$$r\frac{\mathrm{d}p}{\mathrm{d}x} + \frac{\mathrm{d}(r\tau)}{\mathrm{d}r} = 0 \qquad (7.53)$$

Now substituting $\tau = -\mu(\mathrm{d}u/\mathrm{d}r)$ and taking μ = constant provides the desired two-dimensional linear partial differentials (PDE) equation (7.54) as:

$$\frac{\mu}{r}\frac{\mathrm{d}}{\mathrm{d}r}\left(r\frac{\mathrm{d}u}{\mathrm{d}r}\right) = \frac{\mathrm{d}p}{\mathrm{d}x} \tag{7.54}$$

Note that the quantity du/dr is negative in pipe flow, and the negative sign is included to obtain positive values for τ (i.e., $(\mathrm{d}u/\mathrm{d}r) = -(\mathrm{d}u/\mathrm{d}y)$ since $y = R - r$).

Utilizing Fig. 7.29 (which is a free-body diagram of a fluid disk element of radius R and length dx in fully developed laminar flow in a horizontal pipe), allows us to justify a solution for partial differential equation (7.54) using the separation of variable method, with given boundary conditions below as:

$$\text{Boundary conditions}: \quad \begin{cases} (\mathrm{d}u/\mathrm{d}r) = 0 & \text{at } r = 0 \\ u = 0 & \text{at } r = R \end{cases} \tag{7.55}$$

Note that as part of justification, we can claim that, the left side of Eq. (7.54) is a function of r, and the right side is a function of x. The equality must hold for any value of r and x, and an equality of the form $f(r) = g(r)$ can be satisfied only if both $f(r)$ and $g(r)$ are equal to the same constant. Thus we conclude that (dp/dx) = constant. This can be verified by writing a force balance on a volume element of radius R and thickness dx (a slice of the pipe), which gives (Fig. 7.29).

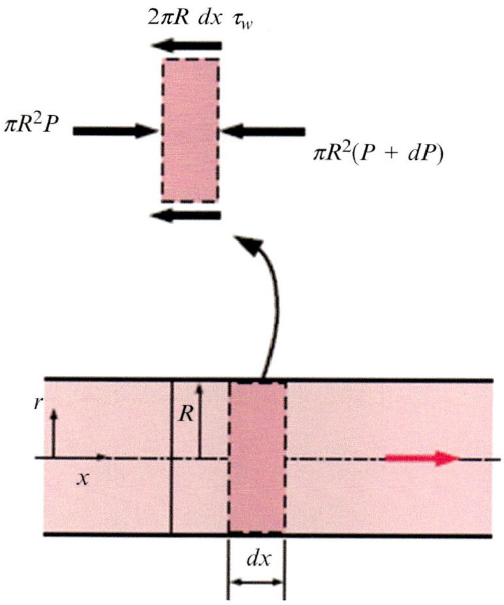

Fig. 7.29 Free-body diagram of a fluid disk element

General solution of such problem is given as:

$$u(r) = \frac{r^2}{4\mu}\left(\frac{\mathrm{d}p}{\mathrm{d}x}\right) + C_1 \ln r + C_2 \tag{7.56}$$

Applying boundary conditions provided by Eq. (7.55) sets causes Eq. (7.56) to yield the following form:

$$u(r) = -\frac{R^2}{4\mu}\left(\frac{\mathrm{d}p}{\mathrm{d}x}\right)\left(1 - \frac{r^2}{R^2}\right) \tag{7.57}$$

Now substitution of Eq. (7.57) into Eq. (7.50) and doing the proper calculus yields to the final form for average velocity V_{avg} as:

$$\begin{aligned} V_{\mathrm{avg}} &= \frac{2}{R^2}\int_0^R u(r)\mathrm{d}r \\ &= -\frac{2}{R^2}\int_0^R \frac{R^2}{4\mu}\left(\frac{\mathrm{d}p}{\mathrm{d}r}\right)\left(1 - \frac{r^2}{R^2}\right) r\mathrm{d}r \\ &= -\frac{2}{8\mu}\left(\frac{\mathrm{d}p}{\mathrm{d}r}\right) \end{aligned} \tag{7.58}$$

Based on the result of Eq. (7.58), the profile velocity $u(r)$ as final form of solution of Eq. (7.57) can also be written as:

$$u(r) = 2V_{\mathrm{avg}}\left(1 - \frac{r^2}{R^2}\right) \tag{7.59}$$

and maximum velocity at centerline of horizontal pipe for a fully developed laminar flow can derived to be $u(r)|_{\max} = 2V_{\mathrm{avg}}$, which is an indication of the remark we made earlier that $V_{\mathrm{avg}} = (1/2)u_{\max}$.

Now that we have established all the above analysis, we can go back to finalizing the analysis of pressure drop or head loss for a fully developed lamina fluid as part of our compact heat exchangers design methodology by paying our attention to Eq. (7.48) and substituting this equation into Eq. (7.58); then the pressure drop can be expressed as:

$$\text{Laminar flow}: \quad \Delta p = p_1 - p_2 = \frac{8\mu L V_{\mathrm{avg}}}{R^2} = \frac{32\mu L V_{\mathrm{avg}}}{D^2} \tag{7.60}$$

Note that: $D = D_{\mathrm{h}} = d_{\mathrm{h}}$ is known as hydraulic diameter.

Study of Eq. (7.60) indicates that the pressure drop is proportional to the viscosity μ of the fluid, and Δp would be zero if there were no friction. Therefore, the drop of pressure from p_1 to p_2 in this case is due entirely to viscous effects, and this equation represents the pressure loss Δp_{L} when a fluid of viscosity m flows through a pipe of constant diameter D and length L at average velocity V_{avg}.

A pressure drop due to viscous effects represents an irreversible pressure loss, and it is called pressure loss Δp_L to emphasize that it is a loss, just like the head loss h_L, which is proportional to it.

In practice, it is found convenient to express the pressure loss for all types of fully developed internal flows (laminar or turbulent flows, circular or noncircular pipes, smooth or rough surfaces, horizontal or inclined pipes) as Fig. 7.30.

Hence pressure drop (loss) for such circumstances can be summarized as:

$$\text{Pressure loss}: \quad \Delta p_L = f \frac{L}{D} \frac{\rho V_{avg}^2}{2} \tag{7.61}$$

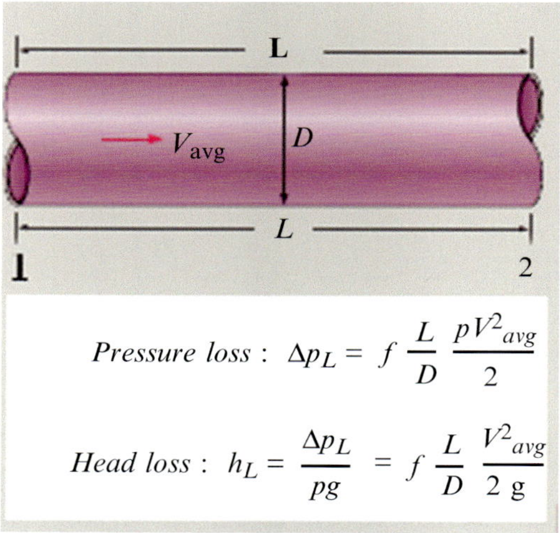

Fig. 7.30 The relation for pressure and head loss valid for laminar and turbulent flows for both circular and noncircular pipes

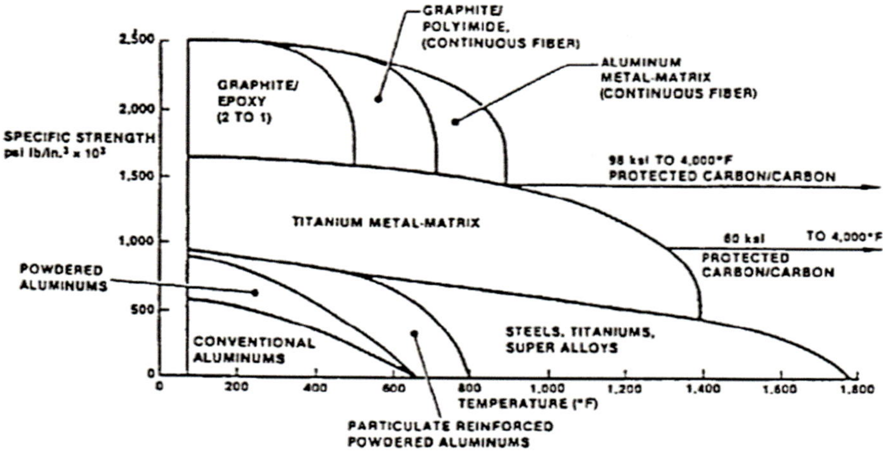

Fig. 7.31 Specific strength vs. temperature [38]

where $\rho V_{\text{avg}}^2/2$ is the *dynamic pressure* and f is the *Darcy friction factor* and can be written as:

$$f = \frac{8\tau_{\text{w}}}{\rho V_{\text{avg}}^2} \tag{7.62}$$

> *Note*: The Darcy friction factor f (some textbooks show that as symbol of f_{d}) should not be confused with the *friction coefficient* f_{c} (some textbooks show that as symbol of C_{f}), which is known as the *Fanning friction factor* and defined as: Fanning friction factor : $C_{\text{f}} = f_{\text{c}} = \frac{2\tau_{\text{w}}}{\rho V_{\text{avg}}^2} = \frac{f}{4}$, where τ_{w} is internal wall shear stress of the passage or pipe.

This friction is also called the Darcy–Weisbach friction factor, named after the Frenchman Henry Darcy (1803–1858) and the German Julius Weisbach (1806–1871).

Setting Eqs. (7.60) and (7.61) equal to each other and solving for f gives the friction factor for fully developed laminar flow in a circular pipe as:

$$\text{Circular pipe, laminar}: \quad f = \frac{64\mu}{\rho D V_{\text{avg}}} = \frac{64}{\text{Re}} \tag{7.63}$$

Eq. (7.63) shows that *in laminar flow, the friction factor is a function of the Reynolds number only and is independent of the roughness of the pipe surface*.

In the analysis of piping or passages on either side of compact heat exchanger systems, pressure losses are commonly expressed in terms of the *equivalent fluid column height*, called the head loss and is symbolically shown as h_{L}. Noting from fluid statics that $\Delta p = \rho g h$ and thus a pressure difference of Δp corresponds to a fluid height of $h = \Delta p / \rho g$, the pipe head loss is obtained by dividing Δp_{L} by ρg to give:

$$\text{Head loss}: \quad h_{\text{L}} = \frac{\Delta p_{\text{L}}}{\rho g} = f \frac{L}{D} \frac{V_{\text{avg}}^2}{2g} \tag{7.64}$$

Again, note that: $D = D_{\text{h}} = d_{\text{h}}$ is known as hydraulic diameter.

The head loss h_{L} represents the additional height that the fluid needs to be raised by a pump in order to overcome the frictional losses in the pipe. The head loss is caused by viscosity, and it is directly related to the wall shear stress. Eqs. (7.61) and (7.64) are valid for both laminar and turbulent flows in both circular and noncircular pipes, but Eq. (7.63) is valid only for fully developed laminar flow in circular pipes.

Finally as part of a design methodology process, it is best to have an Excel spreadsheet template that can be downloaded to make preliminary heat exchanger design calculations, see the "Excel Spreadsheet Templates for Preliminary Heat Exchanger Design" that is given in Fig. 7.19.

7.9 Compact Heat Exchangers in Practice

Factors that are eventually affecting the deployment of compact heat exchangers for their use in open air Brayton combined cycle in support of enhancing the next-generation nuclear power plant's overall efficiency in order to make these reactors more cost-effective are fairly discussed by Hesselgreaves [29].

The decision that can be made for choosing right heat exchanger for intended application (i.e., combined cycle, also see Chap. 5) and going through exercise of design methodology and analysis, fabrication and eventually installation of the unit must be carried out, including all guideline for future Preventive Maintenance (PM) to increase life cycle should also be carried for the heat exchanger as subsystem of big picture (i.e., an operational power plant). Reay and Hesselgreaves [29] have a good chapter on this subject and we refer readers to that reference.

7.10 Heat Exchanger Materials and Comparisons

Material selection is one of the most important things in the high temperature application [34].

There are four main categories of high temperature materials: high temperature nickel-based alloy, high temperature ferritic steels, advanced carbon silicon carbide composite, and ceramics [35].

Ohadi and Buckley [36] extensively reviewed materials for the high temperature applications. High temperature nickel-based material has good potential for helium and molten salts up to 750 °C. High temperature ferrite steels show good performance under fusion and fission neutron irradiation to around 750 °C. Advanced carbon and silicon carbide composite has excellent mechanical strength at temperatures exceeding 1000 °C. It is currently used for high temperature rocket nozzles to eliminate the need for nozzle cooling and for thermal protection of the space shuttle nose and wing leading edges. Many options are available that trade fabrication flexibility and cost, neutron irradiation performance, and coolant compatibility. Table 7.5 compares the properties of most commonly used high temperature materials [36]. It includes nickel-based alloy, ceramic materials as well as carbon and SiC composites. Figure 7.26 shows the specific strength vs. temperature for various composite materials.

Dewson and Li [37] carried out a material selection study of very high temperature reactor (VHTR) intermediate heat exchangers (IHXs) and perhaps the same

Table 7.5 Selected properties of most commonly used high-temperature materials and fabrication technologies [36]

High temp. material/ fabrication technology	Metallic Ni alloys (Inconel 718)	Ceramic oxides of Al, Si, Sr, Ti, Y, Be, Zr, B and SiN, AiN, B4C, BN, WC94/C06	Carbon–carbon composite	Carbon fiber-SiC composite
Temperature range	1200–1250 °C	1500–2500 °C	3300 °C (inert environment) 1400–1650 °C (with SiC layer)	1400–1650 °C
Density	8.19 g/cm^3	1.8–14.95 g/cm^3	2.25 g/cm^3	1.7–2.2 g/cm^3
Hardness	250–410 (Brinell)	400–3000 kgf/mm^2 (V)	0.5–1.0 (Mohs)	2400–3500 (V)
Elongation	<15 %	N/A	N/A	–
Tensile strength	800–1300 MPa	48–2000 MPa	33 (Bulk Mod.)	1400–4500 MPa
Tensile modules	50 GPa	140–600 GPa	4.8 GPa	140–720 GPa
Strength of HE	Strength—adequate, but limited due to creep and thermal exp.	Strength—not adequate, low mechanical parameters for stress. Good thermal and electrical parameters	Strength—poor, oxidation starts at 300 °C	Highest due to carbon fiber and SiC
Electrical conductivity	125 μΩ cm	2E−06–1E+18 Ω cm	1275 μΩ cm	1275 μΩ cm
Thermal conductivity	11.2 W/m K	0.05–300 W/m K	80–240 W/m K	1200 W/m K
Thermal expansion	13E−06 K^{-1}	0.54–10E−06 K^{-1}	0.6–4.3E−06 K^{-1}	–
Comments	Metallic expansion joints are the weak link	Often very expensive fabrication cost for conventional applications. Technology proprietary for the most part. Technologically hard to produce	Life-time is low even protected by SiC (adhesion is poor)	Comparatively less expensive, successful proprietary fabrication technologies available

materials can be looked and used for similar practical purpose and intended operation and application of compact heat exchanges for next generation nuclear plants (NGNP) as well.

They selected and compared the following eight candidate materials based on ASME VIII (Boiler and Pressure Vessel Code): Alloy 617, Alloy 556, Alloy 800H, Alloy 880HT, Alloy 330, Alloy 230, Alloy HX, and 253MA.

Table 7.6 lists the allowable design stress (S) at 898 °C, minimum required mechanical properties (ultimate tensile stress [UTS]), 0.2 % proof stress (PS), and elongation (El) at room temperature, together with the nominal compositions of the

Table 7.6 Candidate materials for VHTR IHXs [37]

Alloys	UNS No.	T_{max} (°C)	S898°C (MPa)	UTS (MPa)	0.2 % PS (MPa)	El (%)	Nominal compositions (wt%)
617	N06617	982	12.4	655	240	30	52Ni-22Cr-13Co-9Mo-1.2A1
556	R30556	898	11.0	690	310	40	21Ni-30Fe-22Cr-18Co-3Mo-3W-0.3A1
800HT	N08811	898	6.3	450	170	30	33Ni-42Fe-21Cr
800H	N08810	898	5.9	450	170	30	33Ni-42Fe-21Cr
330	N08330	898	3.3	483	207	30	Fe-35Ni-19Cr-1.25Si
230	N06230	898	10.3	760	310	40	57Ni-22Cr-14W-2Mo-0.3Al-0.05La
HX	N06002	898	8.3	655	240	35	47Ni-22Cr-9Mo-18Fe
253MA	S30815	898	4.9	600	310	40	Fe-21Cr-11Ni-0.2N

alloys. They extensively compared the mechanical properties, physical properties, and corrosion resistance for the candidate materials, and finally concluded that Alloy 617 and 230 are the most suitable materials for an IHX.

7.11 Guide to Compact Heat Exchangers

As part of our conclusion on subject of compact heat exchangers, we summarize a generic process advantages and limitations of these exchangers design which is applicable to all technologies, such as general advantages and limitations of compact exchangers designs, common applications, fouling, energy efficiency, heat transfer enhancement, exchanger selection and types, specification and operation, process intensification, and finally last but not least software programs.

In general, the final selection of heat exchanger designs is unique for each intended duty and application. In appropriate applications, compact designs offer substantial advantages over more conventional exchanger designs usually shell and tube heat exchangers.

Of course in recent years, the benefits of using compact heat exchanger designs are only realized in appropriate applications as part of combined cycle driven efficiency as part of next generation nuclear plants (NGNP) [2, 4], and researchers around universities, national laboratories, and industry are heavily involved on both aspects of these combined cycle approaches (Open or Closed cycle).

The requirement in most heat transfer applications is to maximize the amount of heat transferred, subject to capital cost and pressure drop constraints. Both cost and heat transfer performance generally increase in proportion to exchange surface area as we discussed in detail above.

7.11.1 Generic Advantages of Compact Design

The main benefits of using compact heat exchangers are:

- Improved heat exchanger thermal effectiveness
- Close approach temperatures
- High heat transfer coefficients and transfer areas per exchanger volume
- Small size
- Multi-stream and multi-pass configurations
- Tighter temperature control
- Energy savings
- Reduced inventory volume and hazard risk
- Process intensification using combined reactors/exchangers

These technological advantages can be converted into reduced operation and capital costs (ROI and TCO), and conserve energy, compared to shell and tube units. The performance characteristics of compact heat exchangers have particular relevance in process integration, for example, in composite process heating and cooling curves. Most details of all above points can be found in different sections of this chapter as well as publication by "Energy Efficiency" [33] and other references that are given at the end of this chapter.

References

1. Mizia, R. E. (2008, April). *Next generation nuclear plant intermediate heat exchanger acquisition strategy*. INL/EXT-08-14054.
2. Zohuri, B. (2015). *Combined cycle driven efficiency for next generation nuclear power plants: An innovative design approach*. New York: Springer.
3. Zohuri, B., & McDaniel, P. (2015). *Thermodynamics in nuclear power plant*. New York: Springer.
4. Zohuri, B. (2014). *Innovative open Air Brayton combined cycle systems for the next generation nuclear power plants*. Albuquerque, NM: University of New Mexico Publications.
5. Zohuri, B., & McDaniel, P. J., de Oliveira, C. R. (2015). Advanced Nuclear Open Air-Brayton Cycles for Highly Efficient Power Conversion. Nuclear Technology, 192(1), 48–60.
6. McDaniel, P. J., Zohuri, B., & de Oliveira, C. R. E. (2014, September). A combined cycle power conversion system for small modular LMFBRs. *ANS Transactions*.
7. Zohuri, B., McDaniel, P., & de Oliveira, C. R. E. (2014, June). A comparison of a recuperated open cycle (Air) Brayton power conversion system with the traditional steam rankine cycle for the next generation nuclear power plant. *ANS Transactions*.
8. McDaniel, P. J., de Oliveira, C. R. E., Zohuri, B., & Cole, J. (2012, November). A combined cycle power conversion system for the next generation nuclear power plant. *ANS Transactions*.
9. Forsberg, C., McDaniel, P., & Zohuri, B. (2015). Variable electricity and steam from salt, helium, and sodium cooled base-load reactors with gas turbines and heat storage. In *Proceedings of ICAPP 2015*, 03–06 May 2015. Nice (France) Paper 15115.
10. Necati Ozisik, M. (1985). *Heat transfer: A basic approach*. New York: McGraw-Hill.

11. McDonald, C. F., & Wilson, D. G. (1996). The utilization of recuperated and regenerated engine cycles for high-efficiency gas turbines in the 21st century. *Applied Thermal Engineering, 16*, 635–653.
12. Kays, W. M., & London, A. L. (1984). *Compact heat exchangers* (3rd ed.). New York: McGraw-Hill.
13. Johnston, A. et al. (1985). Printed circuit heat exchangers for energy conservation. End of Grant Report No. 519, National Energy Research, Development and Demonstration Program, Australia.
14. Johnston, A. et al. (1986, December). Miniaturized heat exchangers for chemical processing. *The Chemical Engineer*.
15. Reay, D. A. (1990). Novel techniques in heat exchanger design including printed circuit units. In: *Symposium on advanced heat exchanger design*. Institute of Chemical Engineering, Leeds, UK.
16. Asadi, M., & Khoshkhoo, R. H. (2013). Investigation into fouling factor in compact heat exchanger. *International Journal of Innovation and Applied Studies, 2*(3), 238–249.
17. Huzayyin, O. A. S. (2011). *Computational modeling of convective heat transfer in compact and enhanced heat exchangers*. University of Cincinnati, 193 pages; 3475151.
18. Zohuri, B. (2011). *Heat pipe design and technology: A practical approach*. Boca Raton, FL: CRC Press.
19. Shah, R. K. *Compact heat exchangers for microturbines*. Rochester Institute of Technology, Dept. of Mechanical Engineering, Rochester, NY.
20. McDonald, C. F. (2003). Recuperator considerations for future higher efficiency microturbines. *Applied Thermal Engineering, 23*, 1463–1487.
21. McDonald, C. F. (2000). *Low cost recuperator concept for microturbine applications*. ASME Paper No. 2000-GT-167, ASME, New York, NY.
22. McDonald, C. F. (2000). Low-cost compact primary surface recuperator concept for microturbines. *Applied Thermal Engineering, 20*, 471–497.
23. Zohuri, B., & Fathi, N. (2015). *Thermal hydraulic analysis of nuclear reactors*. New York: Springer.
24. Stone, K. *Review of literature on heat transfer enhancement in compact heat exchangers*. Air Conditioning & Refrigeration Center, Mechanical & Industrial Engineering Dept., University of Illinois, Urbana, IL.
25. Bergles, A. E., Webb, R. L., & Junkan, G. H. (1979). Energy conservation via heat transfer enhancement. *Energy, 4*, 193–200.
26. Webb, R. L. (1987). Enhancement of single-phase heat transfer. In S. Kakac, R. K. Shah, & A. E. Bergles (Eds.), *Enhancement of single-phase heat transfer* (pp. 17.1–17.62). New York: Wiley.
27. Shah, R. K. (1983). Classification of heat exchangers. In S. Kakac, R. K. Shah, & A. E. Bergles (Eds.), *Low Reynolds number flow heat exchangers* (pp. 9–14). Washington: Hemisphere Publishing Corporation.
28. Shah, R. K., & Sekulic, D. P. (2003). *Fundaments of heat exchangers design*. New York: Wiley.
29. Hesselgreaves, J. E. (2001). *Compact heat exchangers, selection, design and operation*. New York: Pergramon Publication.
30. Kays, W. M., & Crawford, M. E. (1993). *Convective heat and mass transfer* (3rd ed.). New York: McGraw-Hill.
31. London, A. L. (1964). Compact heat exchangers: Part 2—Surface geometry. *Mechanical Engineering, 86*, 31–34.
32. Martin, H. (1999), Economic optimization of compact heat exchangers. In: *Proceedings of the international conference on compact heat exchangers and enhancement technology for the process industries*, Banff. New York: Begell House.
33. Compact heat exchangers—Module 4.1. Energy Efficiency, GIR, Sept 2000.

34. Sabharwall, P., & Kim, E. S. High temperature thermal devices for nuclear process heat transfer applications. USA: Idaho National Laboratory.
35. Sunden, B. (2005). High temperature heat exchangers (HTHE). In: *Proceedings of fifth international conference on enhanced, compact and ultra-compact heat exchangers: Science, engineering and technology*, Hoboken, NJ, USA, Sept 2005.
36. Ohadi, M. M., & Buckey, S. G. (2001). High temperature heat exchangers and microscale combustion systems: applications to thermal system miniaturization. *Experimental Thermal and Fluid Science, 25*, 207–217.
37. Dewson, S., & Li, X. (2005). Selection criteria for the high temperature reactor intermediate heat exchanger. In: *Proceedings of ICAPP'05*, Seoul, Korea, 15–19 May 2005.
38. Brent, A. (1989). *Fundamentals of composites manufacturing materials methods and applications*. Dearborn, MI: SME.
39. Incropera, F. P., DeWitt, D. P., Bergman, T. L., & Lavine, A. S. (2007). *Fundamentals of heat and mass transfer* (6th ed.). New York: Wiley.

Appendix A: Table and Graphs Compilations

The following tables provide samples of engineering data for materials of interest to Thermodynamics. For any detailed design work, more extensive handbooks should be consulted. The thermophysical property data for water, carbon dioxide, and sodium were generated for this book. They represent the best fit to the latest physical measurements. More extensive tables are available from the National Institutes of Standards and Technology (NIST).

A.1 Physical Constants

Acceleration of gravity	$g = 9.80665$ m/s^2 or 32.174 ft/s^2
Sea level atmospheric pressure	$p_{SL} = 101.325$ kPa $= 1.01325$ bar $= 14.696$ psia $= 760$ mmHg (0 °C) $= 29.9213$ in Hg (32 °F) $= 10.3323$ m H_2O (4 °C)
Boltzmann's constant	$k = 1.3806503 \times 10^{-23}$ J/K
Avogadro's number	$N_a = 6.02214199 \times 10^{23}$ 1/g-mol
Electronic charge	$q_e = 1.60217646 \times 10^{-19}$ C
Electron volt	$\varepsilon = 1.60217646 \times 10^{-19}$ J
Atomic mass unit	amu $= 1.6605402 \times 10^{-27}$ kg $= 931.49432$ MeV/c^2
Planck's constant	$h = 6.62606876 \times 10^{-34}$ J-s $\hbar = h/2\pi = 1.054571596 \times 10^{-34}$ J-s
Stefan–Boltzmann constant	$\sigma = 0.17123 \times 10^{-8}$ Btu/(h-ft^2-°R^4) $\sigma = 5.670400 \times 10^{-8}$ W/(m^2-K^4)
Speed of light in vacuum	$c = 2.99792458 \times 10^8$ m/s

(continued)

B. Zohuri, *Application of Compact Heat Exchangers For Combined Cycle Driven Efficiency In Next Generation Nuclear Power Plants*,
DOI 10.1007/978-3-319-23537-0

(continued)

Solar constant	$q_{solar} = 429$ Btu/(h-ft^2) $q^{solar} = 1353$ W/m^2
Universal gas constant	$R = 8.31447$ kJ/(kgmol-K) $= 8.31447$ kPa-m^3/(kgmol-K) $= 0.0831447$ bar-m^3/(kgmol-K) $= 82.05$ L-atm/(kgmol-K) $= 1.98583$ Btu/(lbmol-°R) $= 1545.37$ ft-lbf/(lbmol-°R) $= 10.73$ psia-ft^3/(lbmol-°R)

A.2 Conversion Factors

To convert	To	Multiply by
Distance		
millimeters	meters	10^{-3}
centimeters	meters	10^{-2}
kilometers	meters	10^3
inches	feet	1/12
feet	yards	1/3
feet	miles	1/5280
yards	miles	1/1760
inches	centimeters	2.54
feet	meters	0.3048
Area		
millimeters^2	meters^2	10^{-6}
centimeters^2	meters^2	10^{-4}
kilometers^2	meters^2	10^6
inches^2	feet^2	1/144
inches^2	centimeters^2	6.4516
centimeters^2	inches^2	0.1550
feet^2	meters^2	0.092903
meters^2	feet^2	10.7639
Volume		
milimeters^3	meters^3	10^{-9}
centimeters^3	meters^3	10^{-6}
liters	meters^3	10^{-3}
kilometers^3	meters^3	10^9
inches^3	centimeters^3	16.3871
centimeters^3	inches^3	0.061024
feet^3	meters^3	0.0283168
meters^3	feet^3	35.3147

(continued)

(continued)

To convert	To	Multiply by
Force		
kg-m/s^2	newtons	1.0
pound-mass-ft/s^2	pounds-force	1/32.1745
newtons	pounds-force	4.4482
pounds-force	newtons	0.2248
Mass and density		
grams	kilograms	10^{-3}
metric tonnes	kilograms	10^3
ounces	pounds-mass	1/16
tons	pounds-mass	2000
grams	ounces	1/28.35
ounces	grams	28.35
pounds-mass	kilograms	0.4536
kilograms	pounds-mass	2.2046
tons	metric tonnes	0.9072
metric tonnes	tons	1.1023
grams/cm^3	lbm/ft^3	62.427
grams/cm^3	kg/m^3	10^3
lb/ft^3	kg/m^3	16.0187
Pressure		
psi	in Hg	2.036
psi	in H_2O	27.7
atm	in Hg	29.92
atm	ft H_2O	33.93
atm	Pa	101,320
atm	bar	1.0133
atm	psi	14.69
kPa	psi	0.145
psi	kPa	6.895
Energy		
joules	ergs	107
lbf-ft	joules	1.356
Btu	joules	1055
Btu	ft-lbf	778
Cal	Btu	0.003968
Cal	ft-lbf	3.088
Btu	W-h	0.2930
kW-h	Btu	3412
Cal	joules	4.1868
kJ/kg	Btu/lbm	0.4992
Btu/lbm	kJ/kg	2.3260
Cal/g	Btu/lbm	1.8000

(continued)

(continued)

To convert	To	Multiply by
Power		
hp	ft-lbf/s	550
hp	Btu/h	2545
hp	kW	0.7455
W	Btu/h	3.412
kW	hp	1.341
ton	Btu/h	12,000
ton	kW	3.517
Heat transfer		
W/m/K	Btu/h/ft/°R	0.57779
kcal/h/m/K	W/m/K	1.1630
kcal/h/m/K	Btu/h/ft/°R	0.67197
Btu/h/ft/°R	W/m/K	1.7307
W/m^2/K	Btu/h/ft^2/°R	0.17611
Btu/h/ft^2/°R	W/m^2/K	5.6783
Fluid flow		
m^3/s	ft^3/s	35.3147
g/cm/s (poise)	lbf-s/ft^2	0.002088
lbf-s/ft^2	g/cm/s	478.96
lbf-s/ft^2	kg/m/h	172,400
lbm/ft/s	gm/cm/s	14.882
m^2/s	ft^2/s	10.7639
cm^2/s	ft^2/h	3.875
ft^2/s	m^2/h	334.45

A.3 Standard Atmosphere

A.3.1: SI Units

$p_{SL} = 101{,}325$ Pa			
$T_{SL} = 288.2$ K			
$\rho_{SL} = 1.225$ kg/m^3			
Altitude (km)	p/p_{SL}	T/T_{SL}	ρ/ρ_{SL}
0	1	1	1
1	0.887	0.9774	0.9075
2	0.7846	0.9549	0.8217
3	0.692	0.9324	0.7423
4	0.6085	0.9098	0.6689
5	0.5334	0.8873	0.6012

(continued)

(continued)

$p_{SL} = 101{,}325$ Pa			
$T_{SL} = 288.2$ K			
$\rho_{SL} = 1.225$ kg/m^3			
Altitude (km)	p/p_{SL}	T/T_{SL}	ρ/ρ_{SL}
6	0.466	0.8648	0.5389
7	0.4057	0.8423	0.4817
8	0.3519	0.8198	0.4292
9	0.304	0.7973	0.3813
10	0.2615	0.7748	0.3376
11	0.224	0.7523	0.2978
12	0.1915	0.7519	0.2546
13	0.1636	0.7519	0.2176
14	0.1399	0.7519	0.186
15	0.1195	0.7519	0.159
16	0.1022	0.7519	0.1359

A.3.2: English Units

$p_{SL} = 2116$ lbf/ft^3			
$T_{SL} = 518.7$ °R			
$\rho_{SL} = 0.07647$ lbm/ft^3			
Altitude (kft)	p/p_{SL}	T/T_{SL}	ρ/ρ_{SL}
0	1.0	1.0	1.0
2	0.9298	0.9863	0.9428
4	0.8637	0.9725	0.8881
6	0.8014	0.9588	0.8359
8	0.7429	0.945	0.7861
10	0.6878	0.9313	0.7386
12	0.6362	0.9175	0.6933
14	0.5877	0.9038	0.6502
16	0.5422	0.8901	0.6092
18	0.4997	0.8763	0.5702
20	0.4599	0.8626	0.5332
22	0.4227	0.8489	0.498
24	0.388	0.8352	0.4646
26	0.3557	0.8215	0.433
28	0.3256	0.8077	0.4031
30	0.2975	0.794	0.3747
32	0.2715	0.7803	0.348
34	0.2474	0.7666	0.3227
36	0.225	0.7529	0.2988

(continued)

(continued)

$p_{SL} = 2116$ lbf/ft^3			
$T_{SL} = 518.7$ °R			
$\rho_{SL} = 0.07647$ lbm/ft^3			
Altitude (kft)	p/p_{SL}	T/T_{SL}	ρ/ρ_{SL}
38	0.2044	0.7519	0.2719
40	0.1858	0.7519	0.2471
42	0.1688	0.7519	0.2245
44	0.1534	0.7519	0.204
46	0.1394	0.7519	0.1854
48	0.1267	0.7519	0.1685
50	0.1151	0.7519	0.1531

A.4 Critical State Properties of Gases

		Molar (Mass)	P_{crit} (MPa)	T_{crit} (K)	P_{crit} (atm)	T_{crit} (°R)	v_{crit} (cm^3/g-mol)	Z_{crit} (cm^3/g-mol)
Air		28.996	3.77	133.0	37.21	239.0	88.3	0.300
Ammonia	NH_3	17.031	11.28	405.5	111.32	729.8	72.5	0.243
Argon	Ar	39.950	4.86	151.0	47.96	272.0	75.2	0.291
Carbon dioxide	CO_2	44.010	7.39	304.2	72.93	547.5	94.0	0.293
Carbon monoxide	CO	28.010	3.50	133.0	34.54	240.0	93.1	0.294
Chlorine	Cl_2	70.906	7.99	416.9	78.87	750.4	124.0	0.276
Helium	He	4.003	0.23	5.3	2.27	9.5	57.5	0.303
Hydrogen	H_2	2.016	1.30	33.3	12.83	59.9	65.0	0.304
Iodine	I2	253.809	11.70	819.0	115.47	1474.2	167.0	0.287
Krypton	Kr	83.800	5.50	209.4	54.28	376.9	92.2	0.291
Lithium	Li	6.941	67.00	3223.0	661.24	5801.4	120.0	0.300
Mercury	Hg	200.590	150.97	1763.2	1490.0	3173.7	29.1	0.300
Methane	Ch_4	16.043	4.64	191.1	45.79	343.9	99.0	0.290
Neon	Ne	20.180	2.73	44.5	26.94	227.1	41.7	0.307
Nitrogen	N_2	28.013	3.39	126.2	33.46	227.1	90.1	0.291
Oxygen	O_2	31.999	5.08	154.8	50.14	278.6	73.4	0.288
Potassium	K	39.098	16.42	2105.9	162.02	3790.7	320.0	0.300
Propane	C_3H_8	44.097	4.26	370.0	42.04	665.9	200.0	0.277
Sodium	Na	22.990	25.60	2503.9	252.65	4507.0	244.0	0.300
Water	H_2O	18.015	22.10	647.4	218.11	1165.3	56.0	0.230
Water (Heavy)	D_2O	20.023	21.72	644.7	214.36	1160.4	54.9	0.222
Xenon	Xe	131.290	5.84	289.8	57.65	521.6	118.8	0.290

A.5 Constants for van der Waals Equation of State

Substance		Molar (Mass)	a (kPa-m^6/kmol^2)	b (m^3/kmol)	a (atm-ft^6/lbmol^2)	b (ft^3/lbmol)
Air		28.996	136.83	0.03666	345.21	0.58621
Ammonia	NH_3	17.031	425.09	0.03736	1075.80	0.59826
Argon	Ar	39.95	136.81	0.03229	346.85	0.51752
Carbon dioxide	CO_2	44.0098	365.16	0.04278	924.18	0.68507
Carbon monoxide	CO	28.0104	147.38	0.03949	374.96	0.63407
Chlorine	Cl_2	70.906	634.26	0.05422	1605.62	0.86835
Helium	He	4.0026	3.56	0.02395	8.94	0.38193
Hydrogen	H_2	2.0158	24.87	0.02662	62.88	0.42607
Iodine	I2	253.809	1671.81	0.07275	4232.15	1.16510
Krypton	Kr	83.8	232.51	0.03957	588.59	0.63372
Lithium	Li	6.941	4521.16	0.04999	11,445.24	0.80067
Mercury	Hg	200.59	600.48	0.01214	1520.11	0.19439
Methane	Ch_4	16.043	229.51	0.04280	580.74	0.68534
Neon	Ne	20.1797	21.15	0.01694	430.43	0.76922
Nitrogen	N_2	28.0134	137.00	0.03869	346.63	0.61946
Oxygen	O_2	31.9988	137.56	0.03167	348.12	0.50712
Potassium	K	39.0983	7877.53	0.13331	19,942.70	2.13513
Propane	C_3H_8	44.097	937.13	0.09026	2371.61	1.44542
Sodium	Na	22.9898	7141.57	0.10165	18,078.79	1.62795
Water	H_2O	18.0152	553.04	0.03044	1399.97	0.48757
Water (Heavy)	D_2O	20.023	557.95	0.03084	1412.43	0.49400
Xenon	Xe	131.29	419.20	0.05156	1061.20	0.82572
		$R = 8314$ J/kmol/K = 1545 ft-lbf/lbmol/°R = 0.73023 atm-ft^3/lbmol/°R				

A.6 Constants for Redlich–Kwong Equation of State

Substance		Molar (Mass)	a (kPa-m^6-K$^{1/2}$/kmol^2)	b (m^3/kmol)	a (atm-ft^6-°R$^{1/2}$/lbmol^2)	b (ft^3/lbmol)
Air		28.996	1598.9	0.02541	5407.8	0.40631
Ammonia	NH_3	17.031	8673.7	0.02589	29,448.7	0.41466
Argon	Ar	39.95	1703.5	0.02238	5796.3	0.35870
Carbon dioxide	CO_2	44.0098	6453.4	0.02965	21,912.0	0.47483
Carbon monoxide	CO	28.0104	1722.3	0.02737	5886.1	0.43949
Chlorine	Cl_2	70.906	13,122.4	0.03758	44,568.3	0.60187
Helium	He	4.0026	8.3	0.01660	27.9	0.26473

(continued)

(continued)

Substance		Molar (Mass)	*a* (kPa-m^6-K$^{1/2}$/kmol^2)	*b* (m^3/kmol)	*a* (atm-ft^6-°R$^{1/2}$/lbmol^2)	*b* (ft^3/lbmol)
Hydrogen	H_2	2.0158	145.4	0.01845	493.2	0.29531
Iodine	I2	253.809	48,479.7	0.05042	164,653.6	0.80756
Krypton	Kr	83.8	3409.3	0.02743	11,579.2	0.43925
Lithium	Li	6.941	260,082.9	0.03465	883,330.7	0.55496
Mercury	Hg	200.59	25,549.3	0.00841	86,774.1	0.13473
Methane	Ch_4	16.043	3214.9	0.02967	10,912.6	0.47502
Neon	Ne	20.1797	143.0	0.01174	6572.7	0.53316
Nitrogen	N_2	28.0134	1559.5	0.02682	5293.1	0.42936
Oxygen	O_2	31.9988	1734.2	0.02195	5887.8	0.35150
Potassium	K	39.0983	366,303.6	0.09240	1,244,158.2	1.47990
Propane	C_3H_8	44.097	18,265.5	0.06256	62,012.5	1.00185
Sodium	Na	22.9898	362,104.2	0.07045	1,229,829.9	1.12836
Water	H_2O	18.0152	14,258.5	0.02110	48,424.8	0.33795
Water (Heavy)	D_2O	20.023	14,354.4	0.02138	48,752.6	0.34240
Xenon	Xe	131.29	7230.7	0.03574	24,558.0	0.57232
		$R = 8314$ J/kmol/K = 1545 ft-lbf/lbmol/°R = 0.73023 atm-ft^3/lbmol/°R				

A.7 Constants for the Peng–Robinson Equation of State

Substance		Molar (Mass)	*a* (kPa-m^6/kmol^2)	*b* (m^3/kmol)	*a* (atm-ft^6/lbmol^2)	*b* (ft^3/lbmol)	ω
Air		28.996	148.29	0.02282	374.15	0.36484	0.032
Ammonia	NH_3	17.031	460.72	0.02325	1165.97	0.37234	0.252
Argon	Ar	39.95	148.28	0.02010	375.92	0.32209	−0.004
Carbon dioxide	CO_2	44.0098	395.76	0.02662	1001.65	0.42636	0.225
Carbon monoxide	CO	28.0104	159.73	0.02458	406.39	0.39462	0.049
Chlorine	Cl_2	70.906	687.42	0.03374	1740.20	0.54044	0.073
Helium	He	4.0026	3.86	0.01490	9.69	0.23770	0
Hydrogen	H_2	2.0158	26.96	0.01657	68.16	0.26517	−0.215
Iodine	I_2	253.809	1811.93	0.04528	4586.88	0.72512	0.07
Krypton	Kr	83.8	252.00	0.02463	637.92	0.39441	0.001
Lithium	Li	6.941	4900.11	0.03111	12,404.55	0.49831	−0.153
Mercury	Hg	200.59	650.81	0.00755	1647.52	0.12098	−0.139
Methane	CH_4	16.043	248.75	0.02664	629.41	0.42654	0.008
Neon	Ne	20.1797	22.93	0.01054	466.51	0.47874	−0.041
Nitrogen	N_2	28.0134	148.48	0.02408	375.69	0.38553	0.04

(continued)

(continued)

Substance		Molar (Mass)	*a* (kPa-m^6/kmol^2)	*b* (m^3/kmol)	*a* (atm-ft^6/lbmol^2)	*b* (ft^3/lbmol)	ω
Oxygen	O_2	31.9988	149.09	0.01971	377.30	0.31562	0.021
Potassium	K	39.0983	8537.81	0.08297	21,614.24	1.32884	−0.031
Propane	C_3H_8	44.097	1015.67	0.05618	2570.39	0.89958	0.152
Sodium	Na	22.9898	7740.16	0.06326	19,594.10	1.01318	−0.076
Water	H_2O	18.0152	599.40	0.01895	1517.31	0.30345	0.344
Water (Heavy)	D_2O	20.023	604.71	0.01920	1530.82	0.30745	0.361
Xenon	Xe	131.29	454.34	0.03209	1150.15	0.51390	0.012
		$R = 8314$ J/kmol/K = 1545 ft-lbf/lbmol/°R = 0.73023 atm-ft^3/lbmol/°R					

A.8 Thermophysical Properties of Solids

	Melting			(C_p in J/kg/K)				(*k* in W/m/K)			
	Point (K)	Density									
Material			*T* (K)	200	300	400	600	800	1000	1200	1500
Aluminum (Pure)	933	2702	*k*	237	237	240	231	218			
			C_p	798	903	949	1033	1146			
Aluminum (2024-T6)	775	2770	*k*	163	177	186	186				
			C_p	787	875	925	1042				
Beryllium	1550	1850	*k*	301	200	161	126	106	90.8	78.7	
			C_p	1114	1825	2191	2604	2823	3018	3227	3519
Bismuth	545	9780	*k*	9.69	7.86	7.04					
			C_p	120	122	127					
Boron	2573	2500	*k*	55.5	27	16.8	10.6	9.6	9.85		
			C_p	600	1107	1463	1892	2160	2338		
Cadmium	594	8650	*k*	99.3	96.8	94.7					
			C_p	222	231	242					
Chromium	2118	7160	*k*	111	93.7	90.9	80.7	71.3	65.4	61.9	57.2
			C_p	384	449	484	542	581	616	682	779
Copper (Pure)	1358	8933	*k*	413	401	393	379	366	352	339	
			C_p	356	385	397	417	433	451	480	
Iron (Pure)	1810	7870	*k*	94	80.2	69.5	54.7	43.3	32.8	28.3	32.1
			C_p	384	447	490	574	680	975	609	654
Carbon Steel		7854	*k*		60.5	56.7	48	39.2	31.3		
			C_p		434	487	559	685	1169		
302 Stainless		8055	*k*		15.1	17.3	20	22.8	25.4		
			C_p		480	512	559	585	606		
304 Stainless	1670	7900	*k*	12.6	14.9	16.6	19.8	22.6	25.4	28	31.7
			C_p	402	477	515	557	582	611	640	682

(continued)

(continued)

	Melting			(C_p in J/kg/K)				(k in W/m/K)			
	Point (K)	Density									
Material			T (K)	200	300	400	600	800	1000	1200	1500
316 Stainless		8238	k		13.4	15.2	18.3	21.3	24.2		
			C_p		468	504	550	576	602		
347 Stainless		7978	k		14.2	15.8	18.9	21.9	24.7		
			C_p		480	513	559	585	606		
Lead	601	11,340	k	36.7	35.3	34	31.4				
			C_p	125	129	132	142				
Lithium	454	534	k		84.8						
			C_p		3580						
Magnesium	923	1740	k	159	156	153	149	146			
			C_p	934	1024	1074	1170	1267			
Molybdenum	2894	10,240	k	143	138	134	126	118	112	105	98
			C_p	224	251	261	275	285	295	308	330
Nickel (Pure)	1728	8900	k	107	90.7	80.2	65.6	67.6	71.8	76.2	82.6
			C_p	383	444	485	592	530	562	594	616
Inconel X-750	1665	8510	k	10.3	11.7	13.5	17	20.5	24	27.6	33
			C_p	372	439	473	510	546	626		
Niobium	2741	8570	k	52.6	53.7	55.2	58.2	61.3	64.4	67.5	72.1
			C_p	249	265	274	283	292	301	310	324
Palladium	1827	12,020	k	71.6	71.8	73.6	79.7	86.9	94.2	102	110
			C_p	227	244	251	261	271	281	291	307
Platinum (Pure)	2045	21,450	k	72.6	71.6	71.8	73.2	75.6	78.7	82.6	89.5
			C_p	125	133	136	141	146	152	157	165
60% Pt–40% Rh	1800	16,630	k		47	52	59	65	69	73	76
			C_p		162						
Plutonium	913	19,860	k		25.1						
			C_p		104.4						
Potassium	337	862	k		102.5						
			C_p		757.1						
Rhenium	3453	21,100	k	51	47.9	46.1	44.2	44.1	44.6	45.7	47.8
			C_p	127	136	139	145	151	156	162	171
Rhodium	2236	12,450	k	154	150	146	136	127	121	116	110
			C_p	220	243	253	274	293	311	327	349
Silver	1235	10,500	k	430	429	425	412	396	379	361	
			C_p	225	235	239	250	2623	277	292	
Sodium	371	986	k		142						
			C_p		1084						
Tantalum	3269	16,600	k	57.5	57.5	57.8	58.6	59.4	60.2	61	62.2
			C_p	133	140	144	146	149	152	155	160
Thorium	2023	11,700	k	54.6	54	54.5	55.8	56.9	56.9	58.7	
			C_p	112	118	124	134	145	156	167	
Tin	505	7310	k	73.3	66.6	62.2					
			C_p	215	227	243					
Titanium	1953	4500	k	24.5	21.9	20.4	19.4	19.7	20.7	22	24.5
			C_p	465	522	551	591	633	675	620	686

(continued)

(continued)

	Melting			(C_p in J/kg/K)				(k in W/m/K)			
	Point (K)	Density									
Material			T (K)	200	300	400	600	800	1000	1200	1500
Tungsten	3660	19,300	k	186	174	159	137	125	118	113	107
			C_p	122	132	137	142	145	148	152	157
Uranium	1406	19,070	k	25.1	27.6	29.6	34	38.8	43.9	49	
			C_p	108	116	125	146	176	180	161	
Vanadium	2192	6100	k	31.3	30.7	31.3	33.3	35.7	38.2	40.8	44.6
			C_p	430	489	515	540	563	597	645	714
Zinc	693	7140	k	118	116	111	103				
			C_p	367	389	402	436				
Zirconium	2125	6570	k	25.2	22.7	21.6	20.7	21.6	23.7	26	28.8
			C_p	264	278	300	322	342	362	344	344
Beryllium oxide	2725	3000	k		272	196	111	70	47	33	21.5
			C_p		1030	1350	1690	1865	1975	2055	2145
Carbon	1500	1950	k	1.18	1.6	1.89	2.19	2.37	2.53	2.84	3.48
Amorphous			C_p		509						
Graphite	2273	2210	k		5.7	16.8	9.23	4.09	2.68	2.01	1.6
Pyrolytic			C_p	411	709	992	1406	1650	1793	1890	1974
Silicon carbide	3100	3160	k		490				87	58	30
			C_p		675	880	1050	1135	1195	1243	1310
Silicon dioxide	1883	2220	k	1.14	1.38	1.51	1.75	2.17	2.87	4	
			C_p		745	905	1040	1105	1155	1195	
Silicon nitride	2173	2400	k		16	13.9	11.3	9.88	8.76	8	7.16
			C_p	578	691	778	937	1063	1155	1226	1306
Thorium dioxide	3573	9110	k		13	10.2	6.6	4.7	3.68	3.12	2.73
			C_p		235	255	274	285	295	303	315
Uranium dioxide	3138	10,980	k	13.1	10.05	8.17	5.95	4.67	3.849	3.27	2.67
			C_p	278	277.4	277.3	277	277	277.2	277	277
Plutonium dioxide	2673	11,460	k	16.3	11.17	8.489	5.73	4.33	3.477	2.91	2.33
			C_p		276.1						
Concrete		2300	k		1.4						
			C_p		880						
Glass		2500	k		1.4						
			C_p		750						
Ice	273	920	k	2.03	1.88						
			C_p	1945	2040						
Paraffin		900	k		0.24						
			C_p		2890						
Paper		930	k		0.18						
			C_p		1340						
Sand		1515	k		0.27						
			C_p		800						
Soil		2050	k		0.52						
			C_p		1840						

A.9 Thermophysical Properties of Liquids

	Temp	**273**	**280**	**300**	**320**	**340**	**360**	**400**
Engine oil	ρ (kg/m^3)	899.1	895.3	884.1	871.8	859.9	847.8	825.1
	C_p (J/kg/K)	1796	1827	1909	1993	2076	2161	2337
	μ (N-s/m^2)	3.85	2.17	0.486	0.141	0.00531	0.00252	0.00087
	k (W/m/K)	0.147	0.144	0.145	0.143	0.139	0.138	0.134
	Temp	**273**	**300**	**350**	**400**	**450**	**500**	**600**
Water (liquid)	ρ (kg/m^3)	1000.0	997.0	973.7	937.2	890.5	831.3	648.9
$T_m = 273$	C_p (J/kg/K)	4217.0	4179.0	4195.0	4256.0	4400.0	4660.0	7000.0
$T_b = 373$	μ (N-s/m^2)	1.75E−03	8.55E−04	3.43E−04	2.17E−04	1.52E−04	1.18E−04	8.10E−05
	k (W/m/K)	0.569	0.613	0.668	0.688	0.678	0.642	0.497
	Temp	**273**	**300**	**350**	**400**	**450**	**500**	**600**
Water (vapor)	ρ (kg/m^3)	0.0048	0.0256	0.2600	1.3680	4.8077	13.05	72.99
$T_m = 273$	C_p (J/kg/K)	1854	1872	1954	2158	2560	3270	8750
$T_b = 373$	μ (N-s/m^2)	8.02E−06	9.09E−06	1.11E−05	1.31E−05	1.49E−05	1.66E−05	2.27E−05
	k (W/m/K)	0.0182	0.0196	0.023	0.0272	0.0331	0.0423	0.103
	Temp	**273**	**353**					
Heavy water (liquid)	ρ (kg/m^3)	1105.4	1078.2					
$T_m = 277$	C_p (J/kg/K)	4207.7	4178.4					
$T_b = 376$	μ (N-s/m^2)							
	k (W/m/K)	0.5931	0.632					
	Temp	**313.2**	**300**	**350**	**400**	**450**	**500**	**600**
Heavy water (vapor)	ρ (kg/m^3)	0.0058	0.0250	0.2717	1.4771	5.2966	14.75	83.96
$T_m = 277$	C_p (J/kg/K)	1694.8	1712	1743.4	1779.4	1817.1	1856.4	1938.9
$T_b = 376$	μ (N-s/m^2)							
	k (W/m/K)	0.0187	0.0202	0.0237	0.02884	0.0309	0.0371	0.0482

(continued)

(continued)

	Temp	**473**	**673**	**873**	**1073**	**1273**		
Lithium	ρ (kg/m^3)	508.9	491.2	475.1	457.5	441.4		
$T_m = 452$ K	C_p (kJ/kg/K)	5861.5	4563.6	3810	3056.4	2302.7		
$T_b = 1590$ K	μ (N-s/m^2)	5.65E−04	4.56E−04	4.56E−04	4.56E−04	4.56E−04		
	k (W/m/K)	46.33	39.49	25.93	11.12	10.38		
	Temp	**477.8**	**588.9**	**700.0**	**811.1**	**922.2**	**1033.3**	**1154.8**
Sodium	ρ (kg/m^3)	904.4	878.1	851.6	824.9	798.1	771.1	741.6
$T_m = 371$	C_p (J/kg/K)	1.338	1.300	1.274	1.259	1.255	1.263	1.284
$T_b = 1156$ K	μ (N-s/m^2)	4.52E−04	3.33E−04	2.66E−04	2.26E−04	1.97E−04	1.74E−04	1.57E−04
	k (W/m/K)	81.52	75.81	70.27	65.08	60.23	55.56	50.88
	Temp	**600**	**800**	**1000**	**1200**	**1392.6**	**1400**	**1800**
		588.9	700.0	811.1	922.2	1029.2	1033.3	1255.6
Potassium	ρ (kg/m^3)	771.9	745.8	719.1	692.0	665.4	664.2	607.7
$T_m = 337$	C_p (J/kg/K)	0.783	0.791	0.804	0.825	0.854	0.854	0.934
$T_b = 1033$ K	μ (N-s/m^2)	2.54E−04	2.06E−04	1.71E−04	1.47E−04	1.31E−04	1.31E−04	1.09E−04
	k (W/m/K)	43.61	39.98	36.69	33.75	30.81	30.81	25.10
	Temp	**366**	**644**	**977**				
NaK (45/50)	ρ (kg/m^3)	887.4	821.7	740.1				
$T_m = 292$ K	C_p (J/kg/K)	1130	1055	1043				
$T_b = 1098$ K	μ (N-s/m^2)	5.79E−04	2.36E−04	1.61E−04				
	k (W/m/K)	25.6	27.5	28.9				
	Temp	**366**	**672**	**1033**				
NaK (22/78)	ρ (kg/m^3)	849	775.3	690.4				
$T_m = 262$ K	C_p (J/kg/K)	946	879	883				
$T_b = 1057$ K	μ (N-s/m^2)	4.92E−04	2.07E−04	1.46E−04				
	k (W/m/K)	24.4	26.7	29.4				
	Temp	**644**	**755**	**977**				

(continued)

(continued)

Lead	ρ (kg/m^3)	10,540	10,412	10,140				
$T_m = 600$ K	C_p (J/kg/K)	159	155	151				
$T_b = 2010$ K	μ (N-s/m^2)	2.39E−03	1.93E−03	1.37E−03				
	k (W/m/K)	16.1	15.6	14.9				
	Temp	**273**	**300**	**350**	**400**	**450**	**500**	**600**
Mercury	ρ (kg/m^3)	13,595	13,529	13,407	13,287	13,167	13,048	12,809
$T_m = 234$ K	C_p (J/kg/K)	140.4	139.3	137.7	136.5	135.7	135.3	135.5
$T_b = 630$ K	μ (N-s/m^2)	1.69E−03	1.52E−03	1.31E−03	1.17E−03	1.08E−03	1.01E−03	9.11E−04
	k (W/m/K)	8.18	8.54	9.18	9.8	10.4	10.95	11.95
	Temp	**589**	**811**	**1033**				
Bismuth	ρ (kg/m^3)	10,011	9739	9467				
$T_m = 544$ K	C_p (J/kg/K)	144.4	154.5	164.5				
$T_b = 1750$ K	μ (N-s/m^2)	1.62E−03	1.10E−03	7.90E−04				
	k (W/m/K)	16.4	15.6	15.6				
	Temp	**422**	**644**	**922**				
PbBi (44.5/55.5)	ρ (kg/m^3)	10,524	10,236	9835				
$T_m = 398$ K	C_p (J/kg/K)	147	147	147				
$T_b = 1943$ K	μ (N-s/m^2)	1.85E−03	1.53E−03	1.15E−03				
	k (W/m/K)	9.05	11.86	15.3788				

A.10 Thermophysical Properties of Gases

	Temp (K)	**250**	**350**	**500**	**650**	**800**	**1000**	**1200**	**1400**	**1600**	**1800**
Air	μ (N-s/m^2)	1.60E−05	2.08E−05	2.70E−05	3.23E−05	3.70E−05	4.24E−05	4.73E−05	5.30E−05	5.84E−05	6.37E−05
	k (W/m/K)	0.0223	0.0300	0.0407	0.0497	0.0573	0.0667	0.0763	0.0910	0.1060	0.1200
	Temp (K)	**280**	**300**	**350**	**400**	**450**	**500**	**550**	**600**	**650**	**700**
Carbon dioxide	μ (N-s/m^2)	1.40E−05	1.49E−05	1.69E−05	1.90E−03	2.10E−05	2.31E−05	2.51E−05	2.70E−05	2.88E−05	3.05E−05
	k (W/m/K)	0.0152	0.0166	0.0205	0.0243	0.0283	0.0325	0.0366	0.0407	0.0445	0.0481
	Temp (K)	**250**	**300**	**350**	**400**	**450**	**500**	**550**	**600**	**650**	**700**
Carbon monoxide	μ (N-s/m^2)	1.52E−05	1.75E−05	1.98E−05	2.18E−05	2.37E−05	2.54E−05	2.71E−05	2.86E−05	3.01E−05	3.15E−05
	k (W/m/K)	0.0214	0.0250	0.0285	0.0318	0.0350	0.0381	0.0411	0.0440	0.0470	0.0500
	Temp (K)	**250**	**300**	**350**	**400**	**450**	**500**	**600**	**700**	**800**	**900**
Helium	μ (N-s/m^2)	1.70E−05	1.99E−05	2.21E−05	2.43E−05	2.63E−05	2.83E−05	3.20E−05	3.50E−05	3.82E−05	4.14E−05
	k (W/m/K)	0.1335	0.1520	0.1700	0.1870	0.2040	0.2200	0.2520	0.2780	0.3040	0.3300
	Temp (K)	**250**	**300**	**350**	**400**	**500**	**600**	**800**	**1000**	**1200**	**1400**
Hydrogen	μ (N-s/m^2)	7.89E−06	8.96E−06	9.88E−06	1.08E−05	1.26E−05	1.42E−05	1.72E−05	2.01E−05	2.26E−05	2.51E−05
	k (W/m/K)	0.1570	0.1830	0.2040	0.2260	0.2660	0.3050	0.3780	0.4480	0.5280	0.6100
	Temp (K)	**250**	**300**	**350**	**400**	**500**	**600**	**700**	**800**	**900**	**1100**
Nitrogen	μ (N-s/m^2)	1.55E−05	1.78E−05	2.00E−05	2.22E−05	2.58E−05	2.91E−05	3.21E−05	3.49E−05	3.75E−05	4.23E−05
	k (W/m/K)	0.0222	0.0259	0.0293	0.0327	0.0389	0.0446	0.4990	0.0548	0.0597	0.0700
	Temp (K)	**250**	**300**	**350**	**400**	**500**	**600**	**700**	**800**	**900**	**1100**
Oxygen	μ (N-s/m^2)	1.79E−05	2.07E−05	2.34E−05	2.58E−05	3.03E−05	3.44E−05	3.81E−05	4.15E−05	4.47E−05	5.06E−05
	k (W/m/K)	0.0226	0.0268	0.0296	0.0330	0.0412	0.0473	0.0528	0.0589	0.0649	0.0758
	Temp (K)	**380**	**400**	**450**	**500**	**550**	**600**	**650**	**700**	**750**	**800**
Water	μ (N-s/m^2)	1.27E−05	1.34E−05	1.53E−05	1.70E−05	1.88E−05	2.07E−05	2.25E−05	2.43E−05	2.60E−05	2.79E−05
	k (W/m/K)	0.0246	0.0261	0.0299	0.0339	0.0379	0.0422	0.0464	0.0505	0.0549	0.0592

A.11 Ideal Gas Heat Capacities for Selected Gases

	τ	C_0^*	C_1	C_2	C_3	C_4	Range (°R)	Range (K)	Max error (%)
Air	0.0140	9.0761	−3.4743	1.7804	−0.3134	0.0188	100–6200	56–3444	0.77
100 % Air + CP(CH_1)	0.0400	9.5147	−3.9901	2.1908	−0.3898	0.0229	300–4000	167–2222	0.32
100 % Air + CP(CH_2)	0.0400	9.7159	−4.0842	2.1394	−0.3635	0.0202	300–4000	167–2222	0.35
100 % Air + CP(CH_3)	0.0400	9.8608	−4.1581	2.1088	−0.3464	0.0184	300–4000	167–2222	0.37
Carbon dioxide	1.0000	−5.8291	12.0893	−2.7155	0.2747	−0.0102	300–5800	167–3222	0.18
Carbon monoxide	0.0750	12.1297	−7.2191	3.4222	−0.6162	0.0387	300–5800	167–3222	0.36
Hydrogen	0.5000	12.2786	−4.7650	1.4175	−0.1407	0.0037	300–5800	167–3222	0.48
Nitrogen	0.1000	12.2568	−7.0276	3.1766	−0.5465	0.0328	300–6400	167–3556	0.38
Oxygen	0.0005	8.5200	−3.1865	1.9424	−0.3855	0.0261	300–6400	167–3556	0.98
Water	0.1500	15.8257	−10.0486	4.3395	−0.6669	0.0355	300–6400	167–3556	0.28
Hydroxyl ion	0.0025	11.4039	−4.7972	1.7247	−0.2252	0.0099	540–6300	300–3500	0.62
Nitrous oxide	7.5500	−21.2671	28.5044	−9.0836	1.2789	−0.0676	540–6300	300–3500	0.71
Nitrogen dioxide	1.5000	−16.0807	21.1941	−5.8523	0.7382	−0.0356	540–6300	300–3500	0.81
Methane	2.8500	−21.3124	17.0304	−1.4849	−0.0505	0.0043	540–6300	300–3500	0.20

The ideal gas constant pressure specific heat is given by

$$\theta = \left(\frac{T(\mathrm{K})}{100}\right) = \left(\frac{T(^\circ\mathrm{R})}{180}\right) \qquad C_\mathrm{p} = \sum_{i=0}^{i=4} C_i(\theta)^{i/2} = \frac{\mathrm{kcal}}{\mathrm{kgmol-K}} = \frac{\mathrm{Btu}}{\mathrm{lbm}-^\circ\mathrm{R}} \qquad C_0 = C_0^*\left(\frac{\theta^2}{\theta^2+\tau}\right)$$

A.12 Enthalpy of Formation and Enthalpy of Vaporization

				25 °C (77 °F), 1 atm			
		h_f	g_f	s_o	h_f	g_f	s_o
		kJ/kmol	kJ/kmol	kJ/kmol/K	Btu/lbmol	Btu/lbmol	Btu/lbmol/°R
Carbon	C(s)	0	0	5.74	0	0	1.371
Hydrogen	H_2(g)	0	0	130.68	0	0	31.215
Nitrogen	N_2(g)	0	0	191.61	0	0	45.768
Oxygen	O_2(g)	0	0	205.04	0	0	48.976
Carbon monoxide	CO(g)	−110,530	−137,150	197.65	−47,523	−58,968	47.211
Carbon dioxide	$CO_{2(g)}$	−393,520	−394,360	213.8	−169,195	−169,556	51.069
Steam	H_2O(g)	−241,820	−228,590	188.83	−103,971	−98,283	45.104
Water	H_2O(l)	−285,830	−237,180	69.92	−122,893	−101,976	16.701
Ammonia	NH_3(g)	−46,190	−16,590	192.33	−19,860	−7133	45.940
Methane	CH_4(g)	−74,850	−50,790	186.16	−32,182	−21,837	44.467
Propane	C_3H_8(g)	−103,850	−23,490	269.91	−44,651	−10,100	64.471
n-Octane	C_8H_{18}(g)	−208,450	16,530	466.73	−89,624	7107	111.484
n-Octane	C_8H_{18}(l)	−249,950	6610	360.79	−107,467	2842	86.179
Methyl alcohol	CH_3OH (g)	−200,670	−162,000	239.7	−86,279	−69,652	57.255
Methyl alcohol	CH_3OH (l)	−277,690	−166,360	126.8	−119,394	−71,527	30.288
Monatomic oxygen	O(g)	249,190	231,770	161.06	107,140	99,650	38.471
Monatomic hydrogen	H(g)	218,000	203,290	114.72	93,730	87,405	27.402
Monatomic nitrogen	N(g)	472,650	455,510	153.3	203,217	195,848	36.618
Hydroxyl ion	OH(g)	39,460	34,280	183.7	16,966	14,739	43.879

	h_{fg}	C_p	h_{fg}	C_p
	kJ/kmol	kJ/kmol/K	Btu/lbmol	Btu/lbmol/°R
Water	44,010	35.3	18,922	8.43
Propane	15,060	122.2	6475	29.18
n-Octane	41,460	254.7	17,826	60.84
Methyl alcohol	37,900	83.5	16,295	19.95
Sodium	100,819	29.9	43,347	7.14
Potassium	83,934	30.6	36,088	7.31
Mercury	61,303	27.9	26,358	6.66

A.13 Gas Property Tables for Selected Gases

A.13.1 Air Properties (SI Units)

Temperature		*h*	*u*			s_o	
K	°C	kg/kgmol	kJ/kgmol	Pr	Vr	kJ/kgmol/K	Gamma
100	−173.1	1604.8	1142.6	0.03	6018.5	23.8878	1.400
125	−148.1	2331.0	1663.3	0.07	1916.6	30.3695	1.401
150	−123.1	3056.4	2182.4	0.12	1217.2	35.6598	1.402
175	−98.1	3781.8	2701.0	0.21	829.17	40.1327	1.401
200	−73.1	4507.5	3219.7	0.34	594.50	44.0087	1.401
225	−48.1	5233.5	3738.5	0.51	443.22	47.4293	1.401
250	−23.1	5959.9	4257.7	0.73	340.76	50.4908	1.401
275	1.9	6686.9	4777.2	1.02	268.59	53.2620	1.400
300	26.9	7414.4	5297.2	1.39	216.07	55.7943	1.400
325	51.9	8142.9	5818.2	1.84	176.81	58.1266	1.399
350	76.9	8872.6	6340.2	2.38	146.79	60.2895	1.398
375	101.9	9603.7	6863.8	3.04	123.39	62.3073	1.396
400	126.9	10,336.8	7389.2	3.82	104.82	64.1997	1.395
425	151.9	11,072.0	7916.7	4.73	89.874	65.9826	1.393
450	176.9	11,809.8	8446.8	5.79	77.685	67.6695	1.391
475	201.9	12,550.5	8979.8	7.02	67.631	69.2711	1.389
500	226.9	13,294.2	9515.8	8.44	59.253	70.7971	1.387
525	251.9	14,041.4	10,055.3	10.06	52.205	72.2555	1.384
550	276.9	14,792.3	10,598.4	11.90	46.231	73.6525	1.381
575	301.9	15,547.0	11,145.4	13.98	41.128	74.9945	1.379
600	326.9	16,305.7	11,696.3	16.33	36.742	76.2859	1.376
625	351.9	17,068.5	12,251.4	18.97	32.948	77.5314	1.373
650	376.9	17,835.5	12,810.7	21.92	29.648	78.7349	1.370
675	401.9	18,606.9	13,374.1	25.22	26.765	79.8992	1.367
700	426.9	19,382.6	13,942.1	28.89	24.233	81.0276	1.365
725	451.9	20,162.6	14,514.4	32.95	22.002	82.1225	1.362
750	476.9	20,946.8	15,090.8	37.45	20.027	83.1861	1.359
775	501.9	21,735.5	15,671.7	42.41	18.274	84.2203	1.357
800	526.8	22,528.4	16,256.8	47.87	16.711	85.2274	1.354
825	551.8	23,325.6	16,846.1	53.87	15.315	86.2087	1.351
850	576.8	24,126.8	17,439.7	60.44	14.064	87.1653	1.349
875	601.8	24,932.1	18,037.2	67.62	12.940	88.0990	1.347
900	626.8	25,741.4	18,638.5	75.46	11.926	89.0111	1.344
925	651.8	26,554.5	19,243.9	84.00	11.012	89.9020	1.342
950	676.8	27,371.3	19,852.8	93.28	10.184	90.7733	1.340
975	701.8	28,191.9	20,465.7	103.36	9.433	91.6261	1.338
1000	726.8	29,016.1	21,082.2	114.27	8.751	92.4606	1.336
1025	751.8	29,843.6	21,701.8	126.08	8.130	93.2783	1.334

(continued)

(continued)

Temperature		h	u			s_o	
K	°C	kg/kgmol	kJ/kgmol	Pr	Vr	kJ/kgmol/K	Gamma
1050	776.8	30,674.4	22,324.6	138.83	7.563	94.0789	1.333
1075	801.8	31,508.7	22,951.0	152.57	7.046	94.8638	1.331
1100	826.8	32,346.0	23,580.6	167.37	6.572	95.6337	1.329
1125	851.8	33,186.2	24,212.8	183.29	6.138	96.3890	1.328
1150	876.8	34,029.3	24,848.2	200.39	5.739	97.1304	1.326
1175	901.8	34,875.2	25,486.4	218.70	5.373	97.8575	1.325
1200	926.8	35,723.9	26,127.4	238.35	5.035	98.5727	1.324
1225	951.8	36,574.9	26,770.7	259.35	4.723	99.2747	1.322
1250	976.8	37,428.9	27,416.6	281.80	4.436	99.9647	1.321
1275	1001.8	38,285.6	28,065.2	305.76	4.170	100.6433	1.320
1300	1026.8	39,143.5	28,715.8	331.30	3.924	101.3101	1.319
1325	1051.8	40,004.7	29,369.0	358.49	3.696	101.9659	1.318
1350	1076.8	40,868.0	30,024.6	387.42	3.485	102.6112	1.317
1375	1101.8	41,733.0	30,681.8	418.17	3.288	103.2462	1.316
1400	1126.8	42,600.6	31,341.7	450.85	3.105	103.8718	1.315
1425	1151.8	43,470.4	32,003.0	485.52	2.935	104.4876	1.314
1450	1176.8	44,340.9	32,665.9	522.22	2.777	105.0935	1.313
1475	1201.8	45,214.2	33,331.4	561.10	2.629	105.6904	1.312
1500	1226.8	46,089.4	33,999.0	602.25	2.491	106.2789	1.311
1525	1251.8	46,966.6	34,668.5	645.68	2.362	106.8578	1.310
1550	1276.8	47,845.5	35,339.7	691.69	2.241	107.4300	1.309
1575	1301.8	48,726.2	36,011.9	740.16	2.128	107.9931	1.309
1600	1326.8	49,607.8	36,685.8	791.40	2.022	108.5495	1.308
1625	1351.8	50,491.7	37,362.0	845.22	1.923	109.0966	1.307
1650	1376.8	51,376.2	38,038.9	902.11	1.829	109.6381	1.306
1675	1401.8	52,262.5	38,717.5	961.77	1.742	110.1705	1.306
1700	1426.8	53,152.0	39,397.5	1024.76	1.659	110.6979	1.305
1725	1451.8	54,042.3	40,080.2	1090.79	1.581	111.2171	1.304
1750	1476.8	54,932.8	40,763.0	1160.12	1.508	111.7294	1.304
1775	1501.8	55,825.3	41,449.4	1233.09	1.439	112.2365	1.303
1800	1526.8	56,720.1	42,134.9	1309.56	1.375	112.7367	1.303
1825	1551.8	57,613.3	42,820.4	1389.69	1.313	113.2305	1.302
1850	1576.8	58,509.4	43,508.9	1473.67	1.255	113.7182	1.302
1875	1601.8	59,409.7	44,201.5	1561.58	1.201	114.2000	1.301
1900	1626.8	60,306.9	44,891.0	1653.65	1.149	114.6762	1.300
1925	1651.8	61,207.5	45,582.2	1749.90	1.100	115.1466	1.300
1950	1676.8	62,110.5	46,279.2	1850.59	1.054	115.6117	1.299
1975	1701.8	63,012.8	46,972.2	1955.91	1.010	116.0718	1.299
2000	1726.8	63,915.9	47,669.2	2066.18	0.968	116.5278	1.298
2025	1751.8	64,819.8	48,365.4	2181.02	0.928	116.9775	1.298
2050	1776.8	65,726.4	49,062.8	2300.86	0.891	117.4222	1.298
2075	1801.8	66,632.6	49,762.9	2425.62	0.855	117.8612	1.297

(continued)

(continued)

Temperature		h	u			s_o	
K	°C	kg/kgmol	kJ/kgmol	Pr	Vr	kJ/kgmol/K	Gamma
2100	1826.8	67,543.5	50,464.5	2555.79	0.822	118.2958	1.297
2125	1851.8	68,450.4	51,165.3	2691.76	0.789	118.7267	1.296
2150	1876.8	69,360.0	51,865.6	2833.21	0.759	119.1525	1.296
2175	1901.8	70,272.1	52,568.3	2979.99	0.730	119.5725	1.295
2200	1926.8	71,182.2	53,272.4	3133.78	0.702	119.9908	1.295
2225	1951.8	72,097.2	53,978.1	3293.20	0.676	120.4033	1.295
2250	1976.8	73,011.9	54,686.7	3458.58	0.651	120.8107	1.294
2275	2001.8	73,923.4	55,388.9	3630.89	0.627	121.2149	1.294
2300	2026.8	74,842.5	56,098.7	3809.81	0.604	121.6148	1.294
2325	2051.9	75,754.3	56,804.4	3995.67	0.582	122.0108	1.293
2350	2076.9	76,674.1	57,518.2	4189.73	0.561	122.4051	1.293
2375	2101.9	77,592.9	58,227.6	4389.97	0.541	122.7932	1.293
2400	2126.9	78,507.2	58,935.8	4597.72	0.522	123.1776	1.292
2425	2151.9	79,426.7	59,646.1	4813.26	0.504	123.5585	1.292
2450	2176.9	80,349.7	60,363.0	5037.23	0.486	123.9366	1.292
2475	2201.9	81,273.4	61,077.4	5270.24	0.470	124.3125	1.291
2500	2226.9	82,192.5	61,790.4	5509.22	0.454	124.6812	1.291
2525	2251.9	83,114.8	62,506.7	5757.86	0.439	125.0482	1.291
2550	2276.9	84,041.1	63,223.6	6015.53	0.424	125.4122	1.290
2575	2301.9	84,969.4	63,942.5	6280.78	0.410	125.7709	1.290
2600	2326.9	85,886.2	64,650.1	6557.81	0.396	126.1298	1.290
2625	2351.9	86,807.6	65,368.9	6844.26	0.384	126.4852	1.289
2650	2376.9	87,741.2	66,093.0	7139.73	0.371	126.8366	1.289
2675	2401.9	88,664.3	66,806.7	7447.82	0.359	127.1878	1.289
2700	2426.9	89,596.2	67,529.3	7761.68	0.348	127.5310	1.289
2725	2451.9	90,518.9	68,249.2	8089.94	0.337	127.8753	1.288
2750	2476.9	91,450.9	68,965.3	8423.94	0.326	128.2117	1.288
2775	2501.9	92,381.9	69,687.0	8769.42	0.316	128.5458	1.288
2800	2526.9	93,311.9	70,414.3	9131.02	0.307	128.8818	1.288
2825	2551.9	94,248.4	71,141.4	9502.36	0.297	129.2132	1.287
2850	2576.9	95,182.0	71,865.7	9888.05	0.288	129.5439	1.287
2875	2601.9	96,115.2	72,589.6	10,282.3	0.280	129.8690	1.287
2900	2626.9	97,050.2	73,321.8	10,686.3	0.271	130.1894	1.286
2925	2651.9	97,967.6	74,029.8	11,103.6	0.263	130.5078	1.286
2950	2676.9	98,908.9	74,761.9	11,543.9	0.256	130.8311	1.286
2975	2701.9	99,845.8	75,489.4	11,986.1	0.248	131.1437	1.286
3000	2726.9	100,781.8	76,216.1	12,444.3	0.241	131.4555	1.285
3025	2751.9	101,706.6	76,944.6	12,933.0	0.234	131.7758	1.285
3050	2776.9	102,662.5	77,678.1	13,408.5	0.227	132.0760	1.285
3075	2801.9	103,596.5	78,402.8	13,907.1	0.221	132.3795	1.285
3100	2826.9	104,539.8	79,149.9	14,431.7	0.215	132.6873	1.284
3125	2851.9	105,465.8	79,866.5	14,965.8	0.209	132.9895	1.284

(continued)

(continued)

Temperature		h	u			s_o	
K	°C	kg/kgmol	kJ/kgmol	Pr	Vr	kJ/kgmol/K	Gamma
3150	2876.9	106,407.4	80,611.9	15,519.7	0.203	133.2916	1.284
3175	2901.9	107,353.1	81,348.2	16,072.2	0.198	133.5824	1.284
3200	2926.9	108,280.3	82,053.1	16,655.2	0.192	133.8786	1.284
3225	2951.9	109,232.0	82,808.5	17,257.4	0.187	134.1739	1.283
3250	2976.9	110,178.3	83,545.5	17,879.8	0.182	134.4685	1.283
3275	3001.9	111,123.2	84,267.9	18,497.0	0.177	134.7506	1.283
3300	3026.9	112,069.4	85,004.8	19,152.7	0.172	135.0402	1.283
3325	3051.9	113,008.3	85,747.5	19,806.8	0.168	135.3194	1.283
3350	3076.9	113,938.4	86,468.3	20,500.1	0.163	135.6055	1.283
3375	3101.9	114,883.0	87,203.5	21,213.6	0.159	135.8899	1.282
3400	3126.9	115,838.3	87,949.5	21,938.6	0.155	136.1693	1.283

A.13.2 Air Properties (English Units)

Temperature		h	u			s_o	
°R	°F	Btu/lbmol	Btu/lbmol	Pr	Vr	Btu/lbmol/°R	Gamma
100	−359.7	690.0	491.3	0.004	26,215.8	5.7059	1.402
140	−319.7	968.4	692.7	0.012	11,288.7	8.0473	1.399
180	−279.7	1246.6	892.9	0.030	6018.51	9.7953	1.400
220	−239.7	1524.1	1091.9	0.060	3648.55	11.1878	1.401
260	−199.7	1801.4	1290.4	0.108	2406.77	12.3457	1.402
300	−159.7	2078.6	1488.6	0.178	1685.34	13.3375	1.402
340	−119.7	2355.9	1686.8	0.276	1233.92	14.2051	1.401
380	−79.7	2633.3	1885.1	0.406	935.173	14.9765	1.401
420	−39.7	2910.8	2083.4	0.576	728.608	15.6709	1.401
460	0.3	3188.5	2281.8	0.792	580.625	16.3024	1.401
500	40.3	3466.4	2480.4	1.061	471.455	16.8816	1.400
540	80.3	3744.5	2679.2	1.388	388.917	17.4166	1.400
580	120.3	4022.8	2878.2	1.784	325.178	17.9140	1.399
620	160.3	4301.6	3077.7	2.254	275.068	18.3788	1.398
660	200.3	4580.9	3277.6	2.808	235.036	18.8153	1.397
700	240.3	4860.7	3478.1	3.455	202.608	19.2269	1.396
740	280.3	5141.3	3679.3	4.204	176.016	19.6167	1.394
780	320.3	5422.6	3881.3	5.066	153.972	19.9869	1.393
820	360.3	5704.9	4084.2	6.051	135.513	20.3398	1.391
860	400.3	5988.1	4288.0	7.171	119.925	20.6771	1.389
900	440.3	6272.5	4493.0	8.438	106.655	21.0003	1.387
940	480.3	6558.0	4699.1	9.866	95.277	21.3106	1.384
980	520.3	6844.7	4906.4	11.468	85.459	21.6094	1.382

(continued)

(continued)

Temperature		h	u			s_o	
°R	°F	Btu/lbmol	Btu/lbmol	Pr	Vr	Btu/lbmol/°R	Gamma
1020	560.3	7132.7	5115.0	13.257	76.939	21.8974	1.380
1060	600.3	7422.1	5325.0	15.252	69.500	22.1757	1.377
1100	640.3	7712.8	5536.3	17.466	62.978	22.4449	1.375
1140	680.3	8004.9	5749.1	19.918	57.234	22.7058	1.372
1180	720.3	8298.6	5963.3	22.626	52.152	22.9589	1.370
1220	760.3	8593.7	6178.9	25.609	47.640	23.2048	1.367
1260	800.3	8890.2	6396.0	28.886	43.620	23.4439	1.365
1300	840.3	9188.2	6614.7	32.479	40.025	23.6768	1.362
1340	880.3	9487.6	6834.7	36.411	36.802	23.9037	1.360
1380	920.3	9788.6	7056.3	40.703	33.904	24.1250	1.357
1420	960.3	10,091.0	7279.2	45.382	31.290	24.3410	1.355
1460	1000.3	10,394.8	7503.7	50.467	28.930	24.5520	1.353
1500	1040.3	10,700.1	7729.5	55.993	26.789	24.7583	1.351
1540	1080.3	11,006.8	7956.8	61.980	24.847	24.9600	1.349
1580	1120.3	11,314.8	8185.4	68.460	23.079	25.1575	1.346
1620	1160.3	11,624.2	8415.3	75.463	21.468	25.3509	1.344
1660	1200.3	11,934.8	8646.5	83.016	19.996	25.5403	1.343
1700	1240.3	12,246.8	8879.1	91.154	18.650	25.7260	1.341
1740	1280.3	12,560.0	9112.8	99.909	17.416	25.9082	1.339
1780	1320.3	12,874.3	9347.9	109.313	16.283	26.0868	1.337
1820	1360.3	13,190.1	9584.1	119.404	15.242	26.2621	1.336
1860	1400.3	13,506.8	9821.5	130.218	14.284	26.4343	1.334
1900	1440.3	13,824.8	10,059.9	141.791	13.400	26.6034	1.332
1940	1480.3	14,143.7	10,299.3	154.172	12.583	26.7696	1.331
1980	1520.3	14,463.8	10,540.2	167.374	11.830	26.9328	1.329
2020	1560.3	14,784.7	10,781.6	181.464	11.132	27.0933	1.328
2060	1600.3	15,106.9	11,024.3	196.483	10.484	27.2512	1.327
2100	1640.3	15,429.8	11,267.8	212.465	9.884	27.4065	1.325
2140	1680.3	15,753.9	11,512.5	229.464	9.326	27.5594	1.324
2180	1720.3	16,078.8	11,757.9	247.495	8.808	27.7096	1.323
2220	1760.3	16,404.3	12,004.1	266.670	8.325	27.8578	1.322
2260	1800.3	16,731.0	12,251.4	286.998	7.875	28.0036	1.321
2300	1840.3	17,058.3	12,499.3	308.512	7.455	28.1472	1.320
2340	1880.3	17,386.4	12,748.0	331.296	7.063	28.2887	1.319
2380	1920.3	17,715.5	12,997.7	355.367	6.697	28.4280	1.318
2420	1960.3	18,045.4	13,247.9	380.859	6.354	28.5655	1.317
2460	2000.3	18,375.8	13,498.9	407.717	6.034	28.7009	1.316
2500	2040.3	18,706.8	13,750.4	436.076	5.733	28.8344	1.315
2540	2080.3	19,038.8	14,003.3	465.990	5.451	28.9661	1.314
2580	2120.3	19,371.5	14,256.2	497.469	5.186	29.0960	1.313
2620	2160.3	19,704.9	14,510.2	530.594	4.938	29.2240	1.313
2660	2200.3	20,038.5	14,764.7	565.557	4.703	29.3507	1.312

(continued)

(continued)

Temperature		h	u			s_o	
°R	°F	Btu/lbmol	Btu/lbmol	Pr	Vr	Btu/lbmol/°R	Gamma
2700	2240.3	20,372.8	15,019.5	602.250	4.483	29.4755	1.311
2740	2280.3	20,708.4	15,275.7	640.811	4.276	29.5988	1.310
2780	2320.3	21,044.0	15,531.8	681.229	4.081	29.7202	1.310
2820	2360.3	21,380.2	15,788.6	723.753	3.896	29.8405	1.309
2860	2400.3	21,716.9	16,045.9	768.283	3.723	29.9591	1.308
2900	2440.3	22,054.3	16,303.8	814.954	3.558	30.0762	1.308
2940	2480.3	22,392.2	16,562.5	863.846	3.403	30.1919	1.307
2980	2520.3	22,730.5	16,821.4	915.164	3.256	30.3065	1.306
3020	2560.3	23,069.5	17,080.2	968.642	3.118	30.4192	1.306
3060	2600.3	23,409.4	17,340.7	1024.76	2.986	30.5311	1.305
3100	2640.3	23,749.1	17,601.7	1083.25	2.862	30.6413	1.305
3140	2680.3	24,090.0	17,863.1	1144.37	2.744	30.7503	1.304
3180	2720.3	24,430.2	18,123.9	1208.42	2.632	30.8585	1.303
3220	2760.3	24,772.7	18,386.8	1275.04	2.525	30.9650	1.303
3260	2800.3	25,113.9	18,647.9	1344.65	2.424	31.0706	1.302
3300	2840.3	25,456.2	18,911.5	1417.230	2.328	31.1750	1.302
3340	2880.3	25,799.0	19,174.8	1492.974	2.237	31.2784	1.301
3380	2920.3	26,141.9	19,438.3	1571.546	2.151	31.3802	1.301
3420	2960.3	26,485.7	19,702.6	1653.649	2.068	31.4814	1.300
3460	3000.3	26,829.7	19,967.2	1739.133	1.989	31.5814	1.300
3500	3040.3	27,175.2	20,233.2	1828.024	1.915	31.6804	1.300
3540	3080.3	27,519.7	20,498.2	1920.853	1.843	31.7788	1.299
3580	3120.3	27,863.3	20,762.4	2016.791	1.775	31.8756	1.299
3620	3160.3	28,210.6	21,030.3	2116.627	1.710	31.9715	1.298
3660	3200.3	28,555.4	21,295.6	2220.483	1.648	32.0667	1.298
3700	3240.3	28,902.8	21,563.5	2327.995	1.589	32.1606	1.297
3740	3280.3	29,248.9	21,830.9	2439.306	1.533	32.2533	1.297
3780	3320.3	29,597.1	22,098.9	2555.792	1.479	32.3459	1.297
3820	3360.3	29,943.8	22,366.9	2675.917	1.428	32.4371	1.296
3860	3400.3	30,290.4	22,633.4	2801.559	1.378	32.5283	1.296
3900	3440.3	30,640.2	22,904.4	2930.566	1.331	32.6177	1.295
3940	3480.3	30,986.3	23,170.4	3064.678	1.286	32.7065	1.295
3980	3520.3	31,337.5	23,442.8	3203.413	1.242	32.7944	1.295
4020	3560.3	31,685.2	23,710.3	3347.464	1.201	32.8818	1.295
4060	3600.3	32,034.6	23,981.0	3496.259	1.161	32.9682	1.294
4100	3640.3	32,384.9	24,251.2	3650.195	1.123	33.0537	1.294
4140	3680.3	32,735.3	24,521.4	3809.811	1.087	33.1387	1.294
4180	3720.3	33,082.6	24,791.3	3975.642	1.051	33.2233	1.293
4220	3760.3	33,434.7	25,063.3	4144.901	1.018	33.3061	1.293
4260	3800.3	33,783.8	25,333.7	4322.743	0.985	33.3895	1.293
4300	3840.3	34,136.5	25,606.2	4503.871	0.955	33.4711	1.292
4340	3880.3	34,488.3	25,877.8	4692.876	0.925	33.5527	1.292

(continued)

(continued)

Temperature		h	u			s_o	
°R	°F	Btu/lbmol	Btu/lbmol	Pr	Vr	Btu/lbmol/°R	Gamma
4380	3920.3	34,839.6	26,150.3	4886.152	0.896	33.6328	1.292
4420	3960.3	35,191.0	26,421.6	5089.733	0.868	33.7139	1.291
4460	4000.3	35,543.2	26,695.1	5296.056	0.842	33.7928	1.291
4500	4040.3	35,895.4	26,968.5	5509.218	0.817	33.8712	1.291
4540	4080.3	36,248.5	27,241.5	5729.551	0.792	33.9491	1.291
4580	4120.3	36,602.1	27,513.4	5957.094	0.769	34.0264	1.290
4620	4160.3	36,952.2	27,787.6	6193.744	0.746	34.1038	1.290
4660	4200.3	37,307.9	28,061.8	6433.935	0.724	34.1793	1.290
4700	4240.3	37,662.5	28,337.6	6686.213	0.703	34.2557	1.290
4740	4280.3	38,016.2	28,612.6	6939.447	0.683	34.3295	1.289
4780	4320.3	38,370.7	28,885.5	7206.479	0.663	34.4045	1.289
4820	4360.3	38,721.3	29,157.4	7476.027	0.645	34.4774	1.289
4860	4400.3	39,078.7	29,436.0	7761.679	0.626	34.5519	1.289
4900	4440.3	39,433.7	29,712.2	8050.20	0.609	34.6244	1.288
4940	4480.3	39,787.4	29,987.2	8348.87	0.592	34.6967	1.288
4980	4520.3	40,143.3	30,264.4	8651.44	0.576	34.7674	1.288
5020	4560.3	40,499.7	30,539.2	8967.89	0.560	34.8387	1.288
5060	4600.3	40,853.1	30,813.9	9294.72	0.544	34.9098	1.287
5100	4640.3	41,209.2	31,091.2	9628.23	0.530	34.9798	1.287
5140	4680.3	41,569.5	31,369.9	9975.90	0.515	35.0503	1.287
5180	4720.3	41,920.0	31,641.7	10,326.3	0.502	35.1188	1.287
5220	4760.3	42,283.6	31,926.5	10,686.3	0.488	35.1869	1.286
5260	4800.3	42,634.6	32,198.8	11,058.6	0.476	35.2549	1.286
5300	4840.3	42,998.2	32,480.8	11,436.7	0.463	35.3216	1.286
5340	4880.3	43,352.6	32,753.6	11,837.8	0.451	35.3901	1.286
5380	4920.3	43,707.3	33,029.6	12,237.7	0.440	35.4561	1.285
5420	4960.3	44,063.5	33,307.1	12,652.2	0.428	35.5222	1.285
5460	5000.3	44,425.7	33,590.6	13,080.4	0.417	35.5883	1.285
5500	5040.3	44,778.9	33,864.9	13,523.8	0.407	35.6545	1.285
5540	5080.3	45,140.7	34,148.0	13,970.3	0.397	35.7190	1.285
5580	5120.3	45,503.7	34,432.3	14,431.7	0.387	35.7835	1.284
5620	5160.3	45,860.1	34,704.3	14,902.8	0.377	35.8473	1.284
5660	5200.3	46,219.4	34,984.8	15,386.9	0.368	35.9108	1.284
5700	5240.3	46,574.9	35,261.6	15,895.4	0.359	35.9754	1.284
5740	5280.3	46,937.0	35,550.6	16,402.2	0.350	36.0377	1.284
5780	5320.3	47,297.3	35,826.5	16,926.3	0.341	36.1002	1.283
5820	5360.3	47,652.4	36,102.9	17,458.6	0.333	36.1617	1.283
5860	5400.3	48,016.0	36,387.7	18,015.0	0.325	36.2240	1.283
5900	5440.3	48,376.2	36,669.2	18,567.9	0.318	36.2840	1.283
5940	5480.3	48,741.1	36,949.6	19,152.7	0.310	36.3456	1.283
5980	5520.3	49,098.1	37,233.5	19,748.3	0.303	36.4064	1.283
6020	5560.3	49,454.6	37,511.3	20,352.3	0.296	36.4662	1.282

(continued)

(continued)

Temperature		h	u			s_o	
°R	°F	Btu/lbmol	Btu/lbmol	Pr	Vr	Btu/lbmol/°R	Gamma
6060	5600.3	49,816.3	37,794.3	20,969.2	0.289	36.5255	1.283
6100	5640.3	50,180.0	38,073.6	21,615.6	0.282	36.5858	1.282
6140	5680.3	50,540.0	38,354.8	22,238.5	0.276	36.6422	1.282
6180	5720.3	50,901.3	38,637.4	22,929.3	0.270	36.7030	1.282

A.13.3 H_2O Properties (SI Units)

Temperature		h	u			s_o	
K	°C	kJ/kgmol	kJ/kgmol	Pr	Vr	kJ/kgmol/K	Gamma
200	−73.1	5510.1	4124.4	0.290	1259.160	169.2769	1.332
225	−48.1	6344.3	4757.4	0.460	490.529	173.2069	1.332
250	−23.1	7179.1	5389.8	0.700	356.982	176.7249	1.331
275	1.9	8015.2	6022.5	1.030	267.626	179.9124	1.330
300	26.9	8853.5	6656.6	1.460	205.544	182.8300	1.329
325	51.9	9694.8	7293.2	2.020	161.051	185.5235	1.327
350	76.9	10,539.8	7933.1	2.730	128.322	188.0283	1.325
375	101.9	11,389.1	8576.9	3.620	103.711	190.3722	1.323
400	126.9	12,243.3	9225.3	4.710	84.853	192.5772	1.320
425	151.9	13,102.8	9878.8	6.060	70.166	194.6613	1.318
450	176.9	13,967.9	10,537.7	7.680	58.563	196.6392	1.315
475	201.9	14,838.9	11,202.3	9.640	49.284	198.5230	1.312
500	226.9	15,716.2	11,873.1	11.970	41.779	200.3229	1.309
525	251.9	16,600.0	12,550.2	14.730	35.649	202.0476	1.306
550	276.9	17,490.4	13,233.8	17.970	30.599	203.7044	1.303
575	301.9	18,387.7	13,924.2	21.780	26.404	205.2997	1.300
600	326.9	19,291.9	14,621.5	26.210	22.895	206.8390	1.297
625	351.9	20,203.2	15,325.8	31.340	19.941	208.3270	1.294
650	376.9	21,121.8	16,037.4	37.280	17.438	209.7682	1.291
675	401.9	22,047.9	16,756.2	44.100	15.306	211.1660	1.288
700	426.9	22,981.3	17,482.5	51.930	13.481	212.5239	1.285
725	451.9	23,922.3	18,216.3	60.870	11.911	213.8447	1.282
750	476.9	24,870.9	18,957.7	71.050	10.556	215.1311	1.279
775	501.9	25,827.4	19,706.8	82.620	9.380	216.3855	1.276
800	526.8	26,791.6	20,463.6	95.730	8.357	217.6099	1.273
825	551.8	27,763.5	21,228.2	110.550	7.463	218.8063	1.271
850	576.8	28,743.5	22,000.8	127.260	6.679	219.9765	1.268
875	601.8	29,731.3	22,781.2	146.060	5.991	221.1218	1.265
900	626.8	30,727.1	23,569.5	167.160	5.384	222.2440	1.262
925	651.8	31,731.0	24,365.9	190.810	4.848	223.3439	1.260

(continued)

(continued)

Temperature		h	u			s_o	
K	°C	kJ/kgmol	kJ/kgmol	Pr	Vr	kJ/kgmol/K	Gamma
950	676.8	32,742.7	25,170.1	217.260	4.373	224.4232	1.257
975	701.8	33,762.4	25,982.4	246.790	3.951	225.4828	1.255
1000	726.8	34,790.1	26,802.5	279.710	3.575	226.5235	1.252
1025	751.8	35,825.8	27,630.7	316.320	3.240	227.5464	1.250
1050	776.8	36,869.4	28,466.6	357.010	2.941	228.5522	1.248
1075	801.8	37,920.9	29,310.7	402.140	2.673	229.5419	1.245
1100	826.8	38,980.1	30,162.4	452.140	2.433	230.5162	1.243
1125	851.8	40,047.3	31,022.0	507.430	2.217	231.4753	1.241
1150	876.8	41,122.3	31,889.3	568.520	2.023	232.4204	1.239
1175	901.8	42,204.8	32,764.1	635.900	1.848	233.3515	1.237
1200	926.8	43,294.8	33,646.7	710.130	1.690	234.2695	1.235
1225	951.8	44,392.7	34,536.8	791.850	1.547	235.1750	1.233
1250	976.8	45,497.7	35,434.4	881.630	1.418	236.0679	1.231
1275	1002	46,610.2	36,338.9	980.230	1.301	236.9493	1.229
1300	1027	47,729.9	37,251.2	1088.310	1.195	237.8188	1.227
1325	1052	48,856.9	38,170.4	1206.690	1.098	238.6773	1.225
1350	1077	49,991.0	39,096.9	1336.320	1.010	239.5256	1.224
1375	1102	51,131.9	40,030.1	1477.870	0.930	240.3626	1.222
1400	1127	52,279.6	40,969.9	1632.520	0.858	241.1900	1.220
1425	1152	53,434.1	41,916.8	1801.100	0.791	242.0070	1.219
1450	1177	54,595.2	42,870.1	1984.990	0.730	242.8153	1.217
1475	1202	55,762.9	43,830.1	2185.000	0.675	243.6134	1.216
1500	1227	56,936.6	44,796.6	2402.600	0.624	244.4027	1.214
1525	1252	58,116.8	45,768.6	2638.960	0.578	245.1828	1.213
1550	1277	59,302.9	46,747.5	2895.720	0.535	245.9547	1.212
1575	1302	60,495.3	47,732.2	3174.180	0.496	246.7180	1.210
1600	1327	61,693.9	48,722.7	3475.590	0.460	247.4722	1.209
1625	1352	62,898.1	49,719.2	3802.260	0.427	248.2190	1.208
1650	1377	64,107.6	50,721.0	4155.830	0.397	248.9582	1.207
1675	1402	65,323.3	51,729.0	4537.600	0.369	249.6889	1.206
1700	1427	66,543.8	52,741.4	4950.400	0.343	250.4128	1.205
1725	1452	67,769.7	53,760.0	5395.200	0.320	251.1281	1.204
1750	1477	69,000.6	54,783.2	5875.270	0.298	251.8368	1.203
1775	1502	70,237.3	55,811.5	6392.490	0.278	252.5382	1.202
1800	1527	71,478.6	56,845.0	6949.430	0.259	253.2327	1.201
1825	1552	72,724.0	57,883.6	7548.220	0.242	253.9198	1.200
1850	1577	73,975.0	58,926.9	8192.660	0.226	254.6010	1.199
1875	1602	75,230.7	59,974.9	8884.150	0.211	255.2746	1.198
1900	1627	76,490.6	61,026.3	9627.270	0.197	255.9425	1.197
1925	1652	77,754.6	62,083.5	10,424.460	0.185	256.6039	1.196
1950	1677	79,024.1	63,144.5	11,278.480	0.173	257.2585	1.195
1975	1702	80,296.8	64,209.4	12,194.130	0.162	257.9074	1.195

(continued)

(continued)

Temperature		h	u			s_o	
K	°C	kJ/kgmol	kJ/kgmol	Pr	Vr	kJ/kgmol/K	Gamma
2000	1727	81,574.7	65,279.6	13,173.870	0.152	258.5499	1.194
2025	1752	82,855.8	66,353.0	14,222.670	0.142	259.1868	1.193
2050	1777	84,140.9	67,430.4	15,346.000	0.134	259.8188	1.193
2075	1802	85,430.3	68,512.2	16,540.930	0.125	260.4421	1.192
2100	1827	86,723.8	69,598.0	17,823.400	0.118	261.0630	1.191
2125	1852	88,019.4	70,685.9	19,185.650	0.111	261.6753	1.191
2150	1877	89,320.2	71,778.9	20,643.450	0.104	262.2841	1.190
2175	1902	90,626.4	72,875.9	22,198.160	0.098	262.8878	1.189
2200	1927	91,933.1	73,976.5	23,853.41	0.092	263.4857	1.189
2225	1952	93,244.3	75,078.4	25,619.01	0.087	264.0793	1.188
2250	1977	94,559.1	76,185.5	27,490.67	0.082	264.6656	1.188
2275	2002	95,876.4	77,295.1	29,486.28	0.077	265.2482	1.187
2300	2027	97,196.6	78,407.7	31,609.24	0.073	265.8262	1.186
2325	2052	98,522.0	79,526.9	33,860.18	0.069	266.3981	1.186
2350	2077	99,849.5	80,646.8	36,256.23	0.065	266.9665	1.185
2375	2102	101,181.8	81,769.7	38,795.49	0.061	267.5293	1.185
2400	2127	102,514.1	82,894.4	41,493.24	0.058	268.0882	1.184
2425	2152	103,850.5	84,023.1	44,350.89	0.055	268.6419	1.184
2450	2177	105,189.8	85,154.6	47,387.27	0.052	269.1924	1.183
2475	2202	106,531.3	86,288.5	50,591.75	0.049	269.7364	1.183
2500	2227	107,878.4	87,427.9	53,995.10	0.046	270.2777	1.183
2525	2252	109,226.4	88,568.2	57,590.81	0.044	270.8137	1.182
2550	2277	110,576.9	89,711.0	61,398.02	0.042	271.3459	1.182
2575	2302	111,929.5	90,855.9	65,430.73	0.039	271.8747	1.181
2600	2327	113,285.2	92,005.5	69,692.74	0.037	272.3994	1.181
2625	2352	114,646.5	93,157.5	74,183.61	0.035	272.9185	1.180
2650	2377	116,004.6	94,309.6	78,938.94	0.034	273.4351	1.180
2675	2402	117,370.5	95,466.2	83,955.33	0.032	273.9473	1.180
2700	2427	118,734.1	96,623.8	89,240.35	0.030	274.4548	1.179
2725	2452	120,104.6	97,784.9	94,835.10	0.029	274.9603	1.179
2750	2477	121,475.6	98,946.5	100,725.1	0.027	275.4613	1.178
2775	2502	122,851.6	100,113.2	106,929.9	0.026	275.9583	1.178
2800	2527	124,226.5	101,282.1	113,469.0	0.025	276.4517	1.178
2825	2552	125,603.7	102,449.9	120,392.1	0.023	276.9441	1.177
2850	2577	126,984.2	103,624.4	127,621.4	0.022	277.4289	1.177
2875	2602	128,364.0	104,798.1	135,249.5	0.021	277.9115	1.177
2900	2627	129,749.0	105,973.8	143,271.5	0.020	278.3906	1.176
2925	2652	131,137.9	107,153.4	151,752.6	0.019	278.8687	1.176
2950	2677	132,529.4	108,338.8	160,597.6	0.018	279.3397	1.176
2975	2702	133,916.9	109,520.3	169,925.2	0.018	279.8090	1.176
3000	2727	135,309.9	110,704.0	179,739.8	0.017	280.2758	1.175
3025	2752	136,707.7	111,892.5	190,023.6	0.016	280.7384	1.175

(continued)

(continued)

Temperature		h	u			s_o	
K	°C	kJ/kgmol	kJ/kgmol	Pr	Vr	kJ/kgmol/K	Gamma
3050	2777	138,101.4	113,080.1	200,807.2	0.015	281.1973	1.175
3075	2802	139,504.2	114,270.3	212,141.3	0.014	281.6538	1.174
3100	2827	140,902.4	115,465.7	224,116.0	0.014	282.1103	1.174
3125	2852	142,306.1	116,666.6	236,620.6	0.013	282.5617	1.174
3150	2877	143,714.6	117,859.2	249,705.6	0.013	283.0092	1.173
3175	2902	145,122.6	119,057.9	263,395.2	0.012	283.4529	1.173
3200	2927	146,534.1	120,260.1	277,801.660	0.012	283.8956	1.173

A.13.4 H_2O Properties (English Units)

Temperature		h	u			s_o	
°R	°F	Btu/lbmol	Btu/lbmol	Pr	Vr	Btu/lbmol/°R	Gamma
300	−159.7	2369.1	1773.3	0.138	2180.3	40.4340	1.332
340	−119.7	2687.7	2016.1	0.227	1495.8	41.4308	1.332
380	−79.7	3006.4	2258.2	0.355	1069.9	42.3172	1.332
420	−39.7	3325.3	2500.0	0.531	791.26	43.1150	1.332
460	0.3	3644.4	2741.7	0.765	601.31	43.8408	1.331
500	40.3	3964.1	2983.5	1.070	467.29	44.5071	1.330
540	80.3	4284.5	3226.0	1.460	369.98	45.1236	1.329
580	120.3	4606.0	3469.2	1.949	297.60	45.6979	1.327
620	160.3	4928.7	3713.5	2.555	242.62	46.2359	1.326
660	200.3	5252.8	3959.1	3.298	200.12	46.7425	1.324
700	240.3	5578.5	4206.2	4.198	166.75	47.2216	1.321
740	280.3	5905.9	4455.0	5.279	140.19	47.6765	1.319
780	320.3	6235.2	4705.5	6.566	118.80	48.1098	1.317
820	360.3	6566.5	4957.9	8.088	101.38	48.5240	1.314
860	400.3	6899.8	5212.3	9.877	87.07	48.9208	1.312
900	440.3	7235.2	5468.8	11.968	75.20	49.3020	1.309
940	480.3	7572.8	5727.4	14.397	65.29	49.6690	1.306
980	520.3	7912.7	5988.3	17.207	56.95	50.0231	1.304
1020	560.3	8254.8	6251.4	20.444	49.89	50.3653	1.301
1060	600.3	8599.4	6516.8	24.155	43.88	50.6966	1.298
1100	640.3	8946.3	6784.7	28.397	38.74	51.0179	1.296
1140	680.3	9295.7	7054.9	33.228	34.31	51.3299	1.293
1180	720.3	9647.5	7327.6	38.712	30.48	51.6332	1.290
1220	760.3	10,001.9	7602.8	44.919	27.16	51.9285	1.288
1260	800.3	10,358.8	7880.6	51.925	24.27	52.2164	1.285
1300	840.3	10,718.3	8160.8	59.814	21.73	52.4972	1.282
1340	880.3	11,080.3	8443.7	68.674	19.51	52.7716	1.280

(continued)

(continued)

Temperature		h	u			s_o	
°R	°F	Btu/lbmol	Btu/lbmol	Pr	Vr	Btu/lbmol/°R	Gamma
1380	920.3	11,445.0	8729.1	78.604	17.56	53.0397	1.277
1420	960.3	11,812.3	9017.2	89.707	15.83	53.3021	1.275
1460	1000.3	12,182.4	9308.0	102.10	14.30	53.5591	1.272
1500	1040.3	12,555.0	9601.4	115.90	12.94	53.8109	1.270
1540	1080.3	12,930.4	9897.5	131.25	11.73	54.0578	1.267
1580	1120.3	13,308.4	10,196.2	148.29	10.66	54.3002	1.265
1620	1160.3	13,689.1	10,497.7	167.16	9.69	54.5382	1.262
1660	1200.3	14,072.6	10,801.9	188.05	8.83	54.7719	1.260
1700	1240.3	14,458.7	11,108.7	211.13	8.05	55.0019	1.258
1740	1280.3	14,847.6	11,418.3	236.59	7.35	55.2280	1.256
1780	1320.3	15,239.2	11,730.6	264.64	6.73	55.4504	1.253
1820	1360.3	15,633.5	12,045.6	295.50	6.16	55.6695	1.251
1860	1400.3	16,030.5	12,363.2	329.42	5.65	55.8852	1.249
1900	1440.3	16,430.2	12,683.6	366.64	5.182	56.0979	1.247
1940	1480.3	16,832.6	13,006.6	407.45	4.761	56.3074	1.245
1980	1520.3	17,237.5	13,332.3	452.14	4.379	56.5141	1.243
2020	1560.3	17,645.2	13,660.6	501.01	4.032	56.7179	1.241
2060	1600.3	18,055.5	13,991.6	554.41	3.716	56.9190	1.239
2100	1640.3	18,468.5	14,325.1	612.69	3.428	57.1175	1.237
2140	1680.3	18,883.9	14,661.4	676.25	3.165	57.3135	1.235
2180	1720.3	19,302.0	15,000.1	745.49	2.924	57.5071	1.234
2220	1760.3	19,722.6	15,341.4	820.84	2.705	57.6983	1.232
2260	1800.3	20,145.8	15,685.2	902.78	2.503	57.8873	1.230
2300	1840.3	20,571.4	16,031.4	991.74	2.319	58.0739	1.229
2340	1880.3	20,999.5	16,380.1	1088.31	2.150	58.2584	1.227
2380	1920.3	21,430.1	16,731.3	1193.04	1.995	58.4409	1.225
2420	1960.3	21,863.0	17,084.8	1306.48	1.852	58.6213	1.224
2460	2000.3	22,298.3	17,440.8	1429.28	1.721	58.7997	1.222
2500	2040.3	22,735.9	17,799.0	1562.15	1.600	58.9762	1.221
2540	2080.3	23,175.9	18,159.7	1705.67	1.489	59.1507	1.220
2580	2120.3	23,618.1	18,522.4	1860.70	1.387	59.3235	1.218
2620	2160.3	24,062.5	18,887.5	2027.89	1.292	59.4944	1.217
2660	2200.3	24,509.1	19,254.7	2208.25	1.205	59.6636	1.216
2700	2240.3	24,958.0	19,624.3	2402.60	1.124	59.8311	1.214
2740	2280.3	25,408.8	19,995.7	2611.83	1.049	59.9969	1.213
2780	2320.3	25,861.9	20,369.3	2836.81	0.980	60.1610	1.212
2820	2360.3	26,316.9	20,745.1	3078.89	0.916	60.3236	1.211
2860	2400.3	26,774.0	21,122.7	3338.68	0.857	60.4844	1.210
2900	2440.3	27,233.1	21,502.4	3617.71	0.802	60.6438	1.209
2940	2480.3	27,694.1	21,884.1	3917.13	0.751	60.8017	1.208
2980	2520.3	28,157.1	22,267.6	4238.07	0.703	60.9581	1.207
3020	2560.3	28,621.8	22,652.9	4581.75	0.659	61.1130	1.206

(continued)

(continued)

Temperature		h	u			s_o	
°R	°F	Btu/lbmol	Btu/lbmol	Pr	Vr	Btu/lbmol/°R	Gamma
3060	2600.3	29,088.6	23,040.2	4950.40	0.618	61.2666	1.205
3100	2640.3	29,556.9	23,429.2	5344.50	0.580	61.4188	1.204
3140	2680.3	30,027.3	23,820.2	5765.36	0.545	61.5693	1.203
3180	2720.3	30,498.9	24,212.7	6216.06	0.512	61.7188	1.202
3220	2760.3	30,972.8	24,607.1	6696.83	0.481	61.8667	1.201
3260	2800.3	31,447.9	25,002.8	7210.55	0.452	62.0135	1.200
3300	2840.3	31,925.1	25,400.5	7758.22	0.425	62.1589	1.199
3340	2880.3	32,403.6	25,799.5	8342.94	0.400	62.3031	1.199
3380	2920.3	32,883.8	26,200.3	8964.83	0.377	62.4459	1.198
3420	2960.3	33,365.2	26,602.3	9627.27	0.355	62.5875	1.197
3460	3000.3	33,848.5	27,006.1	10,332.8	0.335	62.7279	1.196
3500	3040.3	34,333.3	27,411.5	11,083.4	0.316	62.8672	1.196
3540	3080.3	34,819.3	27,818.0	11,881.8	0.298	63.0053	1.195
3580	3120.3	35,307.1	28,226.7	12,730.8	0.281	63.1424	1.194
3620	3160.3	35,795.5	28,635.7	13,632.1	0.266	63.2782	1.194
3660	3200.3	36,286.0	29,046.7	14,587.9	0.251	63.4128	1.193
3700	3240.3	36,777.7	29,459.3	15,603.8	0.237	63.5465	1.192
3740	3280.3	37,270.8	29,872.9	16,681.3	0.224	63.6791	1.192
3780	3320.3	37,765.0	30,287.7	17,823.4	0.212	63.8106	1.191
3820	3360.3	38,260.4	30,703.7	19,031.2	0.201	63.9408	1.191
3860	3400.3	38,757.5	31,121.3	20,314.7	0.190	64.0704	1.190
3900	3440.3	39,256.3	31,540.6	21,672.0	0.180	64.1988	1.189
3940	3480.3	39,754.2	31,959.8	23,104.2	0.171	64.3259	1.189
3980	3520.3	40,254.5	32,380.6	24,626.0	0.162	64.4526	1.188
4020	3560.3	40,756.8	32,802.8	26,228.5	0.153	64.5778	1.188
4060	3600.3	41,259.2	33,225.7	27,925.1	0.145	64.7022	1.187
4100	3640.3	41,763.0	33,650.8	29,712.9	0.138	64.8255	1.187
4140	3680.3	42,267.9	34,075.5	31,609.3	0.131	64.9483	1.186
4180	3720.3	42,774.2	34,502.4	33,603.3	0.124	65.0698	1.186
4220	3760.3	43,281.6	34,930.3	35,710.0	0.118	65.1906	1.185
4260	3800.3	43,789.6	35,358.9	37,931.6	0.112	65.3104	1.185
4300	3840.3	44,299.2	35,789.0	40,277.3	0.107	65.4296	1.185
4340	3880.3	44,809.7	36,220.1	42,746.5	0.102	65.5477	1.184
4380	3920.3	45,320.6	36,652.2	45,340.8	0.097	65.6647	1.184
4420	3960.3	45,832.4	37,084.5	48,086.0	0.092	65.7815	1.183
4460	4000.3	46,346.1	37,518.8	50,963.2	0.088	65.8969	1.183
4500	4040.3	46,860.5	37,953.7	53,994.9	0.083	66.0116	1.183
4540	4080.3	47,375.0	38,388.8	57,190.8	0.079	66.1258	1.182
4580	4120.3	47,891.9	38,826.2	60,542.9	0.076	66.2389	1.182
4620	4160.3	48,409.0	39,263.8	64,063.8	0.072	66.3512	1.181
4660	4200.3	48,927.8	39,702.5	67,771.4	0.069	66.4629	1.181
4700	4240.3	49,445.7	40,141.6	71,653.6	0.066	66.5735	1.181

(continued)

(continued)

Temperature		h	u			s_o	
°R	°F	Btu/lbmol	Btu/lbmol	Pr	Vr	Btu/lbmol/°R	Gamma
4740	4280.3	49,964.5	40,581.7	75,740.5	0.063	66.6837	1.180
4780	4320.3	50,485.9	41,023.0	80,033.8	0.060	66.7932	1.180
4820	4360.3	51,006.9	41,465.2	84,536.6	0.057	66.9019	1.180
4860	4400.3	51,527.9	41,907.5	89,240.3	0.054	67.0094	1.179
4900	4440.3	52,051.3	42,350.8	94,201.3	0.052	67.1168	1.179
4940	4480.3	52,575.6	42,794.9	99,387.8	0.050	67.2233	1.179
4980	4520.3	53,100.7	43,241.2	104,828	0.048	67.3291	1.178
5020	4560.3	53,625.6	43,687.4	110,520	0.045	67.4341	1.178
5060	4600.3	54,151.9	44,133.5	116,483	0.043	67.5385	1.178
5100	4640.3	54,679.0	44,581.9	122,741	0.042	67.6424	1.177
5140	4680.3	55,207.7	45,030.4	129,266	0.040	67.7452	1.177
5180	4720.3	55,735.8	45,478.4	136,123	0.038	67.8479	1.177
5220	4760.3	56,263.8	45,927.6	143,271	0.036	67.9495	1.176
5260	4800.3	56,795.0	46,378.6	150,767	0.035	68.0508	1.176
5300	4840.3	57,325.7	46,830.5	158,623	0.033	68.1516	1.176
5340	4880.3	57,857.0	47,281.7	166,780	0.032	68.2512	1.176
5380	4920.3	58,388.1	47,734.1	175,311	0.031	68.3503	1.175
5420	4960.3	58,921.9	48,189.1	184,290	0.029	68.4495	1.175
5460	5000.3	59,454.2	48,641.3	193,579	0.028	68.5471	1.175
5500	5040.3	59,989.7	49,098.0	203,309	0.027	68.6445	1.175
5540	5080.3	60,523.7	49,551.9	213,526	0.026	68.7419	1.174
5580	5120.3	61,059.2	50,008.6	224,116	0.025	68.8380	1.174
5620	5160.3	61,597.8	50,465.0	235,132	0.024	68.9333	1.174
5660	5200.3	62,135.0	50,924.1	246,738	0.023	69.0290	1.173
5700	5240.3	62,670.5	51,380.8	258,763	0.022	69.1235	1.173
5740	5280.3	63,210.0	51,838.8	271,347	0.021	69.2178	1.173
5780	5320.3	63,746.5	52,299.4	284,416	0.020	69.3112	1.173

A.13.5 CO_2 Properties (SI Units)

Temperature		h	u			s_o	
K	°C	kJ/kgmol	kJ/kgmol	Pr	Vr	kJ/kgmol/K	Gamma
200	−73.1	4898.9	3513.4	0.28	1292.15	194.0903	1.345
225	−48.1	5723.6	4168.1	0.44	506.17	197.9744	1.329
250	−23.1	6579.3	4847.3	0.69	364.53	201.5795	1.313
275	1.9	7465.7	5552.2	1.03	267.07	204.9580	1.300
300	26.9	8382.0	6283.1	1.51	198.55	208.1462	1.288
325	51.9	9326.8	7039.4	2.17	149.50	211.1705	1.277
350	76.9	10,298.7	7820.2	3.07	113.86	214.0510	1.267

(continued)

(continued)

Temperature		h	u			s_o	
K	°C	kJ/kgmol	kJ/kgmol	Pr	Vr	kJ/kgmol/K	Gamma
375	101.9	11,296.2	8624.6	4.28	87.607	216.8035	1.259
400	126.9	12,318.0	9451.5	5.88	68.045	219.4409	1.252
425	151.9	13,362.7	10,299.9	7.97	53.308	221.9741	1.245
450	176.9	14,429.0	11,168.8	10.69	42.099	224.4118	1.239
475	201.9	15,516.0	12,057.2	14.18	33.493	226.7625	1.234
500	226.9	16,622.5	12,964.2	18.63	26.832	229.0325	1.229
525	251.9	17,747.4	13,889.0	24.27	21.635	231.2279	1.224
550	276.9	18,890.0	14,830.6	31.34	17.551	233.3538	1.220
575	301.9	20,049.4	15,788.5	40.15	14.320	235.4150	1.217
600	326.9	21,224.9	16,762.0	51.08	11.746	237.4160	1.213
625	351.9	22,415.5	17,750.2	64.54	9.684	239.3600	1.210
650	376.9	23,620.8	18,752.7	81.02	8.023	241.2508	1.207
675	401.9	24,840.1	19,768.7	101.10	6.677	243.0914	1.204
700	426.9	26,072.5	20,797.6	125.43	5.581	244.8847	1.202
725	451.9	27,317.8	21,839.1	154.78	4.684	246.6325	1.199
750	476.9	28,575.4	22,892.7	190.02	3.947	248.3377	1.197
775	501.9	29,844.5	23,957.4	232.14	3.339	250.0022	1.195
800	526.8	31,124.9	25,033.3	282.27	2.834	251.6279	1.193
825	551.8	32,415.8	26,119.5	341.74	2.414	253.2172	1.191
850	576.8	33,717.0	27,215.8	411.95	2.063	254.7706	1.189
875	601.8	35,027.9	28,321.6	494.59	1.769	256.2906	1.188
900	626.8	36,348.2	29,436.6	591.50	1.522	257.7783	1.186
925	651.8	37,677.6	30,560.6	704.80	1.312	259.2352	1.185
950	676.8	39,015.2	31,692.8	836.76	1.135	260.6620	1.183
975	701.8	40,361.3	32,833.2	990.10	0.985	262.0610	1.182
1000	726.8	41,715.3	33,981.6	1167.6	0.856	263.4322	1.181
1025	751.8	43,076.4	35,137.1	1372.6	0.747	264.7767	1.180
1050	776.8	44,445.2	36,299.4	1608.6	0.653	266.0959	1.179
1075	801.8	45,820.7	37,468.8	1879.7	0.572	267.3905	1.178
1100	826.8	47,202.7	38,644.7	2190.0	0.502	268.6610	1.177
1125	851.8	48,591.4	39,827.4	2544.8	0.442	269.9092	1.176
1150	876.8	49,985.1	41,015.0	2949.0	0.390	271.1346	1.175
1175	901.8	51,385.5	42,209.0	3408.9	0.345	272.3395	1.174
1200	926.8	52,791.4	43,408.4	3930.4	0.305	273.5230	1.173
1225	951.8	54,201.5	44,612.9	4521.0	0.271	274.6868	1.172
1250	976.8	55,618.1	45,822.6	5188.2	0.241	275.8313	1.172
1275	1001.8	57,039.4	47,037.0	5940.2	0.215	276.9566	1.171
1300	1026.8	58,464.9	48,255.7	6786.6	0.192	278.0639	1.170
1325	1051.8	59,895.8	49,479.7	7737.8	0.171	279.1545	1.170
1350	1076.8	61,329.7	50,707.6	8803.1	0.153	280.2268	1.169
1375	1101.8	62,768.5	51,939.5	9994.9	0.138	281.2824	1.169
1400	1126.8	64,211.8	53,175.9	11,326.2	0.124	282.3220	1.168

(continued)

(continued)

Temperature		h	u			s_o	
K	°C	kJ/kgmol	kJ/kgmol	Pr	Vr	kJ/kgmol/K	Gamma
1425	1151.8	65,658.4	54,415.7	12,812.8	0.111	283.3473	1.168
1450	1176.8	67,108.3	55,658.6	14,466.2	0.100	284.3563	1.167
1475	1201.8	68,563.2	56,906.7	16,304.6	0.090	285.3509	1.167
1500	1226.8	70,020.9	58,157.6	18,344.6	0.082	286.3310	1.166
1525	1251.8	71,481.7	59,410.6	20,600.9	0.074	287.2954	1.166
1550	1276.8	72,945.4	60,668.3	23,101.9	0.067	288.2480	1.165
1575	1301.8	74,412.2	61,927.4	25,868.1	0.061	289.1882	1.165
1600	1326.8	75,883.0	63,192.1	28,916.1	0.055	290.1142	1.164
1625	1351.8	77,355.4	64,456.8	32,270.5	0.050	291.0267	1.164
1650	1376.8	78,831.3	65,726.7	35,969.4	0.046	291.9289	1.164
1675	1401.8	80,312.5	67,000.2	40,024.4	0.042	292.8169	1.163
1700	1426.8	81,791.3	68,272.9	44,496.2	0.038	293.6975	1.163
1725	1451.8	83,278.7	69,551.0	49,369.3	0.035	294.5615	1.163
1750	1476.8	84,762.2	70,828.5	54,730.5	0.032	295.4185	1.162
1775	1501.8	86,252.2	72,110.8	60,579.4	0.029	296.2627	1.162
1800	1526.8	87,742.6	73,395.1	66,965.8	0.027	297.0959	1.162
1825	1551.8	89,235.2	74,680.0	73,954.8	0.025	297.9212	1.162
1850	1576.8	90,731.9	75,972.3	81,574.9	0.023	298.7365	1.161
1875	1601.8	92,231.5	77,262.6	89,849.9	0.021	299.5398	1.161
1900	1626.8	93,730.0	78,555.0	98,864.0	0.019	300.3346	1.161
1925	1651.8	95,230.1	79,845.8	108,640.8	0.018	301.1186	1.160
1950	1676.8	96,736.7	81,146.4	119,300.4	0.016	301.8968	1.160
1975	1701.8	98,244.5	82,444.8	130,839.1	0.015	302.6643	1.160
2000	1726.8	99,751.9	83,746.2	143,305.3	0.014	303.4209	1.160
2025	1751.8	101,259.1	85,044.1	156,882.2	0.013	304.1735	1.160
2050	1776.8	102,767.9	86,346.8	171,483.8	0.012	304.9133	1.159
2075	1801.8	104,283.4	87,653.0	187,278.1	0.011	305.6458	1.159
2100	1826.8	105,797.3	88,960.8	204,390.5	0.010	306.3727	1.159
2125	1851.8	107,314.4	90,271.8	222,790.0	0.010	307.0894	1.159
2150	1876.8	108,833.1	91,581.2	242,630.1	0.009	307.7986	1.159
2175	1901.8	110,353.8	92,895.8	264,062.5	0.008	308.5023	1.158
2200	1926.8	111,871.4	94,204.1	287,095.1	0.008	309.1976	1.158
2225	1951.8	113,391.6	95,521.6	311,872.0	0.007	309.8858	1.158
2250	1976.8	114,915.7	96,836.3	338,455.4	0.007	310.5658	1.158
2275	2001.8	116,442.1	98,153.4	367,187.4	0.006	311.2432	1.158
2300	2026.8	117,966.3	99,471.5	397,703.9	0.006	311.9070	1.158
2325	2051.9	119,491.0	100,793.5	430,530.6	0.005	312.5663	1.157
2350	2076.9	121,023.1	102,109.7	465,834.2	0.005	313.2215	1.157
2375	2101.9	122,555.9	103,439.7	503,490.9	0.005	313.8678	1.157
2400	2126.9	124,083.5	104,758.0	544,061.8	0.004	314.5121	1.157
2425	2151.9	125,613.1	106,078.3	587,250.6	0.004	315.1472	1.157
2450	2176.9	127,156.1	107,418.5	633,421.1	0.004	315.7764	1.157

(continued)

(continued)

Temperature		h	u			s_o	
K	°C	kJ/kgmol	kJ/kgmol	Pr	Vr	kJ/kgmol/K	Gamma
2475	2201.9	128,681.3	108,734.4	682,687.7	0.004	316.3991	1.157
2500	2226.9	130,225.1	110,068.8	735,216.5	0.003	317.0154	1.156
2525	2251.9	131,764.7	111,399.1	791,136.3	0.003	317.6248	1.156
2550	2276.9	133,298.1	112,723.1	851,306.1	0.003	318.2342	1.156
2575	2301.9	134,830.4	114,052.7	914,842.4	0.003	318.8326	1.156
2600	2326.9	136,372.1	115,385.0	982,925.9	0.003	319.4294	1.156
2625	2351.9	137,910.9	116,721.1	1,054,793.8	0.002	320.0161	1.156
2650	2376.9	139,455.6	118,056.4	1,132,149.1	0.002	320.6044	1.156
2675	2401.9	140,996.3	119,387.8	1,213,719.6	0.002	321.1828	1.156
2700	2426.9	142,542.7	120,731.5	1,299,797.3	0.002	321.7525	1.155
2725	2451.9	144,091.2	122,070.6	1,393,380.3	0.002	322.3305	1.155
2750	2476.9	145,633.1	123,403.2	1,491,186.5	0.002	322.8945	1.155
2775	2501.9	147,180.6	124,741.3	1,594,517.1	0.002	323.4515	1.155
2800	2526.9	148,735.4	126,086.8	1,705,346.0	0.002	324.0101	1.155
2825	2551.9	150,272.8	127,428.0	1,821,001.6	0.002	324.5557	1.155
2850	2576.9	151,831.0	128,776.9	1,945,136.3	0.001	325.1039	1.155
2875	2601.9	153,381.2	130,104.7	2,077,527.0	0.001	325.6513	1.155
2900	2626.9	154,935.0	131,462.2	2,215,282.5	0.001	326.1851	1.155
2925	2651.9	156,479.4	132,797.3	2,363,265.3	0.001	326.7227	1.155
2950	2676.9	158,032.4	134,141.0	2,516,307.0	0.001	327.2444	1.154
2975	2701.9	159,594.8	135,494.0	2,681,335.5	0.001	327.7725	1.154
3000	2726.9	161,139.8	136,842.8	2,854,293.8	0.001	328.2921	1.154
3025	2751.9	162,703.6	138,197.3	3,035,087.0	0.001	328.8027	1.154
3050	2776.9	164,262.6	139,533.9	3,231,907.8	0.001	329.3251	1.154
3075	2801.9	165,809.9	140,884.9	3,433,278.0	0.001	329.8276	1.154
3100	2826.9	167,368.2	142,233.9	3,647,235.0	0.001	330.3302	1.154
3125	2851.9	168,937.7	143,594.1	3,874,568.3	0.001	330.8329	1.154
3150	2876.9	170,504.6	144,951.7	4,115,300.8	0.001	331.3340	1.154
3175	2901.9	172,053.7	146,304.5	4,366,548.0	0.001	331.8267	1.154
3200	2926.9	173,619.6	147,648.0	4,633,662.0	0.001	332.3203	1.154

A.13.6 CO_2 Properties (English Units)

Temperature		h	u			s_o	
°R	°F	Btu/lbmol	Btu/lbmol	Pr	Vr	Btu/lbmol/°R	Gamma
300	−159.7	2106.3	1510.6	0.139	2152.4	46.3610	1.369
340	−119.7	2405.9	1749.9	0.223	1521.8	47.2980	1.353
380	−79.7	2715.2	1995.7	0.345	1103.0	48.1580	1.338
420	−39.7	3035.0	2249.4	0.515	814.94	48.9579	1.323

(continued)

(continued)

Temperature		h	u			s_o	
°R	°F	Btu/lbmol	Btu/lbmol	Pr	Vr	Btu/lbmol/°R	Gamma
460	0.3	3365.3	2511.7	0.752	611.48	49.7090	1.310
500	40.3	3706.0	2782.8	1.076	464.84	50.4190	1.298
540	80.3	4056.8	3062.7	1.511	357.39	51.0939	1.288
580	120.3	4417.3	3351.2	2.090	277.56	51.7378	1.278
620	160.3	4787.1	3648.1	2.850	217.53	52.3541	1.269
660	200.3	5165.6	3953.0	3.839	171.90	52.9458	1.262
700	240.3	5552.6	4265.8	5.114	136.89	53.5150	1.255
740	280.3	5947.6	4586.0	6.741	109.78	54.0636	1.249
780	320.3	6350.1	4913.3	8.802	88.62	54.5933	1.243
820	360.3	6759.9	5247.5	11.392	71.98	55.1057	1.238
860	400.3	7176.6	5588.2	14.626	58.80	55.6018	1.233
900	440.3	7599.8	5935.3	18.635	48.30	56.0828	1.229
940	480.3	8029.4	6288.3	23.574	39.87	56.5498	1.225
980	520.3	8465.0	6647.1	29.626	33.08	57.0036	1.221
1020	560.3	8906.3	7011.6	37.001	27.57	57.4450	1.218
1060	600.3	9353.2	7381.3	45.939	23.07	57.8746	1.215
1100	640.3	9805.3	7756.2	56.722	19.39	58.2933	1.212
1140	680.3	10,262.6	8136.0	69.673	16.36	58.7017	1.209
1180	720.3	10,724.8	8520.6	85.149	13.86	59.1001	1.206
1220	760.3	11,191.6	8909.7	103.578	11.78	59.4891	1.204
1260	800.3	11,662.9	9303.3	125.433	10.05	59.8693	1.202
1300	840.3	12,138.6	9701.1	151.242	8.596	60.2409	1.199
1340	880.3	12,618.4	10,102.9	181.629	7.378	60.6045	1.197
1380	920.3	13,102.2	10,508.7	217.266	6.352	60.9602	1.195
1420	960.3	13,589.9	10,918.2	258.923	5.484	61.3086	1.194
1460	1000.3	14,081.3	11,331.3	307.466	4.748	61.6498	1.192
1500	1040.3	14,576.3	11,748.0	363.861	4.122	61.9842	1.190
1540	1080.3	15,074.6	12,168.0	429.178	3.588	62.3121	1.189
1580	1120.3	15,576.2	12,591.3	504.618	3.131	62.6336	1.187
1620	1160.3	16,080.9	13,017.6	591.502	2.739	62.9491	1.186
1660	1200.3	16,588.8	13,447.0	691.365	2.401	63.2589	1.185
1700	1240.3	17,099.6	13,879.2	805.734	2.110	63.5629	1.184
1740	1280.3	17,613.1	14,314.1	936.428	1.858	63.8614	1.182
1780	1320.3	18,129.4	14,751.8	1085.49	1.640	64.1548	1.181
1820	1360.3	18,648.2	15,192.0	1255.08	1.450	64.4430	1.180
1860	1400.3	19,169.6	15,634.8	1447.55	1.285	64.7264	1.179
1900	1440.3	19,693.5	16,079.8	1665.6	1.141	65.0049	1.178
1940	1480.3	20,219.4	16,527.1	1912.1	1.015	65.2791	1.177
1980	1520.3	20,747.9	16,976.7	2190.0	0.904	65.5486	1.177
2020	1560.3	21,278.2	17,428.2	2503.0	0.807	65.8138	1.176
2060	1600.3	21,810.7	17,882.0	2854.7	0.722	66.0749	1.175
2100	1640.3	22,345.2	18,337.8	3248.9	0.646	66.3318	1.174

(continued)

(continued)

Temperature		h	u			s_o	
°R	°F	Btu/lbmol	Btu/lbmol	Pr	Vr	Btu/lbmol/°R	Gamma
2140	1680.3	22,881.5	18,795.1	3690.6	0.58	66.5849	1.173
2180	1720.3	23,420.2	19,254.7	4183.8	0.521	66.8340	1.173
2220	1760.3	23,959.9	19,715.7	4734.5	0.469	67.0796	1.172
2260	1800.3	24,501.8	20,178.5	5347.8	0.423	67.3215	1.172
2300	1840.3	25,045.1	20,643.0	6029.7	0.381	67.5598	1.171
2340	1880.3	25,590.1	21,108.9	6786.6	0.345	67.7946	1.170
2380	1920.3	26,136.7	21,576.7	7626.4	0.312	68.0263	1.170
2420	1960.3	26,684.6	22,045.6	8555.0	0.283	68.2545	1.169
2460	2000.3	27,234.1	22,515.9	9583.4	0.257	68.4799	1.169
2500	2040.3	27,785.2	22,988.0	10,716.0	0.233	68.7017	1.168
2540	2080.3	28,337.1	23,461.1	11,966.3	0.212	68.9209	1.168
2580	2120.3	28,890.4	23,935.0	13,342.3	0.193	69.1370	1.167
2620	2160.3	29,445.2	24,411.0	14,855.7	0.176	69.3504	1.167
2660	2200.3	30,001.2	24,887.9	16,518.6	0.161	69.5611	1.166
2700	2240.3	30,558.6	25,366.2	18,344.6	0.147	69.7693	1.166
2740	2280.3	31,116.4	25,845.2	20,339.8	0.135	69.9743	1.166
2780	2320.3	31,676.4	26,325.8	22,525.3	0.123	70.1770	1.165
2820	2360.3	32,236.2	26,806.9	24,913.5	0.113	70.3771	1.165
2860	2400.3	32,797.9	27,289.1	27,520.3	0.104	70.5747	1.165
2900	2440.3	33,360.1	27,772.5	30,363.9	0.096	70.7700	1.164
2940	2480.3	33,923.7	28,256.7	33,463.0	0.088	70.9630	1.164
2980	2520.3	34,488.2	28,742.4	36,840.5	0.081	71.1540	1.164
3020	2560.3	35,053.5	29,228.3	40,502.2	0.075	71.3421	1.163
3060	2600.3	35,619.3	29,715.4	44,496.3	0.069	71.5289	1.163
3100	2640.3	36,186.2	30,202.8	48,807.2	0.064	71.7125	1.163
3140	2680.3	36,754.6	30,691.8	53,499.3	0.059	71.8948	1.162
3180	2720.3	37,323.8	31,182.2	58,565.9	0.054	72.0745	1.162
3220	2760.3	37,894.3	31,673.2	64,062.9	0.05	72.2526	1.162
3260	2800.3	38,464.2	32,163.7	70,006.4	0.047	72.4288	1.162
3300	2840.3	39,035.2	32,655.3	76,418.2	0.043	72.6029	1.161
3340	2880.3	39,606.3	33,147.6	83,339.1	0.04	72.7750	1.161
3380	2920.3	40,180.1	33,642.6	90,818.4	0.037	72.9457	1.161
3420	2960.3	40,752.4	34,136.2	98,864.0	0.035	73.1143	1.161
3460	3000.3	41,326.5	34,630.1	107,520.6	0.032	73.2809	1.161
3500	3040.3	41,900.9	35,125.8	116,863.0	0.030	73.4464	1.160
3540	3080.3	42,476.5	35,622.6	126,858.2	0.028	73.6094	1.160
3580	3120.3	43,051.5	36,117.4	137,653.7	0.026	73.7716	1.160
3620	3160.3	43,629.7	36,616.9	149,204.5	0.024	73.9316	1.160
3660	3200.3	44,205.8	37,112.8	161,585.4	0.023	74.0899	1.160
3700	3240.3	44,783.2	37,611.5	174,927.8	0.021	74.2474	1.159
3740	3280.3	45,361.9	38,111.4	189,169.0	0.020	74.4029	1.159
3780	3320.3	45,940.8	38,610.2	204,391.3	0.018	74.5565	1.159

(continued)

(continued)

Temperature		h	u			s_o	
°R	°F	Btu/lbmol	Btu/lbmol	Pr	Vr	Btu/lbmol/°R	Gamma
3820	3360.3	46,519.2	39,109.8	220,690.0	0.017	74.7089	1.159
3860	3400.3	47,100.7	39,611.2	238,098.2	0.016	74.8597	1.159
3900	3440.3	47,680.2	40,112.0	256,730.5	0.015	75.0093	1.158
3940	3480.3	48,261.0	40,614.0	276,623.2	0.014	75.1575	1.158
3980	3520.3	48,843.2	41,117.5	297,891.2	0.013	75.3046	1.158
4020	3560.3	49,423.9	41,617.9	320,487.7	0.013	75.4498	1.158
4060	3600.3	50,007.8	42,123.1	344,603.6	0.012	75.5939	1.158
4100	3640.3	50,588.8	42,624.0	370,381.8	0.011	75.7371	1.158
4140	3680.3	51,172.8	43,129.3	397,704.7	0.010	75.8785	1.158
4180	3720.3	51,756.2	43,633.9	426,811.9	0.010	76.0187	1.157
4220	3760.3	52,341.1	44,140.0	457,729.2	0.009	76.1576	1.157
4260	3800.3	52,925.3	44,642.7	490,688.2	0.009	76.2957	1.157
4300	3840.3	53,509.1	45,150.5	525,804.1	0.008	76.4329	1.157
4340	3880.3	54,095.4	45,655.2	562,836.0	0.008	76.5681	1.157
4380	3920.3	54,680.1	46,161.2	602,004.0	0.007	76.7017	1.157
4420	3960.3	55,268.8	46,668.4	644,330.8	0.007	76.8366	1.157
4460	4000.3	55,853.4	47,174.2	688,207.0	0.006	76.9675	1.157
4500	4040.3	56,443.5	47,685.6	735,215.0	0.006	77.0987	1.156
4540	4080.3	57,030.7	48,194.0	785,035.6	0.006	77.2289	1.156
4580	4120.3	57,616.4	48,700.9	837,485.5	0.005	77.3573	1.156
4620	4160.3	58,205.2	49,208.2	893,302.4	0.005	77.4854	1.156
4660	4200.3	58,793.7	49,717.9	952,144.0	0.005	77.6121	1.156
4700	4240.3	59,383.1	50,228.6	1,014,058.7	0.005	77.7372	1.156
4740	4280.3	59,971.4	50,738.1	1,080,090.8	0.004	77.8625	1.156
4780	4320.3	60,560.5	51,248.5	1,149,724.5	0.004	77.9866	1.156
4820	4360.3	61,147.8	51,757.1	1,223,372.8	0.004	78.1099	1.156
4860	4400.3	61,739.5	52,270.0	1,299,799.6	0.004	78.2302	1.155
4900	4440.3	62,331.9	52,783.6	1,382,923.1	0.004	78.3533	1.155
4940	4480.3	62,921.9	53,292.1	1,468,143.8	0.003	78.4720	1.155
4980	4520.3	63,513.1	53,807.4	1,559,496.6	0.003	78.5919	1.155
5020	4560.3	64,102.9	54,312.8	1,655,452.1	0.003	78.7105	1.155
5060	4600.3	64,698.0	54,829.1	1,755,773.1	0.003	78.8273	1.155
5100	4640.3	65,286.3	55,338.7	1,862,446	0.003	78.9445	1.155
5140	4680.3	65,878.2	55,851.9	1,973,642	0.003	79.0596	1.155
5180	4720.3	66,474.8	56,369.7	2,091,677	0.002	79.1750	1.155
5220	4760.3	67,067.6	56,883.7	2,215,282	0.002	79.2890	1.155
5260	4800.3	67,664.1	57,401.5	2,345,174	0.002	79.4021	1.154
5300	4840.3	68,253.5	57,912.1	2,480,459	0.002	79.5135	1.154
5340	4880.3	68,847.6	58,427.5	2,625,082	0.002	79.6260	1.154
5380	4920.3	69,441.6	58,937.1	2,776,344	0.002	79.7373	1.154
5420	4960.3	70,032.3	59,449.0	2,934,292	0.002	79.8472	1.154
5460	5000.3	70,626.9	59,964.9	3,101,107	0.002	79.9570	1.154

(continued)

(continued)

Temperature		h	u			s_o	
°R	°F	Btu/lbmol	Btu/lbmol	Pr	Vr	Btu/lbmol/°R	Gamma
5500	5040.3	71,225.6	60,490.4	3,274,133	0.002	80.0648	1.154
5540	5080.3	71,821.0	61,001.5	3,457,955	0.002	80.1733	1.154
5580	5120.3	72,413.3	61,515.0	3,647,232	0.002	80.2791	1.154
5620	5160.3	73,007.7	62,030.7	3,851,144	0.001	80.3871	1.154
5660	5200.3	73,615.6	62,554.3	4,063,894	0.001	80.4939	1.154
5700	5240.3	74,205.5	63,071.0	4,279,592	0.001	80.5966	1.154
5740	5280.3	74,808.4	63,589.5	4,513,548	0.001	80.7023	1.154
5780	5320.3	75,401.4	64,103.8	4,753,843	0.001	80.8053	1.153

A.13.7 Nitrogen Properties (SI Units)

Temperature		h	u			s_o	
K	°C	kJ/kgmol	kJ/kgmol	Pr	Vr	kJ/kgmol/K	Gamma
200	−73.1	4842.2	3456.7	0.34	1072.9	174.5226	1.399
225	−48.1	5570.4	3981.6	0.51	443.82	177.9535	1.399
250	−23.1	6298.6	4505.5	0.73	340.92	181.0223	1.400
275	1.9	7026.5	5028.6	1.02	268.58	183.7977	1.400
300	26.9	7754.4	5551.2	1.39	216.03	186.3310	1.400
325	51.9	8482.4	6073.4	1.84	176.82	188.6617	1.400
350	76.9	9210.7	6595.7	2.38	146.87	190.8208	1.399
375	101.9	9939.8	7118.5	3.04	123.53	192.8329	1.398
400	126.9	10,670.1	7642.3	3.81	105.04	194.7179	1.397
425	151.9	11,401.8	8167.4	4.71	90.150	196.4925	1.396
450	176.9	12,135.5	8694.3	5.77	78.014	198.1698	1.394
475	201.9	12,871.4	9223.4	6.99	68.000	199.7613	1.393
500	226.9	13,609.9	9754.9	8.38	59.654	201.2764	1.391
525	251.9	14,351.2	10,289.2	9.97	52.632	202.7233	1.388
550	276.9	15,095.7	10,826.5	11.78	46.675	204.1086	1.386
575	301.9	15,843.5	11,367.2	13.83	41.584	205.4382	1.384
600	326.9	16,594.9	11,911.4	16.13	37.204	206.7174	1.381
625	351.9	17,350.1	12,459.2	18.71	33.413	207.9503	1.378
650	376.9	18,109.0	13,010.8	21.59	30.112	209.1411	1.376
675	401.9	18,872.0	13,566.4	24.79	27.225	210.2928	1.373
700	426.9	19,639.0	14,126.0	28.35	24.688	211.4086	1.370
725	451.9	20,410.0	14,689.6	32.30	22.448	212.4909	1.368
750	476.9	21,185.3	15,257.4	36.65	20.464	213.5421	1.365
775	501.9	21,964.7	15,829.3	41.44	18.700	214.5643	1.362
800	526.8	22,748.2	16,405.3	46.71	17.125	215.5594	1.360
825	551.8	23,535.9	16,985.5	52.49	15.717	216.5289	1.357

(continued)

(continued)

Temperature		h	u			s_o	
K	°C	kJ/kgmol	kJ/kgmol	Pr	Vr	kJ/kgmol/K	Gamma
850	576.8	24,327.7	17,569.7	58.81	14.452	217.4743	1.355
875	601.8	25,123.5	18,158.0	65.72	13.314	218.3971	1.352
900	626.8	25,923.4	18,750.3	73.24	12.288	219.2984	1.350
925	651.8	26,727.3	19,346.6	81.43	11.359	220.1794	1.348
950	676.8	27,534.9	19,946.6	90.32	10.518	221.0410	1.345
975	701.8	28,346.4	20,550.5	99.96	9.754	221.8840	1.343
1000	726.8	29,161.6	21,158.1	110.40	9.058	222.7097	1.341
1025	751.8	29,980.4	21,769.3	121.68	8.424	223.5184	1.339
1050	776.8	30,802.9	22,384.0	133.85	7.844	224.3111	1.337
1075	801.8	31,628.7	23,002.2	146.97	7.314	225.0885	1.335
1100	826.8	32,458.0	23,623.8	161.09	6.828	225.8510	1.334
1125	851.8	33,290.4	24,248.6	176.26	6.383	226.5993	1.332
1150	876.8	34,126.0	24,876.5	192.54	5.973	227.3338	1.330
1175	901.8	34,964.7	25,507.5	210.00	5.595	228.0554	1.329
1200	926.8	35,806.4	26,141.4	228.69	5.247	228.7641	1.327
1225	951.8	36,650.9	26,778.3	248.67	4.926	229.4606	1.326
1250	976.8	37,498.2	27,417.9	270.02	4.629	230.1453	1.324
1275	1002	38,348.2	28,060.2	292.80	4.355	230.8186	1.323
1300	1027	39,200.7	28,704.9	317.08	4.100	231.4809	1.322
1325	1052	40,055.8	29,352.3	342.93	3.864	232.1324	1.321
1350	1077	40,913.3	30,002.1	370.42	3.645	232.7735	1.319
1375	1102	41,773.1	30,654.2	399.63	3.441	233.4045	1.318
1400	1127	42,635.2	31,308.5	430.64	3.251	234.0258	1.317
1425	1152	43,499.3	31,964.9	463.53	3.074	234.6377	1.316
1450	1177	44,365.7	32,623.6	498.38	2.909	235.2404	1.315
1475	1202	45,234.0	33,284.2	535.27	2.756	235.8341	1.314
1500	1227	46,104.4	33,946.7	574.30	2.612	236.4192	1.313
1525	1252	46,976.4	34,611.0	615.54	2.477	236.9958	1.312
1550	1277	47,850.4	35,277.3	659.11	2.352	237.5643	1.312
1575	1302	48,726.0	35,945.2	705.07	2.234	238.1247	1.311
1600	1327	49,603.5	36,614.8	753.54	2.123	238.6775	1.310
1625	1352	50,482.5	37,285.9	804.59	2.020	239.2224	1.309
1650	1377	51,363.2	37,958.9	858.34	1.922	239.7601	1.309
1675	1402	52,245.1	38,633.2	914.92	1.831	240.2907	1.308
1700	1427	53,128.6	39,308.8	974.39	1.745	240.8144	1.307
1725	1452	54,013.4	39,985.9	1036.90	1.664	241.3313	1.307
1750	1477	54,899.8	40,664.5	1102.50	1.587	241.8413	1.306
1775	1502	55,787.6	41,344.7	1171.36	1.515	242.3450	1.305
1800	1527	56,676.6	42,025.8	1243.57	1.447	242.8423	1.305
1825	1552	57,566.8	42,708.1	1319.26	1.383	243.3335	1.304
1850	1577	58,458.1	43,391.7	1398.50	1.323	243.8184	1.304
1875	1602	59,350.7	44,076.7	1481.52	1.266	244.2979	1.303

(continued)

(continued)

Temperature		h	u			s_o	
K	°C	kJ/kgmol	kJ/kgmol	Pr	Vr	kJ/kgmol/K	Gamma
1900	1627	60,244.3	44,762.3	1568.34	1.211	244.7713	1.303
1925	1652	61,139.2	45,449.6	1659.15	1.160	245.2393	1.302
1950	1677	62,035.0	46,137.7	1754.00	1.112	245.7015	1.302
1975	1702	62,932.4	46,827.0	1853.15	1.066	246.1586	1.301
2000	1727	63,830.5	47,517.4	1956.68	1.022	246.6106	1.301
2025	1752	64,729.4	48,208.6	2064.67	0.981	247.0572	1.300
2050	1777	65,629.4	48,900.8	2177.38	0.942	247.4991	1.300
2075	1802	66,530.4	49,593.8	2294.82	0.904	247.9358	1.300
2100	1827	67,432.2	50,287.9	2417.25	0.869	248.3679	1.299
2125	1852	68,335.2	50,983.2	2544.70	0.835	248.7951	1.299
2150	1877	69,238.7	51,679.0	2677.57	0.803	249.2182	1.298
2175	1902	70,143.8	52,376.0	2815.68	0.772	249.6364	1.298
2200	1927	71,049.1	53,073.7	2959.36	0.743	250.0501	1.298
2225	1952	71,955.4	53,772.2	3108.89	0.716	250.4599	1.297
2250	1977	72,862.1	54,471.2	3264.18	0.689	250.8652	1.297
2275	2002	73,770.2	55,171.6	3425.64	0.664	251.2666	1.297
2300	2027	74,679.0	55,872.4	3593.33	0.640	251.6639	1.296
2325	2052	75,588.3	56,574.0	3767.45	0.617	252.0573	1.296
2350	2077	76,498.6	57,276.6	3948.23	0.595	252.4469	1.296
2375	2102	77,409.8	57,979.6	4135.53	0.574	252.8322	1.295
2400	2127	78,321.5	58,683.6	4329.93	0.554	253.2141	1.295
2425	2152	79,233.6	59,388.1	4531.69	0.535	253.5928	1.295
2450	2177	80,147.2	60,093.6	4740.30	0.517	253.9669	1.295
2475	2202	81,061.3	60,799.6	4956.64	0.499	254.3380	1.294
2500	2227	81,974.9	61,506.3	5180.90	0.483	254.7058	1.294
2525	2252	82,889.7	62,213.5	5412.57	0.467	255.0695	1.294
2550	2277	83,806.5	62,921.7	5653.13	0.451	255.4310	1.293
2575	2302	84,722.2	63,629.7	5901.50	0.436	255.7885	1.293
2600	2327	85,639.7	64,339.5	6158.56	0.422	256.1430	1.293
2625	2352	86,557.4	65,048.8	6424.15	0.409	256.4940	1.293
2650	2377	87,474.6	65,759.0	6698.76	0.396	256.8420	1.292
2675	2402	88,394.1	66,470.0	6982.77	0.383	257.1872	1.292
2700	2427	89,313.6	67,181.9	7276.13	0.371	257.5293	1.292
2725	2452	90,232.9	67,893.5	7579.28	0.360	257.8687	1.292
2750	2477	91,153.4	68,606.3	7891.81	0.348	258.2046	1.292
2775	2502	92,074.3	69,318.7	8214.82	0.338	258.5381	1.291
2800	2527	92,995.6	70,033.0	8547.82	0.328	258.8685	1.291
2825	2552	93,917.5	70,746.5	8891.44	0.318	259.1961	1.291
2850	2577	94,840.0	71,461.3	9246.23	0.308	259.5214	1.291
2875	2602	95,762.3	72,176.7	9612.03	0.299	259.8440	1.291
2900	2627	96,685.2	72,891.1	9988.31	0.290	260.1632	1.290
2925	2652	97,609.5	73,607.7	10,376.35	0.282	260.4801	1.290

(continued)

(continued)

Temperature		h	u			s_o	
K	°C	kJ/kgmol	kJ/kgmol	Pr	Vr	kJ/kgmol/K	Gamma
2950	2677	98,533.2	74,323.8	10,777.49	0.274	260.7954	1.290
2975	2702	99,457.2	75,039.2	11,188.71	0.266	261.1068	1.290
3000	2727	100,382.7	75,757.0	11,613.83	0.258	261.4168	1.290
3025	2752	101,307.1	76,473.7	12,049.96	0.251	261.7233	1.290
3050	2777	102,233.1	77,192.0	12,500.42	0.244	262.0284	1.290
3075	2802	103,158.5	77,909.7	12,963.57	0.237	262.3308	1.289
3100	2827	104,084.9	78,628.4	13,439.74	0.231	262.6307	1.289
3125	2852	105,011.0	79,346.8	13,930.07	0.224	262.9286	1.289
3150	2877	105,937.8	80,065.2	14,434.00	0.218	263.2241	1.289
3175	2902	106,865.5	80,786.8	14,951.77	0.212	263.5171	1.289
3200	2927	107,793.2	81,505.2	15,484.30	0.207	263.8080	1.289
3225	2952	108,720.8	82,225.1	16,030.68	0.201	264.0963	1.289
3250	2977	109,649.0	82,945.6	16,593.23	0.196	264.3831	1.289
3275	3002	110,577.2	83,666.1	17,171.93	0.191	264.6681	1.288
3300	3027	111,504.7	84,386.0	17,764.55	0.186	264.9502	1.288
3325	3052	112,434.6	85,108.1	18,374.79	0.181	265.2310	1.288
3350	3077	113,363.7	85,829.6	18,999.02	0.176	265.5087	1.288
3375	3102	114,292.6	86,550.8	19,643.30	0.172	265.7859	1.288
3400	3127	115,224.0	87,272.8	20,301.83	0.167	266.0601	1.288
3425	3152	116,154.8	87,995.9	20,979.58	0.163	266.3331	1.288
3450	3177	117,084.1	88,719.2	21,673.51	0.159	266.6036	1.287
3475	3202	118,014.7	89,442.1	22,386.46	0.155	266.8727	1.287
3500	3227	118,947.9	90,166.0	23,118.61	0.151	267.1403	1.287
3525	3252	119,880.8	90,891.2	23,867.95	0.148	267.4055	1.287
3550	3277	120,813.0	91,615.7	24,637.25	0.144	267.6692	1.287

A.13.8 Nitrogen Properties (English Units)

Temperature		h	u			s_o	
°R	°F	Btu/lbmol	Btu/lbmol	Pr	Vr	Btu/lbmol/°R	Gamma
300	−159.7	2081.9	1486.2	0.177	1692.8	41.6870	1.400
340	−119.7	2360.0	1687.3	0.275	1237.9	42.5571	1.400
380	−79.7	2638.2	1888.1	0.406	937.06	43.3308	1.399
420	−39.7	2916.5	2088.5	0.576	729.35	44.0272	1.399
460	0.3	3194.8	2288.7	0.792	580.81	44.6601	1.400
500	40.3	3473.0	2488.5	1.061	471.43	45.2400	1.400
540	80.3	3751.2	2688.2	1.389	388.86	45.7752	1.400
580	120.3	4029.4	2887.8	1.784	325.19	46.2722	1.400
620	160.3	4307.7	3087.4	2.253	275.18	46.7363	1.399

(continued)

(continued)

Temperature		h	u			s_o	
°R	°F	Btu/lbmol	Btu/lbmol	Pr	Vr	Btu/lbmol/°R	Gamma
660	200.3	4586.3	3287.2	2.805	235.26	47.1717	1.399
700	240.3	4865.2	3487.2	3.449	202.95	47.5819	1.398
740	280.3	5144.5	3687.6	4.194	176.46	47.9700	1.397
780	320.3	5424.5	3888.5	5.048	154.50	48.3384	1.396
820	360.3	5705.1	4090.1	6.024	136.12	48.6893	1.394
860	400.3	5986.5	4292.4	7.131	120.60	49.0243	1.392
900	440.3	6268.8	4495.6	8.382	107.38	49.3451	1.391
940	480.3	6552.0	4699.8	9.788	96.040	49.6531	1.389
980	520.3	6836.4	4904.9	11.362	86.252	49.9493	1.387
1020	560.3	7121.8	5111.2	13.119	77.752	50.2348	1.384
1060	600.3	7408.4	5318.6	15.072	70.330	50.5104	1.382
1100	640.3	7696.3	5527.3	17.237	63.815	50.7770	1.380
1140	680.3	7985.5	5737.2	19.631	58.073	51.0352	1.378
1180	720.3	8275.9	5948.5	22.269	52.988	51.2856	1.375
1220	760.3	8567.8	6161.0	25.171	48.469	51.5288	1.373
1260	800.3	8861.0	6375.0	28.354	44.438	51.7653	1.370
1300	840.3	9155.6	6590.3	31.839	40.831	51.9955	1.368
1340	880.3	9451.6	6807.1	35.645	37.593	52.2198	1.366
1380	920.3	9749.0	7025.2	39.795	34.677	52.4385	1.363
1420	960.3	10,047.9	7244.7	44.312	32.046	52.6520	1.361
1460	1000.3	10,348.1	7465.6	49.217	29.664	52.8605	1.359
1500	1040.3	10,649.8	7688.0	54.537	27.504	53.0643	1.356
1540	1080.3	10,952.8	7911.7	60.297	25.540	53.2637	1.354
1580	1120.3	11,257.2	8136.8	66.523	23.751	53.4588	1.352
1620	1160.3	11,563.0	8363.2	73.244	22.118	53.6499	1.350
1660	1200.3	11,870.1	8591.0	80.487	20.624	53.8372	1.348
1700	1240.3	12,178.6	8820.1	88.285	19.256	54.0208	1.346
1740	1280.3	12,488.3	9050.5	96.664	18.001	54.2009	1.344
1780	1320.3	12,799.3	9282.2	105.660	16.846	54.3776	1.342
1820	1360.3	13,111.6	9515.1	115.308	15.784	54.5511	1.340
1860	1400.3	13,425.1	9749.2	125.636	14.805	54.7215	1.339
1900	1440.3	13,739.7	9984.5	136.685	13.901	54.8889	1.337
1940	1480.3	14,055.5	10,221.0	148.490	13.065	55.0534	1.335
1980	1520.3	14,372.5	10,458.6	161.090	12.291	55.2151	1.334
2020	1560.3	14,690.6	10,697.3	174.521	11.575	55.3741	1.332
2060	1600.3	15,009.8	10,937.1	188.827	10.909	55.5306	1.331
2100	1640.3	15,330.0	11,177.9	204.048	10.292	55.6845	1.329
2140	1680.3	15,651.2	11,419.8	220.229	9.717	55.8361	1.328
2180	1720.3	15,973.4	11,662.6	237.408	9.183	55.9852	1.327
2220	1760.3	16,296.6	11,906.4	255.635	8.684	56.1321	1.325
2260	1800.3	16,620.7	12,151.2	274.959	8.219	56.2768	1.324
2300	1840.3	16,945.7	12,396.7	295.426	7.785	56.4194	1.323

(continued)

(continued)

Temperature		h	u			s_o	
°R	°F	Btu/lbmol	Btu/lbmol	Pr	Vr	Btu/lbmol/°R	Gamma
2340	1880.3	17,271.6	12,643.3	317.078	7.380	56.5599	1.322
2380	1920.3	17,598.4	12,890.6	339.970	7.001	56.6983	1.321
2420	1960.3	17,925.9	13,138.8	364.158	6.645	56.8348	1.320
2460	2000.3	18,254.3	13,387.8	389.695	6.313	56.9694	1.319
2500	2040.3	18,583.4	13,637.5	416.625	6.001	57.1021	1.318
2540	2080.3	18,913.3	13,888.0	445.019	5.708	57.2330	1.317
2580	2120.3	19,243.9	14,139.2	474.922	5.432	57.3622	1.316
2620	2160.3	19,575.2	14,391.1	506.397	5.174	57.4896	1.315
2660	2200.3	19,907.2	14,643.7	539.497	4.931	57.6153	1.314
2700	2240.3	20,239.9	14,897.0	574.296	4.701	57.7395	1.313
2740	2280.3	20,573.1	15,150.8	610.851	4.486	57.8620	1.313
2780	2320.3	20,907.0	15,405.4	649.229	4.282	57.9830	1.312
2820	2360.3	21,241.5	15,660.4	689.473	4.090	58.1024	1.311
2860	2400.3	21,576.5	15,916.1	731.685	3.909	58.2204	1.310
2900	2440.3	21,912.2	16,172.2	775.912	3.738	58.3370	1.310
2940	2480.3	22,248.3	16,429.0	822.200	3.576	58.4520	1.309
2980	2520.3	22,585.0	16,686.3	870.690	3.423	58.5658	1.308
3020	2560.3	22,922.3	16,944.2	921.370	3.278	58.6782	1.308
3060	2600.3	23,260.0	17,202.4	974.393	3.140	58.7893	1.307
3100	2640.3	23,598.0	17,461.2	1029.80	3.010	58.8991	1.307
3140	2680.3	23,936.7	17,720.4	1087.65	2.887	59.0077	1.306
3180	2720.3	24,275.9	17,980.2	1148.04	2.770	59.1150	1.306
3220	2760.3	24,615.5	18,240.3	1211.08	2.659	59.2211	1.305
3260	2800.3	24,955.4	18,500.9	1276.77	2.553	59.3260	1.305
3300	2840.3	25,295.9	18,761.9	1345.29	2.453	59.4298	1.304
3340	2880.3	25,636.7	19,023.3	1416.60	2.358	59.5324	1.304
3380	2920.3	25,977.8	19,285.1	1490.97	2.267	59.6340	1.303
3420	2960.3	26,319.4	19,547.2	1568.34	2.181	59.7345	1.303
3460	3000.3	26,661.4	19,809.7	1648.83	2.098	59.8339	1.302
3500	3040.3	27,003.8	20,072.7	1732.57	2.020	59.9322	1.302
3540	3080.3	27,346.5	20,336.0	1819.63	1.945	60.0296	1.302
3580	3120.3	27,689.5	20,599.7	1910.11	1.874	60.1260	1.301
3620	3160.3	28,032.8	20,863.5	2004.12	1.806	60.2214	1.301
3660	3200.3	28,376.6	21,127.9	2101.77	1.741	60.3158	1.300
3700	3240.3	28,720.7	21,392.7	2203.02	1.680	60.4093	1.300
3740	3280.3	29,065.2	21,657.7	2308.13	1.620	60.5018	1.300
3780	3320.3	29,409.8	21,922.9	2417.25	1.564	60.5936	1.299
3820	3360.3	29,754.8	22,188.4	2530.38	1.510	60.6844	1.299
3860	3400.3	30,100.3	22,454.5	2647.55	1.458	60.7743	1.299
3900	3440.3	30,445.8	22,720.7	2769.04	1.408	60.8634	1.298
3940	3480.3	30,791.8	22,987.2	2894.83	1.361	60.9516	1.298
3980	3520.3	31,137.9	23,253.9	3025.20	1.316	61.0391	1.298

(continued)

(continued)

Temperature		h	u			s_o	
°R	°F	Btu/lbmol	Btu/lbmol	Pr	Vr	Btu/lbmol/°R	Gamma
4020	3560.3	31,484.6	23,521.1	3159.98	1.272	61.1256	1.297
4060	3600.3	31,831.1	23,788.4	3299.54	1.230	61.2115	1.297
4100	3640.3	32,178.1	24,055.9	3443.92	1.191	61.2965	1.297
4140	3680.3	32,525.6	24,324.0	3593.33	1.152	61.3808	1.296
4180	3720.3	32,873.0	24,591.9	3747.78	1.115	61.4644	1.296
4220	3760.3	33,220.9	24,860.3	3907.33	1.080	61.5472	1.296
4260	3800.3	33,569.1	25,129.1	4072.25	1.046	61.6293	1.296
4300	3840.3	33,917.3	25,398.0	4242.58	1.014	61.7107	1.295
4340	3880.3	34,266.0	25,667.3	4418.55	0.982	61.7914	1.295
4380	3920.3	34,614.7	25,936.5	4600.30	0.952	61.8714	1.295
4420	3960.3	34,963.6	26,206.0	4787.81	0.923	61.9508	1.294
4460	4000.3	35,313.2	26,476.1	4981.34	0.895	62.0295	1.294
4500	4040.3	35,662.5	26,746.3	5180.90	0.869	62.1075	1.294
4540	4080.3	36,012.3	27,016.7	5386.68	0.843	62.1848	1.294
4580	4120.3	36,362.4	27,287.3	5598.82	0.818	62.2615	1.294
4620	4160.3	36,712.3	27,557.7	5817.90	0.794	62.3377	1.293
4660	4200.3	37,062.9	27,828.8	6043.10	0.771	62.4132	1.293
4700	4240.3	37,413.4	28,099.9	6275.62	0.749	62.4881	1.293
4740	4280.3	37,764.0	28,371.1	6514.80	0.728	62.5624	1.293
4780	4320.3	38,115.1	28,642.7	6761.13	0.707	62.6361	1.292
4820	4360.3	38,466.5	28,914.7	7015.17	0.687	62.7094	1.292
4860	4400.3	38,817.8	29,186.5	7276.14	0.668	62.7819	1.292
4900	4440.3	39,169.1	29,458.4	7544.44	0.649	62.8538	1.292
4940	4480.3	39,520.6	29,730.4	7821.43	0.632	62.9254	1.292
4980	4520.3	39,873.0	30,003.4	8105.91	0.614	62.9963	1.291
5020	4560.3	40,224.8	30,275.7	8398.19	0.598	63.0667	1.291
5060	4600.3	40,577.1	30,548.5	8699.26	0.582	63.1366	1.291
5100	4640.3	40,929.5	30,821.5	9008.12	0.566	63.2059	1.291
5140	4680.3	41,281.6	31,094.5	9326.59	0.551	63.2749	1.291
5180	4720.3	41,634.7	31,368.2	9653.38	0.537	63.3433	1.291
5220	4760.3	41,987.2	31,641.2	9988.32	0.523	63.4110	1.290
5260	4800.3	42,340.2	31,914.7	10,333.02	0.509	63.4784	1.290
5300	4840.3	42,693.8	32,188.9	10,686.94	0.496	63.5453	1.290
5340	4880.3	43,046.5	32,462.5	11,050.74	0.483	63.6118	1.290
5380	4920.3	43,399.9	32,736.1	11,423.45	0.471	63.6776	1.290
5420	4960.3	43,753.4	33,010.2	11,806.61	0.459	63.7432	1.290
5460	5000.3	44,107.0	33,284.6	12,198.95	0.448	63.8081	1.290
5500	5040.3	44,460.7	33,558.8	12,602.40	0.436	63.8727	1.290
5540	5080.3	44,814.5	33,832.9	13,015.39	0.426	63.9367	1.289
5580	5120.3	45,168.7	34,108.0	13,439.72	0.415	64.0004	1.289
5620	5160.3	45,522.8	34,382.3	13,874.76	0.405	64.0637	1.289
5660	5200.3	45,876.8	34,657.2	14,319.60	0.395	64.1264	1.289

(continued)

(continued)

Temperature		h	u			s_o	
°R	°F	Btu/lbmol	Btu/lbmol	Pr	Vr	Btu/lbmol/°R	Gamma
5700	5240.3	46,231.2	34,931.8	14,777.22	0.386	64.1888	1.289
5740	5280.3	46,586.3	35,207.4	15,245.71	0.376	64.2508	1.289
5780	5320.3	46,940.0	35,482.3	15,725.32	0.368	64.3123	1.289
5820	5360.3	47,294.6	35,757.5	16,217.27	0.359	64.3735	1.289
5860	5400.3	47,648.8	36,032.3	16,721.14	0.350	64.4342	1.288
5900	5440.3	48,004.0	36,308.0	17,236.29	0.342	64.4945	1.288
5940	5480.3	48,358.9	36,583.4	17,764.58	0.334	64.5545	1.288
5980	5520.3	48,714.6	36,859.7	18,306.96	0.327	64.6142	1.288
6020	5560.3	49,069.8	37,135.4	18,859.23	0.319	64.6732	1.288
6060	5600.3	49,424.7	37,410.9	19,427.11	0.312	64.7321	1.288
6100	5640.3	49,780.2	37,687.0	20,007.33	0.305	64.7906	1.288
6140	5680.3	50,135.4	37,962.7	20,600.29	0.298	64.8486	1.288
6180	5720.3	50,491.1	38,238.9	21,209.51	0.291	64.9064	1.288
6220	5760.3	50,847.7	38,516.1	21,830.38	0.285	64.9637	1.287
6260	5800.3	51,202.9	38,791.8	22,466.66	0.279	65.0208	1.287
6300	5840.3	51,559.1	39,068.6	23,118.64	0.273	65.0776	1.287
6340	5880.3	51,915.2	39,345.2	23,782.05	0.267	65.1338	1.287
6380	5920.3	52,271.8	39,622.4	24,462.81	0.261	65.1898	1.287

A.13.9 Oxygen Properties (SI Units)

Temperature		h	u			s_o	
K	°C	kJ/kgmol	kJ/kgmol	Pr	Vr	kJ/kgmol/K	Gamma
200	−73.1	4844.7	3459.2	0.330	1074.6	188.0074	1.399
225	−48.1	5573.2	3979.9	0.510	444.49	191.4393	1.399
250	−23.1	6302.3	4501.1	0.730	341.27	194.5121	1.398
275	1.9	7032.8	5023.8	1.020	268.54	197.2970	1.397
300	26.9	7765.6	5548.8	1.390	215.56	199.8474	1.395
325	51.9	8501.6	6077.0	1.850	175.89	202.2038	1.392
350	76.9	9241.6	6609.2	2.410	145.49	204.3974	1.389
375	101.9	9986.3	7146.0	3.080	121.74	206.4524	1.385
400	126.9	10,736.2	7688.1	3.890	102.89	208.3882	1.381
425	151.9	11,491.8	8235.9	4.850	87.693	210.2205	1.377
450	176.9	12,253.4	8789.6	5.980	75.307	211.9616	1.373
475	201.9	13,021.2	9349.5	7.300	65.100	213.6220	1.369
500	226.9	13,795.1	9915.7	8.830	56.611	215.2100	1.365
525	251.9	14,575.5	10,488.2	10.610	49.493	216.7328	1.361
550	276.9	15,362.2	11,067.0	12.650	43.479	218.1965	1.357
575	301.9	16,155.0	11,652.1	14.990	38.366	219.6063	1.353

(continued)

(continued)

Temperature		h	u			s_o	
K	°C	kJ/kgmol	kJ/kgmol	Pr	Vr	kJ/kgmol/K	Gamma
600	326.9	16,954.0	12,243.2	17.650	33.992	220.9665	1.350
625	351.9	17,759.0	12,840.3	20.670	30.230	222.2808	1.346
650	376.9	18,569.7	13,443.2	24.090	26.980	223.5526	1.343
675	401.9	19,386.0	14,051.6	27.940	24.158	224.7849	1.340
700	426.9	20,207.7	14,665.5	32.260	21.697	225.9803	1.337
725	451.9	21,034.5	15,284.5	37.100	19.544	227.1409	1.334
750	476.9	21,866.4	15,908.5	42.490	17.653	228.2688	1.332
775	501.9	22,702.9	16,537.2	48.480	15.986	229.3661	1.329
800	526.8	23,544.0	17,170.5	55.130	14.512	230.4342	1.327
825	551.8	24,389.3	17,807.9	62.470	13.206	231.4744	1.325
850	576.8	25,238.9	18,449.7	70.580	12.043	232.4888	1.323
875	601.8	26,092.4	19,095.3	79.500	11.006	233.4786	1.321
900	626.8	26,949.3	19,744.4	89.300	10.079	234.4443	1.319
925	651.8	27,810.0	20,397.4	100.030	9.248	235.3876	1.318
950	676.8	28,674.0	21,053.5	111.750	8.501	236.3092	1.316
975	701.8	29,541.2	21,712.9	124.550	7.828	237.2105	1.314
1000	726.8	30,411.7	22,375.4	138.470	7.222	238.0915	1.313
1025	751.8	31,284.8	23,040.8	153.610	6.673	238.9542	1.312
1050	776.8	32,160.9	23,708.9	170.030	6.175	239.7984	1.310
1075	801.8	33,039.5	24,379.7	187.820	5.723	240.6257	1.309
1100	826.8	33,921.0	25,053.3	207.050	5.313	241.4361	1.308
1125	851.8	34,804.5	25,729.1	227.810	4.938	242.2303	1.307
1150	876.8	35,690.4	26,407.2	250.190	4.597	243.0094	1.306
1175	901.8	36,578.8	27,087.6	274.280	4.284	243.7737	1.305
1200	926.8	37,469.3	27,770.3	300.160	3.998	244.5233	1.304
1225	951.8	38,362.2	28,455.3	327.960	3.735	245.2596	1.303
1250	976.8	39,256.8	29,142.2	357.750	3.494	245.9826	1.302
1275	1002	40,153.4	29,830.7	389.660	3.272	246.6929	1.301
1300	1027	41,051.8	30,521.4	423.800	3.068	247.3910	1.300
1325	1052	41,952.2	31,214.1	460.240	2.879	248.0768	1.300
1350	1077	42,854.5	31,908.4	499.160	2.705	248.7517	1.299
1375	1102	43,758.2	32,604.3	540.600	2.543	249.4148	1.298
1400	1127	44,664.2	33,302.6	584.770	2.394	250.0678	1.297
1425	1152	45,572.1	34,002.5	631.780	2.256	250.7106	1.297
1450	1177	46,480.9	34,703.6	681.690	2.127	251.3426	1.296
1475	1202	47,391.8	35,406.7	734.760	2.007	251.9659	1.295
1500	1227	48,304.3	36,111.1	791.040	1.896	252.5795	1.295
1525	1252	49,218.7	36,817.5	850.640	1.793	253.1835	1.294
1550	1277	50,133.9	37,525.4	913.810	1.696	253.7789	1.293
1575	1302	51,051.3	38,234.7	980.670	1.606	254.3660	1.293
1600	1327	51,970.3	38,946.0	1051.430	1.522	254.9453	1.292
1625	1352	52,890.2	39,657.4	1126.040	1.443	255.5152	1.291

(continued)

(continued)

Temperature		h	u			s_o	
K	°C	kJ/kgmol	kJ/kgmol	Pr	Vr	kJ/kgmol/K	Gamma
1650	1377	53,811.9	40,372.2	1205.050	1.369	256.0790	1.291
1675	1402	54,735.2	41,087.9	1288.250	1.300	256.6340	1.290
1700	1427	55,660.0	41,804.1	1375.960	1.236	257.1816	1.290
1725	1452	56,586.7	42,523.1	1468.570	1.175	257.7232	1.289
1750	1477	57,513.6	43,242.3	1565.810	1.118	258.2562	1.288
1775	1502	58,444.0	43,964.2	1668.390	1.064	258.7838	1.288
1800	1527	59,374.3	44,687.6	1776.380	1.013	259.3051	1.287
1825	1552	60,306.7	45,412.4	1889.640	0.966	259.8190	1.287
1850	1577	61,239.9	46,137.1	2008.650	0.921	260.3268	1.286
1875	1602	62,175.3	46,864.8	2133.750	0.879	260.8291	1.286
1900	1627	63,111.9	47,593.7	2264.990	0.839	261.3253	1.285
1925	1652	64,050.0	48,323.3	2402.790	0.801	261.8163	1.284
1950	1677	64,989.3	49,054.8	2547.020	0.766	262.3010	1.284
1975	1702	65,929.9	49,786.9	2697.750	0.732	262.7790	1.283
2000	1727	66,870.6	50,521.6	2856.260	0.700	263.2536	1.283
2025	1752	67,815.3	51,258.6	3021.750	0.670	263.7219	1.282
2050	1777	68,759.6	51,993.6	3195.660	0.641	264.1871	1.282
2075	1802	69,706.0	52,732.3	3376.570	0.615	264.6449	1.281
2100	1827	70,653.6	53,472.2	3565.950	0.589	265.0986	1.281
2125	1852	71,602.0	54,212.9	3763.990	0.565	265.5479	1.280
2150	1877	72,552.5	54,955.7	3970.630	0.541	265.9922	1.280
2175	1902	73,505.1	55,699.0	4186.520	0.520	266.4324	1.279
2200	1927	74,456.5	56,444.3	4412.320	0.499	266.8691	1.279
2225	1952	75,410.8	57,190.9	4646.560	0.479	267.2992	1.278
2250	1977	76,366.4	57,938.9	4892.130	0.460	267.7273	1.278
2275	2002	77,324.9	58,688.0	5147.180	0.442	268.1498	1.277
2300	2027	78,282.5	59,439.6	5413.640	0.425	268.5695	1.277
2325	2052	79,242.9	60,190.7	5690.300	0.409	268.9838	1.276
2350	2077	80,202.2	60,942.3	5978.470	0.393	269.3945	1.276
2375	2102	81,165.1	61,697.5	6278.510	0.378	269.8016	1.275
2400	2127	82,128.8	62,451.9	6591.310	0.364	270.2058	1.275
2425	2152	83,092.6	63,209.4	6915.330	0.351	270.6048	1.274
2450	2177	84,060.2	63,967.9	7253.860	0.338	271.0021	1.274
2475	2202	85,028.5	64,726.9	7605.280	0.325	271.3954	1.273
2500	2227	85,997.5	65,489.8	7969.600	0.314	271.7845	1.273
2525	2252	86,964.4	66,250.6	8348.850	0.302	272.1710	1.273
2550	2277	87,935.4	67,015.6	8740.540	0.292	272.5521	1.272
2575	2302	88,909.4	67,780.3	9149.560	0.281	272.9323	1.272
2600	2327	89,883.7	68,545.3	9573.050	0.272	273.3085	1.271
2625	2352	90,857.1	69,312.6	10,011.66	0.262	273.6809	1.271
2650	2377	91,831.8	70,077.9	10,468.70	0.253	274.0521	1.270
2675	2402	92,811.6	70,851.8	10,942.12	0.244	274.4198	1.270

(continued)

(continued)

Temperature		*h*	*u*			s_o	
K	°C	kJ/kgmol	kJ/kgmol	Pr	Vr	kJ/kgmol/K	Gamma
2700	2427	93,788.0	71,618.8	11,430.69	0.236	274.7829	1.269
2725	2452	94,769.9	72,391.4	11,939.27	0.228	275.1448	1.269
2750	2477	95,754.4	73,166.6	12,466.23	0.221	275.5039	1.269
2775	2502	96,733.4	73,939.4	13,008.47	0.213	275.8579	1.268
2800	2527	97,720.0	74,716.8	13,570.96	0.206	276.2098	1.268
2825	2552	98,698.0	75,492.0	14,159.12	0.200	276.5625	1.267
2850	2577	99,688.5	76,269.9	14,760.84	0.193	276.9085	1.267
2875	2602	100,674.1	77,052.6	15,385.88	0.187	277.2533	1.267
2900	2627	101,660.0	77,829.3	16,033.14	0.181	277.5959	1.266
2925	2652	102,649.4	78,609.3	16,702.24	0.175	277.9358	1.266
2950	2677	103,642.2	79,392.9	17,392.24	0.170	278.2724	1.265
2975	2702	104,632.1	80,176.6	18,109.35	0.164	278.6083	1.265
3000	2727	105,629.1	80,964.3	18,842.33	0.159	278.9381	1.265
3025	2752	106,618.5	81,747.7	19,605.69	0.154	279.2683	1.264
3050	2777	107,617.3	82,537.1	20,394.44	0.150	279.5962	1.264
3075	2802	108,613.8	83,324.3	21,208.77	0.145	279.9217	1.264
3100	2827	109,604.7	84,112.4	22,050.11	0.141	280.2451	1.263
3125	2852	110,603.4	84,901.8	22,912.96	0.136	280.5643	1.263
3150	2877	111,600.9	85,690.0	23,813.95	0.132	280.8849	1.263
3175	2902	112,598.4	86,484.7	24,740.19	0.128	281.2021	1.262
3200	2927	113,599.7	87,276.7	25,688.12	0.125	281.5147	1.262
3225	2952	114,600.5	88,068.1	26,669.83	0.121	281.8265	1.262
3250	2977	115,604.2	88,862.5	27,674.83	0.117	282.1341	1.261
3275	3002	116,609.5	89,658.5	28,729.51	0.114	282.4450	1.261
3300	3027	117,612.1	90,458.3	29,793.74	0.111	282.7474	1.261
3325	3052	118,627.2	91,257.6	30,908.52	0.108	283.0528	1.260
3350	3077	119,629.9	92,057.5	32,048.53	0.105	283.3539	1.260
3375	3102	120,639.1	92,857.4	33,222.66	0.102	283.6530	1.260
3400	3127	121,644.6	93,653.5	34,440.44	0.099	283.9523	1.259
3425	3152	122,653.0	94,459.1	35,676.52	0.096	284.2455	1.259
3450	3177	123,666.0	95,262.8	36,961.71	0.093	284.5397	1.259
3475	3202	124,674.9	96,062.4	38,297.42	0.091	284.8348	1.259
3500	3227	125,690.8	96,869.0	39,663.05	0.088	285.1261	1.258
3525	3252	126,703.0	97,671.8	41,036.38	0.086	285.4091	1.258
3550	3277	127,718.6	98,484.7	42,482.84	0.084	285.6971	1.258

A.13.10 Oxygen Properties (English Units)

Temperature		h	u			s_o	
°R	°F	Btu/lbmol	Btu/lbmol	Pr	Vr	Btu/lbmol/°R	Gamma
300	−159.7	2083.0	1487.3	0.177	1696.0	44.9080	1.400
340	−119.7	2361.2	1686.1	0.274	1240.0	45.7784	1.399
380	−79.7	2639.5	1885.0	0.405	938.55	46.5524	1.399
420	−39.7	2918.0	2084.0	0.575	730.37	47.2491	1.399
460	0.3	3196.7	2283.4	0.791	581.31	47.8831	1.398
500	40.3	3476.1	2483.2	1.061	471.30	48.4653	1.397
540	80.3	3756.2	2683.9	1.392	388.01	49.0043	1.395
580	120.3	4037.4	2885.7	1.792	323.61	49.5066	1.392
620	160.3	4319.9	3088.8	2.272	272.87	49.9777	1.390
660	200.3	4604.0	3293.5	2.841	232.28	50.4217	1.387
700	240.3	4889.8	3499.8	3.511	199.35	50.8421	1.383
740	280.3	5177.5	3708.1	4.294	172.33	51.2418	1.380
780	320.3	5467.1	3918.3	5.203	149.92	51.6229	1.376
820	360.3	5758.8	4130.6	6.252	131.17	51.9876	1.372
860	400.3	6052.6	4344.9	7.456	115.34	52.3375	1.369
900	440.3	6348.6	4561.5	8.832	101.90	52.6738	1.365
940	480.3	6646.7	4780.1	10.398	90.403	52.9979	1.361
980	520.3	6946.9	5001.0	12.171	80.517	53.3106	1.358
1020	560.3	7249.3	5223.9	14.174	71.965	53.6131	1.355
1060	600.3	7553.8	5448.9	16.425	64.536	53.9058	1.351
1100	640.3	7860.3	5676.0	18.949	58.052	54.1897	1.348
1140	680.3	8168.8	5905.1	21.768	52.371	54.4651	1.345
1180	720.3	8479.2	6136.0	24.909	47.372	54.7328	1.342
1220	760.3	8791.5	6368.9	28.397	42.962	54.9931	1.340
1260	800.3	9105.7	6603.7	32.262	39.055	55.2465	1.337
1300	840.3	9421.6	6840.1	36.531	35.586	55.4932	1.335
1340	880.3	9739.2	7078.3	41.237	32.495	55.7339	1.332
1380	920.3	10,058.4	7318.1	46.411	29.734	55.9686	1.330
1420	960.3	10,379.2	7559.4	52.088	27.262	56.1977	1.328
1460	1000.3	10,701.4	7802.3	58.302	25.042	56.4216	1.326
1500	1040.3	11,025.2	8046.6	65.092	23.044	56.6403	1.324
1540	1080.3	11,350.2	8292.2	72.494	21.243	56.8542	1.322
1580	1120.3	11,676.6	8539.1	80.549	19.615	57.0634	1.321
1620	1160.3	12,004.2	8787.4	89.298	18.142	57.2682	1.319
1660	1200.3	12,333.1	9036.8	98.787	16.804	57.4687	1.318
1700	1240.3	12,663.1	9287.3	109.057	15.588	57.6651	1.316
1740	1280.3	12,994.2	9539.1	120.159	14.481	57.8577	1.315
1780	1320.3	13,326.4	9791.8	132.140	13.471	58.0464	1.314
1820	1360.3	13,659.6	10,045.5	145.054	12.547	58.2316	1.312
1860	1400.3	13,993.7	10,300.2	158.939	11.703	58.4131	1.311

(continued)

(continued)

Temperature		h	u			s_o	
°R	°F	Btu/lbmol	Btu/lbmol	Pr	Vr	Btu/lbmol/°R	Gamma
1900	1440.3	14,328.8	10,555.9	173.867	10.928	58.5914	1.310
1940	1480.3	14,664.9	10,812.5	189.878	10.217	58.7663	1.309
1980	1520.3	15,001.7	11,069.9	207.053	9.563	58.9383	1.308
2020	1560.3	15,339.4	11,328.1	225.424	8.961	59.1071	1.307
2060	1600.3	15,677.9	11,587.3	245.068	8.406	59.2730	1.306
2100	1640.3	16,017.1	11,847.0	266.042	7.893	59.4361	1.305
2140	1680.3	16,357.1	12,107.6	288.437	7.419	59.5966	1.304
2180	1720.3	16,697.9	12,368.9	312.265	6.981	59.7542	1.304
2220	1760.3	17,039.2	12,630.9	337.650	6.575	59.9094	1.303
2260	1800.3	17,381.4	12,893.6	364.668	6.197	60.0623	1.302
2300	1840.3	17,724.2	13,157.0	393.349	5.847	60.2126	1.301
2340	1880.3	18,067.6	13,420.9	423.798	5.522	60.3607	1.300
2380	1920.3	18,411.9	13,685.7	456.060	5.219	60.5064	1.300
2420	1960.3	18,756.6	13,951.0	490.289	4.936	60.6501	1.299
2460	2000.3	19,101.8	14,216.9	526.500	4.672	60.7916	1.298
2500	2040.3	19,447.6	14,483.3	564.786	4.426	60.9310	1.298
2540	2080.3	19,794.1	14,750.4	605.319	4.196	61.0686	1.297
2580	2120.3	20,141.4	15,017.9	648.093	3.981	61.2042	1.296
2620	2160.3	20,489.0	15,286.3	693.189	3.780	61.3378	1.296
2660	2200.3	20,837.0	15,554.9	740.793	3.591	61.4697	1.295
2700	2240.3	21,185.9	15,824.3	791.037	3.413	61.6000	1.295
2740	2280.3	21,535.4	16,094.4	843.836	3.247	61.7283	1.294
2780	2320.3	21,885.1	16,364.7	899.451	3.091	61.8551	1.293
2820	2360.3	22,235.5	16,635.4	957.983	2.944	61.9803	1.293
2860	2400.3	22,586.5	16,907.2	1019.44	2.805	62.1038	1.292
2900	2440.3	22,938.1	17,179.4	1084.06	2.675	62.2258	1.292
2940	2480.3	23,290.0	17,451.8	1151.88	2.552	62.3463	1.291
2980	2520.3	23,642.0	17,724.4	1223.12	2.436	62.4655	1.291
3020	2560.3	23,994.9	17,997.9	1297.84	2.327	62.5832	1.290
3060	2600.3	24,348.5	18,272.0	1375.96	2.224	62.6993	1.290
3100	2640.3	24,702.3	18,546.3	1458.02	2.126	62.8143	1.289
3140	2680.3	25,057.2	18,821.8	1543.78	2.034	62.9278	1.289
3180	2720.3	25,412.4	19,097.5	1633.68	1.947	63.0402	1.288
3220	2760.3	25,767.9	19,373.6	1727.83	1.864	63.1515	1.287
3260	2800.3	26,123.7	19,649.9	1826.07	1.785	63.2613	1.287
3300	2840.3	26,480.3	19,927.1	1928.68	1.711	63.3699	1.286
3340	2880.3	26,836.9	20,204.6	2036.13	1.640	63.4776	1.286
3380	2920.3	27,194.4	20,482.6	2147.96	1.574	63.5837	1.285
3420	2960.3	27,552.4	20,761.2	2264.99	1.510	63.6891	1.285
3460	3000.3	27,910.7	21,039.7	2387.13	1.449	63.7934	1.285
3500	3040.3	28,269.7	21,319.2	2514.14	1.392	63.8963	1.284
3540	3080.3	28,629.2	21,599.6	2646.86	1.337	63.9985	1.284

(continued)

(continued)

Temperature		h	u			s_o	
°R	°F	Btu/lbmol	Btu/lbmol	Pr	Vr	Btu/lbmol/°R	Gamma
3580	3120.3	28,988.8	21,879.8	2785.34	1.285	64.0997	1.283
3620	3160.3	29,348.9	22,160.4	2929.13	1.236	64.1997	1.283
3660	3200.3	29,710.1	22,442.2	3078.65	1.189	64.2986	1.282
3700	3240.3	30,071.2	22,723.8	3235.02	1.144	64.3970	1.282
3740	3280.3	30,433.1	23,006.3	3397.46	1.101	64.4942	1.281
3780	3320.3	30,795.0	23,288.7	3565.94	1.060	64.5904	1.281
3820	3360.3	31,157.5	23,571.8	3741.25	1.021	64.6857	1.280
3860	3400.3	31,520.6	23,855.4	3924.16	0.984	64.7805	1.280
3900	3440.3	31,884.3	24,139.6	4114.02	0.948	64.8743	1.279
3940	3480.3	32,247.9	24,424.7	4310.84	0.914	64.9671	1.279
3980	3520.3	32,612.7	24,709.1	4515.11	0.881	65.0590	1.278
4020	3560.3	32,978.3	24,995.2	4727.45	0.850	65.1503	1.278
4060	3600.3	33,343.0	25,280.5	4947.04	0.821	65.2404	1.278
4100	3640.3	33,708.5	25,566.6	5175.92	0.792	65.3303	1.277
4140	3680.3	34,075.1	25,854.4	5413.64	0.765	65.4194	1.277
4180	3720.3	34,441.8	26,141.0	5659.10	0.739	65.5075	1.276
4220	3760.3	34,809.4	26,429.1	5914.01	0.714	65.5950	1.276
4260	3800.3	35,176.7	26,717.0	6177.31	0.690	65.6815	1.275
4300	3840.3	35,545.1	27,005.9	6450.56	0.667	65.7674	1.275
4340	3880.3	35,913.5	27,294.8	6734.85	0.644	65.8531	1.275
4380	3920.3	36,282.0	27,584.6	7026.65	0.623	65.9373	1.274
4420	3960.3	36,650.8	27,874.7	7330.08	0.603	66.0213	1.274
4460	4000.3	37,021.3	28,163.6	7644.56	0.583	66.1047	1.273
4500	4040.3	37,392.1	28,455.7	7969.59	0.565	66.1874	1.273
4540	4080.3	37,762.7	28,746.1	8305.63	0.547	66.2694	1.273
4580	4120.3	38,133.7	29,038.3	8653.25	0.529	66.3508	1.272
4620	4160.3	38,504.2	29,330.1	9013.26	0.513	66.4318	1.272
4660	4200.3	38,876.5	29,623.7	9383.72	0.497	66.5117	1.271
4700	4240.3	39,249.3	29,916.3	9767.59	0.481	66.5914	1.271
4740	4280.3	39,622.2	30,210.4	10,162.9	0.466	66.6702	1.271
4780	4320.3	39,995.4	30,503.5	10,573.0	0.452	66.7487	1.270
4820	4360.3	40,369.8	30,797.7	10,995.9	0.438	66.8266	1.270
4860	4400.3	40,741.7	31,090.9	11,430.7	0.425	66.9036	1.269
4900	4440.3	41,117.8	31,386.8	11,879.5	0.412	66.9801	1.269
4940	4480.3	41,492.4	31,682.7	12,343.5	0.400	67.0562	1.269
4980	4520.3	41,867.4	31,977.5	12,823.6	0.388	67.1319	1.268
5020	4560.3	42,244.0	32,275.3	13,321.3	0.377	67.2076	1.268
5060	4600.3	42,619.0	32,570.2	13,831.8	0.366	67.2822	1.268
5100	4640.3	42,995.1	32,867.5	14,356.2	0.355	67.3561	1.267
5140	4680.3	43,373.6	33,165.5	14,897.9	0.345	67.4297	1.267
5180	4720.3	43,750.6	33,464.1	15,455.8	0.335	67.5027	1.267
5220	4760.3	44,126.3	33,761.1	16,033.1	0.326	67.5755	1.266

(continued)

(continued)

Temperature		h	u			s_o	
°R	°F	Btu/lbmol	Btu/lbmol	Pr	Vr	Btu/lbmol/°R	Gamma
5260	4800.3	44,505.8	34,060.4	16,624.9	0.316	67.6475	1.266
5300	4840.3	44,883.6	34,359.5	17,240.2	0.307	67.7197	1.265
5340	4880.3	45,262.6	34,658.3	17,867.5	0.299	67.7906	1.265
5380	4920.3	45,641.6	34,958.6	18,513.8	0.291	67.8612	1.265
5420	4960.3	46,019.8	35,258.0	19,178.5	0.283	67.9312	1.265
5460	5000.3	46,400.1	35,556.8	19,868.9	0.275	68.0015	1.264
5500	5040.3	46,779.6	35,860.3	20,578.2	0.267	68.0711	1.264
5540	5080.3	47,162.7	36,159.1	21,300.1	0.260	68.1396	1.264
5580	5120.3	47,542.1	36,462.5	22,050.1	0.253	68.2083	1.263
5620	5160.3	47,924.1	36,762.9	22,821.5	0.246	68.2766	1.263
5660	5200.3	48,306.8	37,066.9	23,609.7	0.240	68.3440	1.263
5700	5240.3	48,689.4	37,370.7	24,424.0	0.233	68.4114	1.262
5740	5280.3	49,071.1	37,670.9	25,258.6	0.227	68.4781	1.262
5780	5320.3	49,452.4	37,973.4	26,122.2	0.221	68.5449	1.262
5820	5360.3	49,836.4	38,278.6	27,005.5	0.216	68.6109	1.262
5860	5400.3	50,219.4	38,583.0	27,911.8	0.210	68.6765	1.261
5900	5440.3	50,604.0	38,886.0	28,843.0	0.205	68.7416	1.261
5940	5480.3	50,984.9	39,190.9	29,793.6	0.199	68.8060	1.261
5980	5520.3	51,369.0	39,496.3	30,780.5	0.194	68.8707	1.260
6020	5560.3	51,755.9	39,801.7	31,791.3	0.189	68.9349	1.260
6060	5600.3	52,142.4	40,106.5	32,828.2	0.185	68.9986	1.260
6100	5640.3	52,526.9	40,412.4	33,889.5	0.180	69.0618	1.260
6140	5680.3	52,910.4	40,717.1	34,992.4	0.175	69.1254	1.259
6180	5720.3	53,295.7	41,023.6	36,107.9	0.171	69.1877	1.259
6220	5760.3	53,684.3	41,333.4	37,269.0	0.167	69.2506	1.259
6260	5800.3	54,071.3	41,638.9	38,424.6	0.163	69.3112	1.259
6300	5840.3	54,458.4	41,947.2	39,662.9	0.159	69.3742	1.258
6340	5880.3	54,843.1	42,253.2	40,889.6	0.155	69.4347	1.258
6380	5920.3	55,229.9	42,561.3	42,160.0	0.151	69.4955	1.258

A.13.11 Hydrogen Properties (SI Units)

Temperature		h	u			s_o	
K	°C	kJ/kgmol	kJ/kgmol	Pr	Vr	kJ/kgmol/K	Gamma
200	−73.1	4805.7	3420.1	0.350	1029.2	114.446	1.438
225	−48.1	5495.8	3924.5	0.520	435.08	117.697	1.425
250	−23.1	6198.6	4437.6	0.740	338.55	120.659	1.416
275	1.9	6910.5	4956.6	1.020	268.69	123.373	1.409
300	26.9	7628.8	5479.7	1.380	216.99	125.873	1.405

(continued)

(continued)

Temperature		h	u			s_o	
K	°C	kJ/kgmol	kJ/kgmol	Pr	Vr	kJ/kgmol/K	Gamma
325	51.9	8351.6	6005.5	1.830	177.96	128.187	1.402
350	76.9	9077.5	6532.9	2.370	147.95	130.338	1.400
375	101.9	9805.5	7061.1	3.010	124.49	132.347	1.399
400	126.9	10,534.8	7589.8	3.780	105.88	134.230	1.398
425	151.9	11,265.0	8118.5	4.670	90.912	136.001	1.398
450	176.9	11,995.8	8647.1	5.720	78.734	137.672	1.397
475	201.9	12,727.0	9175.5	6.910	68.713	139.253	1.397
500	226.9	13,458.4	9703.7	8.280	60.384	140.754	1.397
525	251.9	14,190.1	10,231.7	9.830	53.397	142.182	1.397
550	276.9	14,922.0	10,759.6	11.580	47.487	143.544	1.396
575	301.9	15,654.3	11,287.4	13.550	42.449	144.846	1.396
600	326.9	16,386.9	11,815.4	15.740	38.124	146.093	1.396
625	351.9	17,120.1	12,343.6	18.180	34.387	147.290	1.395
650	376.9	17,853.8	12,872.2	20.870	31.138	148.441	1.395
675	401.9	18,588.3	13,401.3	23.850	28.299	149.550	1.394
700	426.9	19,323.5	13,931.1	27.130	25.804	150.619	1.394
725	451.9	20,059.8	14,461.6	30.720	23.601	151.653	1.393
750	476.9	20,797.1	14,993.2	34.640	21.649	152.653	1.392
775	501.9	21,535.7	15,525.8	38.930	19.910	153.621	1.391
800	526.8	22,275.6	16,059.6	43.580	18.356	154.561	1.390
825	551.8	23,016.8	16,594.8	48.640	16.962	155.473	1.389
850	576.8	23,759.7	17,131.3	54.120	15.707	156.360	1.388
875	601.8	24,504.3	17,669.5	60.040	14.575	157.224	1.387
900	626.8	25,250.5	18,209.4	66.430	13.549	158.065	1.385
925	651.8	25,998.7	18,751.0	73.310	12.617	158.884	1.384
950	676.8	26,748.8	19,294.5	80.720	11.769	159.685	1.383
975	701.8	27,500.9	19,840.0	88.670	10.995	160.466	1.381
1000	726.8	28,255.2	20,387.5	97.210	10.287	161.230	1.380
1025	751.8	29,011.6	20,937.1	106.340	9.638	161.977	1.378
1050	776.8	29,770.2	21,489.0	116.120	9.042	162.708	1.377
1075	801.8	30,531.1	22,043.0	126.570	8.493	163.425	1.375
1100	826.8	31,294.6	22,599.6	137.720	7.987	164.126	1.373
1125	851.8	32,060.3	23,158.3	149.620	7.519	164.815	1.372
1150	876.8	32,828.6	23,719.8	162.280	7.087	165.490	1.370
1175	901.8	33,599.4	24,283.3	175.750	6.686	166.153	1.368
1200	926.8	34,372.8	24,849.8	190.080	6.313	166.8049	1.367
1225	951.8	35,148.7	25,418.6	205.290	5.967	167.4448	1.365
1250	976.8	35,927.5	25,990.4	221.430	5.645	168.0741	1.363
1275	1001.8	36,708.8	26,564.4	238.530	5.345	168.6927	1.362
1300	1026.8	37,493.0	27,141.3	256.670	5.065	169.3019	1.360
1325	1051.8	38,279.6	27,720.8	275.850	4.803	169.9012	1.358
1350	1076.8	39,068.9	28,303.0	296.150	4.559	170.4915	1.357

(continued)

(continued)

Temperature		h	u			s_o	
K	°C	kJ/kgmol	kJ/kgmol	Pr	Vr	kJ/kgmol/K	Gamma
1375	1101.8	39,861.4	28,888.0	317.590	4.329	171.0727	1.355
1400	1126.8	40,656.2	29,475.7	340.260	4.115	171.6458	1.353
1425	1151.8	41,453.9	30,066.2	364.180	3.913	172.2106	1.352
1450	1176.8	42,254.3	30,659.3	389.400	3.724	172.7673	1.350
1475	1201.8	43,057.5	31,254.8	415.990	3.546	173.3165	1.348
1500	1226.8	43,863.2	31,853.6	444.010	3.378	173.8584	1.347
1525	1251.8	44,671.7	32,454.5	473.490	3.221	174.3928	1.345
1550	1276.8	45,483.5	33,058.9	504.510	3.072	174.9204	1.344
1575	1301.8	46,297.5	33,665.3	537.150	2.932	175.4416	1.342
1600	1326.8	47,113.8	34,274.7	571.450	2.800	175.9562	1.341
1625	1351.8	47,933.5	34,886.7	607.450	2.675	176.4641	1.339
1650	1376.8	48,755.8	35,501.8	645.290	2.557	176.9665	1.338
1675	1401.8	49,580.5	36,118.8	684.920	2.446	177.4620	1.336
1700	1426.8	50,407.3	36,738.3	726.580	2.340	177.9529	1.335
1725	1451.8	51,237.3	37,361.0	770.150	2.240	178.4370	1.333
1750	1476.8	52,069.7	37,985.7	815.840	2.145	178.9162	1.332
1775	1501.8	52,903.8	38,612.9	863.700	2.055	179.3901	1.331
1800	1526.8	53,740.8	39,241.4	913.720	1.970	179.8581	1.330
1825	1551.8	54,581.4	39,875.1	966.120	1.889	180.3218	1.328
1850	1576.8	55,424.0	40,510.0	1020.85	1.812	180.7799	1.327
1875	1601.8	56,267.7	41,146.0	1078.02	1.739	181.2329	1.326
1900	1626.8	57,114.5	41,785.2	1137.85	1.670	181.6819	1.325
1925	1651.8	57,963.9	42,427.7	1200.31	1.604	182.1262	1.324
1950	1676.8	58,816.2	43,072.4	1265.35	1.541	182.5649	1.322
1975	1701.8	59,669.1	43,717.5	1333.32	1.481	183.0000	1.321
2000	1726.8	60,524.9	44,365.6	1404.24	1.424	183.4308	1.320
2025	1751.8	61,382.1	45,016.0	1478.15	1.370	183.8572	1.319
2050	1776.8	62,242.3	45,667.7	1555.15	1.318	184.2794	1.318
2075	1801.8	63,105.5	46,323.1	1635.33	1.269	184.6973	1.317
2100	1826.8	63,969.2	46,979.2	1718.79	1.222	185.1112	1.316
2125	1851.8	64,836.1	47,638.4	1805.64	1.177	185.5210	1.315
2150	1876.8	65,704.2	48,300.4	1896.32	1.134	185.9284	1.314
2175	1901.8	66,575.0	48,963.5	1990.34	1.093	186.3307	1.313
2200	1926.8	67,446.9	49,627.8	2088.08	1.054	186.729	1.312
2225	1951.8	68,324.0	50,295.5	2189.74	1.016	187.124	1.311
2250	1976.8	69,197.7	50,963.2	2295.26	0.980	187.516	1.310
2275	2001.8	70,077.9	51,634.0	2405.30	0.946	187.905	1.310
2300	2026.8	70,956.1	52,304.6	2519.14	0.913	188.290	1.309
2325	2051.9	71,839.3	52,980.1	2637.34	0.882	188.671	1.308
2350	2076.9	72,722.6	53,657.3	2759.89	0.851	189.048	1.307
2375	2101.9	73,608.8	54,335.9	2887.56	0.822	189.424	1.306
2400	2126.9	74,496.5	55,015.8	3019.41	0.795	189.795	1.305

(continued)

(continued)

Temperature		h	u			s_o	
K	°C	kJ/kgmol	kJ/kgmol	Pr	Vr	kJ/kgmol/K	Gamma
2425	2151.9	75,383.7	55,697.0	3156.42	0.768	190.164	1.305
2450	2176.9	76,275.0	56,378.9	3298.22	0.743	190.530	1.304
2475	2201.9	77,168.5	57,063.2	3444.71	0.718	190.891	1.303
2500	2226.9	78,065.4	57,754.0	3597.46	0.695	191.252	1.302
2525	2251.9	78,958.7	58,441.2	3755.16	0.672	191.608	1.302
2550	2276.9	79,858.3	59,131.5	3918.30	0.651	191.962	1.301
2575	2301.9	80,757.3	59,824.4	4087.42	0.630	192.313	1.300
2600	2326.9	81,660.9	60,518.7	4262.98	0.610	192.663	1.299
2625	2351.9	82,562.7	61,211.2	4443.41	0.591	193.008	1.299
2650	2376.9	83,468.5	61,911.0	4630.75	0.572	193.351	1.298
2675	2401.9	84,374.3	62,607.4	4824.38	0.554	193.691	1.297
2700	2426.9	85,280.6	63,307.7	5023.60	0.537	194.028	1.297
2725	2451.9	86,190.2	64,007.9	5230.92	0.521	194.364	1.296
2750	2476.9	87,100.1	64,711.8	5444.85	0.505	194.697	1.295
2775	2501.9	88,012.9	65,418.5	5666.24	0.490	195.029	1.295
2800	2526.9	88,928.4	66,124.7	5893.76	0.475	195.356	1.294
2825	2551.9	89,842.6	66,829.6	6130.21	0.461	195.683	1.293
2850	2576.9	90,761.3	67,545.5	6370.52	0.447	196.003	1.293
2875	2601.9	91,678.1	68,252.9	6621.13	0.434	196.323	1.292
2900	2626.9	92,599.3	68,964.8	6879.94	0.422	196.642	1.291
2925	2651.9	93,523.4	69,679.6	7146.94	0.409	196.959	1.291
2950	2676.9	94,447.0	70,393.8	7424.87	0.397	197.276	1.290
2975	2701.9	95,365.3	71,109.4	7707.28	0.386	197.586	1.290
3000	2726.9	96,299.9	71,828.1	8001.78	0.375	197.898	1.289
3025	2751.9	97,225.9	72,551.3	8304.25	0.364	198.206	1.289
3050	2776.9	98,155.4	73,271.5	8614.29	0.354	198.511	1.288
3075	2801.9	99,082.3	73,989.1	8934.31	0.344	198.814	1.287
3100	2826.9	100,015.5	74,712.9	9264.16	0.335	199.116	1.287
3125	2851.9	100,948.4	75,443.1	9604.56	0.325	199.416	1.286
3150	2876.9	101,874.6	76,160.0	9951.58	0.317	199.711	1.286
3175	2901.9	102,824.7	76,900.7	10,313.1	0.308	200.008	1.285
3200	2926.9	103,754.9	77,628.2	10,687.4	0.299	200.3040	1.284

A.13.12 Hydrogen Properties (English Units)

Temperature		h	u			s_o	
°R	°F	Btu/lbmol	Btu/lbmol	Pr	Vr	Btu/lbmol/°R	Gamma
300	−159.7	2066.2	1470.5	0.195	1541.3	27.3370	1.467
340	−119.7	2319.9	1656.3	0.290	1171.3	28.1307	1.446

(continued)

(continued)

Temperature		h	u			s_o	
°R	°F	Btu/lbmol	Btu/lbmol	Pr	Vr	Btu/lbmol/°R	Gamma
380	−79.7	2580.6	1847.0	0.418	908.85	28.8554	1.432
420	−39.7	2846.5	2041.2	0.584	718.63	29.5205	1.421
460	0.3	3116.2	2238.0	0.796	577.94	30.1338	1.414
500	40.3	3388.7	2436.7	1.060	471.90	30.7019	1.409
540	80.3	3663.4	2636.7	1.383	390.58	31.2303	1.405
580	120.3	3939.5	2837.6	1.772	327.23	31.7236	1.403
620	160.3	4216.8	3039.0	2.237	277.16	32.1859	1.401
660	200.3	4494.8	3240.8	2.784	237.05	32.6204	1.399
700	240.3	4773.4	3442.8	3.422	204.54	33.0302	1.398
740	280.3	5052.3	3644.9	4.160	177.90	33.4177	1.398
780	320.3	5331.5	3847.0	5.005	155.84	33.7851	1.397
820	360.3	5610.8	4049.0	5.968	137.41	34.1344	1.397
860	400.3	5890.3	4250.9	7.056	121.88	34.4671	1.397
900	440.3	6169.8	4452.8	8.280	108.69	34.7848	1.397
940	480.3	6449.4	4654.5	9.650	97.410	35.0888	1.397
980	520.3	6729.1	4856.3	11.175	87.693	35.3802	1.397
1020	560.3	7009.0	5058.0	12.867	79.276	35.6601	1.396
1060	600.3	7288.9	5259.8	14.734	71.940	35.9293	1.396
1100	640.3	7569.0	5461.6	16.790	65.514	36.1887	1.396
1140	680.3	7849.2	5663.5	19.045	59.857	36.4389	1.395
1180	720.3	8129.7	5865.6	21.512	54.854	36.6807	1.395
1220	760.3	8410.5	6067.8	24.202	50.410	36.9147	1.394
1260	800.3	8691.5	6270.3	27.128	46.447	37.1414	1.394
1300	840.3	8972.9	6473.1	30.303	42.899	37.3612	1.393
1340	880.3	9254.6	6676.1	33.742	39.713	37.5747	1.392
1380	920.3	9536.7	6879.6	37.457	36.842	37.7821	1.391
1420	960.3	9819.3	7083.4	41.465	34.246	37.9839	1.391
1460	1000.3	10,102.3	7287.7	45.779	31.892	38.1805	1.390
1500	1040.3	10,385.8	7492.4	50.417	29.752	38.3721	1.389
1540	1080.3	10,669.9	7697.6	55.392	27.802	38.5590	1.388
1580	1120.3	10,954.6	7903.4	60.724	26.019	38.7415	1.387
1620	1160.3	11,239.8	8109.8	66.427	24.388	38.9198	1.385
1660	1200.3	11,525.7	8316.7	72.521	22.890	39.0941	1.384
1700	1240.3	11,812.3	8524.3	79.026	21.512	39.2647	1.383
1740	1280.3	12,099.5	8732.6	85.961	20.242	39.4317	1.382
1780	1320.3	12,387.4	8941.5	93.341	19.070	39.5953	1.380
1820	1360.3	12,676.1	9151.2	101.193	17.985	39.7557	1.379
1860	1400.3	12,965.5	9361.5	109.533	16.981	39.9129	1.378
1900	1440.3	13,255.7	9572.7	118.388	16.049	40.0673	1.376
1940	1480.3	13,546.7	9784.6	127.776	15.183	40.2189	1.375
1980	1520.3	13,838.5	9997.4	137.723	14.377	40.3677	1.373
2020	1560.3	14,131.1	10,210.8	148.254	13.625	40.5141	1.372

(continued)

(continued)

Temperature		h	u			s_o	
°R	°F	Btu/lbmol	Btu/lbmol	Pr	Vr	Btu/lbmol/°R	Gamma
2060	1600.3	14,424.6	10,425.3	159.396	12.924	40.6580	1.370
2100	1640.3	14,718.8	10,640.4	171.165	12.269	40.7994	1.369
2140	1680.3	15,014.1	10,856.5	183.598	11.656	40.9387	1.367
2180	1720.3	15,310.2	11,073.4	196.724	11.082	41.0758	1.366
2220	1760.3	15,607.1	11,291.2	210.553	10.544	41.2107	1.364
2260	1800.3	15,905.1	11,510.0	225.140	10.038	41.3437	1.363
2300	1840.3	16,203.8	11,729.5	240.496	9.564	41.4747	1.361
2340	1880.3	16,503.5	11,950.1	256.667	9.117	41.6040	1.360
2380	1920.3	16,804.0	12,171.5	273.665	8.697	41.7313	1.358
2420	1960.3	17,105.6	12,393.8	291.537	8.301	41.8569	1.357
2460	2000.3	17,408.0	12,617.1	310.317	7.927	41.9809	1.355
2500	2040.3	17,711.4	12,841.2	330.038	7.575	42.1033	1.354
2540	2080.3	18,015.7	13,066.4	350.727	7.242	42.2240	1.352
2580	2120.3	18,321.1	13,292.4	372.431	6.927	42.3433	1.351
2620	2160.3	18,627.3	13,519.5	395.188	6.630	42.4610	1.350
2660	2200.3	18,934.6	13,747.4	419.029	6.348	42.5774	1.348
2700	2240.3	19,242.4	13,976.2	444.012	6.081	42.6924	1.347
2740	2280.3	19,551.5	14,206.0	470.143	5.828	42.8059	1.345
2780	2320.3	19,861.4	14,436.7	497.506	5.588	42.9183	1.344
2820	2360.3	20,172.2	14,668.2	526.107	5.360	43.0293	1.343
2860	2400.3	20,483.9	14,900.6	555.982	5.144	43.1389	1.341
2900	2440.3	20,796.5	15,133.9	587.240	4.938	43.2476	1.340
2940	2480.3	21,110.2	15,368.3	619.862	4.743	43.3549	1.339
2980	2520.3	21,424.6	15,603.4	653.922	4.557	43.4611	1.337
3020	2560.3	21,739.9	15,839.4	689.489	4.380	43.5663	1.336
3060	2600.3	22,056.0	16,076.3	726.583	4.211	43.6704	1.335
3100	2640.3	22,373.2	16,314.2	765.241	4.051	43.7733	1.334
3140	2680.3	22,690.9	16,552.8	805.452	3.898	43.8750	1.332
3180	2720.3	23,009.8	16,792.3	847.464	3.752	43.9760	1.331
3220	2760.3	23,329.4	17,032.4	891.139	3.613	44.0758	1.330
3260	2800.3	23,650.2	17,273.7	936.736	3.480	44.1749	1.329
3300	2840.3	23,971.1	17,515.6	983.986	3.354	44.2726	1.328
3340	2880.3	24,293.4	17,758.7	1033.37	3.232	44.3698	1.327
3380	2920.3	24,616.2	18,002.0	1084.55	3.116	44.4658	1.326
3420	2960.3	24,939.8	18,246.3	1137.85	3.006	44.5611	1.325
3460	3000.3	25,264.2	18,491.2	1193.24	2.900	44.6555	1.324
3500	3040.3	25,589.6	18,737.5	1250.77	2.798	44.7490	1.323
3540	3080.3	25,916.0	18,984.8	1310.49	2.701	44.8416	1.322
3580	3120.3	26,242.3	19,231.6	1372.46	2.608	44.9334	1.321
3620	3160.3	26,570.1	19,479.9	1436.67	2.520	45.0242	1.320
3660	3200.3	26,898.3	19,729.1	1503.57	2.434	45.1146	1.319
3700	3240.3	27,227.4	19,978.7	1572.68	2.353	45.2038	1.318

(continued)

(continued)

Temperature		h	u			s_o	
°R	°F	Btu/lbmol	Btu/lbmol	Pr	Vr	Btu/lbmol/°R	Gamma
3740	3280.3	27,557.4	20,229.3	1644.49	2.274	45.2925	1.317
3780	3320.3	27,887.0	20,479.4	1718.79	2.199	45.3802	1.316
3820	3360.3	28,218.9	20,731.8	1795.98	2.127	45.4675	1.315
3860	3400.3	28,550.0	20,983.5	1875.66	2.058	45.5537	1.314
3900	3440.3	28,882.4	21,237.2	1958.71	1.991	45.6397	1.314
3940	3480.3	29,215.3	21,490.6	2044.45	1.927	45.7248	1.313
3980	3520.3	29,549.5	21,745.3	2132.74	1.866	45.8087	1.312
4020	3560.3	29,883.8	22,000.2	2224.60	1.807	45.8925	1.311
4060	3600.3	30,219.1	22,256.8	2319.38	1.750	45.9753	1.310
4100	3640.3	30,555.0	22,513.2	2417.69	1.696	46.0578	1.310
4140	3680.3	30,891.1	22,769.1	2519.14	1.643	46.1394	1.309
4180	3720.3	31,227.8	23,027.1	2623.96	1.593	46.2204	1.308
4220	3760.3	31,566.3	23,286.1	2731.96	1.545	46.3005	1.307
4260	3800.3	31,904.5	23,544.9	2844.50	1.498	46.3806	1.307
4300	3840.3	32,243.1	23,804.0	2959.95	1.453	46.4596	1.306
4340	3880.3	32,582.4	24,064.6	3079.53	1.409	46.5383	1.305
4380	3920.3	32,922.2	24,325.6	3203.30	1.367	46.6165	1.304
4420	3960.3	33,263.4	24,586.7	3330.60	1.327	46.6939	1.304
4460	4000.3	33,604.6	24,849.2	3461.81	1.288	46.7706	1.303
4500	4040.3	33,947.7	25,112.1	3597.46	1.251	46.8470	1.302
4540	4080.3	34,288.8	25,374.5	3737.92	1.215	46.9230	1.302
4580	4120.3	34,632.8	25,638.3	3882.32	1.180	46.9983	1.301
4620	4160.3	34,976.3	25,901.6	4030.11	1.146	47.0725	1.300
4660	4200.3	35,318.8	26,166.8	4183.69	1.114	47.1468	1.300
4700	4240.3	35,664.8	26,432.6	4342.71	1.082	47.2208	1.299
4740	4280.3	36,010.7	26,698.3	4504.54	1.052	47.2935	1.298
4780	4320.3	36,354.3	26,963.2	4672.92	1.023	47.3664	1.298
4820	4360.3	36,702.3	27,231.0	4846.26	0.995	47.4387	1.297
4860	4400.3	37,049.9	27,499.9	5023.59	0.967	47.5101	1.297
4900	4440.3	37,397.7	27,767.5	5207.93	0.941	47.5816	1.296
4940	4480.3	37,745.1	28,036.2	5396.59	0.915	47.6523	1.295
4980	4520.3	38,094.5	28,305.4	5591.04	0.891	47.7226	1.295
5020	4560.3	38,442.8	28,575.0	5790.58	0.867	47.7922	1.294
5060	4600.3	38,793.1	28,846.5	5998.17	0.844	47.8622	1.294
5100	4640.3	39,141.9	29,116.6	6207.91	0.822	47.9304	1.293
5140	4680.3	39,495.4	29,388.5	6425.75	0.800	47.9989	1.293
5180	4720.3	39,846.4	29,660.7	6649.98	0.779	48.0670	1.292
5220	4760.3	40,196.6	29,932.2	6879.93	0.759	48.1345	1.291
5260	4800.3	40,549.5	30,203.6	7117.84	0.739	48.2021	1.291
5300	4840.3	40,902.4	30,477.7	7361.83	0.720	48.2690	1.290
5340	4880.3	41,256.2	30,752.8	7610.55	0.702	48.3350	1.290
5380	4920.3	41,610.1	31,027.9	7868.25	0.684	48.4011	1.289

(continued)

(continued)

Temperature		h	u			s_o	
°R	°F	Btu/lbmol	Btu/lbmol	Pr	Vr	Btu/lbmol/°R	Gamma
5420	4960.3	41,960.6	31,299.7	8132.56	0.666	48.4667	1.289
5460	5000.3	42,316.4	31,576.7	8406.08	0.650	48.5324	1.288
5500	5040.3	42,674.0	31,852.7	8681.65	0.634	48.5965	1.288
5540	5080.3	43,026.8	32,129.6	8971.08	0.618	48.6616	1.287
5580	5120.3	43,385.2	32,403.6	9264.15	0.602	48.7254	1.287
5620	5160.3	43,743.0	32,685.5	9566.16	0.587	48.7891	1.286
5660	5200.3	44,099.9	32,960.9	9875.62	0.573	48.8523	1.286
5700	5240.3	44,455.9	33,238.0	10,197.0	0.559	48.9160	1.285
5740	5280.3	44,814.2	33,517.6	10,519.0	0.546	48.9777	1.285
5780	5320.3	45,171.6	33,796.3	10,854.5	0.532	49.0400	1.284

A.14 Thermodynamic Properties for Water

A.14.1 The Saturation Temperature vs. Pressure (SI Units)

Press (kPa)	Temp (K)	Specific Volume (m**3/kg)		Internal Energy (kJ/kg)			Enthalpy (kJ/kg)			Entropy (kJ/kg/K)		
		vf	vg	uf	ufg	ug	hf	hfg	hg	sf	sfg	sg
0.61	273.2	0.001000	205.9975	0.00	2374.9	2374.9	0.00	2500.9	2500.9	0.00000	9.15549	9.15549
0.80	276.9	0.001000	159.6461	15.81	2364.3	2380.1	15.81	2492.0	2507.8	0.05748	8.99925	9.05672
1.00	280.1	0.001000	129.1833	29.30	2355.2	2384.5	29.30	2484.4	2513.7	0.10591	8.86902	8.97493
1.20	282.8	0.001000	108.6740	40.57	2347.6	2388.2	40.57	2478.0	2518.6	0.14595	8.76236	8.90831
1.40	285.1	0.001001	93.9033	50.28	2341.1	2391.4	50.28	2472.5	2522.8	0.18016	8.67199	8.85214
1.60	287.2	0.001001	82.7463	58.83	2335.3	2394.2	58.84	2467.7	2526.6	0.21005	8.59355	8.80360
1.80	289.0	0.001001	74.0143	66.49	2330.2	2396.7	66.49	2463.4	2529.9	0.23663	8.52424	8.76087
2.00	290.6	0.001001	66.9896	73.43	2325.5	2398.9	73.43	2459.5	2532.9	0.26058	8.46214	8.72272
2.50	294.2	0.001002	54.2421	88.43	2315.4	2403.8	88.43	2451.0	2539.4	0.31186	8.33030	8.64215
3.00	297.2	0.001003	45.6550	100.99	2306.9	2407.9	100.99	2443.9	2544.9	0.35433	8.22223	8.57656
3.50	299.8	0.001003	39.4678	111.83	2299.6	2411.4	111.84	2437.7	2549.6	0.39066	8.13060	8.52126
4.00	302.1	0.001004	34.7925	121.40	2293.1	2414.5	121.40	2432.3	2553.7	0.42245	8.05104	8.47349
5.00	306.0	0.001005	28.1863	137.76	2282.1	2419.8	137.76	2423.0	2560.8	0.47625	7.91766	8.39391
6.00	309.3	0.001006	23.7342	151.49	2272.8	2424.3	151.49	2415.2	2566.7	0.52087	7.80827	8.32915
7.00	312.2	0.001007	20.5252	163.36	2264.7	2428.1	163.37	2408.4	2571.8	0.55908	7.71549	8.27456
8.00	314.7	0.001008	18.0994	173.84	2257.6	2431.4	173.85	2402.4	2576.2	0.59253	7.63488	8.22741
9.00	316.9	0.001009	16.1997	183.25	2251.2	2434.5	183.26	2397.0	2580.3	0.62233	7.56359	8.18592
10.00	319.0	0.001010	14.6706	191.80	2245.4	2437.2	191.81	2392.1	2583.9	0.64922	7.49968	8.14889
15.00	327.1	0.001014	10.0204	225.92	2222.1	2448.0	225.94	2372.4	2598.3	0.75484	7.25228	8.00712
20.00	333.2	0.001017	7.6482	251.38	2204.6	2456.0	251.40	2357.5	2608.9	0.83195	7.07528	7.90724
25.00	338.1	0.001020	6.2034	271.90	2190.5	2462.4	271.93	2345.5	2617.4	0.89309	6.93708	7.83016
30.00	342.2	0.001022	5.2286	289.20	2178.5	2467.7	289.23	2335.3	2624.6	0.94394	6.82351	7.76745
40.00	349.0	0.001026	3.9931	317.53	2158.8	2476.3	317.57	2318.5	2636.1	1.02590	6.64307	7.66897
50.00	354.5	0.001030	3.2401	340.42	2142.8	2483.2	340.48	2304.7	2645.2	1.09101	6.50196	7.59296
60.00	359.1	0.001033	2.7318	359.77	2129.2	2488.9	359.84	2293.0	2652.9	1.14524	6.38586	7.53110
70.00	363.1	0.001036	2.3649	376.61	2117.3	2493.9	376.68	2282.7	2659.4	1.19186	6.28709	7.47895
80.00	366.6	0.001038	2.0872	391.56	2106.6	2498.2	391.64	2273.5	2665.2	1.23283	6.20106	7.43389
90.00	369.8	0.001041	1.8695	405.03	2097.0	2502.1	405.13	2265.2	2670.3	1.26944	6.12479	7.39423
101.33	373.1	0.001043	1.6733	418.88	2087.1	2506.0	418.99	2256.5	2675.5	1.30672	6.04766	7.35439
120.00	377.9	0.001047	1.4284	439.17	2072.5	2511.6	439.30	2243.8	2683.1	1.36075	5.93688	7.29763
140.00	382.4	0.001051	1.2366	458.22	2058.6	2516.9	458.37	2231.6	2690.0	1.41085	5.83517	7.24602
160.00	386.4	0.001054	1.0914	475.17	2046.2	2521.4	475.34	2220.7	2696.0	1.45494	5.74643	7.20137
180.00	390.1	0.001058	0.9775	490.48	2035.0	2525.5	490.67	2210.7	2701.4	1.49437	5.66765	7.16203
200.00	393.4	0.001061	0.8857	504.47	2024.6	2529.1	504.68	2201.6	2706.2	1.53010	5.59676	7.12686

Press (kPa)	Temp (K)	Specific Volume (m**3/kg)		Internal Energy (kJ/kg)			Enthalpy (kJ/kg)			Entropy (kJ/kg/K)		
		vf	vg	uf	ufg	ug	hf	hfg	hg	sf	sfg	sg
300.00	406.7	0.001073	0.6058	561.13	1982.0	2543.2	561.46	2163.4	2724.9	1.67176	5.31980	6.99157
400.00	416.8	0.001084	0.4624	604.29	1948.8	2553.1	604.72	2133.3	2738.1	1.77660	5.11882	6.89542
500.00	425.0	0.001093	0.3748	639.64	1921.1	2560.7	640.19	2107.9	2748.1	1.86060	4.95998	6.82058
600.00	432.0	0.001101	0.3156	669.84	1897.0	2566.8	670.50	2085.6	2756.1	1.93110	4.82807	6.75917
700.00	438.1	0.001108	0.2728	696.37	1875.4	2571.8	697.14	2065.6	2762.7	1.99208	4.71490	6.70698
800.00	443.6	0.001115	0.2403	720.13	1855.9	2576.0	721.02	2047.3	2768.3	2.04599	4.61555	6.66154
900.00	448.5	0.001121	0.2149	741.72	1837.9	2579.7	742.72	2030.3	2773.0	2.09440	4.52683	6.62124
1000.00	453.0	0.001127	0.1943	761.56	1821.2	2582.8	762.68	2014.4	2777.1	2.13843	4.44655	6.58498
1200.00	461.1	0.001139	0.1632	797.13	1790.7	2587.9	798.50	1985.3	2783.8	2.21630	4.30539	6.52169
1400.00	468.2	0.001149	0.1408	828.52	1763.3	2591.8	830.13	1958.8	2788.9	2.28388	4.18364	6.46752
1600.00	474.5	0.001159	0.1237	856.76	1738.2	2594.9	858.61	1934.3	2792.9	2.34381	4.07621	6.42002
1800.00	480.3	0.001168	0.1104	882.51	1714.8	2597.3	884.61	1911.4	2796.0	2.39779	3.97980	6.37760
2000.00	485.5	0.001177	0.0996	906.27	1693.0	2599.2	908.62	1889.8	2798.4	2.44702	3.89214	6.33916
2500.00	497.1	0.001197	0.0799	958.99	1643.2	2602.2	961.98	1840.1	2802.0	2.55443	3.70155	6.25597
3000.00	507.0	0.001217	0.0667	1004.72	1598.6	2603.3	1008.37	1794.9	2803.3	2.64562	3.54017	6.18579
4000.00	523.5	0.001253	0.0498	1082.42	1519.4	2601.8	1087.43	1713.5	2800.9	2.79665	3.27306	6.06971
5000.00	537.1	0.001286	0.0394	1148.07	1448.9	2597.0	1154.50	1639.7	2794.2	2.92075	3.05296	5.97370
6000.00	548.7	0.001319	0.0324	1205.82	1384.1	2589.9	1213.73	1570.8	2784.6	3.02744	2.86263	5.89007
7000.00	559.0	0.001352	0.0274	1257.97	1322.9	2580.9	1267.44	1505.1	2772.6	3.12199	2.69264	5.81463
8000.00	568.2	0.001385	0.0235	1306.00	1264.4	2570.4	1317.08	1441.5	2758.6	3.20765	2.53720	5.74485
9000.00	576.5	0.001418	0.0205	1350.89	1207.6	2558.4	1363.65	1379.2	2742.9	3.28657	2.39244	5.67901
10000.00	584.1	0.001453	0.0180	1393.34	1151.8	2545.1	1407.87	1317.6	2725.5	3.36029	2.25560	5.61589
11000.00	591.2	0.001489	0.0160	1433.90	1096.6	2530.5	1450.28	1256.1	2706.4	3.42995	2.12458	5.55453
12000.00	597.8	0.001526	0.0143	1473.01	1041.3	2514.4	1491.33	1194.3	2685.6	3.49646	1.99766	5.49412
13000.00	604.0	0.001566	0.0128	1511.04	985.6	2496.7	1531.40	1131.5	2662.9	3.56058	1.87331	5.43388
14000.00	609.8	0.001610	0.0115	1548.34	928.9	2477.3	1570.88	1067.2	2638.1	3.62300	1.75005	5.37305
15000.00	615.3	0.001657	0.0103	1585.30	870.5	2455.8	1610.15	1000.7	2610.9	3.68445	1.62636	5.31080
16000.00	620.5	0.001710	0.0093	1622.32	809.6	2431.9	1649.67	931.1	2580.8	3.74568	1.50059	5.24627
17000.00	625.4	0.001769	0.0084	1659.98	745.2	2405.1	1690.05	857.4	2547.4	3.80770	1.37081	5.17850
18000.00	630.1	0.001840	0.0075	1698.93	675.6	2374.6	1732.04	777.5	2509.5	3.87170	1.23384	5.10554
19000.00	634.6	0.001925	0.0036	1740.32	279.6	2019.9	1776.91	310.7	2087.6	3.93967	0.48827	4.42795
20000.00	638.9	0.002039	0.0034	1786.35	235.7	2022.1	1827.12	263.1	2090.2	4.01541	0.41112	4.42653
21000.00	643.0	0.002212	0.0033	1842.97	180.8	2023.8	1889.43	203.0	2092.4	4.10930	0.31555	4.42485
22064.00	647.1	0.003209	0.0032	2034.47	0.0	2034.5	2105.27	0.0	2105.3	4.43941	0.00000	4.43941

A.14.2 The Saturation Pressure vs. Temperature (SI Units)

Temp	Press	Specific Volume		Internal Energy			Enthalpy			Entropy		
(K)	(kPa)	(m**3/kg)		(kJ/kg)			(kJ/kg)			(kJ/kg/K)		
		vf	vg	uf	ufg	ug	hf	hfg	hg	sf	sfg	sg
273.1	0.61	0.001000	206.1397	0.00	2374.9	2374.9	0.00	2500.9	2500.9	0.00000	9.15591	9.15591
275.0	0.70	0.001000	181.6044	7.76	2369.7	2377.4	7.76	2496.5	2504.3	0.02831	9.07831	9.10662
280.0	0.99	0.001000	130.1941	28.79	2355.5	2384.3	28.80	2484.7	2513.5	0.10412	8.87382	8.97794
285.0	1.39	0.001001	94.6073	49.78	2341.4	2391.2	49.78	2472.8	2522.6	0.17840	8.67661	8.85501
290.0	1.92	0.001001	69.6305	70.73	2327.3	2398.1	70.73	2461.0	2531.7	0.25128	8.48623	8.73751
295.0	2.62	0.001002	51.8694	91.66	2313.2	2404.9	91.66	2449.2	2540.8	0.32283	8.30229	8.62511
300.0	3.54	0.001003	39.0821	112.57	2299.1	2411.7	112.57	2437.3	2549.9	0.39312	8.12441	8.51754
305.0	4.72	0.001005	29.7669	133.47	2285.0	2418.4	133.48	2425.4	2558.9	0.46222	7.95228	8.41451
310.0	6.23	0.001007	22.9051	154.37	2270.8	2425.2	154.38	2413.5	2567.9	0.53018	7.78559	8.31577
315.0	8.14	0.001009	17.7969	175.26	2256.6	2431.9	175.27	2401.6	2576.8	0.59704	7.62405	8.22110
320.0	10.55	0.001011	13.9557	196.16	2242.4	2438.6	196.17	2389.6	2585.7	0.66285	7.46741	8.13026
325.0	13.53	0.001013	11.0396	217.06	2228.1	2445.2	217.07	2377.5	2594.6	0.72765	7.31540	8.04305
330.0	17.21	0.001015	8.8056	237.96	2213.8	2451.8	237.98	2365.4	2603.3	0.79148	7.16779	7.95928
335.0	21.72	0.001018	7.0792	258.87	2199.4	2458.3	258.89	2353.2	2612.1	0.85438	7.02436	7.87875
340.0	27.19	0.001021	5.7341	279.80	2185.0	2464.8	279.82	2340.9	2620.7	0.91638	6.88491	7.80128
345.0	33.78	0.001024	4.6778	300.73	2170.5	2471.2	300.77	2328.5	2629.3	0.97751	6.74921	7.72673
350.0	41.68	0.001027	3.8420	321.69	2155.9	2477.6	321.73	2316.0	2637.7	1.03781	6.61710	7.65492
355.0	51.08	0.001030	3.1760	342.66	2141.2	2483.9	342.71	2303.4	2646.1	1.09731	6.48839	7.58570
360.0	62.19	0.001034	2.6415	363.66	2126.4	2490.1	363.72	2290.7	2654.4	1.15604	6.36289	7.51894
365.0	75.26	0.001037	2.2099	384.68	2111.5	2496.2	384.75	2277.8	2662.5	1.21403	6.24046	7.45449
370.0	90.54	0.001041	1.8591	405.72	2096.5	2502.3	405.81	2264.8	2670.6	1.27129	6.12094	7.39223
375.0	108.30	0.001045	1.5723	426.79	2081.4	2508.2	426.91	2251.6	2678.5	1.32787	6.00417	7.33204
380.0	128.85	0.001049	1.3365	447.90	2066.1	2514.0	448.03	2238.2	2686.3	1.38378	5.89001	7.27379
385.0	152.52	0.001053	1.1415	469.04	2050.7	2519.8	469.20	2224.7	2693.9	1.43904	5.77834	7.21738
390.0	179.64	0.001058	0.9794	490.21	2035.2	2525.4	490.40	2210.9	2701.3	1.49369	5.66901	7.16270
395.0	210.59	0.001062	0.8440	511.43	2019.5	2530.9	511.65	2197.0	2708.6	1.54775	5.56190	7.10964
400.0	245.75	0.001067	0.7303	532.68	2003.5	2536.2	532.95	2182.8	2715.7	1.60122	5.45690	7.05812
405.0	285.55	0.001072	0.6345	553.99	1987.5	2541.4	554.29	2168.3	2722.6	1.65415	5.35389	7.00804
410.0	330.42	0.001077	0.5533	575.33	1971.2	2546.5	575.69	2153.6	2729.3	1.70654	5.25277	6.95931
415.0	380.82	0.001082	0.4842	596.73	1954.7	2551.4	597.15	2138.7	2735.8	1.75843	5.15343	6.91186
420.0	437.24	0.001087	0.4253	618.19	1938.0	2556.1	618.66	2123.4	2742.1	1.80982	5.05578	6.86560
425.0	500.18	0.001093	0.3747	639.70	1921.0	2560.7	640.24	2107.9	2748.1	1.86074	4.95972	6.82046
430.0	570.18	0.001098	0.3311	661.27	1903.8	2565.1	661.90	2092.0	2753.9	1.91121	4.86516	6.77637
435.0	647.77	0.001104	0.2935	682.91	1886.4	2569.3	683.62	2075.8	2759.4	1.96124	4.77202	6.73327
440.0	733.55	0.001110	0.2609	704.61	1868.7	2573.3	705.43	2059.3	2764.7	2.01086	4.68022	6.69108
445.0	828.10	0.001117	0.2326	726.39	1850.7	2577.1	727.32	2042.4	2769.7	2.06009	4.58966	6.64975

Temp	Press	Specific Volume		Internal Energy			Enthalpy			Entropy		
(K)	(kPa)	(m**3/kg)		(kJ/kg)			(kJ/kg)			(kJ/kg/K)		
		vf	vg	uf	ufg	ug	hf	hfg	hg	sf	sfg	sg
455.0	1046.02	0.001130	0.1862	770.18	1813.9	2584.1	771.37	2007.4	2778.8	2.15744	4.41199	6.56943
460.0	1170.68	0.001137	0.1672	792.21	1795.0	2587.2	793.54	1989.4	2782.9	2.20560	4.32472	6.53032
465.0	1306.72	0.001144	0.1504	814.33	1775.8	2590.1	815.82	1970.8	2786.7	2.25345	4.23839	6.49183
470.0	1454.84	0.001152	0.1356	836.55	1756.2	2592.7	838.22	1951.9	2790.1	2.30099	4.15292	6.45392
475.0	1615.75	0.001159	0.1226	858.87	1736.3	2595.1	860.74	1932.4	2793.2	2.34826	4.06826	6.41652
480.0	1790.19	0.001167	0.1110	881.30	1715.9	2597.2	883.39	1912.5	2795.9	2.39527	3.98431	6.37958
485.0	1978.94	0.001176	0.1006	903.85	1695.2	2599.0	906.18	1892.0	2798.2	2.44204	3.90101	6.34305
490.0	2182.77	0.001185	0.0914	926.53	1674.0	2600.6	929.11	1871.0	2800.1	2.48859	3.81829	6.30688
495.0	2402.48	0.001194	0.0832	949.34	1652.4	2601.8	952.20	1849.3	2801.5	2.53495	3.73606	6.27101
500.0	2638.90	0.001203	0.0758	972.29	1630.3	2602.6	975.46	1827.1	2802.6	2.58113	3.65426	6.23539
505.0	2892.85	0.001213	0.0691	995.40	1607.8	2603.2	998.90	1804.3	2803.2	2.62717	3.57281	6.19997
510.0	3165.22	0.001223	0.0632	1018.66	1584.7	2603.3	1022.53	1780.7	2803.3	2.67307	3.49163	6.16470
515.0	3456.86	0.001233	0.0578	1042.10	1561.0	2603.1	1046.37	1756.5	2802.8	2.71888	3.41064	6.12952
520.0	3768.70	0.001245	0.0529	1065.73	1536.8	2602.5	1070.42	1731.5	2801.9	2.76461	3.32977	6.09437
525.0	4101.65	0.001256	0.0485	1089.55	1511.9	2601.4	1094.70	1705.7	2800.4	2.81029	3.24892	6.05921
530.0	4456.65	0.001268	0.0445	1113.58	1486.4	2599.9	1119.23	1679.1	2798.3	2.85595	3.16802	6.02396
535.0	4834.69	0.001281	0.0409	1137.84	1460.1	2598.0	1144.04	1651.5	2795.6	2.90162	3.08696	5.98857
540.0	5236.75	0.001294	0.0376	1162.35	1433.1	2595.5	1169.13	1623.1	2792.2	2.94734	3.00564	5.95298
545.0	5663.85	0.001308	0.0345	1187.12	1405.4	2592.5	1194.53	1593.6	2788.1	2.99314	2.92397	5.91711
550.0	6117.05	0.001323	0.0318	1212.18	1376.7	2588.9	1220.27	1563.0	2783.3	3.03906	2.84182	5.88087
555.0	6597.43	0.001339	0.0292	1237.54	1347.2	2584.7	1246.37	1531.3	2777.6	3.08515	2.75905	5.84420
560.0	7106.12	0.001355	0.0269	1263.25	1316.6	2579.9	1272.88	1498.3	2771.2	3.13146	2.67553	5.80700
565.0	7644.26	0.001373	0.0248	1289.32	1285.0	2574.3	1299.82	1464.0	2763.8	3.17805	2.59110	5.76915
570.0	8213.06	0.001392	0.0228	1315.80	1252.2	2568.0	1327.23	1428.2	2755.4	3.22497	2.50557	5.73054
575.0	8813.76	0.001412	0.0210	1342.73	1218.0	2560.8	1355.17	1390.8	2745.9	3.27231	2.41873	5.69103
580.0	9447.69	0.001433	0.0193	1370.16	1182.5	2552.7	1383.70	1351.6	2735.3	3.32014	2.33033	5.65048
585.0	10116.21	0.001457	0.0178	1398.14	1145.4	2543.5	1412.88	1310.5	2723.3	3.36858	2.24011	5.60869
590.0	10820.77	0.001482	0.0163	1426.75	1106.4	2533.2	1442.79	1267.1	2709.9	3.41772	2.14771	5.56544
595.0	11562.92	0.001510	0.0150	1456.07	1065.5	2521.6	1473.53	1221.4	2694.9	3.46773	2.05273	5.52046
600.0	12344.30	0.001540	0.0137	1486.21	1022.2	2508.5	1505.22	1172.8	2678.0	3.51877	1.95463	5.47340
605.0	13166.69	0.001573	0.0126	1517.30	976.3	2493.6	1538.01	1120.9	2658.9	3.57108	1.85272	5.42380
610.0	14032.02	0.001611	0.0114	1549.53	927.1	2476.6	1572.14	1065.1	2637.3	3.62498	1.74610	5.37108
615.0	14942.40	0.001654	0.0104	1583.17	873.9	2457.1	1607.89	1004.6	2612.5	3.68092	1.63352	5.31444
620.0	15900.20	0.001704	0.0094	1618.61	815.8	2434.4	1645.70	938.2	2583.9	3.73955	1.51329	5.25284
630.0	17969.08	0.001837	0.0075	1697.70	677.9	2375.6	1730.71	780.1	2510.8	3.86968	1.23822	5.10789
640.0	20265.91	0.002076	0.0034	1799.92	222.7	2022.7	1842.00	248.9	2090.9	4.03783	0.38844	4.42627
647.1	22063.97	0.003209	0.0032	2034.52	0.0	2034.5	2105.33	0.0	2105.3	4.43950	0.00000	4.43950

A.14.3 Superheated Steam Table (SI Units)

Press = 0.010 MPa Tsat= 319.0 K

T(K) =	325.	350.	400.	450.	500.	550.	600.	650.	700.	800.	900.	1000.
v-m3/kg	14.9541	16.1208	18.4417	20.7557	23.0669	25.3768	27.6860	29.9947	32.3032	36.9196	41.5356	46.1514
u-kj/kg	2446.0	2482.1	2554.3	2627.2	2701.3	2776.8	2853.8	2932.4	3012.6	3177.9	3350.3	3529.6
h-kj/kg	2595.6	2643.3	2738.7	2834.7	2932.0	3030.6	3130.7	3232.3	3335.6	3547.1	3765.6	3991.2
s-kj/kg/K	8.18513	8.32676	8.58137	8.80763	9.01251	9.20046	9.37463	9.53734	9.69038	9.97277	10.23002	10.46758

Press = 0.015 MPa Tsat= 327.1 K

T(K) =	350.	400.	450.	500.	550.	600.	650.	700.	800.	900.	1000.	1100.
v-m3/kg	10.7362	12.2880	13.8328	15.3748	16.9155	18.4555	19.9950	21.5343	24.6123	27.6899	30.7672	33.8444
u-kj/kg	2481.4	2553.9	2627.0	2701.2	2776.7	2853.7	2932.3	3012.5	3177.9	3350.2	3529.6	3716.1
h-kj/kg	2642.5	2738.2	2834.5	2931.8	3030.4	3130.6	3232.2	3335.5	3547.1	3765.6	3991.1	4223.8
s-kj/kg/K	8.13764	8.39336	8.62004	8.82509	9.01314	9.18737	9.35011	9.50318	9.78559	10.04286	10.28043	10.50212

Press = 0.025 MPa Tsat= 338.1 K

T(K) =	350.	400.	450.	500.	550.	600.	650.	700.	800.	900.	1000.	1100.
v-m3/kg	6.4284	7.3650	8.2944	9.2211	10.1464	11.0710	11.9952	12.9191	14.7664	16.6132	18.4599	20.3063
u-kj/kg	2480.0	2553.2	2626.6	2700.9	2776.5	2853.6	2932.2	3012.4	3177.8	3350.2	3529.6	3716.1
h-kj/kg	2640.7	2737.3	2833.9	2931.4	3030.2	3130.4	3232.1	3335.4	3547.0	3765.5	3991.1	4223.7
s-kj/kg/K	7.89785	8.15585	8.38335	8.58878	8.77701	8.95134	9.11416	9.26727	9.54974	9.80704	10.04463	10.26633

Press = 0.040 MPa Tsat= 349.0 K

T(K) =	350.	400.	450.	500.	550.	600.	650.	700.	800.	900.	1000.	1100.
v-m3/kg	4.0050	4.5957	5.1791	5.7596	6.3388	6.9173	7.4953	8.0731	9.2281	10.3826	11.5370	12.6912
u-kj/kg	2477.8	2552.1	2625.9	2700.5	2776.2	2853.3	2932.0	3012.3	3177.7	3350.1	3529.5	3716.0
h-kj/kg	2638.0	2736.0	2833.1	2930.8	3029.8	3130.0	3231.8	3335.2	3546.9	3765.4	3991.0	4223.7
s-kj/kg/K	7.67465	7.93627	8.16504	8.37102	8.55954	8.73403	8.89695	9.05013	9.33269	9.59003	9.82764	10.04936

Press = 0.060 MPa Tsat= 359.1 K

T(K) =	400.	450.	500.	550.	600.	650.	700.	800.	900.	1000.	1100.	1200.
v-m3/kg	3.05719	3.44835	3.83661	4.22350	4.60967	4.99541	5.38086	6.15124	6.92120	7.69092	8.46050	9.22998
u-kj/kg	2550.7	2625.1	2699.9	2775.8	2853.0	2931.8	3012.0	3177.6	3350.0	3529.4	3716.0	3909.4
h-kj/kg	2734.1	2832.0	2930.1	3029.2	3129.6	3231.5	3334.9	3546.7	3765.3	3990.9	4223.6	4463.2
s-kj/kg/K	7.74553	7.97603	8.18276	8.37166	8.54637	8.70943	8.86270	9.14537	9.40277	9.64043	9.86217	10.07059

Press = 0.101 MPa Tsat= 373.1 K

T(K) =	400.	450.	500.	550.	600.	650.	700.	800.	900.	1000.	1100.	1200.
v-m3/kg	1.80206	2.03654	2.26798	2.49803	2.72735	2.95623	3.18482	3.64148	4.09772	4.55372	5.00959	5.46535
u-kj/kg	2547.7	2623.3	2698.7	2774.9	2852.4	2931.2	3011.6	3177.3	3349.8	3529.3	3715.8	3909.3
h-kj/kg	2730.3	2829.7	2928.5	3028.1	3128.7	3230.8	3334.3	3546.3	3765.0	3990.7	4223.4	4463.0
s-kj/kg/K	7.49608	7.73027	7.93859	8.12828	8.30345	8.46679	8.62025	8.90315	9.16068	9.39842	9.62020	9.82866

Press = 0.200 MPa Tsat= 393.4 K

T(K) =	400.	450.	500.	550.	600.	650.	700.	800.	900.	1000.	1100.	1200.
v-m3/kg	0.90251	1.02507	1.14427	1.26200	1.37897	1.49549	1.61172	1.84366	2.07517	2.30645	2.53758	2.76862
u-kj/kg	2540.0	2619.0	2695.9	2772.9	2850.8	2930.0	3010.6	3176.6	3349.2	3528.8	3715.5	3909.0
h-kj/kg	2720.5	2824.0	2924.8	3025.3	3126.6	3229.1	3332.9	3545.3	3764.3	3990.1	4223.0	4462.7
s-kj/kg/K	7.16292	7.40676	7.61907	7.81074	7.98701	8.15104	8.30496	8.58842	8.84625	9.08417	9.30606	9.51459

Press = 0.300 MPa Tsat= 406.7 K

T(K) =	450.	500.	550.	600.	650.	700.	800.	900.	1000.	1100.	1200.	1300.
v-m3/kg	0.67872	0.75960	0.83891	0.91743	0.99550	1.07327	1.22829	1.38288	1.53724	1.69145	1.84557	1.99962
u-kj/kg	2614.4	2693.0	2770.8	2849.2	2928.7	3009.6	3175.8	3348.7	3528.4	3715.1	3908.7	4108.8
h-kj/kg	2818.1	2920.8	3022.5	3124.5	3227.4	3331.5	3544.3	3763.5	3989.6	4222.6	4462.4	4708.7
s-kj/kg/K	7.20939	7.42604	7.61979	7.79722	7.96195	8.11634	8.40036	8.65851	8.89661	9.11862	9.32722	9.52438

Press = 0.400 MPa Tsat= 416.8 K

T(K) =	450.	500.	550.	600.	650.	700.	800.	900.	1000.	1100.	1200.	1300.
v-m3/kg	0.50545	0.56722	0.62735	0.68666	0.74550	0.80405	0.92061	1.03674	1.15263	1.26838	1.38404	1.49963
u-kj/kg	2609.7	2690.0	2768.7	2847.6	2927.4	3008.5	3175.1	3348.1	3528.0	3714.8	3908.4	4108.6
h-kj/kg	2811.9	2916.9	3019.6	3122.3	3225.6	3330.1	3543.3	3762.8	3989.0	4222.2	4462.0	4708.5
s-kj/kg/K	7.06594	7.28723	7.48316	7.66177	7.82722	7.98208	8.26667	8.52513	8.76341	8.98553	9.19421	9.39142

Press = 0.500 MPa Tsat= 425.0 K

T(K) =	450.	500.	550.	600.	650.	700.	800.	900.	1000.	1100.	1200.	1300.
v-m3/kg	0.40140	0.45177	0.50040	0.54819	0.59550	0.64251	0.73599	0.82905	0.92187	1.01455	1.10712	1.19964
u-kj/kg	2604.8	2686.9	2766.6	2846.0	2926.2	3007.5	3174.3	3347.6	3527.6	3714.5	3908.1	4108.4
h-kj/kg	2805.5	2912.8	3016.8	3120.1	3223.9	3328.7	3542.3	3762.1	3988.5	4221.7	4461.7	4708.2
s-kj/kg/K	6.95182	7.17807	7.37626	7.55608	7.72226	7.87760	8.16276	8.42153	8.66000	8.88223	9.09098	9.28825

Press = 0.600 MPa Tsat= 432.0 K

T(K) =	450.	500.	550.	600.	650.	700.	800.	900.	1000.	1100.	1200.	1300.
v-m3/kg	0.33195	0.37477	0.41576	0.45587	0.49549	0.53482	0.61292	0.69059	0.76803	0.84532	0.92251	0.99965
u-kj/kg	2599.7	2683.8	2764.4	2844.4	2924.9	3006.4	3173.6	3347.0	3527.1	3714.1	3907.8	4108.1
h-kj/kg	2798.8	2908.7	3013.9	3117.9	3222.2	3327.3	3541.4	3761.4	3988.0	4221.3	4461.3	4707.9
s-kj/kg/K	6.85603	7.08760	7.28815	7.46921	7.63613	7.79195	8.07769	8.33677	8.57542	8.79777	9.00659	9.20392

Press = 0.700 MPa Tsat= 438.1 K

T(K) =	450.	500.	550.	600.	650.	700.	800.	900.	1000.	1100.	1200.	1300.
v-m3/kg	0.28228	0.31976	0.35529	0.38992	0.42406	0.45789	0.52501	0.59170	0.65814	0.72444	0.79065	0.85679
u-kj/kg	2594.3	2680.6	2762.2	2842.7	2923.6	3005.4	3172.9	3346.5	3526.7	3713.8	3907.5	4107.9
h-kj/kg	2791.9	2904.4	3010.9	3115.7	3220.4	3325.9	3540.4	3760.7	3987.4	4220.9	4461.0	4707.6
s-kj/kg/K	6.77268	7.00998	7.21299	7.39533	7.56300	7.71931	8.00562	8.26502	8.50385	8.72631	8.93521	9.13258

Press = 0.800 MPa Tsat= 443.6 K

T(K) =	450.	500.	550.	600.	650.	700.	800.	900.	1000.	1100.	1200.	1300.
v-m3/kg	0.24496	0.27847	0.30993	0.34046	0.37049	0.40020	0.45908	0.51752	0.57572	0.63379	0.69175	0.74965
u-kj/kg	2588.7	2677.4	2760.0	2841.1	2922.3	3004.3	3172.1	3345.9	3526.3	3713.4	3907.3	4107.6
h-kj/kg	2784.7	2900.1	3007.9	3113.5	3218.7	3324.5	3539.4	3759.9	3986.9	4220.4	4460.7	4707.4
s-kj/kg/K	6.69816	6.94173	7.14729	7.33094	7.49938	7.65617	7.94306	8.20278	8.44180	8.66437	8.87334	9.07077

Press = 0.900 MPa Tsat= 448.5 K

T(K) =	450.	500.	550.	600.	650.	700.	800.	900.	1000.	1100.	1200.	1300.
v-m3/kg	0.21586	0.24634	0.27465	0.30199	0.32881	0.35533	0.40779	0.45983	0.51162	0.56327	0.61483	0.66632
u-kj/kg	2582.7	2674.0	2757.8	2839.4	2921.0	3003.3	3171.4	3345.4	3525.9	3713.1	3907.0	4107.4
h-kj/kg	2777.0	2895.7	3004.9	3111.2	3216.9	3323.1	3538.4	3759.2	3986.3	4220.0	4460.3	4707.1
s-kj/kg/K	6.63002	6.88060	7.08881	7.27380	7.44301	7.60031	7.88778	8.14781	8.38701	8.60970	8.81874	9.01622

Press = 1.000 MPa Tsat= 453.0 K

T(K) =	500.	550.	600.	650.	700.	800.	900.	1000.	1100.	1200.	1300.	1400.
v-m3/kg	0.22063	0.24641	0.27121	0.29547	0.31943	0.36677	0.41368	0.46034	0.50687	0.55329	0.59966	0.64598
u-kj/kg	2670.6	2755.5	2837.8	2919.7	3002.2	3170.6	3344.8	3525.4	3712.7	3906.7	4107.1	4313.9
h-kj/kg	2891.3	3001.9	3109.0	3215.2	3321.6	3537.4	3758.5	3985.8	4219.6	4460.0	4706.8	4959.9
s-kj/kg/K	6.82505	7.03601	7.22237	7.39237	7.55017	7.83822	8.09857	8.33796	8.56076	8.76988	8.96741	9.15493

Press = 1.200 MPa Tsat= 461.1 K

T(K) =	500.	550.	600.	650.	700.	800.	900.	1000.	1100.	1200.	1300.	1400.
v-m3/kg	0.18201	0.20405	0.22503	0.24546	0.26557	0.30523	0.34445	0.38342	0.42225	0.46099	0.49966	0.53829
u-kj/kg	2663.7	2750.9	2834.4	2917.1	3000.1	3169.2	3343.7	3524.6	3712.0	3906.1	4106.7	4313.5
h-kj/kg	2882.1	2995.7	3104.4	3211.6	3318.8	3535.4	3757.1	3984.7	4218.7	4459.3	4706.2	4959.4
s-kj/kg/K	6.72659	6.94336	7.13256	7.30417	7.46298	7.75221	8.01319	8.25296	8.47598	8.68525	8.88289	9.07048

Press = 1.400 MPa Tsat= 468.2 K

T(K) =	500.	550.	600.	650.	700.	800.	900.	1000.	1100.	1200.	1300.	1400.
v-m3/kg	0.15437	0.17376	0.19203	0.20974	0.22711	0.26127	0.29500	0.32848	0.36181	0.39506	0.42823	0.46137
u-kj/kg	2656.4	2746.2	2831.0	2914.4	2998.0	3167.7	3342.6	3523.7	3711.3	3905.5	4106.2	4313.1
h-kj/kg	2872.5	2989.4	3099.8	3208.1	3315.9	3533.4	3755.6	3983.6	4217.9	4458.6	4705.7	4959.0
s-kj/kg/K	6.64047	6.86348	7.05567	7.22894	7.38878	7.67920	7.94082	8.18096	8.40421	8.61362	8.81137	8.99904

Press = 1.600 MPa Tsat= 474.5 K

T(K) =	500.	550.	600.	650.	700.	800.	900.	1000.	1100.	1200.	1300.	1400.
v-m3/kg	0.13360	0.15103	0.16728	0.18294	0.19825	0.22830	0.25791	0.28727	0.31649	0.34561	0.37466	0.40368
u-kj/kg	2648.8	2741.3	2827.5	2911.8	2995.8	3166.2	3341.5	3522.9	3710.7	3905.0	4105.7	4312.6
h-kj/kg	2862.6	2983.0	3095.2	3204.5	3313.0	3531.5	3754.2	3982.5	4217.0	4457.9	4705.1	4958.5
s-kj/kg/K	6.56317	6.79289	6.98820	7.16319	7.32408	7.61571	7.87796	8.11847	8.34195	8.55151	8.74936	8.93711

Press = 1.800 MPa Tsat= 480.3 K

T(K) =	500.	550.	600.	650.	700.	800.	900.	1000.	1100.	1200.	1300.	1400.
v-m3/kg	0.11740	0.13334	0.14802	0.16209	0.17581	0.20266	0.22906	0.25522	0.28123	0.30715	0.33300	0.35881
u-kj/kg	2640.9	2736.4	2824.0	2909.1	2993.7	3164.7	3340.4	3522.0	3710.0	3904.4	4105.2	4312.2
h-kj/kg	2852.2	2976.4	3090.5	3200.9	3310.1	3529.5	3752.7	3981.4	4216.2	4457.2	4704.6	4958.1
s-kj/kg/K	6.49238	6.72934	6.92792	7.10466	7.26663	7.55947	7.82237	8.06326	8.28696	8.49667	8.69463	8.88246

Press = 2.000 MPa Tsat= 485.5 K

T(K) =	500.	550.	600.	650.	700.	800.	900.	1000.	1100.	1200.	1300.	1400.
v-m3/kg	0.10439	0.11917	0.13261	0.14541	0.15786	0.18215	0.20599	0.22958	0.25303	0.27638	0.29967	0.32291
u-kj/kg	2632.6	2731.3	2820.5	2906.4	2991.5	3163.2	3339.3	3521.2	3709.3	3903.8	4104.7	4311.8
h-kj/kg	2841.4	2969.7	3085.7	3197.2	3307.2	3527.5	3751.3	3980.3	4215.3	4456.6	4704.0	4957.6
s-kj/kg/K	6.42645	6.67128	6.87328	7.05184	7.21490	7.50897	7.77251	8.01377	8.23771	8.44757	8.64563	8.83353

Press = 2.500 MPa Tsat= 497.1 K

T(K) =	500.	550.	600.	650.	700.	800.	900.	1000.	1100.	1200.	1300.	1400.
v-m3/kg	0.08080	0.09361	0.10484	0.11538	0.12553	0.14522	0.16445	0.18342	0.20226	0.22099	0.23967	0.25830
u-kj/kg	2609.7	2718.1	2811.4	2899.5	2986.1	3159.4	3336.5	3519.0	3707.5	3902.4	4103.5	4310.8
h-kj/kg	2811.7	2952.2	3073.5	3188.0	3299.9	3522.5	3747.6	3977.6	4213.2	4454.8	4702.7	4956.5
s-kj/kg/K	6.27535	6.54361	6.75484	6.93818	7.10407	7.40126	7.66643	7.90864	8.13314	8.34338	8.54170	8.72980

Press = 3.000 MPa Tsat= 507.0 K

T(K) =	550.	600.	650.	700.	800.	900.	1000.	1100.	1200.	1300.	1400.	1500.
v-m3/kg	0.07650	0.08631	0.09534	0.10397	0.12060	0.13675	0.15265	0.16841	0.18407	0.19967	0.21522	0.23074
u-kj/kg	2704.1	2801.9	2892.5	2980.5	3155.6	3333.7	3516.8	3705.8	3900.9	4102.3	4309.7	4523.0
h-kj/kg	2933.7	3060.9	3178.5	3292.5	3517.4	3744.0	3974.8	4211.0	4453.1	4701.3	4955.4	5215.2
s-kj/kg/K	6.43310	6.65462	6.84310	7.01196	7.31236	7.57918	7.82233	8.04742	8.25802	8.45662	8.64491	8.82414

Press = 3.500 MPa Tsat= 515.7 K

T(K) =	550.	600.	650.	700.	800.	900.	1000.	1100.	1200.	1300.	1400.	1500.
v-m3/kg	0.06422	0.07304	0.08102	0.08857	0.10301	0.11697	0.13068	0.14424	0.15770	0.17110	0.18445	0.19777
u-kj/kg	2689.3	2792.2	2885.4	2974.9	3151.8	3330.9	3514.7	3704.1	3899.5	4101.0	4308.7	4522.1
h-kj/kg	2914.0	3047.8	3168.9	3284.9	3512.4	3740.3	3972.1	4208.9	4451.4	4699.9	4954.3	5214.3
s-kj/kg/K	6.33371	6.56677	6.76077	6.93274	7.23644	7.50492	7.74903	7.97469	8.18567	8.38453	8.57302	8.75240

Press = 4.000 MPa Tsat= 523.5 K

T(K) =	550.	600.	650.	700.	800.	900.	1000.	1100.	1200.	1300.	1400.	1500.
v-m3/kg	0.05494	0.06306	0.07027	0.07701	0.08982	0.10213	0.11419	0.12610	0.13792	0.14967	0.16138	0.17305
u-kj/kg	2673.4	2782.0	2878.1	2969.3	3148.0	3328.1	3512.5	3702.3	3898.0	4099.8	4307.6	4521.2
h-kj/kg	2893.1	3034.3	3159.1	3277.3	3507.3	3736.7	3969.3	4206.7	4449.7	4698.5	4953.1	5213.4
s-kj/kg/K	6.24171	6.48774	6.68767	6.86291	7.17000	7.44017	7.68524	7.91147	8.12284	8.32196	8.51064	8.69017

Press = 4.500 MPa Tsat= 530.6 K

T(K) =	550.	600.	650.	700.	800.	900.	1000.	1100.	1200.	1300.	1400.	1500.
v-m3/kg	0.04765	0.05528	0.06189	0.06802	0.07955	0.09059	0.10137	0.11200	0.12254	0.13300	0.14343	0.15382
u-kj/kg	2656.3	2771.5	2870.6	2963.5	3144.1	3325.3	3510.4	3700.6	3896.6	4098.6	4306.6	4520.3
h-kj/kg	2870.7	3020.3	3149.1	3269.6	3502.1	3733.0	3966.5	4204.6	4448.0	4697.1	4952.0	5212.5
s-kj/kg/K	6.15444	6.41525	6.62155	6.80022	7.11079	7.38267	7.62871	7.85553	8.06727	8.26666	8.45554	8.63521

Press = 5.000 MPa Tsat= 537.1 K

T(K) =	550.	600.	650.	700.	800.	900.	1000.	1100.	1200.	1300.	1400.	1500.
v-m3/kg	0.04175	0.04904	0.05519	0.06082	0.07134	0.08136	0.09112	0.10072	0.11023	0.11967	0.12907	0.13843
u-kj/kg	2637.7	2760.6	2862.9	2957.7	3140.3	3322.5	3508.2	3698.8	3895.1	4097.4	4305.5	4519.4
h-kj/kg	2846.5	3005.8	3138.9	3261.8	3497.0	3729.3	3963.8	4202.4	4446.2	4695.7	4950.9	5211.6
s-kj/kg/K	6.06984	6.34772	6.56087	6.74313	7.05728	7.33089	7.57790	7.80531	8.01743	8.21709	8.40617	8.58598

Press = 6.000 MPa Tsat= 548.7 K

T(K) =	550.	600.	650.	700.	800.	900.	1000.	1100.	1200.	1300.	1400.	1500.
v-m3/kg	0.03267	0.03961	0.04510	0.05001	0.05902	0.06751	0.07573	0.08380	0.09177	0.09967	0.10753	0.11536
u-kj/kg	2594.6	2737.5	2847.1	2945.8	3132.5	3316.8	3503.8	3695.3	3892.2	4094.9	4303.5	4517.6
h-kj/kg	2790.6	2975.2	3117.7	3245.9	3486.6	3721.9	3958.2	4198.1	4442.8	4693.0	4948.7	5209.7
s-kj/kg/K	5.90115	6.22319	6.45166	6.64164	6.96328	7.24041	7.48939	7.71798	7.93087	8.13107	8.32053	8.50064

Press = 7.000 MPa Tsat= 559.0 K

T(K) =	600.	650.	700.	800.	900.	1000.	1100.	1200.	1300.	1400.	1500.	1600.
v-m3/kg	0.03280	0.03787	0.04227	0.05022	0.05762	0.06474	0.07171	0.07858	0.08539	0.09215	0.09888	0.10558
u-kj/kg	2712.3	2830.5	2933.5	3124.5	3311.1	3499.4	3691.8	3889.3	4092.5	4301.4	4515.8	4735.4
h-kj/kg	2941.9	3095.6	3229.5	3476.1	3714.4	3952.6	4193.8	4439.4	4690.2	4946.4	5207.9	5474.5
s-kj/kg/K	6.10772	6.35409	6.55258	6.88213	7.16288	7.41385	7.64363	7.85730	8.05804	8.24789	8.42829	8.60030

Press = 8.000 MPa Tsat= 568.2 K

T(K) =	600.	650.	700.	800.	900.	1000.	1100.	1200.	1300.	1400.	1500.	1600.
v-m3/kg	0.02761	0.03241	0.03646	0.04361	0.05020	0.05650	0.06264	0.06869	0.07468	0.08061	0.08652	0.09239
u-kj/kg	2684.6	2813.1	2920.9	3116.5	3305.3	3495.0	3688.3	3886.4	4090.0	4299.3	4514.0	4733.8
h-kj/kg	2905.5	3072.4	3212.6	3465.4	3706.9	3947.0	4189.4	4435.9	4687.4	4944.2	5206.1	5473.0
s-kj/kg/K	5.99692	6.26455	6.47242	6.81035	7.09482	7.34781	7.57880	7.79324	7.99453	8.18477	8.36547	8.53770

Press = 9.000 MPa Tsat= 576.5 K

T(K) =	600.	650.	700.	800.	900.	1000.	1100.	1200.	1300.	1400.	1500.	1600.
v-m3/kg	0.02349	0.02815	0.03193	0.03847	0.04442	0.05009	0.05559	0.06100	0.06634	0.07164	0.07690	0.08214
u-kj/kg	2653.8	2794.8	2907.9	3108.4	3299.5	3490.6	3684.7	3883.4	4087.5	4297.2	4512.2	4732.3
h-kj/kg	2865.2	3048.1	3195.2	3454.6	3699.3	3941.4	4185.1	4432.5	4684.6	4941.9	5204.3	5471.5
s-kj/kg/K	5.88729	6.18066	6.39887	6.74568	7.03398	7.28902	7.52122	7.73645	7.93828	8.12892	8.30991	8.48236

Press = 10.000 MPa Tsat= 584.1 K

T(K) =	600.	650.	700.	800.	900.	1000.	1100.	1200.	1300.	1400.	1500.	1600.
v-m3/kg	0.02009	0.02471	0.02829	0.03436	0.03980	0.04496	0.04995	0.05485	0.05968	0.06446	0.06921	0.07393
u-kj/kg	2618.9	2775.4	2894.4	3100.1	3293.7	3486.2	3681.2	3880.5	4085.1	4295.1	4510.4	4730.7
h-kj/kg	2819.8	3022.5	3177.3	3443.7	3691.7	3935.8	4180.7	4429.0	4681.9	4939.7	5202.5	5470.0
s-kj/kg/K	5.77538	6.10069	6.33038	6.68660	6.97883	7.23595	7.46938	7.68539	7.88777	8.07881	8.26009	8.43276

Press = 12.000 MPa Tsat= 597.8 K

T(K) =	600.	650.	700.	750.	800.	850.	900.	950.	1000.	1050.
v-m3/kg	0.01459	0.01947	0.02280	0.02562	0.02818	0.03058	0.03287	0.03510	0.03727	0.03940
u-kj/kg	2528.8	2733.1	2866.2	2979.0	3083.3	3183.5	3281.9	3379.6	3477.2	3575.3
h-kj/kg	2703.9	2966.8	3139.8	3286.5	3421.4	3550.4	3676.4	3800.7	3924.4	4048.1
s-kj/kg/K	5.52472	5.94753	6.20430	6.40683	6.58105	6.73752	6.88148	7.01595	7.14285	7.26349

Press = 14.000 MPa Tsat= 609.8 K

T(K) =	650.	700.	750.	800.	850.	900.	950.	1000.	1050.
v-m3/kg	0.01563	0.01885	0.02145	0.02376	0.02589	0.02792	0.02987	0.03177	0.03363
u-kj/kg	2684.9	2835.9	2957.0	3066.0	3169.3	3269.9	3369.2	3468.2	3567.3
h-kj/kg	2903.7	3099.8	3257.3	3398.6	3531.8	3660.8	3787.5	3913.0	4038.1
s-kj/kg/K	5.79671	6.08799	6.30540	6.48786	6.64947	6.79693	6.93391	7.06267	7.18475

Press = 16.000 MPa Tsat= 620.5 K

T(K) =	650.	700.	750.	800.	850.	900.	950.	1000.	1050.
v-m3/kg	0.01263	0.01586	0.01831	0.02044	0.02238	0.02421	0.02596	0.02765	0.02930
u-kj/kg	2628.3	2803.4	2933.9	3048.2	3154.8	3257.8	3358.8	3459.1	3559.3
h-kj/kg	2830.4	3057.1	3226.9	3375.1	3512.9	3645.1	3774.1	3901.5	4028.1
s-kj/kg/K	5.64059	5.97752	6.21205	6.40350	6.57059	6.72171	6.86128	6.99195	7.11548

Press = 18.000 MPa Tsat= 630.1 K

T(K) =	650.	700.	750.	800.	850.	900.	950.	1000.	1050.
v-m3/kg	0.01014	0.01350	0.01586	0.01785	0.01964	0.02132	0.02291	0.02445	0.02594
u-kj/kg	2559.0	2768.3	2909.9	3029.8	3140.1	3245.4	3348.3	3449.9	3551.2
h-kj/kg	2741.5	3011.2	3195.3	3351.1	3493.7	3629.2	3760.7	3890.0	4018.1
s-kj/kg/K	5.46893	5.87012	6.12446	6.32572	6.49863	6.65356	6.79580	6.92843	7.05344

Press = 20.000 MPa Tsat= 638.9 K

T(K) =	650.	700.	750.	800.	850.	900.	950.	1000.	1050.
v-m3/kg	0.00790	0.01158	0.01388	0.01577	0.01745	0.01901	0.02047	0.02188	0.02325
u-kj/kg	2466.8	2730.1	2884.7	3011.0	3125.0	3232.9	3337.6	3440.7	3543.0
h-kj/kg	2624.9	2961.6	3162.4	3326.5	3474.1	3613.1	3747.1	3878.4	4008.0
s-kj/kg/K	5.26183	5.76355	6.04101	6.25298	6.43206	6.59096	6.73595	6.87059	6.99709

Press = 22.000 MPa Tsat= 646.9 K

T(K) =	650.	700.	750.	800.	850.	900.	950.	1000.	1050.
v-m3/kg	0.00147	0.00997	0.01226	0.01407	0.01566	0.01711	0.01848	0.01979	0.02105
u-kj/kg	1596.4	2688.3	2858.4	2991.6	3109.7	3220.3	3326.9	3431.4	3534.9
h-kj/kg	1698.6	2907.8	3128.1	3301.2	3454.2	3596.8	3733.5	3866.7	3997.9
s-kj/kg/K	3.69088	5.65591	5.96051	6.18419	6.36978	6.53282	6.68065	6.81732	6.94536

Press = 24.000 MPa

T(K) =	650.	700.	750.	800.	850.	900.	950.	1000.	1050.
v-m3/kg	0.00146	0.00861	0.01090	0.01265	0.01416	0.01554	0.01682	0.01804	0.01921
u-kj/kg	1590.2	2642.3	2830.8	2971.7	3094.1	3207.5	3316.1	3422.0	3526.6
h-kj/kg	1695.8	2848.9	3092.4	3275.4	3434.0	3580.4	3719.7	3855.0	3987.8
s-kj/kg/K	3.68017	5.54534	5.88208	6.11853	6.31101	6.47834	6.62908	6.76784	6.89743

Press = 26.000 MPa

T(K) =	650.	700.	750.	800.	850.	900.	950.	1000.	1050.
v-m3/kg	0.00145	0.00742	0.00974	0.01145	0.01290	0.01420	0.01542	0.01656	0.01766
u-kj/kg	1584.4	2591.2	2801.9	2951.2	3078.2	3194.5	3305.1	3412.6	3518.4
h-kj/kg	1693.3	2784.1	3055.2	3248.9	3413.6	3563.8	3705.9	3843.2	3977.7
s-kj/kg/K	3.67008	5.42987	5.80500	6.05541	6.25513	6.42692	6.58065	6.72153	6.85270

Press = 28.000 MPa

T(K) =	650.	700.	750.	800.	850.	900.	950.	1000.	1050.
v-m3/kg	0.00144	0.00637	0.00874	0.01042	0.01182	0.01306	0.01421	0.01530	0.01634
u-kj/kg	1578.9	2533.7	2771.6	2930.2	3062.0	3181.4	3294.1	3403.1	3510.1
h-kj/kg	1691.1	2712.0	3016.4	3221.9	3392.8	3547.1	3692.0	3831.5	3967.5
s-kj/kg/K	3.66050	5.30724	5.72875	5.99434	6.20169	6.37808	6.53487	6.67792	6.81068

Press = 30.000 MPa

T(K) =	650.	700.	750.	800.	850.	900.	950.	1000.	1050.
v-m3/kg	0.00143	0.00543	0.00787	0.00952	0.01088	0.01207	0.01317	0.01420	0.01519
u-kj/kg	1573.7	2468.6	2739.9	2908.7	3045.6	3168.1	3283.0	3393.6	3501.8
h-kj/kg	1689.1	2631.5	2976.2	3194.4	3371.8	3530.2	3678.1	3819.7	3957.4
s-kj/kg/K	3.65138	5.17540	5.65290	5.93499	6.15032	6.33146	6.49139	6.63664	6.77102

Press = 35.000 MPa

T(K) =	650.	700.	750.	800.	850.	900.	950.	1000.	1050.
v-m3/kg	0.00141	0.00141	0.00613	0.00773	0.00900	0.01010	0.01109	0.01201	0.01289
u-kj/kg	1561.7	1701.7	2654.4	2852.5	3003.4	3134.3	3254.9	3369.6	3480.9
h-kj/kg	1685.0	1899.7	2868.9	3123.2	3318.4	3487.6	3643.0	3790.1	3932.0
s-kj/kg/K	3.63027	3.83785	5.46347	5.79237	6.02927	6.22286	6.39089	6.54179	6.68027

Press = 40.000 MPa

T(K) =	650.	700.	750.	800.	850.	900.	950.	1000.	1050.
v-m3/kg	0.00139	0.00139	0.00483	0.00640	0.00760	0.00862	0.00953	0.01038	0.01117
u-kj/kg	1550.8	1691.1	2560.8	2793.3	2959.9	3099.8	3226.4	3345.4	3459.9
h-kj/kg	1681.8	1897.7	2754.1	3049.3	3263.9	3444.6	3607.7	3760.4	3906.6
s-kj/kg/K	3.61114	3.81901	5.27412	5.65605	5.91650	6.12322	6.29967	6.45636	6.59905

A.14.4 H_2O Compressed Liquid Table (SI units)

Temperature = 273.1 K		Psat = 0.001 Mpa								
Press (MPa)=	1	2	5	8	10	15	20	25	30	50
v (m^3/kg)	0.001	0.001	0.001	0.001	0.001	0.00099	0.00099	0.00099	0.00099	0.00098
u (kJ/kg)	0	0	0	0.1	0.1	0.2	0.2	0.3	0.3	0.3
h (kJ/kg)	1	2	5	8.1	10.1	15.1	20	25	29.9	49.1
s (kJ/kg/K)	0	0	0.00014	0.00027	0.00034	0.00045	0.00047	0.00041	0.00028	0
Temperature = 280.0 K		Psat = 0.001 MPa								
Press (MPa)=	1	2	5	8	10	15	20	25	30	50
v (m^3/kg)	0.001	0.001	0.001	0.001	0.001	0.00099	0.00099	0.00099	0.00099	0.00098
u (kJ/kg)	28.8	28.8	28.7	28.7	28.7	28.6	28.5	28.4	28.3	27.9
h (kJ/kg)	29.8	30.8	33.7	36.7	38.6	43.5	48.3	53.1	57.9	76.7
s (kJ/kg/K)	0.10407	0.10402	0.10386	0.10368	0.10354	0.10315	0.1027	0.10219	0.10162	0.09879
Temperature = 300.0 K		Psat = 0.004 Mpa								
Press (MPa)=	1	2	5	8	10	15	20	25	30	50
v (m^3/kg)	0.001	0.001	0.001	0.001	0.001	0.001	0.00099	0.00099	0.00099	0.00098
u (kJ/kg)	112.5	112.4	112.2	111.9	111.8	111.4	111	110.6	110.2	108.6
h (kJ/kg)	113.5	114.4	117.2	119.9	121.7	126.3	130.8	135.4	139.9	157.8
s (kJ/kg/K)	0.39285	0.39257	0.39174	0.39089	0.39033	0.3889	0.38744	0.38597	0.38448	0.37831
Temperature = 320.0 K		Psat = 0.011 MPa								
Press (MPa)=	1	2	5	8	10	15	20	25	30	50
v (m^3/kg)	0.00101	0.00101	0.00101	0.00101	0.00101	0.001	0.001	0.001	0.001	0.00099
u (kJ/kg)	196	195.9	195.5	195	194.8	194.1	193.4	192.8	192.1	189.7
h (kJ/kg)	197	197.9	200.5	203.1	204.8	209.2	213.5	217.8	222.1	239.2
s (kJ/kg/K)	0.66242	0.66198	0.66066	0.65934	0.65847	0.65628	0.65409	0.65191	0.64973	0.64104

(continued)

(continued)

Temperature = 340.0 K		Psat = 0.027 MPa							
Press (MPa)=	1	2	5	8	10	15	20	25	30	50
v (m^3/kg)	0.00102	0.00102	0.00102	0.00102	0.00102	0.00101	0.00101	0.00101	0.00101	0.001
u (kJ/kg)	279.6	279.4	278.8	278.3	277.9	276.9	276	275.1	274.2	270.9
h (kJ/kg)	280.6	281.5	283.9	286.4	288	292.2	296.3	300.4	304.5	320.9
s (kJ/kg/K)	0.91582	0.91524	0.91352	0.91181	0.91067	0.90785	0.90504	0.90226	0.89949	0.88862
Temperature = 360.0 K		Psat = 0.062 MPa								
Press (MPa)=	1	2	5	8	10	15	20	25	30	50
v (m^3/kg)	0.00103	0.00103	0.00103	0.00103	0.00103	0.00103	0.00102	0.00102	0.00102	0.00101
u (kJ/kg)	363.4	363.2	362.4	361.7	361.2	360	358.8	357.7	356.5	352.2
h (kJ/kg)	364.5	365.2	367.6	369.9	371.5	375.4	379.3	383.2	387.1	402.8
s (kJ/kg/K)	1.15538	1.15468	1.15259	1.15052	1.14914	1.14573	1.14236	1.13902	1.13572	1.12283
Temperature = 380.0 K		Psat = 0.129 MPa								
Press (MPa)=	1	2	5	8	10	15	20	25	30	50
v (m^3/kg)	0.00105	0.00105	0.00105	0.00104	0.00104	0.00104	0.00104	0.00104	0.00103	0.00102
u (kJ/kg)	447.6	447.3	446.4	445.5	444.9	443.4	441.9	440.5	439.1	433.8
h (kJ/kg)	448.7	449.4	451.6	453.8	455.3	459	462.7	466.4	470.1	485.1
s (kJ/kg/K)	1.38306	1.38224	1.37978	1.37735	1.37574	1.37175	1.36782	1.36394	1.36011	1.34526
Temperature = 400.0 K		Psat = 0.246 MPa								
Press (MPa)=	1	2	5	8	10	15	20	25	30	50
v (m^3/kg)	0.00107	0.00107	0.00106	0.00106	0.00106	0.00106	0.00106	0.00105	0.00105	0.00104
u (kJ/kg)	532.4	532	530.9	529.8	529	527.3	525.5	523.8	522.1	515.8
h (kJ/kg)	533.5	534.1	536.2	538.3	539.7	543.1	546.6	550.1	553.6	567.8
s (kJ/kg/K)	1.60051	1.59955	1.59672	1.59392	1.59207	1.58749	1.58298	1.57854	1.57417	1.5573
Temperature = 420.0 K		Psat = 0.437 MPa								
Press (MPa)=	1	2	5	8	10	15	20	25	30	50
v (m^3/kg)	0.00109	0.00109	0.00108	0.00108	0.00108	0.00108	0.00107	0.00107	0.00107	0.00106

(continued)

(continued)

u (kJ/kg)	617.9	617.5	616.1	614.8	613.9	611.7	609.6	607.6	605.6	598.1
h (kJ/kg)	619	619.6	621.5	623.4	624.7	627.9	631.1	634.4	637.7	650.9
s (kJ/kg/K)	1.8092	1.80811	1.80487	1.80166	1.79954	1.79432	1.78919	1.78415	1.7792	1.76019
Temperature = 440.0 K		Psat = 0.734 MPa								
Press (MPa)=	1	2	5	8	10	15	20	25	30	50
v (m^3/kg)	0.00111	0.00111	0.00111	0.0011	0.0011	0.0011	0.0011	0.00109	0.00109	0.00108
u (kJ/kg)	704.5	703.9	702.3	700.7	699.6	697.1	694.6	692.1	689.8	680.9
h (kJ/kg)	705.6	706.1	707.8	709.5	710.7	713.6	716.5	719.5	722.4	734.7
s (kJ/kg/K)	2.01053	2.00928	2.00556	2.00189	1.99947	1.99352	1.9877	1.98199	1.9764	1.95505
Temperature = 460.0 K		Psat = 1.171 MPa								
Press (MPa)=	1	2	5	8	10	15	20	25	30	50
v (m^3/kg)	0.19849	0.00114	0.00113	0.00113	0.00113	0.00113	0.00112	0.00112	0.00111	0.0011
u (kJ/kg)	2597	791.7	789.7	787.8	786.5	783.4	780.5	777.6	774.8	764.3
h (kJ/kg)	2795.5	793.9	795.4	796.8	797.8	800.3	802.9	805.5	808.2	819.2
s (kJ/kg/K)	6.62522	2.20441	2.20012	2.19591	2.19314	2.18633	2.1797	2.17322	2.16689	2.1429
Temperature = 480.0 K		Psat = 1.790 MPa								
Press (MPa)=	1	2	5	8	10	15	20	25	30	50
v (m^3/kg)	0.20981	0.00117	0.00116	0.00116	0.00116	0.00115	0.00115	0.00114	0.00114	0.00112
u (kJ/kg)	2634.9	881.1	878.7	876.4	874.9	871.2	867.6	864.2	860.9	848.5
h (kJ/kg)	2844.7	883.5	884.6	885.7	886.5	888.5	890.6	892.8	895.1	904.7
s (kJ/kg/K)	6.73004	2.39491	2.38993	2.38504	2.38184	2.374	2.36639	2.35899	2.35179	2.32474
Temperature = 500.0 K		Psat = 2.639 MPa								
Press (MPa)=	1	2	5	8	10	15	20	25	30	50
v (m^3/kg)	0.22063	0.10439	0.0012	0.0012	0.00119	0.00119	0.00118	0.00118	0.00117	0.00115
u (kJ/kg)	2670.6	2632.6	970	967.1	965.3	960.8	956.5	952.3	948.4	933.8
h (kJ/kg)	2891.3	2841.4	976	976.7	977.2	978.6	980.1	981.7	983.5	991.3
s (kJ/kg/K)	6.82505	6.42645	2.57651	2.57075	2.56699	2.55784	2.54901	2.54049	2.53223	2.50153

(continued)

(continued)

Temperature = 520.0 K		Psat = 3.769 MPa								
Press (MPa)=	1	2	5	8	10	15	20	25	30	50
v (m^3/kg)	0.23111	0.11055	0.00124	0.00124	0.00123	0.00123	0.00122	0.00121	0.00121	0.00118
u (kJ/kg)	2705.1	2674	1064.2	1060.7	1058.3	1052.8	1047.5	1042.5	1037.7	1020.3
h (kJ/kg)	2936.2	2895.1	1070.4	1070.6	1070.7	1071.2	1071.9	1072.8	1073.8	1079.4
s (kJ/kg/K)	6.91324	6.53188	2.7617	2.75477	2.75027	2.73938	2.72897	2.719	2.70941	2.67426
Temperature = 540.0 K		Psat = 5.237 MPa								
Press (MPa)=	1	2	5	8	10	15	20	25	30	50
v (m^3/kg)				0.00129	0.00128	0.00127	0.00126	0.00125	0.00125	0.00122
u (kJ/kg)				1158.1	1155.1	1148.1	1141.5	1135.2	1129.3	1108.5
h (kJ/kg)				1168.4	1168	1167.2	1166.7	1166.6	1166.7	1169.3
s (kJ/kg/K)				2.93939	2.93382	2.9205	2.90793	2.89602	2.88469	2.84393
Temperature = 560.0 K		Psat = 7.106 MPa								
Press (MPa)=	1	2	5	8	10	15	20	25	30	50
v (m^3/kg)				0.00135	0.00135	0.00133	0.00132	0.00131	0.00129	0.00126
u (kJ/kg)				1261.4	1257.4	1248	1239.4	1231.5	1224.1	1198.7
h (kJ/kg)				1272.2	1270.8	1268	1265.8	1264.2	1263	1261.6
s (kJ/kg/K)				3.12812	3.12086	3.1038	3.08808	3.07344	3.0597	3.01163
Temperature = 580.0 K		Psat = 9.45 MPa								
Press (MPa)=	1	2	5	8	10	15	20	25	30	50
v (m^3/kg)					0.00143	0.00141	0.00139	0.00137	0.00135	0.00131
u (kJ/kg)					1368.5	1355.1	1343.3	1332.8	1323.2	1291.5
h (kJ/kg)					1382.9	1376.2	1371.1	1367	1363.8	1356.7
s (kJ/kg/K)					3.31731	3.29363	3.27272	3.25386	3.23661	3.17855
Temperature = 600.0 K		Psat = 12.34 MPa								
Press (MPa)=	1	2	5	8	10	15	20	25	30	50
v (m^3/kg)						0.00152	0.00148	0.00145	0.00143	0.00136

(continued)

(continued)

u (kJ/kg)						1474.8	1456.6	1441.5	1428.3	1387.5
h (kJ/kg)						1497.5	1486.3	1477.8	1471.2	1455.6
s (kJ/kg/K)						3.49917	3.46794	3.44163	3.41862	3.34608
Temperature = 620.0 K		Psat = 15.9 Mpa								
Press (MPa)=	1	2	5	8	10	15	20	25	30	50
v (m^3/kg)							0.00163	0.00157	0.00153	0.00143
u (kJ/kg)							1588.6	1562.9	1542.9	1487.6
h (kJ/kg)							1621.2	1602.3	1588.9	1559.1
s (kJ/kg/K)							3.68902	3.64556	3.6115	3.5159
Temperature = 640.0 K		Psat = 20.27 Mpa								
Press (MPa)=	1	2	5	8	10	15	20	25	30	50
v (m^3/kg)								0.01322	0.01322	0.01322
u (kJ/kg)								2616.4	2616.4	2616.4
h (kJ/kg)								2815.1	2815.1	2815.1
s (kJ/kg/K)								5.63646	5.63646	5.63646

A.14.5 The Saturation Temperature vs. Pressure (English Units)

Press (psi)	Temp (R)	Specific Volume (ft**3/lbm)		Internal Energy (Btu/lbm)			Enthalpy (Btu/lbm)			Entropy (Btu/lbm/R)		
		vf	vg	uf	ufg	ug	hf	hfg	hg	sf	sfg	sg
0.1	494.7	0.0161	2951.1577	3.01	1021.2	1024.2	3.02	1075.8	1078.8	0.00611	2.17479	2.18090
0.2	512.8	0.0161	1529.1317	21.25	1008.9	1030.2	21.25	1065.5	1086.8	0.04231	2.07785	2.12017
0.3	524.1	0.0161	1041.6294	32.60	1001.3	1033.9	32.60	1059.1	1091.7	0.06421	2.02075	2.08496
0.4	532.5	0.0161	793.5160	41.00	995.6	1036.6	41.00	1054.4	1095.4	0.08010	1.98002	2.06013
0.6	544.8	0.0161	541.0339	53.35	987.3	1040.6	53.36	1047.4	1100.7	0.10305	1.92229	2.02534
0.8	554.0	0.0161	412.4360	62.52	981.1	1043.6	62.52	1042.1	1104.7	0.11973	1.88108	2.00081
1.0	561.4	0.0162	334.2068	69.88	976.1	1046.0	69.88	1037.9	1107.8	0.13292	1.84897	1.98188
1.5	575.3	0.0162	228.1557	83.83	966.6	1050.4	83.83	1029.9	1113.8	0.15746	1.79023	1.94769
2.5	594.1	0.0163	141.1594	102.58	953.8	1056.4	102.59	1019.1	1121.7	0.18954	1.71547	1.90501
3.0	601.1	0.0163	118.9527	109.62	949.0	1058.6	109.63	1015.0	1124.6	0.20132	1.68857	1.88989
3.5	607.2	0.0164	102.9342	115.73	944.7	1060.5	115.74	1011.4	1127.1	0.21143	1.66571	1.87715
4.0	612.6	0.0164	90.8181	121.13	941.0	1062.1	121.15	1008.2	1129.4	0.22030	1.64584	1.86614
5.0	621.9	0.0164	73.6775	130.43	934.6	1065.0	130.44	1002.7	1133.2	0.23536	1.61246	1.84781
6.0	629.7	0.0165	62.1094	138.27	929.1	1067.4	138.29	998.0	1136.3	0.24789	1.58501	1.83290
7.0	636.5	0.0165	53.7612	145.08	924.3	1069.4	145.11	994.0	1139.1	0.25865	1.56168	1.82033
8.0	642.5	0.0166	47.4442	151.13	920.1	1071.2	151.15	990.3	1141.5	0.26811	1.54137	1.80947
10.0	652.8	0.0166	38.5033	161.54	912.7	1074.3	161.57	984.0	1145.5	0.28418	1.50721	1.79139
12.5	663.6	0.0167	31.2495	172.36	905.0	1077.4	172.40	977.3	1149.7	0.30062	1.47276	1.77337
14.7	671.6	0.0167	26.8597	180.48	899.2	1079.7	180.52	972.2	1152.8	0.31278	1.44757	1.76035
20.0	687.6	0.0169	20.1340	196.61	887.6	1084.2	196.67	962.0	1158.7	0.33651	1.39911	1.73562
25.0	699.7	0.0170	16.3406	208.88	878.6	1087.5	208.96	954.1	1163.1	0.35421	1.36356	1.71777
30.0	710.0	0.0170	13.7771	219.30	870.9	1090.2	219.40	947.3	1166.6	0.36900	1.33420	1.70320
35.0	718.9	0.0171	11.9251	228.41	864.1	1092.5	228.52	941.2	1169.7	0.38175	1.30914	1.69089
40.0	726.9	0.0172	10.5224	236.53	857.9	1094.5	236.66	935.7	1172.4	0.39298	1.28724	1.68022
45.0	734.1	0.0172	9.4220	243.88	852.3	1096.2	244.02	930.7	1174.7	0.40304	1.26777	1.67081
50.0	740.7	0.0173	8.5350	250.61	847.2	1097.8	250.77	926.0	1176.8	0.41217	1.25022	1.66239
55.0	746.7	0.0174	7.8041	256.82	842.4	1099.2	257.00	921.6	1178.6	0.42052	1.23424	1.65476
60.0	752.4	0.0174	7.1912	262.61	837.9	1100.5	262.80	917.5	1180.3	0.42824	1.21955	1.64779
65.0	757.6	0.0175	6.6696	268.02	833.7	1101.7	268.24	913.7	1181.9	0.43542	1.20595	1.64137
70.0	762.6	0.0175	6.2201	273.13	829.6	1102.8	273.35	910.0	1183.3	0.44213	1.19329	1.63542
75.0	767.3	0.0176	5.8286	277.95	825.8	1103.8	278.19	906.5	1184.7	0.44844	1.18143	1.62987
80.0	771.7	0.0176	5.4844	282.53	822.2	1104.7	282.79	903.1	1185.9	0.45439	1.17028	1.62467
85.0	775.9	0.0176	5.1794	286.89	818.7	1105.6	287.17	899.9	1187.0	0.46003	1.15975	1.61977
90.0	779.9	0.0177	4.9071	291.06	815.3	1106.4	291.36	896.8	1188.1	0.46539	1.14977	1.61515
95.0	783.8	0.0177	4.6626	295.05	812.1	1107.1	295.37	893.7	1189.1	0.47049	1.14028	1.61078
100.0	787.5	0.0178	4.4417	298.89	809.0	1107.9	299.22	890.8	1190.1	0.47538	1.13124	1.60661

Press (psi)	Temp (R)	Specific Volume (ft**3/lbm)		Internal Energy (Btu/lbm)			Enthalpy (Btu/lbm)			Entropy (Btu/lbm/R)		
		vf	vg	uf	ufg	ug	hf	hfg	hg	sf	sfg	sg
110.0	794.5	0.0178	4.0581	306.14	803.0	1109.2	306.50	885.3	1191.8	0.48454	1.11432	1.59886
120.0	800.9	0.0179	3.7365	312.90	797.4	1110.3	313.29	880.0	1193.3	0.49302	1.09874	1.59176
130.0	807.0	0.0180	3.4626	319.24	792.1	1111.4	319.67	875.0	1194.7	0.50091	1.08429	1.58520
140.0	812.7	0.0181	3.2267	325.22	787.1	1112.3	325.69	870.2	1195.9	0.50830	1.07080	1.57910
150.0	818.1	0.0181	3.0211	330.89	782.3	1113.2	331.40	865.7	1197.1	0.51525	1.05814	1.57340
160.0	823.2	0.0182	2.8404	336.28	777.7	1114.0	336.82	861.3	1198.1	0.52183	1.04622	1.56805
170.0	828.1	0.0183	2.6802	341.43	773.3	1114.7	342.00	857.0	1199.0	0.52806	1.03494	1.56300
180.0	832.8	0.0183	2.5373	346.35	769.0	1115.4	346.96	852.9	1199.9	0.53399	1.02423	1.55822
190.0	837.2	0.0184	2.4088	351.07	764.9	1116.0	351.72	849.0	1200.7	0.53965	1.01403	1.55367
200.0	841.5	0.0184	2.2928	355.62	760.9	1116.5	356.30	845.1	1201.4	0.54506	1.00429	1.54935
250.0	860.7	0.0187	1.8478	376.11	742.6	1118.7	376.97	827.2	1204.2	0.56915	0.96114	1.53029
300.0	877.0	0.0189	1.5467	393.80	726.3	1120.1	394.85	811.1	1206.0	0.58953	0.92485	1.51438
350.0	891.4	0.0192	1.3290	409.48	711.5	1121.0	410.72	796.3	1207.1	0.60728	0.89335	1.50063
400.0	904.3	0.0194	1.1640	423.66	697.8	1121.5	425.09	782.6	1207.6	0.62308	0.86537	1.48845
450.0	916.0	0.0196	1.0345	436.64	685.0	1121.6	438.27	769.5	1207.8	0.63737	0.84010	1.47747
500.0	926.7	0.0198	0.9301	448.67	672.9	1121.6	450.50	757.1	1207.6	0.65045	0.81698	1.46743
600.0	945.9	0.0202	0.7718	470.48	650.3	1120.8	472.72	733.7	1206.5	0.67378	0.77569	1.44947
700.0	962.8	0.0206	0.6572	490.01	629.3	1119.4	492.68	711.8	1204.5	0.69430	0.73931	1.43361
800.0	977.9	0.0209	0.5704	507.83	609.6	1117.4	510.93	691.0	1201.9	0.71271	0.70654	1.41925
900.0	991.7	0.0213	0.5021	524.31	590.8	1115.1	527.86	670.9	1198.7	0.72950	0.67650	1.40600
1000.0	1004.3	0.0216	0.4470	539.73	572.7	1112.4	543.73	651.4	1195.1	0.74501	0.64861	1.39362
1100.0	1016.0	0.0220	0.4015	554.27	555.1	1109.4	558.75	632.4	1191.1	0.75948	0.62242	1.38190
1200.0	1026.9	0.0224	0.3632	568.10	538.0	1106.1	573.07	613.7	1186.8	0.77310	0.59760	1.37069
1300.0	1037.2	0.0227	0.3306	581.34	521.2	1102.5	586.81	595.2	1182.0	0.78600	0.57389	1.35990
1400.0	1046.8	0.0231	0.3023	594.07	504.6	1098.6	600.06	576.9	1177.0	0.79831	0.55110	1.34941
1500.0	1055.9	0.0235	0.2776	606.38	488.1	1094.5	612.91	558.6	1171.5	0.81012	0.52904	1.33916
1600.0	1064.6	0.0239	0.2558	618.34	471.7	1090.0	625.42	540.4	1165.8	0.82151	0.50757	1.32907
1800.0	1080.7	0.0248	0.2189	641.42	438.9	1080.3	649.67	503.5	1153.2	0.84327	0.46589	1.30916
2000.0	1095.5	0.0257	0.1886	663.74	405.4	1069.2	673.25	465.7	1139.0	0.86408	0.42510	1.28918
2250.0	1112.4	0.0270	0.1573	691.19	361.7	1052.9	702.44	416.0	1118.4	0.88942	0.37393	1.26335
2500.0	1127.8	0.0287	0.1309	719.11	314.2	1033.3	732.37	361.5	1093.9	0.91498	0.32054	1.23552
2750.0	1142.0	0.0308	0.0573	749.09	121.2	870.3	764.78	134.6	899.4	0.94233	0.11756	1.05990
3000.0	1155.1	0.0345	0.0531	785.51	86.3	871.8	804.64	96.7	901.3	0.97571	0.08362	1.05933
3200.0	1164.8	0.0517	0.0517	877.11	0.0	877.1	907.70	0.0	907.7	1.06317	0.00000	1.06317

A.14.6 The Saturation Pressure vs. Temperature (English Units)

Temp (R)	Press (psi)	Specific Volume (ft**3/lbm)		Internal Energy (Btu/lbm)			Enthalpy (Btu/lbm)			Entropy (Btu/lbm/R)		
		vf	vg	uf	ufg	ug	hf	hfg	hg	sf	sfg	sg
491.7	0.09	0.0161	3308.9548	0.00	1023.2	1023.2	0.00	1077.5	1077.5	0.00000	2.19152	2.19152
500.0	0.12	0.0161	2418.8743	8.38	1017.6	1026.0	8.38	1072.8	1081.2	0.01690	2.14557	2.16247
510.0	0.18	0.0161	1686.7268	18.44	1010.8	1029.3	18.44	1067.1	1085.6	0.03681	2.09238	2.12920
520.0	0.26	0.0161	1195.2240	28.47	1004.1	1032.5	28.47	1061.5	1089.9	0.05630	2.04126	2.09755
530.0	0.37	0.0161	859.8069	38.49	997.3	1035.8	38.49	1055.8	1094.3	0.07538	1.99205	2.06744
540.0	0.51	0.0161	627.3451	48.50	990.6	1039.1	48.50	1050.1	1098.6	0.09410	1.94466	2.03876
550.0	0.71	0.0161	463.8765	58.51	983.8	1042.3	58.51	1044.4	1102.9	0.11246	1.89896	2.01142
560.0	0.96	0.0162	347.3375	68.51	977.0	1045.5	68.51	1038.7	1107.2	0.13048	1.85486	1.98535
570.0	1.29	0.0162	263.1730	78.51	970.2	1048.7	78.52	1033.0	1111.5	0.14819	1.81227	1.96045
580.0	1.71	0.0162	201.6415	88.52	963.4	1051.9	88.52	1027.2	1115.8	0.16558	1.77109	1.93668
590.0	2.25	0.0163	156.1334	98.52	956.6	1055.1	98.53	1021.4	1120.0	0.18269	1.73126	1.91395
600.0	2.92	0.0163	122.1055	108.53	949.7	1058.2	108.54	1015.6	1124.2	0.19951	1.69269	1.89220
610.0	3.75	0.0164	96.3962	118.55	942.8	1061.3	118.56	1009.7	1128.3	0.21606	1.65531	1.87137
620.0	4.78	0.0164	76.7798	128.57	935.9	1064.4	128.58	1003.8	1132.4	0.23236	1.61906	1.85142
630.0	6.05	0.0165	61.6720	138.60	928.9	1067.5	138.62	997.8	1136.5	0.24841	1.58387	1.83228
640.0	7.57	0.0165	49.9327	148.64	921.8	1070.5	148.66	991.8	1140.5	0.26422	1.54969	1.81391
650.0	9.42	0.0166	40.7337	158.69	914.8	1073.4	158.72	985.7	1144.4	0.27981	1.51645	1.79626
660.0	11.62	0.0167	33.4671	168.76	907.6	1076.4	168.79	979.5	1148.3	0.29518	1.48411	1.77928
670.0	14.23	0.0167	27.6832	178.84	900.4	1079.2	178.88	973.3	1152.1	0.31034	1.45261	1.76294
680.0	17.31	0.0168	23.0457	188.93	893.1	1082.1	188.99	966.9	1155.9	0.32529	1.42190	1.74720
690.0	20.92	0.0169	19.3017	199.05	885.8	1084.8	199.11	960.5	1159.6	0.34006	1.39195	1.73201
700.0	25.13	0.0170	16.2589	209.18	878.4	1087.5	209.26	953.9	1163.2	0.35464	1.36270	1.71734
710.0	30.02	0.0170	13.7705	219.33	870.8	1090.2	219.43	947.2	1166.7	0.36904	1.33412	1.70316
720.0	35.64	0.0171	11.7232	229.51	863.2	1092.7	229.62	940.4	1170.1	0.38327	1.30616	1.68943
730.0	42.10	0.0172	10.0291	239.71	855.5	1095.2	239.84	933.5	1173.4	0.39734	1.27880	1.67613
740.0	49.48	0.0173	8.6197	249.93	847.7	1097.6	250.09	926.5	1176.6	0.41125	1.25198	1.66323
750.0	57.86	0.0174	7.4409	260.18	839.8	1100.0	260.37	919.3	1179.6	0.42501	1.22569	1.65070
760.0	67.35	0.0175	6.4501	270.46	831.7	1102.2	270.68	911.9	1182.6	0.43863	1.19989	1.63852
770.0	78.05	0.0176	5.6133	280.77	823.6	1104.3	281.03	904.4	1185.4	0.45211	1.17454	1.62665
780.0	90.07	0.0177	4.9034	291.12	815.3	1106.4	291.41	896.7	1188.1	0.46546	1.14963	1.61509
790.0	103.52	0.0178	4.2985	301.50	806.8	1108.3	301.84	888.8	1190.7	0.47869	1.12511	1.60380
800.0	118.52	0.0179	3.7809	311.92	798.2	1110.2	312.31	880.8	1193.1	0.49180	1.10098	1.59277
810.0	135.18	0.0180	3.3362	322.38	789.5	1111.9	322.83	872.5	1195.4	0.50480	1.07719	1.58198
820.0	153.64	0.0181	2.9527	332.89	780.6	1113.5	333.40	864.1	1197.5	0.51769	1.05372	1.57141
830.0	174.03	0.0183	2.6207	343.44	771.5	1115.0	344.03	855.4	1199.4	0.53048	1.03055	1.56104
840.0	196.49	0.0184	2.3323	354.04	762.3	1116.3	354.71	846.4	1201.1	0.54319	1.00766	1.55085
Temp (R)	Press (psi)	Specific Volume (ft**3/lbm)		Internal Energy (Btu/lbm)			Enthalpy (Btu/lbm)			Entropy (Btu/lbm/R)		
		vf	vg	uf	ufg	ug	hf	hfg	hg	sf	sfg	sg
850.0	221.14	0.0185	2.0809	364.69	752.9	1117.6	365.45	837.3	1202.7	0.55580	0.98501	1.54082
860.0	248.15	0.0187	1.8611	375.41	743.2	1118.6	376.26	827.8	1204.1	0.56834	0.96259	1.53093
870.0	277.66	0.0188	1.6683	386.18	733.4	1119.6	387.15	818.1	1205.3	0.58080	0.94038	1.52118
880.0	309.82	0.0190	1.4986	397.02	723.3	1120.3	398.11	808.1	1206.2	0.59320	0.91833	1.51153
890.0	344.79	0.0191	1.3488	407.93	713.0	1120.9	409.15	797.8	1207.0	0.60553	0.89644	1.50198
900.0	382.74	0.0193	1.2163	418.91	702.4	1121.3	420.28	787.2	1207.5	0.61782	0.87468	1.49250
910.0	423.83	0.0195	1.0986	429.98	691.6	1121.6	431.50	776.3	1207.8	0.63006	0.85302	1.48309
920.0	468.23	0.0197	0.9940	441.13	680.5	1121.6	442.83	764.9	1207.8	0.64227	0.83144	1.47371
930.0	516.12	0.0199	0.9006	452.37	669.1	1121.5	454.27	753.2	1207.5	0.65444	0.80992	1.46436
940.0	567.68	0.0201	0.8171	463.72	657.4	1121.1	465.83	741.1	1206.9	0.66660	0.78841	1.45501
950.0	623.10	0.0203	0.7422	475.17	645.3	1120.5	477.51	728.6	1206.1	0.67874	0.76691	1.44565
960.0	682.56	0.0205	0.6749	486.74	632.9	1119.6	489.33	715.6	1204.9	0.69089	0.74537	1.43625
970.0	746.26	0.0207	0.6143	498.44	620.1	1118.5	501.30	702.1	1203.4	0.70304	0.72376	1.42680
980.0	814.41	0.0210	0.5595	510.28	606.9	1117.1	513.44	688.0	1201.5	0.71522	0.70206	1.41728
990.0	887.21	0.0212	0.5100	522.27	593.2	1115.4	525.75	673.4	1199.2	0.72743	0.68022	1.40765
1000.0	964.87	0.0215	0.4651	534.42	579.0	1113.4	538.26	658.2	1196.5	0.73969	0.65819	1.39789
1010.0	1047.63	0.0218	0.4243	546.75	564.3	1111.0	550.98	642.3	1193.3	0.75202	0.63595	1.38796
1020.0	1135.71	0.0221	0.3871	559.29	549.0	1108.3	563.94	625.7	1189.6	0.76443	0.61341	1.37784
1030.0	1229.35	0.0225	0.3531	572.04	533.0	1105.1	577.16	608.3	1185.4	0.77695	0.59054	1.36749
1040.0	1328.81	0.0229	0.3220	585.05	516.4	1101.4	590.67	589.9	1180.6	0.78960	0.56724	1.35685
1050.0	1434.35	0.0233	0.2935	598.34	498.9	1097.2	604.52	570.6	1175.1	0.80242	0.54344	1.34587
1060.0	1546.25	0.0237	0.2672	611.95	480.5	1092.5	618.73	550.2	1168.9	0.81544	0.51904	1.33448
1070.0	1664.83	0.0242	0.2429	625.93	461.1	1087.0	633.38	528.5	1161.9	0.82869	0.49390	1.32260
1080.0	1790.40	0.0247	0.2205	640.33	440.4	1080.8	648.52	505.3	1153.8	0.84225	0.46786	1.31011
1090.0	1923.31	0.0253	0.1995	655.24	418.3	1073.6	664.26	480.4	1144.6	0.85619	0.44069	1.29688
1100.0	2063.95	0.0260	0.1800	670.78	394.5	1065.3	680.71	453.3	1134.0	0.87060	0.41209	1.28270
1110.0	2212.74	0.0268	0.1616	687.10	368.4	1055.5	698.08	423.6	1121.7	0.88566	0.38163	1.26730
1120.0	2370.19	0.0278	0.1441	704.47	339.5	1044.0	716.64	390.6	1107.2	0.90160	0.34870	1.25030
1130.0	2536.83	0.0289	0.1274	723.34	306.7	1030.0	736.92	352.9	1089.8	0.91885	0.31230	1.23114
1140.0	2713.33	0.0305	0.0580	744.45	125.6	870.1	759.76	139.4	899.2	0.93811	0.12195	1.06006
1150.0	2900.52	0.0327	0.0547	769.61	101.6	871.2	787.17	113.4	900.5	0.96110	0.09844	1.05953
1160.0	3099.69	0.0371	0.0517	805.97	65.9	871.9	827.28	74.2	901.5	0.99470	0.06398	1.05868
1164.8	3200.12	0.0515	0.0515	876.57	0.0	876.6	907.08	0.0	907.1	1.06264	0.00000	1.06264

A.14.7 Superheated Steam Table (English Units)

Press = 1.0 psi, Tsat = 561.4 R

T(R) =	600.	650.	700.	750.	800.	850.	900.	1000.	1100.	1200.	1300.	1400.
v-ft3/lbm	357.745	387.798	417.779	447.720	477.638	507.541	537.435	597.207	656.965	716.715	776.461	836.203
u-Btu/lbm	1059.3	1076.5	1093.7	1111.0	1128.5	1146.1	1163.9	1200.1	1237.0	1274.9	1313.6	1353.1
h-Btu/lbm	1125.5	1148.2	1170.9	1193.8	1216.8	1240.0	1263.3	1310.5	1358.6	1407.4	1457.2	1507.8
s-Btu/lbm/R	2.01229	2.04866	2.08238	2.11392	2.14360	2.17168	2.19835	2.24811	2.29388	2.33640	2.37620	2.41372

Press = 2.5 psi, Tsat = 594.1 R

T(R) =	600.	650.	700.	750.	800.	850.	900.	1000.	1100.	1200.	1300.	1400.
v-ft3/lbm	142.701	154.841	166.904	178.926	190.924	202.909	214.883	238.816	262.735	286.646	310.553	334.456
u-Btu/lbm	1058.5	1076.0	1093.3	1110.7	1128.3	1145.9	1163.8	1200.0	1237.0	1274.8	1313.5	1353.1
h-Btu/lbm	1124.5	1147.5	1170.5	1193.5	1216.5	1239.8	1263.1	1310.4	1358.5	1407.4	1457.1	1507.7
s-Btu/lbm/R	1.90967	1.94663	1.98065	2.01236	2.04214	2.07028	2.09699	2.14680	2.19260	2.23513	2.27495	2.31247

Press = 5.0 psi, Tsat = 621.9 R

T(R) =	650.	700.	750.	800.	850.	900.	1000.	1100.	1200.	1300.	1400.	1500.
v-ft3/lbm	77.185	83.277	89.327	95.353	101.364	107.366	119.352	131.325	143.290	155.250	167.206	179.161
u-Btu/lbm	1075.1	1092.7	1110.3	1127.9	1145.7	1163.6	1199.8	1236.9	1274.7	1313.4	1353.0	1393.6
h-Btu/lbm	1146.4	1169.7	1192.9	1216.1	1239.4	1262.8	1310.2	1358.3	1407.2	1457.0	1507.7	1559.2
s-Btu/lbm/R	1.86870	1.90322	1.93521	1.96516	1.99340	2.02019	2.07008	2.11592	2.15848	2.19832	2.23586	2.27143

Press = 7.5 psi, Tsat = 639.6 R

T(R) =	650.	700.	750.	800.	850.	900.	1000.	1100.	1200.	1300.	1400.	1500.
v-ft3/lbm	51.297	55.400	59.460	63.495	67.516	71.527	79.531	87.521	95.504	103.482	111.457	119.429
u-Btu/lbm	1074.2	1092.1	1109.9	1127.6	1145.4	1163.4	1199.7	1236.8	1274.6	1313.4	1353.0	1393.5
h-Btu/lbm	1145.3	1169.0	1192.4	1215.7	1239.1	1262.6	1310.0	1358.2	1407.1	1456.9	1507.6	1559.2
s-Btu/lbm/R	1.82250	1.85755	1.88983	1.91995	1.94831	1.97516	2.02514	2.07103	2.11362	2.15347	2.19103	2.22661

Press = 10.0 psi, Tsat = 652.8 R

T(R) =	700.	750.	800.	850.	900.	1000.	1100.	1200.	1300.	1400.	1500.	1600.
v-ft3/lbm	41.4602	44.5258	47.5662	50.5914	53.6069	59.6202	65.6198	71.6115	77.5984	83.5819	89.5631	95.5426
u-Btu/lbm	1091.5	1109.4	1127.3	1145.2	1163.1	1199.5	1236.7	1274.6	1313.3	1352.9	1393.5	1434.9
h-Btu/lbm	1168.2	1191.8	1215.3	1238.7	1262.3	1309.8	1358.0	1407.0	1456.8	1507.5	1559.1	1611.6
s-Btu/lbm/R	1.82489	1.85746	1.88776	1.91622	1.94315	1.99321	2.03915	2.08177	2.12164	2.15920	2.19480	2.22869

Press = 12.5 psi, Tsat = 663.6 R

T(R) =	700.	750.	800.	850.	900.	1000.	1100.	1200.	1300.	1400.	1500.	1600.
v-ft3/lbm	33.0955	35.5648	38.0085	40.4367	42.8550	47.6737	52.4787	57.2759	62.0681	66.8570	71.6435	76.4284
u-Btu/lbm	1090.9	1109.0	1126.9	1144.9	1162.9	1199.4	1236.5	1274.5	1313.2	1352.9	1393.4	1434.9
h-Btu/lbm	1167.4	1191.2	1214.8	1238.4	1262.0	1309.6	1357.9	1406.9	1456.7	1507.4	1559.0	1611.6
s-Btu/lbm/R	1.79933	1.83221	1.86269	1.89126	1.91827	1.96841	2.01440	2.05704	2.09693	2.13451	2.17011	2.20401

Press = 14.7 psi, Tsat = 671.6 R

T(R) =	700.	750.	800.	850.	900.	1000.	1100.	1200.	1300.	1400.	1500.	1600.
v-ft3/lbm	28.0952	30.2085	32.2956	34.3669	36.4283	40.5332	44.6241	48.7072	52.7854	56.8603	60.9328	65.0036
u-Btu/lbm	1090.3	1108.6	1126.7	1144.7	1162.7	1199.3	1236.4	1274.4	1313.2	1352.8	1393.4	1434.8
h-Btu/lbm	1166.7	1190.7	1214.4	1238.1	1261.8	1309.4	1357.7	1406.8	1456.6	1507.4	1559.0	1611.5
s-Btu/lbm/R	1.78065	1.81380	1.84444	1.87311	1.90018	1.95040	1.99643	2.03910	2.07901	2.11659	2.15220	2.18611

Press = 20.0 psi, Tsat = 687.6 R

T(R) =	700.	750.	800.	850.	900.	1000.	1100.	1200.	1300.	1400.	1500.	1600.
v-ft3/lbm	20.5449	22.1217	23.6710	25.2040	26.7268	29.7540	32.7671	35.7724	38.7726	41.7696	44.7642	47.7572
u-Btu/lbm	1088.9	1107.6	1125.9	1144.1	1162.3	1198.9	1236.2	1274.2	1313.0	1352.7	1393.3	1434.8
h-Btu/lbm	1164.9	1189.4	1213.5	1237.3	1261.2	1309.0	1357.4	1406.5	1456.4	1507.2	1558.8	1611.4
s-Btu/lbm/R	1.74461	1.77846	1.80949	1.83841	1.86564	1.91605	1.96217	2.00490	2.04485	2.08246	2.11809	2.15201

Press = 40.0 psi, Tsat = 726.9 R

T(R) =	750.	800.	850.	900.	1000.	1100.	1200.	1300.	1400.	1500.	1600.	1700.
v-ft3/lbm	10.9125	11.7197	12.5083	13.2855	14.8204	16.3406	17.8527	19.3597	20.8634	22.3648	23.8644	25.3628
u-Btu/lbm	1103.8	1123.1	1142.0	1160.6	1197.8	1235.3	1273.5	1312.4	1352.2	1392.9	1434.4	1476.9
h-Btu/lbm	1184.5	1209.8	1234.5	1258.9	1307.4	1356.2	1405.6	1455.7	1506.6	1558.3	1611.0	1664.5
s-Btu/lbm/R	1.69668	1.72936	1.75928	1.78714	1.83827	1.88478	1.92774	1.96784	2.00555	2.04126	2.07523	2.10768

Press = 60.0 psi, Tsat = 752.4 R

T(R) =	800.	850.	900.	1000.	1100.	1200.	1300.	1400.	1500.	1600.	1700.	1800.
v-ft3/lbm	7.7325	8.2745	8.8041	9.8422	10.8650	11.8794	12.8887	13.8947	14.8983	15.9002	16.9008	17.9005
u-Btu/lbm	1120.1	1139.7	1158.8	1196.5	1234.4	1272.8	1311.9	1351.8	1392.5	1434.1	1476.6	1520.0
h-Btu/lbm	1205.9	1231.5	1256.5	1305.8	1355.0	1404.6	1454.9	1505.9	1557.8	1610.5	1664.1	1718.6
s-Btu/lbm/R	1.68079	1.71180	1.74036	1.79226	1.83917	1.88236	1.92261	1.96043	1.99620	2.03022	2.06272	2.09388

Press = 80.0 psi, Tsat = 771.7 R

T(R) =	800.	850.	900.	1000.	1100.	1200.	1300.	1400.	1500.	1600.	1700.	1800.
v-ft3/lbm	5.7361	6.1561	6.5625	7.3528	8.1271	8.8927	9.6532	10.4103	11.1650	11.9180	12.6698	13.4206
u-Btu/lbm	1117.0	1137.4	1157.0	1195.3	1233.5	1272.1	1311.3	1351.3	1392.1	1433.7	1476.3	1519.8
h-Btu/lbm	1201.8	1228.4	1254.1	1304.1	1353.7	1403.7	1454.1	1505.3	1557.3	1610.1	1663.8	1718.3
s-Btu/lbm/R	1.64497	1.67722	1.70652	1.75924	1.80656	1.84999	1.89039	1.92831	1.96416	1.99823	2.03077	2.06196

Press = 100.0 psi, Tsat = 787.5 R

T(R) =	800.	850.	900.	1000.	1100.	1200.	1300.	1400.	1500.	1600.	1700.	1800.
v-ft3/lbm	4.5358	4.8839	5.2169	5.8589	6.4842	7.1007	7.7119	8.3196	8.9250	9.5287	10.1312	10.7327
u-Btu/lbm	1113.6	1134.9	1155.1	1194.1	1232.6	1271.4	1310.7	1350.8	1391.7	1433.4	1476.0	1519.5
h-Btu/lbm	1197.5	1225.2	1251.6	1302.4	1352.5	1402.7	1453.4	1504.7	1556.7	1609.6	1663.4	1718.0
s-Btu/lbm/R	1.61596	1.64965	1.67975	1.73333	1.78107	1.82474	1.86529	1.90332	1.93924	1.97337	2.00595	2.03717

Press = 120.0 psi, Tsat = 800.9 R

T(R) =	850.	900.	1000.	1100.	1200.	1300.	1400.	1500.	1600.	1700.	1800.	1900.
v-ft3/lbm	4.0347	4.3192	4.8628	5.3889	5.9059	6.4176	6.9259	7.4317	7.9359	8.4388	8.9407	9.4420
u-Btu/lbm	1132.4	1153.2	1192.8	1231.7	1270.7	1310.2	1350.3	1391.3	1433.1	1475.7	1519.2	1563.7
h-Btu/lbm	1221.9	1249.0	1300.7	1351.3	1401.7	1452.6	1504.0	1556.2	1609.2	1663.0	1717.7	1773.2
s-Btu/lbm/R	1.62645	1.65742	1.71191	1.76009	1.80401	1.84471	1.88284	1.91884	1.95302	1.98564	2.01689	2.04692

Press = 140.0 psi, Tsat = 812.7 R

T(R) =	850.	900.	1000.	1100.	1200.	1300.	1400.	1500.	1600.	1700.	1800.	1900.
v-ft3/lbm	3.4272	3.6775	4.1511	4.6064	5.0525	5.4931	5.9303	6.3651	6.7981	7.2299	7.6608	8.0909
u-Btu/lbm	1129.7	1151.2	1191.5	1230.7	1269.9	1309.6	1349.9	1390.9	1432.7	1475.4	1519.0	1563.5
h-Btu/lbm	1218.5	1246.4	1299.0	1350.0	1400.8	1451.8	1503.4	1555.7	1608.8	1662.6	1717.4	1772.9
s-Btu/lbm/R	1.60622	1.63813	1.69358	1.74221	1.78638	1.82724	1.86547	1.90154	1.93578	1.96844	1.99972	2.02977

Press = 160.0 psi, Tsat = 823.2 R

T(R) =	850.	900.	1000.	1100.	1200.	1300.	1400.	1500.	1600.	1700.	1800.	1900.
v-ft3/lbm	2.9707	3.1958	3.6172	4.0195	4.4124	4.7998	5.1836	5.5651	5.9448	6.3233	6.7008	7.0776
u-Btu/lbm	1127.0	1149.1	1190.2	1229.8	1269.2	1309.0	1349.4	1390.5	1432.4	1475.1	1518.7	1563.2
h-Btu/lbm	1214.9	1243.7	1297.2	1348.7	1399.8	1451.0	1502.8	1555.2	1608.3	1662.2	1717.0	1772.7
s-Btu/lbm/R	1.58812	1.62105	1.67750	1.72659	1.77102	1.81204	1.85038	1.88653	1.92082	1.95351	1.98482	2.01490

Press = 180.0 psi, Tsat = 832.8 R

T(R) =	850.	900.	1000.	1100.	1200.	1300.	1400.	1500.	1600.	1700.	1800.	1900.
v-ft3/lbm	2.6148	2.8207	3.2017	3.5630	3.9145	4.2605	4.6029	4.9428	5.2811	5.6181	5.9541	6.2895
u-Btu/lbm	1124.1	1147.0	1188.9	1228.8	1268.5	1308.4	1348.9	1390.1	1432.1	1474.8	1518.5	1563.0
h-Btu/lbm	1211.1	1240.9	1295.4	1347.5	1398.8	1450.3	1502.1	1554.6	1607.9	1661.9	1716.7	1772.4
s-Btu/lbm/R	1.57158	1.60564	1.66314	1.71272	1.75740	1.79858	1.83703	1.87325	1.90759	1.94033	1.97167	2.00177

Press = 200.0 psi, Tsat = 841.5 R

T(R) =	850.	900.	1000.	1100.	1200.	1300.	1400.	1500.	1600.	1700.	1800.	1900.
v-ft3/lbm	2.3291	2.5202	2.8693	3.1977	3.5162	3.8290	4.1382	4.4451	4.7501	5.0540	5.3568	5.6590
u-Btu/lbm	1121.0	1144.8	1187.5	1227.9	1267.8	1307.8	1348.4	1389.7	1431.7	1474.6	1518.2	1562.8
h-Btu/lbm	1207.2	1238.1	1293.6	1346.2	1397.8	1449.5	1501.5	1554.1	1607.4	1661.5	1716.4	1772.1
s-Btu/lbm/R	1.55619	1.59152	1.65013	1.70020	1.74515	1.78649	1.82505	1.86134	1.89574	1.92852	1.95989	1.99002

Press = 250.0 psi, Tsat = 860.7 R

T(R) =	900.	1000.	1100.	1200.	1300.	1400.	1500.	1600.	1700.	1800.	1900.	2000.
v-ft3/lbm	1.9779	2.2703	2.5400	2.7991	3.0523	3.3019	3.5490	3.7944	4.0385	4.2816	4.5241	4.7660
u-Btu/lbm	1139.1	1184.1	1225.5	1265.9	1306.4	1347.2	1388.7	1430.9	1473.8	1517.6	1562.2	1607.7
h-Btu/lbm	1230.6	1289.0	1342.9	1395.4	1447.5	1499.9	1552.8	1606.3	1660.5	1715.6	1771.4	1828.1
s-Btu/lbm/R	1.56033	1.62195	1.67333	1.71896	1.76072	1.79954	1.83603	1.87056	1.90344	1.93489	1.96507	1.99414

Press = 300.0 psi, Tsat = 877.0 R

T(R) =	900.	1000.	1100.	1200.	1300.	1400.	1500.	1600.	1700.	1800.	1900.	2000.
v-ft3/lbm	1.6144	1.8704	2.1013	2.3209	2.5345	2.7443	2.9517	3.1572	3.3615	3.5649	3.7675	3.9696
u-Btu/lbm	1133.0	1180.5	1223.0	1264.1	1304.9	1346.0	1387.7	1430.0	1473.1	1517.0	1561.7	1607.2
h-Btu/lbm	1222.6	1284.2	1339.6	1392.9	1445.5	1498.3	1551.5	1605.2	1659.6	1714.8	1770.7	1827.5
s-Btu/lbm/R	1.53305	1.59813	1.65091	1.69726	1.73944	1.77854	1.81522	1.84989	1.88287	1.91439	1.94463	1.97375

Press = 350.0 psi, Tsat = 891.4 R

T(R) =	900.	1000.	1100.	1200.	1300.	1400.	1500.	1600.	1700.	1800.	1900.	2000.
v-ft3/lbm	1.3527	1.5842	1.7877	1.9793	2.1646	2.3460	2.5250	2.7021	2.8780	3.0529	3.2270	3.4007
u-Btu/lbm	1126.2	1176.7	1220.5	1262.2	1303.4	1344.8	1386.7	1429.2	1472.4	1516.3	1561.1	1606.7
h-Btu/lbm	1213.8	1279.3	1336.2	1390.3	1443.6	1496.7	1550.1	1604.1	1658.7	1713.9	1770.0	1826.8
s-Btu/lbm/R	1.50812	1.57726	1.63155	1.67865	1.72127	1.76065	1.79752	1.83233	1.86541	1.89701	1.92731	1.95647

Press = 400.0 psi, Tsat = 904.3 R

T(R) =	1000.	1100.	1200.	1300.	1400.	1500.	1600.	1700.	1800.	1900.	2000.	2100.
v-ft3/lbm	1.3691	1.5523	1.7230	1.8871	2.0473	2.2049	2.3608	2.5153	2.6689	2.8217	2.9741	3.1259
u-Btu/lbm	1172.8	1217.9	1260.3	1302.0	1343.6	1385.7	1428.3	1471.6	1515.7	1560.5	1606.2	1652.7
h-Btu/lbm	1274.1	1332.7	1387.8	1441.6	1495.1	1548.8	1603.0	1657.7	1713.1	1769.3	1826.2	1884.0
s-Btu/lbm/R	1.55851	1.61441	1.66230	1.70536	1.74503	1.78209	1.81704	1.85022	1.88190	1.91226	1.94146	1.96963

Press = 450.0 psi, Tsat = 916.0 R

T(R) =	1000.	1100.	1200.	1300.	1400.	1500.	1600.	1700.	1800.	1900.	2000.	2100.
v-ft3/lbm	1.2012	1.3691	1.5236	1.6712	1.8149	1.9560	2.0953	2.2332	2.3702	2.5065	2.6422	2.7775
u-Btu/lbm	1168.8	1215.3	1258.4	1300.5	1342.4	1384.7	1427.5	1470.9	1515.0	1560.0	1605.7	1652.3
h-Btu/lbm	1268.8	1329.2	1385.2	1439.5	1493.5	1547.5	1601.8	1656.8	1712.3	1768.6	1825.6	1883.4
s-Btu/lbm/R	1.54131	1.59895	1.64767	1.69119	1.73115	1.76840	1.80349	1.83678	1.86853	1.89895	1.92820	1.95640

Press = 500.0 psi, Tsat = 926.7 R

T(R) =	1000.	1100.	1200.	1300.	1400.	1500.	1600.	1700.	1800.	1900.	2000.	2100.
v-ft3/lbm	1.0665	1.2223	1.3640	1.4985	1.6290	1.7568	1.8828	2.0075	2.1313	2.2543	2.3767	2.4987
u-Btu/lbm	1164.6	1212.6	1256.4	1298.9	1341.2	1383.7	1426.6	1470.2	1514.4	1559.4	1605.2	1651.8
h-Btu/lbm	1263.2	1325.6	1382.6	1437.5	1491.8	1546.1	1600.7	1655.8	1711.5	1767.9	1825.0	1882.9
s-Btu/lbm/R	1.52530	1.58481	1.63438	1.67838	1.71863	1.75608	1.79132	1.82471	1.85654	1.88702	1.91631	1.94454

Press = 550.0 psi, Tsat = 936.7 R

T(R) =	1000.	1100.	1200.	1300.	1400.	1500.	1600.	1700.	1800.	1900.	2000.	2100.
v-ft3/lbm	0.9559	1.1021	1.2333	1.3572	1.4769	1.5939	1.7091	1.8229	1.9358	2.0479	2.1595	2.2707
u-Btu/lbm	1160.2	1209.8	1254.5	1297.4	1340.0	1382.6	1425.7	1469.4	1513.8	1558.8	1604.7	1651.3
h-Btu/lbm	1257.5	1321.9	1379.9	1435.5	1490.2	1544.8	1599.6	1654.8	1710.7	1767.2	1824.4	1882.3
s-Btu/lbm/R	1.51020	1.57171	1.62218	1.66666	1.70722	1.74487	1.78025	1.81374	1.84565	1.87619	1.90552	1.93380

Press = 600.0 psi, Tsat = 945.9 R

T(R) =	1000.	1100.	1200.	1300.	1400.	1500.	1600.	1700.	1800.	1900.	2000.	2100.
v-ft3/lbm	0.8632	1.0017	1.1243	1.2394	1.3501	1.4581	1.5642	1.6690	1.7729	1.8760	1.9785	2.0806
u-Btu/lbm	1155.7	1207.0	1252.5	1295.9	1338.7	1381.6	1424.9	1468.7	1513.1	1558.3	1604.2	1650.9
h-Btu/lbm	1251.5	1318.2	1377.2	1433.4	1488.5	1543.4	1598.5	1653.9	1709.8	1766.4	1823.7	1881.8
s-Btu/lbm/R	1.49578	1.55946	1.61087	1.65586	1.69672	1.73458	1.77010	1.80369	1.83568	1.86628	1.89566	1.92397

Press = 700.0 psi, Tsat = 962.8 R

T(R) =	1000.	1100.	1200.	1300.	1400.	1500.	1600.	1700.	1800.	1900.	2000.	2100.
v-ft3/lbm	0.7162	0.8437	0.9530	1.0541	1.1508	1.2447	1.3366	1.4272	1.5169	1.6057	1.6941	1.7820
u-Btu/lbm	1145.9	1201.2	1248.4	1292.8	1336.2	1379.6	1423.2	1467.2	1511.8	1557.1	1603.2	1650.0
h-Btu/lbm	1238.6	1310.4	1371.8	1429.3	1485.2	1540.7	1596.2	1652.0	1708.2	1765.0	1822.5	1880.7
s-Btu/lbm/R	1.46837	1.53696	1.59037	1.63641	1.67790	1.71617	1.75198	1.78579	1.81793	1.84865	1.87812	1.90650

Press = 800.0 psi, Tsat = 977.9 R

T(R) =	1000.	1100.	1200.	1300.	1400.	1500.	1600.	1700.	1800.	1900.	2000.	2100.
v-ft3/lbm	0.6041	0.7246	0.8243	0.9151	1.0013	1.0846	1.1659	1.2459	1.3249	1.4031	1.4807	1.5580
u-Btu/lbm	1134.9	1195.1	1244.2	1289.6	1333.7	1377.5	1421.4	1465.7	1510.5	1556.0	1602.1	1649.0
h-Btu/lbm	1224.3	1302.3	1366.2	1425.0	1481.9	1538.0	1593.9	1650.0	1706.6	1763.6	1821.2	1879.6
s-Btu/lbm/R	1.44194	1.51644	1.57205	1.61919	1.66134	1.70004	1.73614	1.77016	1.80246	1.83330	1.86286	1.89131

Press = 900.0 psi, Tsat = 991.7 R

T(R) =	1000.	1100.	1200.	1300.	1400.	1500.	1600.	1700.	1800.	1900.	2000.	2100.
v-ft3/lbm	0.5148	0.6315	0.7239	0.8069	0.8850	0.9600	1.0331	1.1048	1.1755	1.2454	1.3148	1.3838
u-Btu/lbm	1122.5	1188.7	1239.9	1286.4	1331.2	1375.4	1419.7	1464.2	1509.2	1554.9	1601.1	1648.1
h-Btu/lbm	1208.2	1293.8	1360.4	1420.7	1478.5	1535.2	1591.6	1648.1	1704.9	1762.2	1820.0	1878.5
s-Btu/lbm/R	1.41552	1.49735	1.55536	1.60367	1.64649	1.68563	1.72203	1.75627	1.78873	1.81969	1.84934	1.87787

Press = 1000.0 psi, Tsat = 1004.3 R

T(R) =	1100.	1200.	1300.	1400.	1500.	1600.	1700.	1800.	1900.	2000.	2100.	2200.
v-ft3/lbm	0.5565	0.6435	0.7203	0.7919	0.8604	0.9269	0.9920	1.0560	1.1193	1.1821	1.2444	1.3063
u-Btu/lbm	1181.9	1235.5	1283.2	1328.6	1373.3	1417.9	1462.7	1507.9	1553.7	1600.1	1647.2	1695.1
h-Btu/lbm	1284.9	1354.5	1416.4	1475.1	1532.5	1589.3	1646.2	1703.2	1760.7	1818.7	1877.4	1936.7
s-Btu/lbm/R	1.47927	1.53992	1.58948	1.63300	1.67258	1.70929	1.74374	1.77636	1.80744	1.83719	1.86579	1.89338

Press = 1100.0 psi, Tsat = 1016.0 R

T(R) =	1100.	1200.	1300.	1400.	1500.	1600.	1700.	1800.	1900.	2000.	2100.	2200.
v-ft3/lbm	0.4947	0.5776	0.6493	0.7157	0.7788	0.8399	0.8996	0.9583	1.0162	1.0735	1.1304	1.1869
u-Btu/lbm	1174.8	1230.9	1279.8	1326.1	1371.2	1416.1	1461.2	1506.6	1552.6	1599.1	1646.3	1694.2
h-Btu/lbm	1275.5	1348.4	1411.9	1471.6	1529.7	1587.0	1644.2	1701.6	1759.3	1817.5	1876.2	1935.7
s-Btu/lbm/R	1.46191	1.52547	1.57634	1.62060	1.66064	1.69765	1.73233	1.76511	1.79631	1.82615	1.85482	1.88247

Press = 1200.0 psi, Tsat = 1026.9 R

T(R) =	1100.	1200.	1300.	1400.	1500.	1600.	1700.	1800.	1900.	2000.	2100.	2200.
v-ft3/lbm	0.4428	0.5224	0.5901	0.6521	0.7109	0.7675	0.8227	0.8768	0.9302	0.9830	1.0353	1.0873
u-Btu/lbm	1167.3	1226.2	1276.5	1323.4	1369.1	1414.4	1459.7	1505.3	1551.4	1598.1	1645.4	1693.4
h-Btu/lbm	1265.6	1342.2	1407.4	1468.2	1526.9	1584.7	1642.3	1699.9	1757.9	1816.2	1875.1	1934.7
s-Btu/lbm/R	1.44502	1.51180	1.56408	1.60909	1.64959	1.68692	1.72182	1.75477	1.78609	1.81602	1.84477	1.87248

Press = 1300.0 psi, Tsat = 1037.2 R

T(R) =	1100.	1200.	1300.	1400.	1500.	1600.	1700.	1800.	1900.	2000.	2100.	2200.
v-ft3/lbm	0.3983	0.4756	0.5400	0.5983	0.6533	0.7062	0.7576	0.8079	0.8575	0.9064	0.9549	1.0031
u-Btu/lbm	1159.3	1221.4	1273.0	1320.8	1367.0	1412.6	1458.2	1504.0	1550.3	1597.0	1644.4	1692.6
h-Btu/lbm	1255.0	1335.8	1402.9	1464.6	1524.0	1582.4	1640.3	1698.2	1756.4	1815.0	1874.0	1933.7
s-Btu/lbm/R	1.42838	1.49877	1.55252	1.59832	1.63931	1.67696	1.71208	1.74519	1.77664	1.80667	1.83549	1.86326

Press = 1400.0 psi, Tsat = 1046.8 R

T(R) =	1100.	1200.	1300.	1400.	1500.	1600.	1700.	1800.	1900.	2000.	2100.	2200.
v-ft3/lbm	0.3596	0.4354	0.4969	0.5522	0.6040	0.6536	0.7017	0.7488	0.7951	0.8408	0.8860	0.9309
u-Btu/lbm	1150.7	1216.4	1269.5	1318.1	1364.8	1410.8	1456.6	1502.7	1549.1	1596.0	1643.5	1691.7
h-Btu/lbm	1243.8	1329.2	1398.2	1461.1	1521.2	1580.0	1638.3	1696.6	1755.0	1813.7	1872.9	1932.8
s-Btu/lbm/R	1.41181	1.48626	1.54158	1.58819	1.62967	1.66764	1.70300	1.73627	1.76785	1.79796	1.82686	1.85470

Press = 1500.0 psi, Tsat = 1055.9 R

T(R) =	1100.	1200.	1300.	1400.	1500.	1600.	1700.	1800.	1900.
v-ft3/lbm	0.3255	0.4003	0.4596	0.5122	0.5613	0.6081	0.6534	0.6976	0.7411
u-Btu/lbm	1141.5	1211.3	1266.0	1315.4	1362.6	1409.0	1455.1	1501.4	1547.9
h-Btu/lbm	1231.7	1322.4	1393.5	1457.5	1518.3	1577.7	1636.4	1694.9	1753.5
s-Btu/lbm/R	1.39510	1.47415	1.53114	1.57860	1.62058	1.65889	1.69447	1.72792	1.75962

Press = 1600.0 psi, Tsat = 1064.6 R

T(R) =	1100.	1200.	1300.	1400.	1500.	1600.	1700.	1800.	1900.
v-ft3/lbm	0.2950	0.3695	0.4268	0.4772	0.5238	0.5682	0.6110	0.6528	0.6938
u-Btu/lbm	1131.4	1206.0	1262.4	1312.7	1360.4	1407.2	1453.6	1500.0	1546.8
h-Btu/lbm	1218.7	1315.3	1388.7	1453.9	1515.5	1575.3	1634.4	1693.2	1752.1
s-Btu/lbm/R	1.37804	1.46238	1.52113	1.56948	1.61198	1.65062	1.68643	1.72005	1.75187

Press = 1700.0 psi, Tsat = 1072.9 R

T(R) =	1100.	1200.	1300.	1400.	1500.	1600.	1700.	1800.	1900.
v-ft3/lbm	0.2673	0.3422	0.3979	0.4462	0.4908	0.5330	0.5737	0.6133	0.6521
u-Btu/lbm	1120.4	1200.5	1258.7	1309.9	1358.2	1405.4	1452.0	1498.7	1545.6
h-Btu/lbm	1204.5	1308.1	1383.8	1450.2	1512.6	1572.9	1632.4	1691.5	1750.6
s-Btu/lbm/R	1.36037	1.45087	1.51150	1.56076	1.60379	1.64277	1.67883	1.71261	1.74456

Press = 1800.0 psi, Tsat = 1080.7 R

T(R) =	1100.	1200.	1300.	1400.	1500.	1600.	1700.	1800.	1900.
v-ft3/lbm	0.2418	0.3178	0.3721	0.4187	0.4614	0.5017	0.5405	0.5781	0.6150
u-Btu/lbm	1108.2	1194.8	1254.9	1307.1	1356.0	1403.5	1450.5	1497.4	1544.4
h-Btu/lbm	1188.7	1300.6	1378.8	1446.5	1509.6	1570.6	1630.4	1689.8	1749.2
s-Btu/lbm/R	1.34174	1.43955	1.50219	1.55240	1.59598	1.63531	1.67160	1.70555	1.73764

Press = 1900.0 psi, Tsat = 1088.3 R

T(R) =	1100.	1200.	1300.	1400.	1500.	1600.	1700.	1800.	1900.
v-ft3/lbm	0.2179	0.2958	0.3490	0.3941	0.4351	0.4738	0.5108	0.5467	0.5818
u-Btu/lbm	1094.2	1189.0	1251.1	1304.3	1353.8	1401.7	1448.9	1496.0	1543.3
h-Btu/lbm	1170.8	1292.9	1373.7	1442.7	1506.7	1568.2	1628.4	1688.1	1747.7
s-Btu/lbm/R	1.32165	1.42838	1.49317	1.54436	1.58849	1.62817	1.66471	1.69884	1.73105

Press = 2000.0 psi, Tsat = 1095.5 R

T(R) =	1100.	1200.	1300.	1400.	1500.	1600.	1700.	1800.	1900.
v-ft3/lbm	0.1949	0.2758	0.3281	0.3719	0.4115	0.4486	0.4840	0.5184	0.5519
u-Btu/lbm	1077.8	1182.8	1247.2	1301.4	1351.5	1399.8	1447.4	1494.7	1542.1
h-Btu/lbm	1149.9	1284.9	1368.5	1439.0	1503.7	1565.8	1626.4	1686.4	1746.2
s-Btu/lbm/R	1.29918	1.41730	1.48438	1.53660	1.58130	1.62134	1.65812	1.69242	1.72476

Press = 2500.0 psi, Tsat = 1127.8 R

T(R) =	1200.	1300.	1400.	1500.	1600.	1700.	1800.	1900.
v-ft3/lbm	0.1980	0.2483	0.2873	0.3214	0.3528	0.3824	0.4109	0.4385
u-Btu/lbm	1148.2	1226.4	1286.5	1340.0	1390.4	1439.5	1487.9	1536.2
h-Btu/lbm	1239.7	1341.3	1419.4	1488.6	1553.6	1616.3	1677.9	1738.9
s-Btu/lbm/R	1.36160	1.44310	1.50107	1.54884	1.59077	1.62881	1.66401	1.69701

Press = 3000.0 psi, Tsat = 1155.1 R

T(R) =	1200.	1300.	1400.	1500.	1600.	1700.	1800.	1900.
v-ft3/lbm	0.1421	0.1943	0.2307	0.2613	0.2889	0.3147	0.3392	0.3629
u-Btu/lbm	1103.1	1203.5	1270.8	1328.0	1380.8	1431.5	1481.1	1530.2
h-Btu/lbm	1181.9	1311.3	1398.8	1473.0	1541.1	1606.1	1669.3	1731.6
s-Btu/lbm/R	1.30046	1.40439	1.46935	1.52058	1.56455	1.60393	1.64007	1.67375

Press = 3500.0 psi

T(R) =	1200.	1300.	1400.	1500.	1600.	1700.	1800.	1900.
v-ft3/lbm	0.0244	0.1550	0.1900	0.2183	0.2433	0.2663	0.2881	0.3089
u-Btu/lbm	720.0	1177.7	1254.1	1315.6	1370.9	1423.3	1474.1	1524.2
h-Btu/lbm	768.0	1278.1	1377.1	1456.9	1528.4	1595.7	1660.6	1724.2
s-Btu/lbm/R	0.91193	1.36649	1.44004	1.49513	1.54131	1.58210	1.61920	1.65358

Press = 4000.0 psi

T(R) =	1200.	1300.	1400.	1500.	1600.	1700.	1800.	1900.
v-ft3/lbm	0.0240	0.1249	0.1594	0.1860	0.2091	0.2301	0.2497	0.2685
u-Btu/lbm	714.7	1148.8	1236.4	1302.7	1360.8	1415.0	1467.1	1518.2
h-Btu/lbm	765.3	1241.1	1354.3	1440.3	1515.5	1585.2	1651.9	1716.8
s-Btu/lbm/R	0.90685	1.32818	1.41226	1.47167	1.52020	1.56247	1.60058	1.63567

Press = 4500.0 psi

T(R) =	1200.	1300.	1400.	1500.	1600.	1700.	1800.	1900.
v-ft3/lbm	0.0237	0.1009	0.1355	0.1609	0.1825	0.2019	0.2200	0.2370
u-Btu/lbm	709.9	1116.0	1217.7	1289.4	1350.5	1406.6	1460.0	1512.1
h-Btu/lbm	763.2	1200.0	1330.4	1423.4	1502.4	1574.6	1643.1	1709.3
s-Btu/lbm/R	0.90232	1.28852	1.38550	1.44969	1.50072	1.54453	1.58367	1.61949

Press = 5000.0 psi

T(R) =	1200.	1300.	1400.	1500.	1600.	1700.	1800.	1900.
v-ft3/lbm	0.0234	0.0815	0.1163	0.1409	0.1613	0.1794	0.1962	0.2119
u-Btu/lbm	705.6	1079.1	1198.0	1275.8	1340.0	1398.1	1452.9	1505.9
h-Btu/lbm	761.3	1154.4	1305.5	1406.0	1489.1	1564.0	1634.3	1701.9
s-Btu/lbm/R	0.89820	1.24703	1.35941	1.42885	1.48252	1.52792	1.56811	1.60468

A.14.8 H_2O Compressed Liquid Tables (English Units)

Temperature = 500.0 °R		Psat = 0.1 psi							
Press (psi)=	250	500	1000	1500	2000	2500	3000	3500	4000	5000
v (ft^3/lbm)	0.016	0.016	0.016	0.016	0.016	0.0159	0.0159	0.0159	0.0159	0.0158
u (Btu/lbm)	8.4	8.4	8.4	8.4	8.3	8.3	8.3	8.3	8.3	8.3
h (Btu/lbm)	9.1	9.9	11.3	12.8	14.3	15.7	17.1	18.6	20	22.9
s (Btu/lbm/°R)	0.0169	0.01689	0.01687	0.01684	0.0168	0.01675	0.0167	0.01664	0.01657	0.01641
Temperature = 550.0 °R		Psat = 0.7 psi								
Press (psi)=	250	500	1000	1500	2000	2500	3000	3500	4000	5000
v (ft^3/lbm)	0.0161	0.0161	0.0161	0.0161	0.016	0.016	0.016	0.016	0.016	0.0159
u (Btu/lbm)	58.4	58.4	58.2	58.1	57.9	57.8	57.6	57.5	57.4	57.1
h (Btu/lbm)	59.2	59.9	61.2	62.5	63.9	65.2	66.5	67.8	69.2	71.8
s (Btu/lbm/°R)	0.11232	0.11219	0.11192	0.11165	0.11137	0.1111	0.11082	0.11054	0.11026	0.1097
Temperature = 600.0 °R		Psat = 2.9 psi								
Press (psi)=	250	500	1000	1500	2000	2500	3000	3500	4000	5000
v (ft^3/lbm)	0.0163	0.0163	0.0163	0.0163	0.0162	0.0162	0.0162	0.0162	0.0161	0.0161
u (Btu/lbm)	108.4	108.3	108	107.8	107.5	107.3	107	106.8	106.5	106.1
h (Btu/lbm)	109.2	109.8	111	112.3	113.5	114.8	116	117.2	118.5	120.9
s (Btu/lbm/°R)	0.19929	0.19907	0.19864	0.1982	0.19777	0.19734	0.19691	0.19648	0.19606	0.19521
Temperature = 650.0 °R		Psat = 9.4 psi								
Press (psi)=	250	500	1000	1500	2000	2500	3000	3500	4000	5000
v (ft^3/lbm)	0.0166	0.0166	0.0166	0.0165	0.0165	0.0165	0.0165	0.0164	0.0164	0.0164
u (Btu/lbm)	158.5	158.3	158	157.6	157.2	156.9	156.5	156.2	155.8	155.2
h (Btu/lbm)	159.3	159.9	161	162.2	163.3	164.5	165.6	166.8	168	170.3

(continued)

(continued)

s (Btu/lbm/°R)	0.27953	0.27924	0.27866	0.27809	0.27752	0.27695	0.27639	0.27584	0.27529	0.27419
Temperature = 700.0 °R		Psat = 25.1 psi								
Press (psi)=	250	500	1000	1500	2000	2500	3000	3500	4000	5000
v (ft^3/lbm)	0.017	0.0169	0.0169	0.0169	0.0169	0.0168	0.0168	0.0168	0.0167	0.0167
u (Btu/lbm)	209	208.7	208.2	207.7	207.2	206.8	206.3	205.8	205.4	204.5
h (Btu/lbm)	209.7	210.3	211.3	212.4	213.5	214.5	215.6	216.7	217.8	219.9
s (Btu/lbm/°R)	0.35431	0.35395	0.35323	0.35252	0.35182	0.35113	0.35044	0.34976	0.34908	0.34775
Temperature = 750.0 °R		Psat = 57.9 psi								
Press (psi)=	250	500	1000	1500	2000	2500	3000	3500	4000	5000
v (ft^3/lbm)	0.0174	0.0174	0.0173	0.0173	0.0173	0.0172	0.0172	0.0172	0.0171	0.0171
u (Btu/lbm)	259.9	259.6	259	258.3	257.7	257.1	256.5	255.9	255.3	254.2
h (Btu/lbm)	260.7	261.2	262.2	263.1	264.1	265.1	266	267	268	270
s (Btu/lbm/°R)	0.42467	0.42423	0.42337	0.42251	0.42167	0.42083	0.42001	0.41919	0.41839	0.41681
Temperature = 800.0 °R		Psat = 118.5 psi								
Press (psi)=	250	500	1000	1500	2000	2500	3000	3500	4000	5000
v (ft^3/lbm)	0.0179	0.0179	0.0178	0.0178	0.0178	0.0177	0.0177	0.0176	0.0176	0.0175
u (Btu/lbm)	311.7	311.3	310.4	309.6	308.8	308.1	307.3	306.5	305.8	304.4
h (Btu/lbm)	312.5	312.9	313.7	314.6	315.4	316.2	317.1	318	318.8	320.6
s (Btu/lbm/°R)	0.49152	0.49099	0.48994	0.48891	0.4879	0.4869	0.48592	0.48495	0.484	0.48212
Temperature = 850.0 °R		Psat = 221.1 psi								
Press (psi)=	250	500	1000	1500	2000	2500	3000	3500	4000	5000
v (ft^3/lbm)	0.0185	0.0185	0.0185	0.0184	0.0184	0.0183	0.0183	0.0182	0.0182	0.0181
u (Btu/lbm)	364.6	364.1	363	362	360.9	359.9	358.9	358	357.1	355.2
h (Btu/lbm)	365.5	365.8	366.4	367.1	367.7	368.4	369.1	369.8	370.5	372
s (Btu/lbm/°R)	0.55573	0.55508	0.5538	0.55255	0.55132	0.55012	0.54894	0.54778	0.54664	0.54442
Temperature = 900.0 °R		Psat = 382.7 psi								
Press (psi)=	250	500	1000	1500	2000	2500	3000	3500	4000	5000
v (ft^3/lbm)		0.0193	0.0192	0.0192	0.0191	0.019	0.019	0.0189	0.0188	0.0187

(continued)

(continued)

u (Btu/lbm)		418.6	417.1	415.8	414.4	413.1	411.8	410.6	409.4	407.1
h (Btu/lbm)		420.4	420.7	421.1	421.5	421.9	422.4	422.9	423.4	424.4
s (Btu/lbm/°R)		0.61744	0.61584	0.61428	0.61277	0.61129	0.60985	0.60844	0.60706	0.6044
Temperature = 950.0 °R		Psat = 623.1 psi								
Press (psi)=	250	500	1000	1500	2000	2500	3000	3500	4000	5000
v (ft^3/lbm)			0.0202	0.0201	0.02	0.0199	0.0198	0.0197	0.0197	0.0195
u (Btu/lbm)			473.7	471.8	470	468.2	466.6	464.9	463.4	460.4
h (Btu/lbm)			477.4	477.4	477.4	477.4	477.6	477.7	477.9	478.4
s (Btu/lbm/°R)			0.67717	0.67515	0.67321	0.67133	0.66951	0.66775	0.66604	0.66277
Temperature = 1000.0 °R		Psat = 964.9 psi								
Press (psi)=	250	500	1000	1500	2000	2500	3000	3500	4000	5000
v (ft^3/lbm)			0.0215	0.0214	0.0212	0.0211	0.0209	0.0208	0.0207	0.0205
u (Btu/lbm)			534.2	531.4	528.8	526.3	524	521.8	519.6	515.7
h (Btu/lbm)			538.2	537.4	536.7	536.1	535.6	535.2	534.9	534.6
s (Btu/lbm/°R)			0.73949	0.73666	0.73399	0.73146	0.72905	0.72675	0.72454	0.72038
Temperature = 1050.0 °R		Psat = 1434.3 psi								
Press (psi)=	250	500	1000	1500	2000	2500	3000	3500	4000	5000
v (ft^3/lbm)				0.0232	0.0229	0.0227	0.0224	0.0222	0.0221	0.0217
u (Btu/lbm)				597.7	593.4	589.5	585.9	582.6	579.5	573.9
h (Btu/lbm)				604.2	601.9	599.9	598.3	597	595.8	594
s (Btu/lbm/°R)				0.80183	0.7976	0.79376	0.79024	0.78696	0.7839	0.7783
Temperature = 1100.0 °R		Psat = 2063.9 psi								
Press (psi)=	250	500	1000	1500	2000	2500	3000	3500	4000	5000
v (ft^3/lbm)						0.0254	0.0249	0.0244	0.024	0.0234
u (Btu/lbm)						663.3	656.4	650.6	645.4	636.7
h (Btu/lbm)						675.1	670.2	666.4	663.2	658.4
s (Btu/lbm/°R)						0.86361	0.85706	0.85149	0.8466	0.83821

A.15 Thermodynamic Property Tables for Carbon Dioxide

A.15.1 CO_2 Saturation Temperature Table (SI Units)

T(K)	P(MPa)	Volume (m**3/kg)			Energy (kJ/kg)			Enthalpy (kJ/kg)			Entropy (kJ/kg/K)		
		vf	vfg	vg	uf	ufg	ug	hf	hfg	hg	sf	sfg	sg
216.5	0.5173	0.000847	0.071931	0.072778	0.00	314.66	314.22	0.00	351.87	351.87	0.00000	1.62497	1.62497
220.0	0.6000	0.000857	0.062281	0.063137	6.49	308.75	315.24	7.00	346.12	353.12	0.03175	1.57326	1.60501
225.0	0.7365	0.000871	0.050940	0.051811	16.88	299.69	316.57	17.52	337.21	354.73	0.07848	1.49872	1.57720
230.0	0.8949	0.000885	0.041984	0.042869	27.28	290.46	317.73	28.07	328.03	356.09	0.12425	1.42620	1.55046
235.0	1.0769	0.000901	0.034828	0.035729	37.55	281.17	318.72	38.52	318.68	357.19	0.16850	1.35607	1.52457
240.0	1.2849	0.000918	0.029049	0.029967	47.72	271.77	319.49	48.90	309.10	358.00	0.21142	1.28791	1.49934
245.0	1.5211	0.000936	0.024333	0.025269	57.89	262.15	320.04	59.32	299.16	358.48	0.25348	1.22106	1.47454
250.0	1.7875	0.000956	0.020449	0.021405	68.15	252.18	320.33	69.85	288.74	358.59	0.29503	1.15495	1.44998
255.0	2.0866	0.000977	0.017222	0.018199	78.52	241.80	320.32	80.56	277.73	358.30	0.33630	1.08914	1.42545
260.0	2.4208	0.001001	0.014516	0.015516	89.06	230.91	319.97	91.49	266.05	357.53	0.37744	1.02326	1.40070
265.0	2.7924	0.001027	0.012227	0.013254	99.79	219.43	319.22	102.66	253.57	356.23	0.41856	0.95688	1.37544
270.0	3.2043	0.001056	0.010272	0.011328	110.77	207.22	317.99	114.16	240.13	354.29	0.45995	0.88937	1.34933
275.0	3.6592	0.001090	0.008586	0.009676	122.13	194.04	316.17	126.12	225.46	351.57	0.50204	0.81985	1.32188
280.0	4.1602	0.001129	0.007113	0.008242	134.02	179.57	313.59	138.72	209.16	347.88	0.54545	0.74701	1.29246
285.0	4.7106	0.001177	0.005805	0.006982	146.67	163.35	310.01	152.21	190.69	342.90	0.59096	0.66910	1.26006
290.0	5.3143	0.001239	0.004616	0.005854	160.36	144.62	304.98	166.94	169.15	336.09	0.63966	0.58327	1.22294
295.0	5.9765	0.001324	0.003484	0.004808	175.75	121.81	297.55	183.66	142.62	326.29	0.69391	0.48347	1.17739
300.0	6.7058	0.001471	0.002278	0.003749	195.40	39.57	234.56	205.06	104.89	309.95	0.76240	0.34960	1.11200
304.2	7.3834	0.002155	0.000000	0.002155	241.37	0.0	241.37	257.31	0.00	257.31	0.93116	0.00000	0.93116

A.15.2 CO_2 Saturation Pressure Table (SI Units)

P(MPa)	T(K)	Volume (m**3/kg)			Energy (kJ/kg)			Enthalpy (kJ/kg)			Entropy (kJ/kg/K)		
		vf	vfg	vg	uf	ufg	ug	hf	hfg	hg	sf	sfg	sg
0.5173	216.5	0.000847	0.071926	0.072774	0.00	314.66	314.23	0.00	351.87	351.87	0.00001	1.62495	1.62496
0.6000	220.0	0.000857	0.062281	0.063137	6.49	308.75	315.24	7.00	346.12	353.12	0.03175	1.57326	1.60501
0.7000	223.7	0.000867	0.053559	0.054425	14.23	302.01	316.24	14.84	339.50	354.34	0.06669	1.51745	1.58413
0.8000	227.1	0.000877	0.046942	0.047819	21.24	295.84	317.08	21.94	333.39	355.33	0.09780	1.46810	1.56590
0.9000	230.2	0.000886	0.041744	0.042630	27.59	290.18	317.77	28.38	327.75	356.13	0.12561	1.42406	1.54967
1.0000	233.0	0.000895	0.037547	0.038442	33.39	284.95	318.34	34.29	322.50	356.78	0.15071	1.38428	1.53499
1.5000	244.6	0.000935	0.024695	0.025630	57.04	262.97	320.00	58.44	300.01	358.45	0.24998	1.22664	1.47662
2.0000	253.6	0.000971	0.018062	0.019033	75.62	244.73	320.36	77.56	280.86	358.42	0.32484	1.10744	1.43229
2.5000	261.1	0.001006	0.013974	0.014981	91.43	228.41	319.84	93.94	263.35	357.29	0.38657	1.00857	1.39514
3.0000	267.6	0.001041	0.011180	0.012221	105.43	213.22	318.65	108.55	246.76	355.32	0.43989	0.92220	1.36208
3.5000	273.3	0.001078	0.009131	0.010209	118.23	198.63	316.86	122.00	230.59	352.59	0.48766	0.84371	1.33137
4.0000	278.5	0.001116	0.007549	0.008665	130.27	184.22	314.48	134.73	214.41	349.15	0.53182	0.77001	1.30183
4.5000	283.1	0.001158	0.006275	0.007433	141.86	169.62	311.48	147.07	197.86	344.93	0.57374	0.69880	1.27254
5.0000	287.5	0.001205	0.005210	0.006415	153.23	154.53	307.77	159.26	180.58	339.84	0.61437	0.62821	1.24258
5.5000	291.5	0.001260	0.004285	0.005545	164.60	138.54	303.14	171.53	162.11	333.64	0.65465	0.55620	1.21085
6.0000	295.2	0.001328	0.003445	0.004773	176.32	120.92	297.23	184.28	141.59	325.87	0.69592	0.47968	1.17560
6.5000	298.6	0.001420	0.002199	0.003619	189.34	77.22	266.56	198.56	90.42	288.95	0.74169	0.30631	1.04800
7.0000	301.9	0.001571	0.000955	0.002526	205.65	14.70	220.35	216.64	39.24	255.88	0.79940	0.13294	0.93234
7.3834	304.2	0.002155	0.000000	0.002155	241.37	0.00	241.37	257.31	0.00	257.31	0.93116	0.00000	0.93116

A.15.3 Superheated CO_2 Table (SI Units)

P= 0.5173 MPa T(K)	216.5	220.	250.	300.	350.	400.	500.	600.	700.	800.	900.	1000.
v,m**3/kg	0.07277	0.07429	0.08691	0.10672	0.12586	0.14469	0.18190	0.21881	0.25559	0.29229	0.32895	0.36557
u,kJ/kg	314.23	316.52	336.19	369.47	404.41	441.28	520.67	606.76	698.42	794.67	894.67	997.77
h,kJ/kg	351.87	354.95	381.15	424.68	469.52	516.13	614.77	719.95	830.64	945.88	1064.84	1186.88
s,kJ/kg/K	1.62496	1.63907	1.75074	1.90943	2.04760	2.17203	2.39184	2.58344	2.75397	2.90778	3.04786	3.17642
P= 0.6000 MPa T(K)	220.0	220.	250.	300.	350.	400.	500.	600.	700.	800.	900.	1000.
v,m**3/kg	0.06314	0.00000	0.07429	0.09161	0.10824	0.12456	0.15673	0.18861	0.22036	0.25202	0.28365	0.31524
u,kJ/kg	315.24	0.00	335.31	368.92	404.01	440.97	520.47	606.60	698.30	794.57	894.58	997.69
h,kJ/kg	353.12	0.00	379.89	423.89	468.96	515.71	614.50	719.77	830.51	945.78	1064.77	1186.83
s,kJ/kg/K	1.60501	0.00000	1.71914	1.87954	2.01844	2.14324	2.36341	2.55516	2.72578	2.87964	3.01975	3.14832
P= 0.7000 MPa T(K)	223.7	250.	300.	350.	400.	500.	600.	700.	800.	900.	1000.	1200.
v,m**3/kg	0.05443	0.06301	0.07811	0.09250	0.10656	0.13424	0.16162	0.18887	0.21604	0.24316	0.27026	0.32439
u,kJ/kg	316.24	334.23	368.24	403.52	440.60	520.21	606.41	698.15	794.44	894.48	997.60	1211.21
h,kJ/kg	354.34	378.34	422.91	468.27	515.19	614.18	719.55	830.36	945.67	1064.69	1186.78	1438.28
s,kJ/kg/K	1.58413	1.68559	1.84813	1.98793	2.11318	2.33378	2.52573	2.69644	2.85036	2.99051	3.11911	3.34826
P= 0.8000 MPa T(K)	227.1	250.	300.	350.	400.	500.	600.	700.	800.	900.	1000.	1200.
v,m**3/kg	0.04782	0.05454	0.06798	0.08069	0.09307	0.11737	0.14138	0.16526	0.18905	0.21280	0.23652	0.28390
u,kJ/kg	317.08	333.12	367.55	403.04	440.22	519.96	606.23	698.00	794.32	894.37	997.50	1211.13
h,kJ/kg	355.33	376.75	421.94	467.59	514.68	613.86	719.33	830.21	945.56	1064.61	1186.72	1438.26
s,kJ/kg/K	1.56590	1.65582	1.82060	1.96130	2.08702	2.30805	2.50019	2.67100	2.82498	2.96516	3.09379	3.32297
P= 0.9000 MPa T(K)	230.2	250.	300.	350.	400.	500.	600.	700.	800.	900.	1000.	1200.
v,m**3/kg	0.04263	0.04794	0.06010	0.07150	0.08258	0.10425	0.12564	0.14689	0.16806	0.18919	0.21028	0.25242
u,kJ/kg	317.77	331.99	366.86	402.55	439.85	519.71	606.04	697.85	794.20	894.27	997.41	1211.06
h,kJ/kg	356.13	375.13	420.95	466.90	514.17	613.54	719.11	830.05	945.45	1064.54	1186.67	1438.23
s,kJ/kg/K	1.54967	1.62888	1.79601	1.93764	2.06383	2.28530	2.47762	2.64854	2.80257	2.94280	3.07145	3.30066
P= 1.0000 MPa T(K)	233.0	250.	300.	350.	400.	500.	600.	700.	800.	900.	1000.	1200.
v,m**3/kg	0.03844	0.04265	0.05379	0.06416	0.07418	0.09376	0.11305	0.13220	0.15127	0.17030	0.18929	0.22723
u,kJ/kg	318.34	330.83	366.16	402.05	439.47	519.46	605.85	697.70	794.07	894.16	997.32	1210.99
h,kJ/kg	356.78	373.47	419.95	466.21	513.65	613.22	718.90	829.90	945.34	1064.46	1186.61	1438.21
s,kJ/kg/K	1.53499	1.60415	1.77374	1.91632	2.04298	2.26489	2.45741	2.62842	2.78251	2.92277	3.05145	3.28069
P= 1.5000 MPa T(K)	244.6	250.	300.	350.	400.	500.	600.	700.	800.	900.	1000.	1200.
v,m**3/kg	0.02563	0.02665	0.03485	0.04211	0.04899	0.06228	0.07527	0.08812	0.10090	0.11362	0.12632	0.15166
u,kJ/kg	320.00	324.48	362.54	399.55	437.58	518.19	604.91	696.95	793.45	893.63	996.86	1210.62
h,kJ/kg	358.45	364.46	414.82	462.71	511.06	611.61	717.81	829.13	944.80	1064.07	1186.34	1438.11
s,kJ/kg/K	1.47662	1.50091	1.68488	1.83254	1.96164	2.18577	2.37925	2.55076	2.70515	2.84559	2.97439	3.20379
P= 2.0000 MPa T(K)	253.6	300.	350.	400.	500.	600.	700.	800.	900.	1000.	1200.	1400.
v,m**3/kg	0.01903	0.02535	0.03107	0.03640	0.04654	0.05638	0.06608	0.07571	0.08529	0.09484	0.11388	0.13287
u,kJ/kg	320.36	358.72	396.99	435.65	516.92	603.97	696.21	792.83	893.10	996.40	1210.25	1431.83
h,kJ/kg	358.42	409.41	459.13	508.45	610.01	716.73	828.37	944.25	1063.68	1186.07	1438.01	1697.57
s,kJ/kg/K	1.43229	1.61745	1.77081	1.90250	2.12891	2.32335	2.49535	2.65003	2.79066	2.91959	3.14914	3.34913
P= 2.5000 MPa T(K)	261.1	300.	350.	400.	500.	600.	700.	800.	900.	1000.	1200.	1400.
v,m**3/kg	0.01498	0.01962	0.02445	0.02884	0.03710	0.04505	0.05286	0.06060	0.06829	0.07595	0.09121	0.10642
u,kJ/kg	319.84	354.65	394.35	433.71	515.65	603.03	695.46	792.22	892.58	995.94	1209.89	1431.53
h,kJ/kg	357.29	403.69	455.48	505.81	608.40	715.66	827.62	943.71	1063.30	1185.81	1437.91	1697.58
s,kJ/kg/K	1.39514	1.56125	1.72105	1.85548	2.08423	2.27964	2.45215	2.60711	2.74793	2.87698	3.10668	3.30675
P= 3.0000 MPa T(K)	267.6	300.	350.	400.	500.	600.	700.	800.	900.	1000.	1200.	1400.
v,m**3/kg	0.01222	0.01577	0.02003	0.02381	0.03081	0.03750	0.04405	0.05050	0.05695	0.06335	0.07609	0.08879
u,kJ/kg	318.65	350.28	391.64	431.73	514.38	602.09	694.72	791.60	892.06	995.49	1209.53	1431.22
h,kJ/kg	355.32	397.58	451.73	503.15	606.80	714.59	826.87	943.18	1062.92	1185.55	1437.81	1697.60
s,kJ/kg/K	1.36208	1.51160	1.67877	1.81611	2.04727	2.24365	2.41666	2.57182	2.71291	2.84208	3.07193	3.27210
P= 3.5000 MPa T(K)	273.3	300.	350.	400.	500.	600.	700.	800.	900.	1000.	1200.	1400.
v,m**3/kg	0.01021	0.01299	0.01687	0.02021	0.02632	0.03211	0.03776	0.04333	0.04886	0.05436	0.06530	0.07620
u,kJ/kg	316.86	345.55	388.85	429.73	513.10	601.15	693.98	790.99	891.53	995.03	1209.17	1430.92
h,kJ/kg	352.59	391.01	447.90	500.47	605.21	713.52	826.13	942.65	1062.55	1185.29	1437.72	1697.61
s,kJ/kg/K	1.33137	1.46584	1.64157	1.78201	2.01563	2.21299	2.38650	2.54204	2.68321	2.81251	3.04251	3.24276
P= 4.0000 MPa T(K)	278.5	300.	350.	400.	500.	600.	700.	800.	900.	1000.	1200.	1400.
v,m**3/kg	0.00867	0.01087	0.01450	0.01751	0.02295	0.02806	0.03304	0.03794	0.04279	0.04761	0.05720	0.06675
u,kJ/kg	314.48	340.37	385.98	427.71	511.82	600.21	693.24	790.38	891.02	994.58	1208.81	1430.62
h,kJ/kg	349.15	383.86	443.96	497.76	603.61	712.47	825.39	942.13	1062.18	1185.03	1437.62	1697.63
s,kJ/kg/K	1.30183	1.42218	1.60799	1.75173	1.98788	2.18624	2.36023	2.51606	2.65742	2.78684	3.01699	3.21732
P= 4.5000 MPa T(K)	283.1	300.	350.	400.	500.	600.	700.	800.	900.	1000.	1200.	1400.
v,m**3/kg	0.00743	0.00919	0.01265	0.01541	0.02033	0.02492	0.02937	0.03374	0.03807	0.04237	0.05091	0.05940
u,kJ/kg	311.48	334.61	383.01	425.66	510.53	599.27	692.50	789.77	890.50	994.13	1208.45	1430.33
h,kJ/kg	344.93	375.94	439.92	495.02	602.02	711.41	824.66	941.61	1061.81	1184.78	1437.53	1697.65
s,kJ/kg/K	1.27254	1.37913	1.57711	1.72436	1.96311	2.16246	2.33695	2.49307	2.63460	2.76414	2.99444	3.19486
P= 5.0000 MPa T(K)	287.5	300.	350.	400.	500.	600.	700.	800.	900.	1000.	1200.	1400.
v,m**3/kg	0.00641	0.00779	0.01116	0.01374	0.01824	0.02241	0.02643	0.03039	0.03429	0.03817	0.04587	0.05353
u,kJ/kg	307.77	328.03	379.95	423.57	509.24	598.34	691.77	789.17	889.98	993.68	1208.09	1430.03
h,kJ/kg	339.84	366.98	435.76	492.26	600.43	710.37	823.93	941.10	1061.45	1184.53	1437.45	1697.67
s,kJ/kg/K	1.24258	1.33514	1.54826	1.69927	1.94068	2.14103	2.31602	2.47242	2.61414	2.74380	2.97424	3.17474

P= 5.5000 MPa	T(K) 291.5	300.	350.	400.	500.	600.	700.	800.	900.	1000.	1200.	1400.
v,m**3/kg	0.00554	0.00659	0.00995	0.01236	0.01653	0.02035	0.02403	0.02764	0.03120	0.03474	0.04175	0.04872
u,kJ/kg	303.14	320.24	376.78	421.47	507.95	597.40	691.03	788.56	889.47	993.23	1207.73	1429.73
h,kJ/kg	333.64	356.46	431.48	489.47	598.84	709.33	823.21	940.59	1061.09	1184.29	1437.36	1697.69
s,kJ/kg/K	1.21085	1.28813	1.52098	1.67602	1.92015	2.12151	2.29699	2.45368	2.59557	2.72535	2.95594	3.15653

P= 6.0000 MPa	T(K) 295.2	300.	350.	400.	500.	600.	700.	800.	900.	1000.	1200.	1400.
v,m**3/kg	0.00477	0.00548	0.00893	0.01122	0.01510	0.01864	0.02203	0.02535	0.02863	0.03188	0.03832	0.04471
u,kJ/kg	297.23	310.38	373.50	419.33	506.66	596.47	690.30	787.96	888.95	992.78	1207.38	1429.44
h,kJ/kg	325.87	343.29	427.06	486.66	597.26	708.29	822.50	940.08	1060.73	1184.04	1437.28	1697.71
s,kJ/kg/K	1.17560	1.23416	1.49490	1.65426	1.90119	2.10356	2.27954	2.43650	2.57858	2.70848	2.93921	3.13988

P= 6.5000 MPa	T(K) 298.6	300.	350.	400.	500.	600.	700.	800.	900.	1000.	1200.	1400.
v,m**3/kg	0.00142	0.00435	0.00806	0.01025	0.01389	0.01719	0.02034	0.02342	0.02645	0.02946	0.03541	0.04132
u,kJ/kg	189.33	295.63	370.09	417.16	505.36	595.54	689.57	787.36	888.44	992.34	1207.02	1429.14
h,kJ/kg	198.55	323.90	422.50	483.81	595.68	707.26	821.79	939.58	1060.38	1183.80	1437.19	1697.73
s,kJ/kg/K	0.74165	1.16131	1.46975	1.63375	1.88354	2.08694	2.26341	2.42065	2.56290	2.69292	2.92380	3.12455

P= 7.0000 MPa	T(K) 301.9	350.	400.	500.	600.	700.	800.	900.	1000.	1200.	1400.	1600.
v,m**3/kg	0.00157	0.00732	0.00943	0.01286	0.01595	0.01889	0.02176	0.02459	0.02738	0.03292	0.03842	0.04389
u,kJ/kg	205.74	366.55	414.96	504.06	594.61	688.84	786.76	887.93	991.90	1206.67	1428.85	1657.11
h,kJ/kg	216.75	417.79	480.95	594.10	706.24	821.08	939.09	1060.03	1183.57	1437.11	1697.76	1964.31
s,kJ/kg/K	0.79975	1.44529	1.61429	1.86703	2.07144	2.24840	2.40593	2.54835	2.67849	2.90951	3.11034	3.28826

P= 7.3834 MPa	T(K) 304.2	350.	400.	500.	600.	700.	800.	900.	1000.	1200.	1400.	1600.
v,m**3/kg	0.00221	0.00682	0.00887	0.01217	0.01511	0.01791	0.02064	0.02333	0.02598	0.03124	0.03645	0.04164
u,kJ/kg	243.72	363.73	413.26	503.07	593.89	688.29	786.30	887.55	991.56	1206.40	1428.62	1656.92
h,kJ/kg	260.02	414.06	478.73	592.89	705.46	820.55	938.71	1059.77	1183.39	1437.05	1697.78	1964.39
s,kJ/kg/K	0.94013	1.42690	1.59999	1.85503	2.06022	2.23756	2.39530	2.53786	2.66808	2.89922	3.10011	3.27807

P= 7.500 MPa		350.	400.	500.	600.	700.	800.	900.	1000.	1200.	1400.	1600.
v,m**3/kg	0.00166	0.00667	0.00871	0.01197	0.01487	0.01764	0.02032	0.02297	0.02558	0.03076	0.03590	0.04101
u,kJ/kg	214.30	362.86	412.74	502.76	593.68	688.12	786.16	887.43	991.46	1206.32	1428.56	1656.86
h,kJ/kg	226.75	412.90	478.05	592.53	705.23	820.38	938.59	1059.69	1183.33	1437.03	1697.79	1964.41
s,kJ/kg/K	0.83007	1.42135	1.59573	1.85148	2.05691	2.23437	2.39217	2.53477	2.66502	2.89619	3.09710	3.27507

P= 8.000 MPa		350.	400.	500.	600.	700.	800.	900.	1000.	1200.	1400.	1600.
v,m**3/kg	0.00148	0.00610	0.00808	0.01119	0.01393	0.01654	0.01907	0.02155	0.02401	0.02887	0.03369	0.03849
u,kJ/kg	200.44	359.01	410.48	501.46	592.75	687.39	785.57	886.92	991.02	1205.97	1428.27	1656.61
h,kJ/kg	212.29	407.83	475.13	590.96	704.22	819.69	938.11	1059.35	1183.10	1436.96	1697.81	1964.51
s,kJ/kg/K	0.78001	1.39775	1.57794	1.83678	2.04324	2.22118	2.37926	2.52203	2.65240	2.88371	3.08470	3.26273

P= 9.000 MPa		350.	400.	500.	600.	700.	800.	900.	1000.	1200.	1400.	1600.
v,m**3/kg	0.00137	0.00515	0.00704	0.00989	0.01237	0.01471	0.01697	0.01920	0.02139	0.02573	0.03002	0.03429
u,kJ/kg	189.73	350.78	405.88	498.85	590.90	685.95	784.39	885.92	990.14	1205.27	1427.69	1656.11
h,kJ/kg	202.07	397.10	469.21	587.83	702.21	818.32	937.15	1058.68	1182.65	1436.81	1697.87	1964.72
s,kJ/kg/K	0.74177	1.35107	1.54429	1.80950	2.01803	2.19696	2.35559	2.49870	2.62930	2.86089	3.06204	3.24016

P= 10.000 MPa		350.	400.	500.	600.	700.	800.	900.	1000.	1200.	1400.	1600.
v,m**3/kg	0.00132	0.00437	0.00620	0.00885	0.01112	0.01325	0.01530	0.01731	0.01929	0.02321	0.02708	0.03093
u,kJ/kg	183.35	341.76	401.15	496.24	589.05	684.52	783.21	884.92	989.27	1204.58	1427.11	1655.62
h,kJ/kg	196.51	385.51	463.19	584.73	700.24	816.97	936.20	1058.02	1182.21	1436.67	1697.94	1964.93
s,kJ/kg/K	0.71907	1.30438	1.51273	1.78460	1.99520	2.17510	2.33428	2.47773	2.60855	2.84042	3.04173	3.21995

P= 12.500 MPa		350.	400.	500.	600.	700.	800.	900.	1000.	1200.	1400.	1600.
v,m**3/kg	0.00124	0.00300	0.00472	0.00699	0.00888	0.01062	0.01229	0.01392	0.01552	0.01868	0.02180	0.02489
u,kJ/kg	173.03	315.85	388.84	489.68	584.47	680.97	780.30	882.45	987.12	1202.87	1425.69	1654.40
h,kJ/kg	188.49	353.29	447.78	577.07	695.41	813.69	933.93	1056.45	1181.16	1436.36	1698.12	1965.48
s,kJ/kg/K	0.68224	1.18642	1.44045	1.73002	1.94585	2.12816	2.28867	2.43295	2.56433	2.79688	2.99857	3.17705

P= 15.000 MPa		350.	400.	500.	600.	700.	800.	900.	1000.	1200.	1400.	1600.
v,m**3/kg	0.00119	0.00222	0.00374	0.00576	0.00739	0.00887	0.01029	0.01166	0.01301	0.01566	0.01827	0.02086
u,kJ/kg	165.95	289.68	376.00	483.13	579.93	677.47	777.44	880.02	985.01	1201.19	1424.28	1653.19
h,kJ/kg	183.79	322.92	432.17	569.58	690.74	810.55	931.75	1054.96	1180.18	1436.10	1698.34	1966.05
s,kJ/kg/K	0.65685	1.08136	1.37520	1.68334	1.90438	2.08905	2.25087	2.39596	2.52788	2.76108	2.96315	3.14185

P= 20.000 MPa		350.	400.	500.	600.	700.	800.	900.	1000.	1200.	1400.	1600.
v,m**3/kg	0.00113	0.00163	0.00262	0.00426	0.00554	0.00670	0.00779	0.00885	0.00988	0.01189	0.01387	0.01582
u,kJ/kg	155.78	255.86	350.57	470.16	571.04	670.64	771.86	875.29	980.90	1197.91	1421.54	1650.83
h,kJ/kg	178.36	288.43	403.03	555.30	681.94	804.67	927.73	1052.23	1178.42	1435.70	1698.87	1967.26
s,kJ/kg/K	0.61997	0.95621	1.26338	1.60545	1.83658	2.02578	2.19009	2.33671	2.46965	2.70411	2.90689	3.08604

A.16 Thermodynamic Property Tables for Sodium

A.16.1 Sodium Temperature Saturation Table: SI units

T(K)	P(MPa)	Volume (m**3/kg)			Energy (kJ/kg)			Enthalpy (kJ.kg)			Entropy (kJ/kg)		
		vf	vfg	vg	uf	ufg	ug	hf	hfg	hg	sf	sfg	sg
800.0	0.001	0.001211	298.0341	298.0353	0.0	4034.4	4034.3	0.0	4315.6	4315.5	0.0000	5.3945	5.3944
850.0	0.002	0.001229	127.4088	127.4100	66.4	3970.8	4037.2	66.4	4267.0	4333.4	0.0804	5.0200	5.1004
900.0	0.005	0.001247	60.0689	60.0702	132.7	3904.9	4037.6	132.7	4215.5	4348.2	0.1563	4.6839	4.8401
950.0	0.011	0.001266	30.7359	30.7372	200.2	3836.0	4036.2	200.2	4160.4	4360.6	0.2292	4.3793	4.6085
1000.0	0.020	0.001286	16.8575	16.8587	268.5	3765.2	4033.7	268.5	4102.8	4371.3	0.2991	4.1028	4.4019
1050.0	0.036	0.001306	9.8135	9.8148	336.6	3694.3	4030.9	336.7	4044.6	4381.2	0.3653	3.8520	4.2172
1100.0	0.060	0.001327	6.0137	6.0150	405.1	3623.4	4028.5	405.2	3986.0	4391.2	0.4285	3.6236	4.0521
1154.6	0.101	0.001351	3.7069	3.7083	479.9	3547.6	4027.5	480.0	3923.2	4403.2	0.4939	3.3979	3.8917
1200.0	0.150	0.001372	2.5685	2.5699	542.4	3486.0	4028.4	542.6	3872.2	4414.9	0.5457	3.2269	3.7725
1250.0	0.224	0.001395	1.7722	1.7736	611.0	3421.1	4032.2	611.3	3818.8	4430.1	0.5996	3.0550	3.6547
1300.0	0.325	0.001419	1.2599	1.2613	682.9	3356.1	4039.0	683.3	3765.0	4448.4	0.6528	2.8962	3.5490
1400.0	0.625	0.001469	0.6885	0.6899	831.8	3232.2	4064.0	832.7	3662.8	4495.5	0.7528	2.6163	3.3691
1500.0	1.101	0.001523	0.4091	0.4106	998.1	3107.4	4105.5	999.7	3557.9	4557.7	0.8494	2.3719	3.2213
1600.0	1.802	0.001581	0.2590	0.2606	1194.0	2970.8	4164.8	1196.8	3437.4	4634.2	0.9463	2.1484	3.0947
1700.0	2.776	0.001642	0.1713	0.1729	1440.8	2804.2	4245.0	1445.4	3279.7	4725.1	1.0487	1.9292	2.9780

A.16.2 Sodium Pressure Saturation Table: SI units

P(MPa)	T(K)	Volume (m**3/kg)			Energy (kJ/kg)			Enthalpy (kJ.kg)			Entropy (kJ/kg)		
		vf	vfg	vg	uf	ufg	ug	hf	hfg	hg	sf	sfg	sg
0.009	941.1	0.001263	34.4471	34.4483	188.2	3848.4	4036.6	188.2	4170.3	4358.5	0.2165	4.4313	4.6479
0.020	999.9	0.001286	16.8773	16.8786	268.2	3765.5	4033.7	268.2	4103.1	4371.3	0.2988	4.1035	4.4023
0.040	1060.5	0.001311	8.8223	8.8236	351.0	3679.4	4030.3	351.0	4032.3	4383.3	0.3788	3.8024	4.1811
0.060	1099.5	0.001327	6.0411	6.0425	404.3	3624.2	4028.6	404.4	3986.7	4391.1	0.4278	3.6258	4.0536
0.080	1129.1	0.001340	4.6196	4.6209	444.9	3582.8	4027.7	445.0	3952.4	4397.4	0.4637	3.5006	3.9643
0.100	1153.1	0.001351	3.7528	3.7541	477.9	3549.6	4027.5	478.0	3924.9	4402.9	0.4921	3.4036	3.8958
0.101	1154.6	0.001351	3.7070	3.7084	480.0	3547.5	4027.5	480.1	3923.1	4403.2	0.4939	3.3978	3.8917
0.200	1235.2	0.001388	1.9714	1.9728	591.3	3439.4	4030.7	591.5	3833.7	4425.3	0.5844	3.1038	3.6882
0.300	1289.0	0.001413	1.3548	1.3562	667.1	3370.0	4037.2	667.6	3776.5	4444.0	0.6414	2.9298	3.5712
0.400	1330.2	0.001434	1.0390	1.0404	726.6	3318.3	4044.9	727.2	3733.9	4461.1	0.6836	2.8070	3.4905
0.500	1364.1	0.001451	0.8459	0.8474	777.0	3276.3	4053.2	777.7	3699.2	4476.9	0.7175	2.7118	3.4293
0.750	1430.6	0.001485	0.5826	0.5841	880.3	3194.6	4074.9	881.4	3631.6	4513.0	0.7825	2.5385	3.3210
1.000	1481.9	0.001513	0.4471	0.4486	966.2	3130.5	4096.8	967.7	3577.6	4545.3	0.8320	2.4142	3.2461
2.000	1623.0	0.001594	0.2348	0.2364	1245.1	2936.1	4181.1	1248.3	3405.6	4653.8	0.9691	2.0983	3.0675
3.000	1719.4	0.001655	0.1586	0.1603	1497.8	2766.0	4263.7	1502.7	3241.9	4744.6	1.0698	1.8855	2.9553

A.16.3 Superheated Sodium Table: SI Units

P= 0.0093 MPa T(K)	941.1	950.	1000.	1050.	1100.	1150.	1200.	1250.	1300.	1350.	1400.	1450.
v,m**3/kg	34.44834	34.9420	37.4751	39.7787	41.9451	44.0527	46.0856	48.0955	50.0854	52.0603	54.0195	55.9839
u,kJ/kg	4036.59	4053.95	4131.43	4187.49	4231.91	4269.71	4303.53	4334.91	4364.73	4393.52	4421.62	4449.24
h,kJ/kg	4358.54	4380.48	4481.67	4559.26	4623.98	4681.33	4734.23	4784.41	4832.83	4880.08	4926.55	4972.47
s,kJ/kg/K	4.64786	4.6711	4.7751	4.8509	4.9111	4.9622	5.0072	5.0481	5.0861	5.1218	5.1555	5.1878

P= 0.0500 MPa T(K)	1081.6	1100.	1150.	1200.	1250.	1300.	1350.	1400.	1450.	1500.	1550.	1600.
v,m**3/kg	7.16339	7.3656	7.8738	8.3370	8.7728	9.1880	9.5898	9.9813	10.3661	10.7461	11.1220	11.4955
u,kJ/kg	4029.30	4067.50	4152.25	4217.57	4270.57	4315.56	4355.23	4391.29	4424.85	4456.64	4487.15	4516.73
h,kJ/kg	4387.47	4435.77	4545.94	4634.47	4709.21	4774.96	4834.71	4890.36	4943.17	4993.95	5043.27	5091.51
s,kJ/kg/K	4.11076	4.1553	4.2536	4.3291	4.3902	4.4418	4.4870	4.5275	4.5645	4.5990	4.6313	4.6619

P= 0.1000 MPa T(K)	1153.1	1200.	1250.	1300.	1350.	1400.	1450.	1500.	1550.	1600.	1700.	1800.
v,m**3/kg	3.75411	4.0095	4.2583	4.4909	4.7099	4.9198	5.1238	5.3224	5.5175	5.7113	6.0886	6.4619
u,kJ/kg	4027.47	4118.13	4194.78	4257.01	4309.31	4354.73	4395.35	4432.52	4467.19	4500.04	4561.93	4620.60
h,kJ/kg	4402.87	4519.07	4620.66	4706.08	4780.30	4846.74	4907.72	4964.76	5018.94	5071.02	5170.79	5266.81
s,kJ/kg/K	3.89575	3.9956	4.0790	4.1462	4.2023	4.2507	4.2936	4.3322	4.3678	4.4009	4.4614	4.5162

P= 0.1013 MPa T(K)	1154.6	1200.	1250.	1300.	1350.	1400.	1450.	1500.	1550.	1600.	1700.	1800.
v,m**3/kg	3.70836	3.9531	4.2086	4.4293	4.6460	4.8546	5.0551	5.2515	5.4449	5.6342	6.0083	6.3769
u,kJ/kg	4027.46	4115.61	4193.14	4255.49	4308.11	4353.79	4394.57	4431.88	4466.67	4499.59	4561.62	4620.37
h,kJ/kg	4403.21	4516.16	4618.74	4704.29	4778.88	4845.62	4906.79	4963.99	5018.31	5070.48	5170.40	5266.51
s,kJ/kg/K	3.89173	3.9887	4.0736	4.1403	4.1967	4.2453	4.2883	4.3271	4.3627	4.3958	4.4564	4.5114

P= 0.2000 MPa T(K)	1235.2	1250.	1300.	1350.	1400.	1450.	1500.	1550.	1600.	1700.	1800.
v,m**3/kg	1.97278	2.0149	2.1492	2.2746	2.3923	2.5043	2.6118	2.7161	2.8176	3.0153	3.2077
u,kJ/kg	4030.70	4060.32	4148.92	4222.58	4284.70	4338.27	4385.53	4428.13	4467.23	4538.19	4602.85
h,kJ/kg	4425.26	4463.26	4578.76	4677.48	4763.15	4839.12	4907.90	4971.33	5030.76	5141.23	5244.38
s,kJ/kg/K	3.68816	3.7197	3.8120	3.8874	3.9501	4.0036	4.0503	4.0920	4.1297	4.1968	4.2557

P= 0.3000 MPa T(K)	1289.0	1300.	1350.	1400.	1450.	1500.	1550.	1600.	1700.	1800.
v,m**3/kg	1.35620	1.3768	1.4673	1.5524	1.6330	1.7094	1.7830	1.8541	1.9910	2.1230
u,kJ/kg	4037.18	4057.97	4145.04	4219.93	4284.39	4340.57	4390.39	4435.34	4514.89	4585.34
h,kJ/kg	4444.04	4471.01	4585.23	4685.67	4774.27	4853.41	4925.30	4991.58	5112.20	5222.25
s,kJ/kg/K	3.57121	3.5931	3.6823	3.7568	3.8197	3.8736	3.9210	3.9632	4.0364	4.0994

P= 0.4000 MPa T(K)	1330.2	1350.	1400.	1450.	1500.	1550.	1600.	1700.	1800.
v,m**3/kg	1.04040	1.0678	1.1349	1.1987	1.2593	1.3172	1.3729	1.4792	1.5807
u,kJ/kg	4044.92	4079.76	4161.92	4234.45	4298.05	4354.22	4404.49	4492.10	4568.10
h,kJ/kg	4461.08	4506.88	4615.92	4713.95	4801.77	4881.10	4953.64	5083.78	5200.44
s,kJ/kg/K	3.49055	3.5271	3.6100	3.6806	3.7410	3.7935	3.8398	3.9190	3.9858

P= 0.5000 MPa T(K)	1364.1	1400.	1450.	1500.	1550.	1600.	1700.	1800.
v,m**3/kg	0.84736	0.8867	0.9396	0.9901	1.0382	1.0845	1.1723	1.2557
u,kJ/kg	4053.23	4112.29	4189.33	4258.40	4319.85	4374.83	4469.86	4551.15
h,kJ/kg	4476.92	4555.62	4659.12	4753.43	4838.98	4917.08	5056.02	5178.99
s,kJ/kg/K	3.42932	3.4912	3.5674	3.6332	3.6903	3.7404	3.8252	3.8957

P= 0.6000 MPa T(K)	1393.2	1400.	1450.	1500.	1550.	1600.	1700.	1800.
v,m**3/kg	0.71675	0.7229	0.7679	0.8113	0.8529	0.8926	0.9679	1.0390
u,kJ/kg	4061.83	4072.44	4149.78	4222.09	4287.59	4346.48	4448.24	4534.51
h,kJ/kg	4491.87	4506.21	4610.55	4708.87	4799.27	4882.03	5028.97	5157.89
s,kJ/kg/K	3.38018	3.3918	3.4718	3.5419	3.6030	3.6564	3.7464	3.8205

P= 0.7000 MPa T(K)	1418.8	1450.	1500.	1550.	1600.	1700.	1800.
v,m**3/kg	0.62230	0.6464	0.6843	0.7208	0.7558	0.8220	0.8842
u,kJ/kg	4070.56	4116.64	4189.55	4257.59	4319.61	4427.27	4518.19
h,kJ/kg	4506.11	4569.10	4668.54	4762.13	4848.67	5002.68	5137.18
s,kJ/kg/K	3.33927	3.3893	3.4624	3.5268	3.5833	3.6782	3.7556

P= 0.8000 MPa T(K)	1441.8	1450.	1500.	1550.	1600.	1700.	1800.
v,m**3/kg	0.55049	0.5561	0.5895	0.6221	0.6535	0.7128	0.7683
u,kJ/kg	4079.29	4090.60	4161.22	4230.19	4294.39	4407.05	4502.25
h,kJ/kg	4519.68	4535.45	4632.88	4727.89	4817.18	4977.26	5116.91
s,kJ/kg/K	3.30409	3.3172	3.3922	3.4592	3.5184	3.6178	3.6984

P= 0.9000 MPa T(K)	1462.7	1500.	1550.	1600.	1700.	1800.
v,m**3/kg	0.49413	0.5164	0.5458	0.5741	0.6280	0.6782
u,kJ/kg	4088.04	4137.55	4205.62	4270.94	4387.58	4486.67
h,kJ/kg	4532.75	4602.32	4696.79	4787.66	4952.69	5097.06
s,kJ/kg/K	3.27341	3.3295	3.3985	3.4599	3.5633	3.6470

P= 1.0000 MPa T(K)	1481.9	1500.	1550.	1600.	1700.	1800.
v,m**3/kg	0.44859	0.4583	0.4849	0.5109	0.5601	0.6062
u,kJ/kg	4096.75	4118.89	4184.13	4249.42	4368.93	4471.50
h,kJ/kg	4545.34	4577.20	4669.06	4760.28	4929.03	5077.67
s,kJ/kg/K	3.24613	3.2730	3.3433	3.4066	3.5136	3.6002

P= 2.0000 MPa T(K)	1623.0	1700.	1800.
v,m**3/kg	0.23636	0.2574	0.2834
u,kJ/kg	4181.14	4241.43	4347.27
h,kJ/kg	4653.85	4756.31	4914.13
s,kJ/kg/K	3.06745	3.1640	3.2716

A.16.4 Sodium Temperature Saturation Table: English Units

T(R)	P(psi)	Volume (ft**3/lbm)			Energy (Btu/lbm)			Enthalpy (Btu/lbm)			Entropy (Btu/lbm)		
		vf	vfg	vg	uf	ufg	ug	hf	hfg	hg	sf	sfg	sg
1440.0	0.1	0.01944	4784.0732	4784.0928	-0.1	8412.9	8412.8	-0.1	8999.3	8999.2	2.77752	3.47195	6.24947
1500.0	0.3	0.01963	2683.4402	2683.4597	91.3	8326.1	8417.4	91.3	8933.5	9024.8	2.70908	3.30871	6.01779
1600.0	0.6	0.01996	1131.0094	1131.0293	245.7	8174.2	8419.9	245.7	8815.3	9061.0	2.61041	3.06085	5.67126
1700.0	1.4	0.02029	527.3803	527.4006	436.5	7979.5	8416.0	436.6	8652.6	9089.1	2.53727	2.82764	5.36491
1800.0	2.9	0.02065	270.5976	270.6182	559.9	7851.7	8411.6	559.9	8555.6	9115.5	2.45895	2.64062	5.09957
1900.0	5.5	0.02101	148.8262	148.8472	717.8	7687.2	8405.0	717.9	8420.5	9138.5	2.40120	2.46215	4.86335
2000.0	9.8	0.02139	87.1279	87.1493	876.3	7523.6	8400.0	876.5	8285.3	9161.8	2.35325	2.30148	4.65473
2100.0	16.4	0.02178	53.8157	53.8375	1035.4	7363.3	8398.6	1035.7	8152.4	9188.1	2.31376	2.15673	4.47049
2200.0	26.2	0.02218	34.8169	34.8391	1195.4	7207.8	8403.2	1195.9	8023.8	9219.7	2.28164	2.02621	4.30784
2300.0	40.1	0.02260	23.4509	23.4735	1357.7	7057.5	8415.3	1358.6	7899.8	9258.4	2.25603	1.90818	4.16421
2400.0	59.2	0.02304	16.3607	16.3838	1523.8	6912.6	8436.4	1525.0	7780.7	9305.7	2.23610	1.80108	4.03717
2500.0	84.7	0.02349	11.7652	11.7887	1698.8	6768.5	8467.3	1700.6	7661.5	9362.1	2.22161	1.70256	3.92417
2600.0	117.8	0.02396	8.6896	8.7136	1882.8	6626.0	8508.8	1885.3	7543.0	9428.3	2.21143	1.61176	3.82319
2800.0	211.7	0.02495	5.0603	5.0853	2298.4	6326.7	8625.1	2303.1	7286.1	9589.2	2.20253	1.44566	3.64818
3000.0	350.7	0.02602	3.1440	3.1700	2819.0	5971.7	8790.7	2827.1	6959.5	9786.6	2.20581	1.28880	3.49461

A.16.5 Sodium Pressure Saturation Table: English Units

P(psi)	T(R)	Volume (ft**3/lbm)			Energy (Btu/lbm)			Enthalpy (Btu/lbm)			Entropy (Btu/lbm)		
		vf	vfg	vg	uf	ufg	ug	hf	hfg	hg	sf	sfg	sg
1.4	1694.0	0.02027	552.9115	552.9318	392.7	8024.8	8417.5	392.7	8696.2	9088.9	0.25091	5.13362	5.38454
2.5	1778.1	0.02057	311.3963	311.4169	525.6	7887.4	8412.9	525.6	8584.7	9110.3	0.32730	4.82810	5.15540
5.0	1884.3	0.02095	162.7399	162.7609	693.0	7713.0	8406.0	693.0	8441.9	9134.9	0.41842	4.48007	4.89849
7.5	1952.7	0.02121	111.4133	111.4345	801.4	7600.7	8402.0	801.4	8349.2	9150.6	0.47446	4.27565	4.75011
10.0	2004.5	0.02140	85.1776	85.1990	883.8	7516.0	8399.8	883.8	8279.1	9162.9	0.51564	4.13034	4.64598
12.5	2046.6	0.02157	69.1843	69.2059	950.6	7448.2	8398.7	950.6	8223.0	9173.6	0.54812	4.01794	4.56606
14.7	2078.3	0.02169	59.5053	59.5270	1001.2	7397.3	8398.5	1001.2	8180.9	9182.1	0.57219	3.93639	4.50858
15.0	2082.4	0.02171	58.3823	58.4040	1007.5	7391.0	8398.5	1007.5	8175.6	9183.2	0.57520	3.92615	4.50135
17.5	2113.6	0.02183	50.5836	50.6055	1057.4	7341.4	8398.9	1057.4	8134.7	9192.1	0.59845	3.84866	4.44711
20.0	2141.5	0.02194	44.6815	44.7034	1101.8	7297.9	8399.7	1101.8	8098.7	9200.5	0.61875	3.78177	4.40052
25.0	2189.8	0.02214	36.3213	36.3435	1179.6	7222.8	8402.4	1179.6	8036.6	9216.2	0.65350	3.66995	4.32345
30.0	2231.0	0.02231	30.6741	30.6964	1246.1	7160.0	8406.0	1246.1	7984.8	9230.9	0.68238	3.57897	4.26135
45.0	2328.6	0.02272	21.0823	21.1050	1405.9	7014.4	8420.3	1405.9	7865.1	9271.0	0.74868	3.37755	4.12623
60.0	2403.4	0.02305	16.1681	16.1911	1531.9	6905.3	8437.2	1531.9	7775.5	9307.4	0.79783	3.23519	4.03302
75.0	2464.9	0.02333	13.1648	13.1881	1638.8	6816.4	8455.2	1638.8	7702.3	9341.2	0.83749	3.12481	3.96230
100.0	2549.2	0.02372	10.1038	10.1275	1790.1	6696.3	8486.4	1790.1	7603.4	9393.5	0.89046	2.98265	3.87311
200.0	2779.2	0.02485	5.3337	5.3586	2256.0	6354.9	8610.9	2256.0	7314.8	9570.8	1.03310	2.63194	3.66503
300.0	2935.0	0.02567	3.6502	3.6758	2640.2	6090.6	8730.7	2640.2	7078.3	9718.5	1.13168	2.41166	3.54334

A.16.6 Superheated Sodium Table: English Units

P= 1.4 psi	T(R) 1694.0	1700.	1800.	1900.	2000.	2100.	2200.	2300.	2400.	2500.	2600.	2700.
v,ft**3/lbm	552.9318	555.9260	601.5211	642.4615	680.8981	717.9449	754.1944	789.7718	825.1837	860.1784	895.1399	929.9736
u,Btu/lbm	8417.5	8431.5	8615.3	8743.4	8843.3	8927.9	9003.8	9074.5	9142.0	9207.5	9271.7	9334.9
h,Btu/lbm	9088.9	9106.5	9345.6	9523.5	9670.0	9799.6	9919.4	10033.4	10143.8	10252.0	10358.5	10464.1
s,Btu/lbm/R	5.3845	5.3949	5.5319	5.6282	5.7034	5.7667	5.8224	5.8731	5.9201	5.9642	6.0060	6.0459
P= 2.5 psi	T(R) 1778.1	1800.	1900.	2000.	2100.	2200.	2300.	2400.	2500.	2600.	2700.	2800.
v,ft**3/lbm	311.4169	317.4174	342.4559	365.0533	386.2528	406.6337	426.4664	446.0121	465.3169	484.4529	503.5057	522.4739
u,Btu/lbm	8412.9	8464.1	8647.7	8780.0	8884.3	8972.8	9051.9	9125.1	9194.5	9261.5	9326.9	9391.0
h,Btu/lbm	9110.3	9174.9	9414.5	9597.4	9749.3	9883.4	10006.9	10123.8	10236.5	10346.3	10454.3	10561.0
s,Btu/lbm/R	5.1554	5.1916	5.3214	5.4153	5.4895	5.5519	5.6068	5.6566	5.7026	5.7457	5.7864	5.8252
P= 5.0 psi	T(R) 1884.3	1900.	2000.	2100.	2200.	2300.	2400.	2500.	2600.	2700.	2800.	2900.
v,ft**3/lbm	162.7609	164.9819	178.0859	189.8740	200.8626	211.3520	221.5073	231.4522	241.2410	250.9281	260.5435	270.0991
u,Btu/lbm	8406.0	8444.5	8644.1	8790.5	8905.9	9002.9	9088.3	9166.3	9239.4	9309.2	9376.8	9442.7
h,Btu/lbm	9134.9	9183.3	9441.7	9640.8	9805.5	9949.4	10080.3	10202.8	10319.8	10433.0	10543.6	10652.3
s,Btu/lbm/R	4.8985	4.9242	5.0571	5.1545	5.2312	5.2952	5.3509	5.4009	5.4468	5.4895	5.5298	5.5679
P= 7.5 psi	T(R) 1952.7	2000.	2100.	2200.	2300.	2400.	2500.	2600.	2700.	2800.	2900.	3000.
v,ft**3/lbm	111.4345	115.8719	124.4683	132.3047	139.6526	146.6837	153.5009	160.1333	166.7368	173.2347	179.6686	186.0625
u,Btu/lbm	8402.0	8513.1	8698.8	8840.1	8954.4	9051.8	9138.2	9217.4	9291.7	9362.6	9431.1	9497.8
h,Btu/lbm	9150.6	9291.4	9534.9	9728.9	9892.6	10037.2	10169.4	10293.4	10411.8	10526.3	10638.1	10747.8
s,Btu/lbm/R	4.7501	4.8217	4.9410	5.0314	5.1043	5.1658	5.2198	5.2684	5.3132	5.3548	5.3940	5.4312
P= 10.0 psi	T(R) 2004.5	2100.	2200.	2300.	2400.	2500.	2600.	2700.	2800.	2900.	3000.	
v,ft**3/lbm	85.1990	91.8104	98.0478	103.8186	109.2796	114.5324	119.6472	124.6441	129.5703	134.5207	139.2899	
u,Btu/lbm	8399.8	8609.7	8775.6	8906.7	9015.7	9110.4	9195.6	9274.3	9348.5	9419.5	9488.2	
h,Btu/lbm	9162.9	9432.0	9653.8	9836.6	9994.5	10136.3	10267.2	10390.7	10509.1	10623.8	10735.8	
s,Btu/lbm/R	4.6460	4.7781	4.8816	4.9630	5.0303	5.0882	5.1396	5.1862	5.2293	5.2697	5.3075	
P= 12.5 psi	T(R) 2046.6	2100.	2200.	2300.	2400.	2500.	2600.	2700.	2800.	2900.	3000.	
v,ft**3/lbm	69.2059	72.2589	77.5131	82.3293	86.8434	91.1549	95.3236	99.3923	103.3869	107.3276	111.2298	
u,Btu/lbm	8398.7	8523.8	8712.5	8859.6	8980.0	9082.8	9173.9	9257.0	9334.5	9408.0	9478.6	
h,Btu/lbm	9173.6	9332.9	9580.3	9781.4	9952.3	10103.4	10241.2	10369.8	10492.0	10609.6	10723.9	
s,Btu/lbm/R	4.5661	4.6436	4.7594	4.8491	4.9220	4.9837	5.0378	5.0863	5.1308	5.1721	5.2108	
P= 14.7 psi	T(R) 2078.3	2100.	2200.	2300.	2400.	2500.	2600.	2700.	2800.	2900.	3000.	
v,ft**3/lbm	59.5270	60.6072	65.2546	69.4941	73.4376	77.1835	80.7923	84.2977	87.7346	91.1159	94.4552	
u,Btu/lbm	8398.5	8451.6	8658.4	8819.0	8949.0	9058.8	9155.0	9241.8	9322.2	9397.9	9470.2	
h,Btu/lbm	9182.1	9249.3	9517.4	9733.7	9915.7	10074.8	10218.4	10351.5	10477.0	10597.2	10713.5	
s,Btu/lbm/R	4.5086	4.5412	4.6670	4.7637	4.8413	4.9063	4.9627	5.0129	5.0586	5.1008	5.1402	
P= 15.0 psi	T(R) 2082.4	2100.	2200.	2300.	2400.	2500.	2600.	2700.	2800.	2900.	3000.	
v,ft**3/lbm	58.4040	59.2735	63.8431	68.0124	71.8927	75.5716	79.1147	82.5565	85.9165	89.2437	92.5195	
u,Btu/lbm	8398.5	8441.9	8651.0	8813.4	8944.8	9055.5	9152.4	9239.7	9320.5	9396.5	9469.0	
h,Btu/lbm	9183.2	9238.1	9508.8	9727.2	9910.7	10070.9	10215.3	10348.9	10474.9	10595.5	10712.1	
s,Btu/lbm/R	4.5013	4.5280	4.6551	4.7526	4.8309	4.8964	4.9531	5.0036	5.0494	5.0917	5.1312	
P= 20.0 psi	T(R) 2141.5	2200.	2300.	2400.	2500.	2600.	2700.	2800.	2900.	3000.		
v,ft**3/lbm	44.7034	46.8068	50.1409	53.2172	56.1055	58.8591	61.5165	64.1042	66.6435	69.1359		
u,Btu/lbm	8399.7	8534.1	8723.8	8875.8	9001.7	9109.8	9205.6	9292.7	9373.6	9450.0		
h,Btu/lbm	9200.5	9372.6	9622.1	9829.1	10006.7	10164.2	10307.6	10441.0	10567.4	10688.5		
s,Btu/lbm/R	4.4005	4.4811	4.5931	4.6817	4.7544	4.8162	4.8704	4.9190	4.9633	5.0044		
P= 25.0 psi	T(R) 2189.8	2200.	2300.	2400.	2500.	2600.	2700.	2800.	2900.	3000.		
v,ft**3/lbm	36.3435	36.6427	39.4482	42.0280	44.4280	46.7121	48.8968	51.0140	53.0731	55.1066		
u,Btu/lbm	8402.4	8426.7	8638.6	8808.9	8949.0	9067.9	9171.9	9265.2	9350.9	9431.1		
h,Btu/lbm	9216.2	9247.2	9521.9	9750.0	9944.0	10113.9	10266.8	10407.5	10539.5	10665.0		
s,Btu/lbm/R	4.3235	4.3379	4.4623	4.5603	4.6398	4.7067	4.7645	4.8157	4.8620	4.9046		
P= 30.0 psi	T(R) 2231.0	2300.	2400.	2500.	2600.	2700.	2800.	2900.	3000.			
v,ft**3/lbm	30.6964	32.3466	34.5823	36.6599	38.6280	40.4872	42.2885	44.0402	45.7548			
u,Btu/lbm	8406.0	8558.4	8744.6	8897.8	9026.9	9138.6	9238.0	9328.4	9412.3			
h,Btu/lbm	9230.9	9427.6	9673.9	9882.9	10064.6	10226.5	10374.3	10511.8	10641.7			
s,Btu/lbm/R	4.2614	4.3505	4.4569	4.5429	4.6146	4.6758	4.7296	4.7779	4.8220			
P= 45.0 psi	T(R) 2328.6	2400.	2500.	2600.	2700.	2800.	2900.	3000.				
v,ft**3/lbm	21.1050	22.2378	23.7421	25.1384	26.4846	27.7569	28.9821	30.1735				
u,Btu/lbm	8420.3	8570.1	8753.6	8908.8	9042.1	9158.4	9262.3	9356.8				
h,Btu/lbm	9271.0	9466.5	9710.6	9922.5	10109.5	10277.2	10430.4	10572.9				
s,Btu/lbm/R	4.1262	4.2126	4.3147	4.3988	4.4701	4.5313	4.5852	4.6336				
P= 60.0 psi	T(R) 2403.4	2500.	2600.	2700.	2800.	2900.	3000.					
v,ft**3/lbm	16.1911	17.3290	18.4446	19.4989	20.5009	21.4601	22.3860					
u,Btu/lbm	8437.2	8627.0	8800.6	8951.1	9082.2	9198.3	9302.7					
h,Btu/lbm	9307.4	9558.3	9791.9	9999.0	10184.0	10351.6	10505.8					
s,Btu/lbm/R	4.0330	4.1415	4.2360	4.3155	4.3834	4.4426	4.4951					
P= 75.0 psi	T(R) 2464.9	2500.	2600.	2700.	2800.	2900.	3000.					
v,ft**3/lbm	13.1881	13.4599	14.4440	15.3240	16.1567	16.9534	17.7190					
u,Btu/lbm	8455.2	8520.2	8703.9	8866.7	9009.9	9136.7	9250.3					
h,Btu/lbm	9341.2	9428.0	9674.2	9896.0	10095.3	10275.6	10440.6					
s,Btu/lbm/R	3.9623	3.9992	4.1034	4.1899	4.2637	4.3277	4.3840					
P= 100.0 psi	T(R) 2549.2	2600.	2700.	2800.	2900.	3000.						
v,ft**3/lbm	10.1275	10.4854	11.1702	11.8280	12.4567	13.0588						
u,Btu/lbm	8486.4	8573.8	8743.2	8899.6	9040.4	9166.8						
h,Btu/lbm	9393.5	9512.9	9743.8	9959.1	10156.1	10336.5						
s,Btu/lbm/R	3.8731	3.9260	4.0204	4.1022	4.1732	4.2353						
P= 200.0 psi	T(R) 2779.2	2800.	2900.	3000.								
v,ft**3/lbm	5.3586	5.4301	5.7768	6.1134								
u,Btu/lbm	8610.9	8633.3	8760.4	8898.8								
h,Btu/lbm	9570.8	9606.2	9795.2	9994.0								
s,Btu/lbm/R	3.6650	3.6835	3.7690	3.8462								
P= 300.0 psi	T(R) 2935.0	3000.										
v,ft**3/lbm	3.6758	3.8323										
u,Btu/lbm	8730.7	8776.9										
h,Btu/lbm	9718.5	9806.6										
s,Btu/lbm/R	3.5433	3.5974										

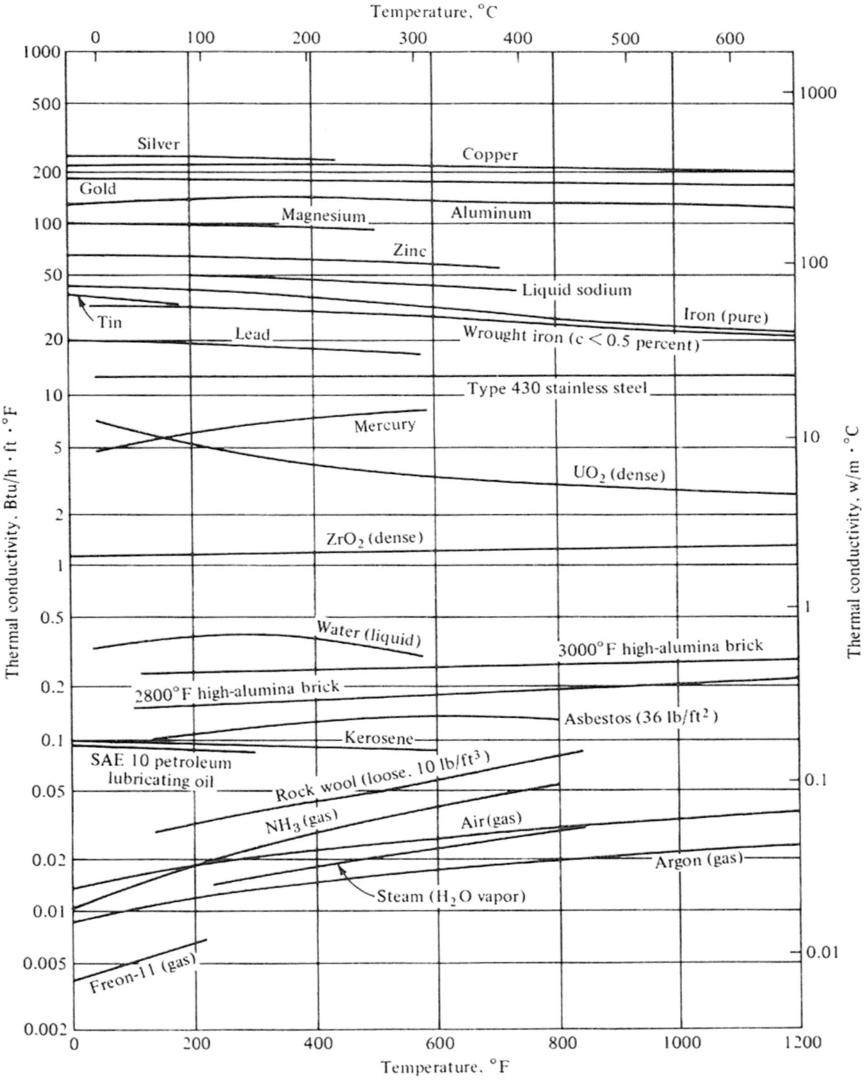
Temperature, °C
0
100
200
300
400
500
600
1000
500
200
100
50
20
10
5
2
1
0.5
0.2
0.1
0.05
0.02
0.01
0.005
0.002
1000
100
10
1
0.1
0.01
Silver
Copper
Gold
Magnesium
Aluminum
Zinc
Liquid sodium
Tin
Iron (pure)
Lead
Wrought iron (c < 0.5 percent)
Type 430 stainless steel
Mercury
UO_2 (dense)
ZrO_2 (dense)
Water (liquid)
3000°F high-alumina brick
2800°F high-alumina brick
Asbestos (36 lb/ft²)
Kerosene
SAE 10 petroleum
lubricating oil
Rock wool (loose, 10 lb/ft³)
NH_3 (gas)
Air (gas)
Argon (gas)
Steam (H_2O vapor)
Freon-11 (gas)
Thermal conductivity, Btu/h · ft · °F
Thermal conductivity, w/m · °C
0
200
400
600
800
1000
1200
Temperature, °F

Appendix B: Nuclear Power Plants

Currently, about half of all nuclear power plants are located in the USA. There are many different kinds of nuclear power plants, and we discuss a few important designs in this text. A nuclear power plant harnesses the energy inside atoms themselves and converts this to electricity. All of us use this electricity. In Sect. B.1 we show you should the idea of the fission process and how it works. A nuclear power plant uses controlled nuclear fission. In this chapter, we explore how a nuclear power plant operates and the manner in which nuclear reactions are controlled. There are several different designs for nuclear reactors. Most of them have the same basic function, but one's implementation of this function separates it from another. There are several classification systems used to distinguish between reactor types. Below is a list of common reactor types and classification systems found throughout the world and they are briefly explained down below according to three types of classifications: (1) Classified by Moderator Material, (2) Classified by Coolant Material, and (3) Classified by Reaction Type.

B.1 Fission Energy Generation

There is strategic as well as economic necessity for nuclear power in the USA and indeed most of the world. The strategic importance lies primarily in the fact that one large nuclear power plant saves more than 50,000 barrels of oil per day. At \$30–\$40 per barrel (1982), such a power plant would pay for its capital cost in a few short years. For those countries that now rely on but do not have oil, or must reduce the importation of foreign oil, these strategic and economic advantages are obvious. For those countries that are oil exporters, nuclear power represents an insurance against the day when oil is depleted. A modest start now will assure that they would not be left behind when the time comes to have to use nuclear technology.

The unit costs per kilowatt-hour for nuclear energy are now comparable to or lower than the unit costs for coal in most parts of the world. Other advantages are

B. Zohuri, *Application of Compact Heat Exchangers For Combined Cycle Driven Efficiency In Next Generation Nuclear Power Plants*,
DOI 10.1007/978-3-319-23537-0

the lack of environmental problems that are associated with coal or oil-fired power plants and the near absence of issues of mine safety, labor problems, and transportation bottlenecks. Natural gas is a good, relatively clean-burning fuel, but it has some availability problems in many countries and should, in any case, be conserved for small-scale industrial and domestic uses. Thus, nuclear power is bound to become the social choice relative to other societal risks and overall health and safety risks.

Nuclear fission is the process of splitting atoms, or fissioning them. This appendix explains.

B.2 The First Chain Reaction

Early in World War II the scientific community in the USA, including those Europeans now calling the USA their safe home, pursued the idea that uranium fission and the production of excess neutrons could be the source of extraordinary new weapons. They knew Lisa Meitner's interpretation, in Sweden, of Hahn's experiments would likely be known in Germany. Clearly there might now be a race commencing for the development and production of a new, super weapon based on the fission of $^{235}U_{92}$ or $^{239}Pu_{94}$.

By early 1942, it was known that the two naturally occurring isotopes of uranium reacted with neutrons as follows:

$$^{235}U + {}^{1}n \rightarrow \text{fission products} + (2.5)^{1}n + 200\,\text{MeV Energy}$$
$$^{238}U_{92} + {}^{1}n \rightarrow {}^{239}U_{92}$$
$$^{239}U_{92} \rightarrow {}^{239}Np_{93} + ß^{-1} t_{1/2} = 23.5\,\text{min}$$
$$^{239}Np_{93} \rightarrow {}^{239}Pu_{94} + ß^{-1} t_{1/2} = 2.33\,\text{days}$$

Each U-235 that undergoes fission produces an average of 2.5 neutrons. In contrast, some U-238 nuclei capture neutrons, become U-239, and subsequently emit two beta particles to produce Pu-239. The plutonium was fissile also and would produce energy by the same mechanism as uranium. A flow sheet for uranium fission is shown in Fig. B.1 below [1].

The answers to two questions were critical to the production of plutonium for atomic bombs:

1. Is it possible, using natural uranium (99.3 % U-238 and 0.7 % U-235), to achieve a controlled chain reaction on a large scale? If so, some of the excess neutrons produced by the fission of U-235 would be absorbed by U-238 and produce fissionable Pu-239.
2. How can we separate (in a reasonable period of time) the relatively small quantities of Pu-239 from the unreacted uranium and the highly radioactive fission-product elements?

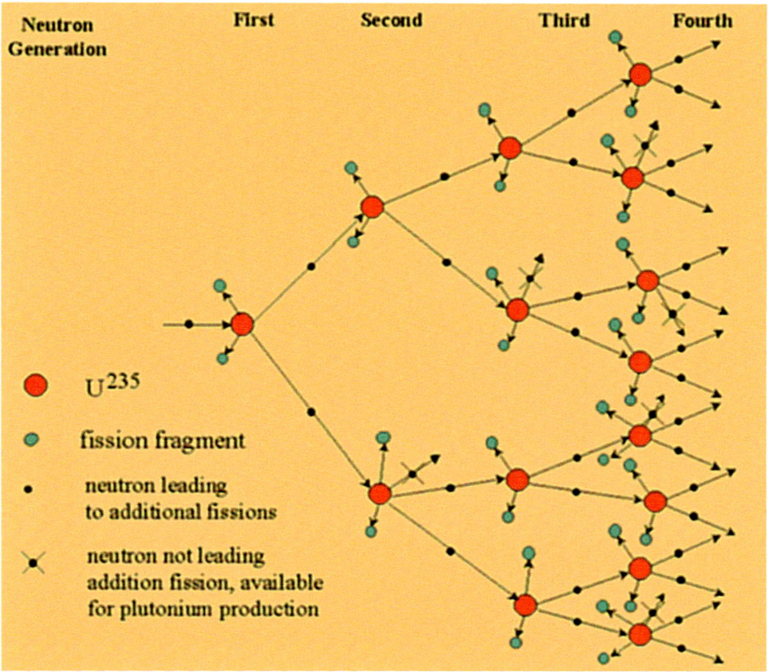

Fig. B.1 The first generations of a nuclear chain reaction [1]

Although fission had been observed on a small scale in many laboratories, no one had carried out a controlled chain reaction that would provide continuous production of plutonium for isolation.

Enrico Fermi thought that he could achieve a controlled chain reaction using natural uranium. He had started this work with Leo Szilard at Columbia University, but moved to the University of Chicago in early 1942.

The first nuclear reactor, called a pile, was a daring and sophisticated experiment that required nearly 50 t of machined and shaped uranium and uranium oxide pellets along with 385 t—the equivalent of four railroad coal hoppers—of graphite blocks, machined on site.

The pile itself was assembled in a squash court under the football field at the University of Chicago from the layered graphite blocks and uranium and uranium oxide lumps (Fermi's term) arranged roughly in a sphere with an anticipated 13 ft radius. Neutron absorbing, cadmium coated control rods were inserted in the pile. By slowly withdrawing the rods, neutron activity within the pile was expected to increase and at some point, Fermi predicted, there would be one neutron produced for each neutron absorbed in either producing fission or by the control rods [1].

On December 2, 1942, with 57 of the anticipated 75 layers in place, Fermi began the first controlled nuclear chain reaction occurred. At around 3:20 p.m. the reactor went critical; that is, it produced one neutron for every neutron absorbed by the uranium nuclei. Fermi allowed the reaction to continue for the next 27 min before

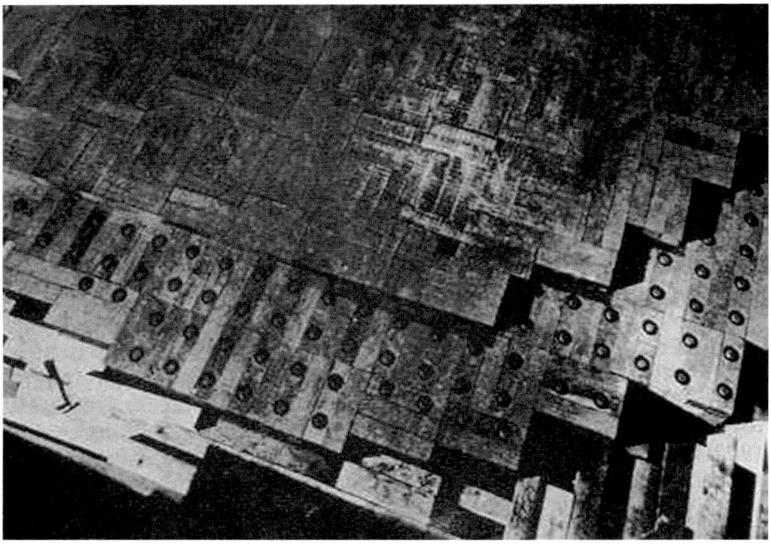

Fig. B.2 CP-1—Graphite blocks with 3 in. diameter uranium cylinders inserted—part of a layer of CP-1, the first nuclear reactor. A layer of graphite blocks without inserted uranium is seen covering the active layer [1]

Fig. B.3 The first controlled chain reaction, Stagg Field, Chicago, Dec. 2, 1942 (Courtesy of the Argonne National Laboratory)

inserting the neutron-absorbing control rods. The energy releasing nuclear chain reaction stopped as Fermi predicted it would. See Fig. B.2.

In addition to excess neutrons and energy, the pile also produced a small amount of Pu-239, the other known fissionable material. See Fig. B.3.

The achievement of the first sustained nuclear reaction was the beginning of a new age in nuclear physics and the study of the atom. Humankind could now use the tremendous potential energy contained in the nucleus of the atom. However, while a controlled chain reaction was achieved with natural uranium, and could produce plutonium, it would be necessary to separate U-235 from U-238 to build a uranium bomb [1].

On December 28, 1942, upon reviewing a report from his advisors, President Franklin Roosevelt recommended building full-scale plants to produce both U-235 and Pu-239.

This changed the effort to develop nuclear weapons from experimental work in academic laboratories administered by the US Office of Scientific Research and Development to a huge effort by private industry. This work, supervised by the US Army Corps of Engineers, was codenamed the Manhattan Project. It spread throughout the entire USA, with the facilities for uranium and plutonium production being located at Oak Ridge, Tennessee, and Hanford, Washington, respectively. Work on plutonium production continued at the University of Chicago, at what became known as the Metallurgical Laboratory or Met Lab. A new laboratory at Los Alamos, New Mexico, became the focal point for development of the uranium and plutonium bombs.

B.3 Concepts in Nuclear Criticality

A nuclear reactor works on the principle of a chain reaction. An initial neutron is absorbed by a fissile nuclide and during the process of fission, additional neutrons are released to replace the neutron that was consumed. If more neutrons are produced than are consumed, then the neutron population grows. If fewer neutrons are produced than are consumed, the neutron population shrinks. The number of fissions caused by the neutron population determines the energy released.

In order to quantify this concept let us define a multiplication factor k. We define k as the ratio of the production to consumption of neutrons:

$$k = \frac{\text{Production}}{\text{Consumption}} \tag{B.1}$$

B.4 Fundamental of Fission Nuclear Reactors

Today many nations are considering an expanded role for nuclear power in their energy portfolios. This expansion is driven by concerns about global warming, growth in energy demand, and relative costs of alternative energy sources. In 2008, 435 nuclear reactors in 30 countries provided 16 % of the world's electricity. In January 2009, 43 reactors were under construction in 11 countries, with several hundred more projected to come on line globally by 2030.

Concerns over energy resource availability, climate change, air quality, and energy security suggest an important role for nuclear power in future energy supplies. While the current Generation II and III nuclear power plant designs

Fig. B.4 A nuclear power plant (Courtesy of R2 Controls)

provide a secure and low-cost electricity supply in many markets, further advances in nuclear energy system design can broaden the opportunities for the use of nuclear energy. To explore these opportunities, the US Department of Energy's Office of Nuclear Energy has engaged governments, industry, and the research community worldwide in a wide ranging discussion on the development of next-generation nuclear energy systems known as "Generation IV." See Sect. B.4 for more information on New Generation of Power Plant know as Gen IV (Fig. B.4).

Nuclear reactors produce energy through a controlled fission chain reaction (See Sect. B.1: The First Chain Reaction). While most reactors generate electric power, some can also produce plutonium for weapons and reactor fuel. Power reactors use the heat from fission to produce steam, which turns turbines to generate electricity. In this respect, they are similar to plants fueled by coal and natural gas. The components common to all nuclear reactors include a fuel assembly, control rods, a coolant, a pressure vessel, a containment structure, and an external cooling facility.

In a nuclear reactor neutrons interact with the nuclei of the surrounding atoms. For some nuclei (e.g., U-235) an interaction with a neutron can lead to fission: the nucleus is split into two parts, giving rise to two new nuclei (the so-called fission products), energy and a number of new highly energetic neutrons. Other possible interactions are absorption (the neutron is removed from the system), and simple

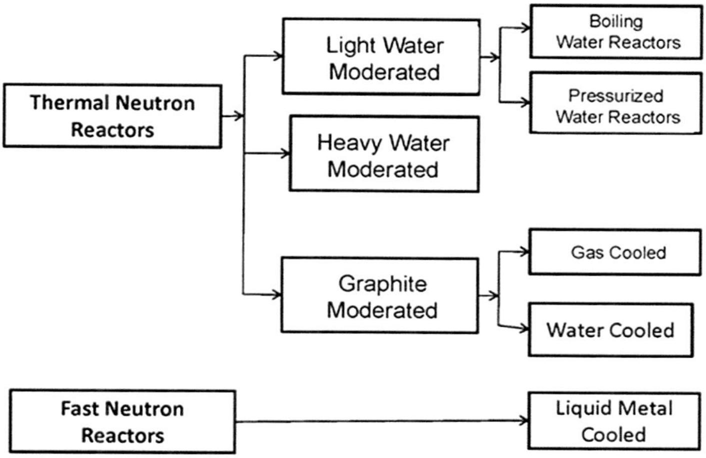

Fig. B.5 Types of nuclear reactors (Courtesy of Chem Cases)

collisions, where the incident neutron transfers energy to the nucleus, either elastically (hard sphere collision) or inelastically [7].

The speed of the neutrons in the chain reaction determines the reactor type (See Fig. B.5). Thermal reactors use slow neutrons to maintain the reaction. These reactors require a moderator to reduce the speed of neutrons produced by fission. Fast neutron reactors, also known as Fast Breeder Reactors (FBR), use high speed, unmoderated neutrons to sustain the chain reaction [1].

Thermal reactors operate on the principle that uranium-235 undergoes fission more readily with slow neutrons than with fast ones. Light water (H_2O), heavy water (D_2O), and carbon in the form of graphite are the most common moderators. Since slow neutron reactors are highly efficient in producing fission in uranium-235, they use fuel assemblies containing either natural uranium (0.7 % U-235) or slightly enriched uranium (0.9–2.0 % U-235) fuel. Rods composed of neutron-absorbing material such as cadmium or boron are inserted into the fuel assembly. The position of these control rods in the reactor core determines the rate of the fission chain reaction. The coolant is a liquid or gas that removes the heat from the core and produces steam to drive the turbines. In reactors using either light water or heavy water, the coolant also serves as the moderator. Reactors employing gaseous coolants (CO_2 or He) use graphite as the moderator. The pressure vessel, made of heavy-duty steel, holds the reactor core containing the fuel assembly, control rods, moderator, and coolant. The containment structure, composed of thick concrete and steel, inhibits the release of radiation in case of an accident and also secures components of the reactor from potential intruders. Finally, the most obvious components of many nuclear power plants are the cooling towers, the external components, which provide cool water for condensing the steam to water for recycling into the containment structure. Cooling towers are also employed with coal and natural gas plants.

B.5 Reactor Fundamentals

It is important to realize that while the U-235 in the fuel assembly of a thermal reactor is undergoing fission, some of the fertile U-238 present in the assembly is also absorbing neutrons to produce fissile Pu-239. Approximately one third of the energy produced by a thermal power reactor comes from fission of this plutonium. Power reactors and those used to produce plutonium for weapons operate in different ways to achieve their goals. Production reactors produce less energy and thus consume less fuel than power reactors. The removal of fuel assemblies from a production reactor is timed to maximize the amount of plutonium in the spent fuel (See Fig. B.6). Fuel rods are removed from production reactors after only several months in order to recover the maximum amount of plutonium-239. The fuel assemblies remain in the core of power reactors for up to 3 years to maximize the energy produced. However, it is possible to recover some plutonium from the spent fuel assemblies of power reactor.

The power output or capacity of a reactor used to generate electricity is measured in megawatts of electricity, MW(e). However, due to the inefficiency of converting heat into electricity, this represents only about one third of the total thermal energy, MW(t), produced by the reactor. Plutonium production is related to MW(t). A production reactor operating at 100 MW(t) can produce 100 g of plutonium per day or enough for one weapon every 2 months.

Another important property of a reactor is its capacity factor. This is the ratio of its actual output of electricity for a period of time to its output if it had been operated at its full capacity. The capacity factor is affected by the time required for maintenance and repair and for the removal and replacement of fuel assemblies. The average capacity factor for US reactors has increased from 50 % in the early 1970s to over 90 % today. This increase in production from existing reactors has kept electricity affordable.

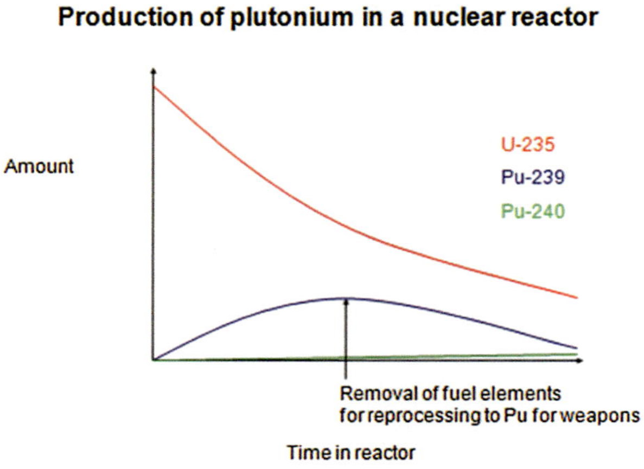

Fig. B.6 The fate of plutonium in a thermal reactor (Courtesy of Chem Cases)

B.6 Thermal Reactors

Currently the majority of nuclear power plants in the world are water-moderated, thermal reactors. They are categorized as either light water or heavy water reactors. Light water reactors use purified natural water (H_2O) as the coolant/moderator, while heavy water reactors employ heavy water, deuterium oxide (D_2O). In light water reactors, the water is either pressured to keep it in superheated form (in a pressurized water reactor, PWR) or allowed to vaporize, forming a mixture of water and steam (in a boiling water reactor, BWR). In a PWR (Fig. B.10), superheated water flowing through tubes in the reactor core transfers the heat generated by fission to a heat exchanger, which produces steam in a secondary loop to generate electricity. None of the water flowing through the reactor core leaves the containment structure. In a BWR (Fig. B.12), the water flowing through the core is directly converted to steam and leaves the containment structure to drive the turbines. Light water reactors use low enriched uranium as fuel. Enriched fuel is required because natural water absorbs some of the neutrons, reducing the number of nuclear fissions. All of the 103 nuclear power plants in the USA are light water reactors; 69 are PWRs and 34 are BWRs.

B.7 Nuclear Power Plants and Their Classifications

A nuclear power plant uses controlled nuclear fission. In this section, we explore how a nuclear power plant operates and the manner in which nuclear reactions are controlled. There are several different designs for nuclear reactors. Most of them have the same basic function, but one's implementation of this function separates it from another. There are several classification systems used to distinguish between reactor types. Below is a list of common reactor types and classification systems found throughout the world and they are briefly explained down below according to three types of classifications:

1. **Classified by Moderator Material** (i.e., Light Water Reactor, or Graphite Moderated Reactor, and Heavy Water Reactor).
2. **Classified by Coolant Material** (i.e., Pressurized Water Reactor, or Boiling Water Reactor, and Gas-Cooled Reactor).
3. **Classified by Reaction Type** (i.e., Fast Neutron Reactor, or Thermal Neutron Reactor, and Liquid Metal Fast Breeder Reactor).

B.8 Classified by Moderator Material

These types of reactors and their general description are presented below.

B.8.1 Light Water Reactors (LWR)

A light water reactor is a type of thermal reactor that uses "light water" (plain water) as a neutron moderator or coolant instead of using deuterium oxide (^2H_2O); light water reactors are the most commonly used among thermal reactors. Light water reactors are contained in highly pressurized steel vessels called reactor vessels. Heat is generated by means of nuclear fission within the core of the reactor. The hundreds into a "fuel assembly" about 12 ft in length and about as thin as a pencil, group the nuclear fuel rods, each. Each fuel rod contains pellets of an oxidized form of Uranium Triuranium Octaoxide (U_3O_8). A light water fuel reactor uses ordinary water to keep the system cool. The water is circulated past the core of the reactor to absorb the generated heat. The heated water then travels away from the reactor where it leaves the system as nothing more than water vapor. This is the method used in all LWRs except the BWR for in that specific system water is boiled directly by the reactor core (Fig. B.7).

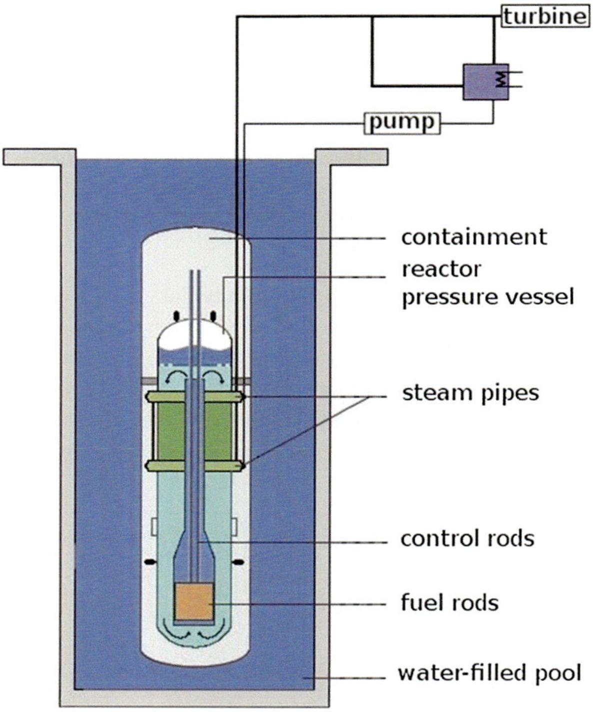

Fig. B.7 A pumpless light water reactor

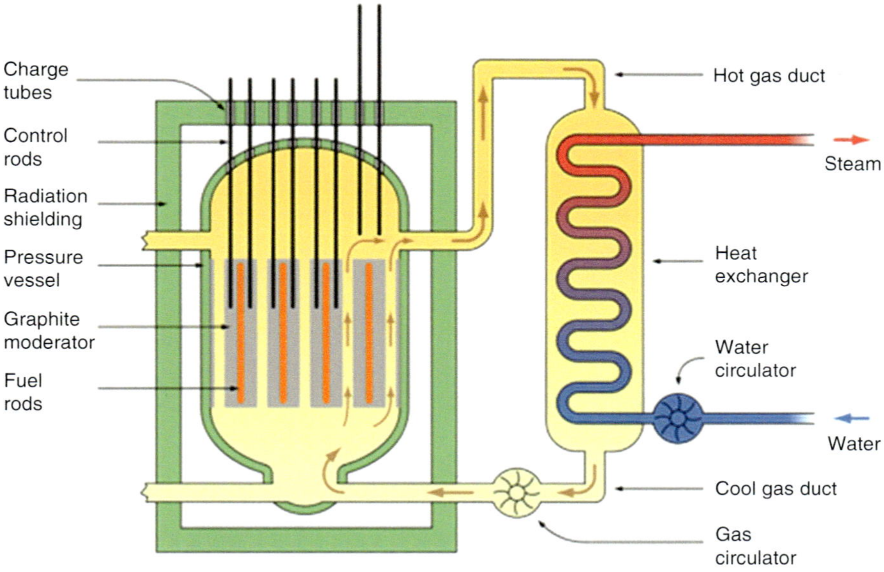

Fig. B.8 A typical core layout of graphite moderated reactor (Courtesy Osterreichisches Ökologie-Institut)

B.8.2 Graphite Moderated Reactors (GMR)

A Graphite Moderated Reactor (GMR) is a type of reactor that is moderated with graphite. The first ongoing nuclear reaction carried out by Enrico Fermi at The University of Chicago was of this type, as well as the reactor associated with the Chernobyl accident. GMRs share a valuable property with heavy water reactors, in that natural un-enriched uranium may be used. Another highlight for the GMR is a low power density, which is ideal if power were to suddenly stop; this would not waste as much power/fuel. The common criticisms for this design are a lack of room for steam suppression and the limited safety precautions available to the design (Fig. B.8).

B.8.3 Heavy Water Reactors (HWR)

Heavy water reactors (HWR) are a class of fission reactor that uses heavy water as a *neutron moderator*. Heavy water is deuterium oxide, D_2O. Neutrons in a nuclear reactor that uses uranium must be slowed down so that they are more likely to split other atoms and get more neutrons released to split other atoms. Light water can be used, as in a light water reactor (LWR), but since it absorbs neutrons, the uranium must be enriched for criticality to be possible. The most common pressurized heavy water reactor (PHWR) is the CANDU reactor.

Usually the heavy water is also used as the coolant but, for example, the Lucens reactor was gas cooled. The advantages of these types of reactors are that they can operate with unenriched uranium fuel. However, the opponents of heavy water reactors suggest that such reactors pose a much greater risk of nuclear proliferation because of two characteristics:

1. They use unenriched uranium as fuel, the acquisition of which is free from supervision of international institutions on uranium enrichment.
2. They produce more plutonium and tritium as by-products than light water reactors, these are hazardous radioactive substances that can be used in the production of modern nuclear weapons such as fission, boosted fission, and neutron bombs as well as the primary stages of thermonuclear weapons. For instance, India produced its plutonium for Operation Smiling Buddha, its first nuclear weapon test, by extraction from the spent fuel of a heavy water research reactor known as "CIRUS (Canada India Research Utility Services)." It is advocated that safeguards need to be established to prevent exploitation of heavy water reactors in such a fashion.

In heavy water reactors, both the moderator and coolant are heavy water (D_2O). A great disadvantage of this type comes from this fact: heavy water is one of the most expensive liquids. However, it is worth its price: this is the best moderator. Therefore, the fuel of HWRs can be slightly (1–2 %) enriched or even natural uranium. Heavy water is not allowed to boil, so in the primary circuit very high pressure, similar to that of PWRs, exists (Fig. B.9).

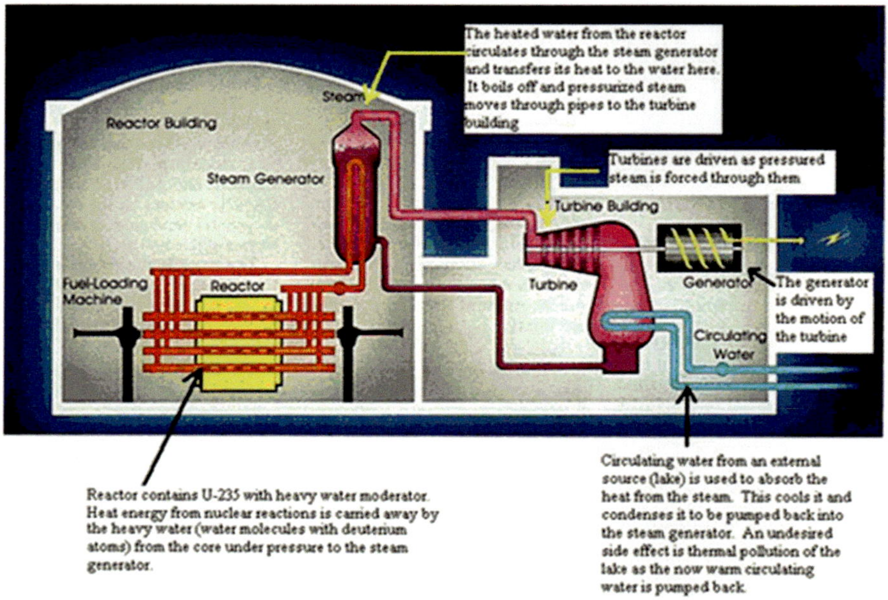

Fig. B.9 A typical outline layout of heavy water reactor (Courtesy Courtesy of Atomic Energy of Canada Limited)

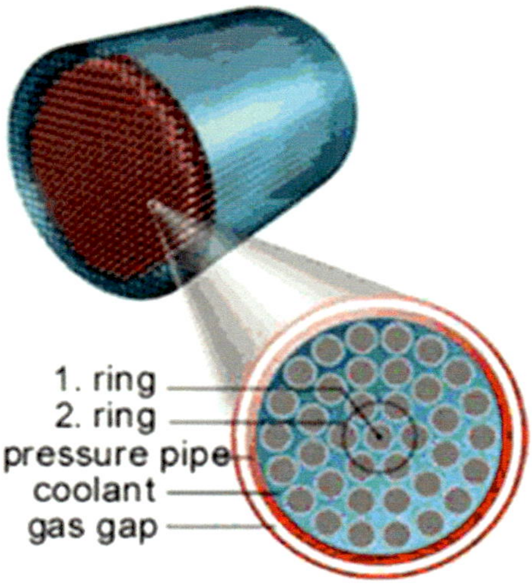

The main representative of the heavy water type is the Canadian CANDU reactor. In these reactors, the moderator and coolant are spatially separated: the moderator is in a large tank (calandria), in which there are pressure tubes surrounding the fuel assemblies. The coolant flows in these tubes only.

The advantage of this construction is that the whole tank need not be kept under high pressure; it is sufficient to pressurize the coolant flowing in the tubes. This arrangement is called pressurized tube reactor. Warming up of the moderator is much less than that of the coolant; it is simply lost for heat generation or steam production. The high temperature and high-pressure coolant, similarly to PWRs, goes to the steam generator where it boils the secondary side light water. Another advantage of this type is that fuel can be replaced during operation and thus there is no need for outages.

The other type of heavy water reactor is the Pressurized Heavy Water Reactor (PHWR). In this type, the moderator and coolant are the same and the reactor pressure vessel has to stand the high pressure.

Heavy water reactors produce CCA. 6 % of the total NPP power of the world; however 13.2 % of the under construction nuclear power plant capacity is accounted for by this type. One reason for this is the safety of the type; another is the high conversion factor, which means that during operation a large amount of fissile material is produced from U-238 by neutron capture.

B.9 Classified by Coolant Material

The descriptions of these types of reactors are as follows.

B.9.1 Pressurized Water Reactors (PWR)

A Pressurized Water Reactor (PWR) is Westinghouse Bettis Atomic Power Laboratory has used a type of light water reactor for decades in designs for military ship applications, now the primary manufacturers are Framatome-ANP and Westinghouse for present day power plant reactors. The pressurized water reactor is unique in that although water passes through the reactor core to act as moderator and coolant it does not flow into the turbine. Instead of the conventional flow cycle, the water passes into a pressurized primary loop. This step in the PWR cycle produces steam in a secondary loop that drives the turbine. Advantages of the PWR include zero fuel leaks of radioactive material into the turbine or environment, and the ability to withstand higher pressures and temperatures to higher the Carnot efficiency. Disadvantages include complex reactor designs and costs. This reactor type accounts for the majority of reactors located in the USA (Fig. B.10).

Pressurized Water Reactor (PWR) is a type of a nuclear power reactor that uses enriched uranium as a fuel, which in turn heats the light water used for producing

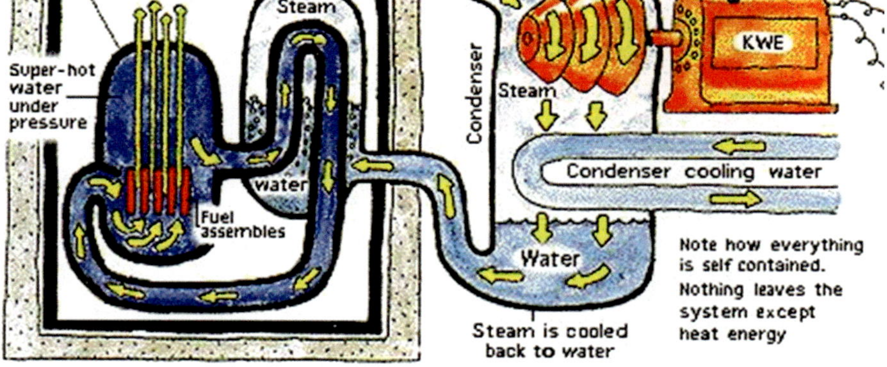

Fig. B.10 A typical pressurized water reactor (Courtesy of the Uranium Information Centre)

steam. The main feature, which differentiates it from a BWR nuclear reactor, is that a PWR has a separate arrangement to make steam in the form of a heat exchanger.

B.9.1.1 Arrangement of PWR

A Pressurized Water Reactor (PWR) is a type of power plant reactor consisting of two basic circuits having light water as the working fluid. In one of the circuits, water is heated to a high temperature and kept at high pressure as well, so that it does not get converted into a gaseous state. This superheated water is used as a coolant and a moderator for the nuclear reactor core hence the name PWR or pressurized water reactor.

The secondary circuit consists of water at high pressure in the gaseous state, i.e., steam which is used to run the turbine-alternator arrangement. The point of interaction between these two circuits is the heat exchanger or the boiler wherein heat from the superheated high-pressure water converts the water in the secondary circuit to steam.

B.9.1.2 Advantages of PWR

- Much fewer control rods are required in a PWR. In fact, for a typical 1000 MW plant just around 5 dozen control rods are sufficient.
- Since the two circuits are independent of each other, it makes it very easy for the maintenance staff to inspect the components of the secondary circuit without having to shut down the power plant entirely.
- A PWR has got a high power density and this, combined with the fact that enriched uranium is used as fuel instead of normal uranium, leads to the construction of very compact core size for a given power output.
- One feature, which makes a PWR reactor very suitable for practical applications, is its positive demand coefficient, which serves to increase the output as a direct proportion to demand of power.
- The water used in the primary circuit is different from that used in the secondary circuit and there is no intermixing between the two, except for heat transfer, which takes place in the boiler or heat exchanger. This means that the water used in the turbine side is free from radioactive steam hence the piping on that side is not required to be clad with special shielding materials.

B.9.1.3 Drawbacks of PWR

- The primary circuit consists of high temperature, high pressure water which accelerates corrosion. This means that the vessel should be constructed of very strong material such as stainless steel, which adds to construction costs of PWR.

- PWR fuel charging requires the plant to be shut down and this certainly requires a long time period of the order of at least a couple of months.
- The pressure in the secondary circuit is relatively quite low as compared to the primary circuit hence the thermodynamic efficiency of PWR reactors is quite low of the order of 20.

B.9.1.4 Drawbacks of PWR

One important point to note here is that despite the changing loads the pressure in the primary circuit needs to be maintained at a constant value. This is achieved by installing a device known as pressure equalizer in the primary circuit. It basically, consists of a dome-shaped structure which has heating coils which are used to increase or decrease pressure as and when required depending on varied load conditions.

Note that in the Pressurized Water Reactor (PWR), the water, which passes over the reactor core to act as moderator and coolant, does not flow to the turbine, but is contained in a pressurized primary loop. The primary loop water produces steam in the secondary loop, which drives the turbine. The obvious advantage to this is that a fuel leak in the core would not pass any radioactive contaminants to the turbine and condenser (Fig. B.11).

Another advantage is that the PWR can operate at higher pressure and temperature, about 160 atm and about 315 °C. This provides a higher *Carnot efficiency* than the BWR, but the reactor is more complicated and more costly to construct. Most of the US reactors are pressurized water reactors.

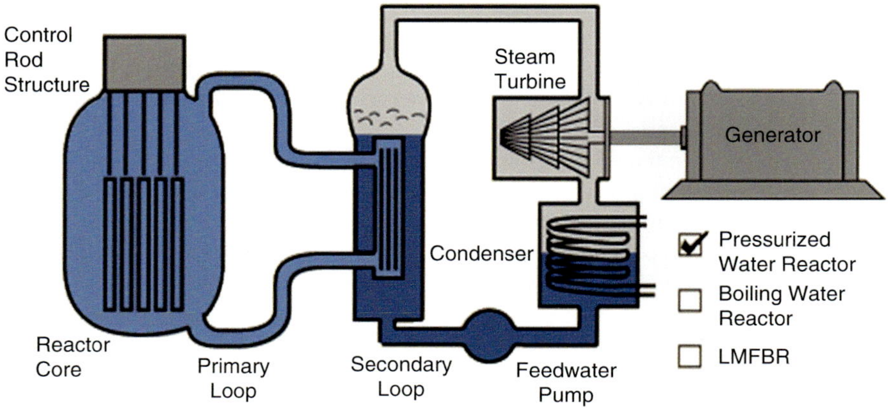

Fig. B.11 A typical outline of pressurized water reactor

B.9.2 Boiling Water Reactor (BWR)

Boiling Water Reactors (BWR) date back to their General Electric introduction in the 1950s. The distinguishing feature in the BWR is the boiling method for steam. In this type of reactor, water passes over the core as a coolant to expand and become steam source for a turbine placed directly above. Advantages of this design type include a simpler reactor design, a smaller reactor system, and lower costs. Disadvantages found are the increase of radioactive materials in the turbine and a greater chance for fuel to burn out as water quickly evaporates to expose fuel rods to an atmosphere absent of a coolant. BWRs have found fame all over the world due to the cheap simple design.

In Fig. B.12, we can see that:

1. The core inside the reactor vessel creates heat.
2. A steam-water mixture is produced when very pure water (reactor coolant) moves upward through the core, absorbing heat.
3. The steam-water mixture leaves the top of the core and enters the two stages of moisture separation where water droplets are removed before the steam is allowed to enter the steam line.
4. The steam line directs the steam to the main turbine, causing it to turn the turbine generator, which produces electricity.

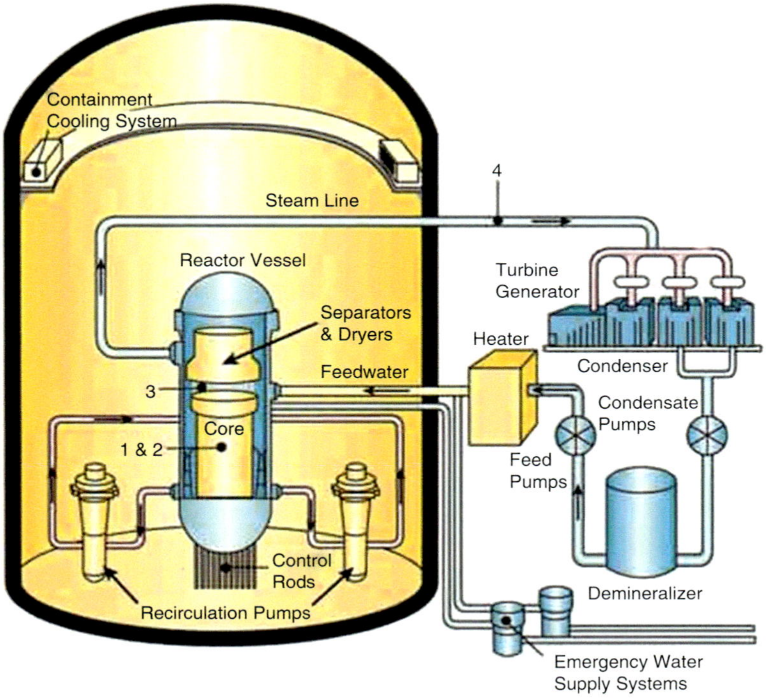

Fig. B.12 A typical boiling water reactor (Courtesy of the US Nuclear Regulatory Commission)

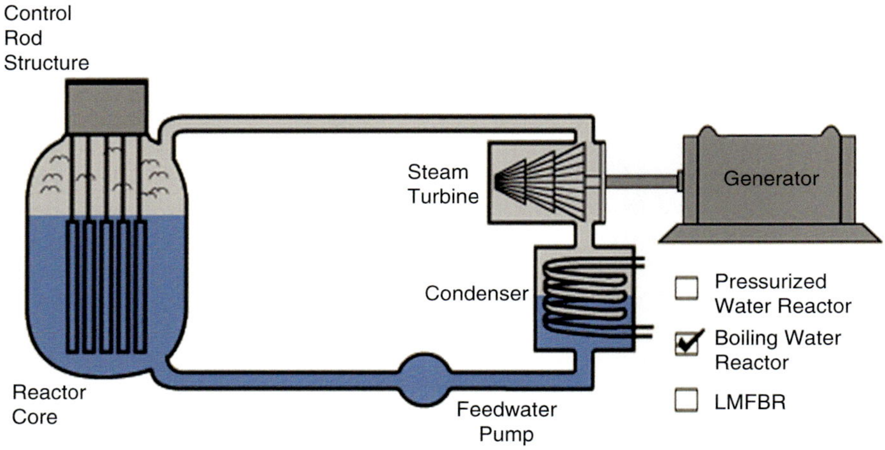

Fig. B.13 A typical layout of boiling water reactor

Note that in the Boiling Water reactor (BWR), the water, which passes over the reactor core to act as moderator and coolant, is also the steam source for the turbine. The disadvantage of this is that any fuel leak might make the water radioactive and that radioactivity would reach the turbine and the rest of the loop (Fig. B.13).

A typical operating pressure for such reactors is about 70 atm at which pressure the water boils at about 285 °C. This operating temperature gives a *Carnot efficiency* of only 42 % with a practical operating efficiency of around 32 %, somewhat less than the PWR.

B.9.3 Gas-Cooled Reactors (GCR)

The Gas-Cooled Reactor (GCR) or the gas-graphite reactors operate using graphite as moderator and some gas (mostly CO_2, lately helium) as coolant. This belongs to the oldest reactor types. The first GGR was the Calder Hall power plant reactor, which was built in 1955 in England. This type is called MAGNOX after the special magnesium alloy (Magnox), of which the fuel cladding was made. The fuel is natural uranium. These reactors account for 1.1 % of the total NPP power of the world and are not built any more (Fig. B.14).

The Advanced Gas-cooled Reactor (AGR) is a development from MAGNOX: the cladding is not Magnox and the fuel is slightly enriched. The moderator is also graphite and the coolant is CO_2. Contribution to total world capacity is 2.5 %. This type is not manufactured any longer (Fig. B.15).

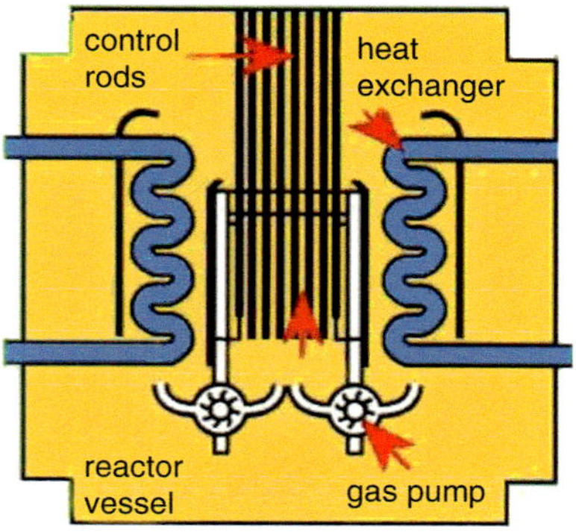

Fig. B.14 A typical core layout of gas-cooled reactor

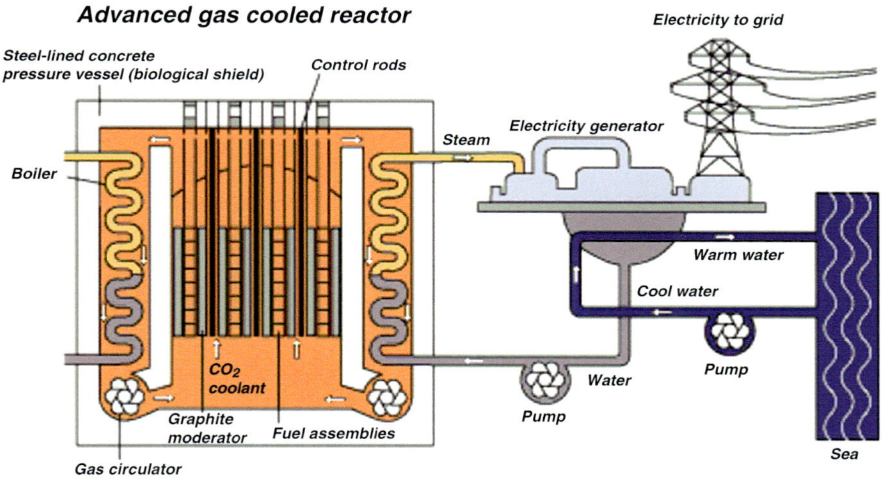

Fig. B.15 A typical outline layout of gas-cooled reactor

The newest gas-cooled reactor type is the High Temperature Gas-cooled Reactor (HTGR), which is cooled by helium and moderated by graphite. In this reactor as high as 950 °C coolant temperature can be achieved. The efficiency of a newly developed type, the Gas Turbine Modular Helium Reactor (GT-MHR) might be as high as almost 50 %.

Gas-Cooled Reactors (GCR) and Advanced Gas-Cooled Reactors (AGR) use carbon dioxide as the coolant to carry the heat to the turbine, and graphite as the moderator. Like heavy water, a graphite moderator allows natural uranium (GCR) or slightly enriched uranium (AGR) to be used as fuel.

B.10 Gas-Cooled Reactors (GCR)

The descriptions of each of these reactors are given as follows.

B.10.1 Fast Neutron Reactor (FNR)

Fast Neutron Reactors (FNR), also known as Fast Breeder Reactors (FBR), use depleted nuclear waste as a form of energy. Uranium, which is composed of 0.7 % Uranium-235 and 99.3 % Uranium-238, is processed in the fast neutron reactors into isotopes of usable plutonium of plutonium 239 and 241. Fast neutron reactors are 60 % more efficient than normal reactors; a fast neutron reactor uses liquid metal as its coolant as opposed to water, which makes the reactor safer to use and its fuel is metallic, which keeps the reactors under control more easily. Some cons of fast neutron reactors though are that they are very unpredictable, making them more tedious to use. Bubbles are more present in processes, so fast neutron reactors tend to heat up more rather than cool down and the coolant that it requires is much more exotic, such as liquid sodium and bismuth eutectic.

Several countries have research and development programs for improved Fast Breeder Reactors (FBR), which are a type of Fast Neutron Reactors. These use the uranium-238 in reactor fuel as well as the fissile U-235 isotope used in most reactors.

Natural uranium contains about 0.7 % U-235 and 99.3 % U-238. In any reactor, the U-238 component is turned into several isotopes of plutonium during its operation. Two of these, Pu 239 and Pu 241, then undergo fission in the same way as U-235 to produce heat. In a fast neutron reactor this process is optimized so that it can "breed" fuel, often using a depleted uranium blanket around the core. FBRs can utilize uranium at least 60 times more efficiently than a normal reactor.

Fast-neutron reactors could extract much more energy from recycled nuclear fuel, minimize the risks of weapons proliferation, and markedly reduce the time nuclear waste must be isolated.

If developed sensibly, nuclear power could be truly sustainable and essentially inexhaustible and could operate without contributing to climate change. In particular, a relatively new form of nuclear technology could overcome the principal drawbacks of current methods—namely, worries about reactor accidents, the potential for diversion of nuclear fuel into highly destructive weapons, the management of dangerous, long-lived radioactive waste, and the depletion of global reserves of economically available uranium. This nuclear fuel cycle would combine two innovations: pyrometallurgical processing (a high-temperature method of recycling reactor waste into fuel) and advanced fast-neutron reactors capable of burning that fuel. With this approach, the radioactivity from the generated waste could drop to safe levels in a few hundred years, thereby eliminating the need to segregate waste for tens of thousands of years.

Fast Reactor Technology:
A Path to Long-Term Energy Sustainability
Position Statement
November 2005

"The American Nuclear Society believes that the development and deployment of advanced nuclear reactors based on fast-neutron fission technology is important to the sustainability, reliability, and security of the world's long-term energy supply. Of the known and proven energy technologies, only nuclear fission can provide the large quantities of energy required by industrial societies in a sustainable and environmentally acceptable manner."

"Natural uranium mined from the earth's crust is composed primarily of two isotopes: 99.3 % is U-238, and 0.7 % is the fissile U-235. Nearly all current power reactors are of the "thermal neutron" design, and their capability to extract the potential energy in the uranium fuel is limited to less than 1 % of that available. The remainder of the potential energy is left unused in the spent fuel and in the uranium, depleted in U-235 that remains from the process of enriching the natural uranium in the isotope U-235 for use in thermal reactors. With known fast reactor technology, this unutilized energy can be harvested, thereby extending by a hundred-fold the amount of energy extracted from the same amount of mined uranium."

"Fast reactors can convert U-238 into fissile material at rates faster than it is consumed making it economically feasible to utilize ores with very low uranium concentrations and potentially even uranium found in the oceans [2–4]. A suitable technology has already been proven on a small scale [5]. Used fuel from thermal reactors and the depleted uranium from the enrichment process can be utilized in fast reactors, and that energy alone would be sufficient to supply the nation's needs for several hundred years."

"Fast reactors in conjunction with fuel recycling can diminish the cost and duration of storing and managing reactor waste with an offsetting increase in the fuel cycle cost due to reprocessing and fuel prefabrications. Virtually all long-lived heavy elements are eliminated during fast reactor operation, leaving a small amount of fission product waste that requires assured isolation from the environment for less than 500 years [5]."

"Although fast reactors do not eliminate the need for international proliferation safeguards, they make the task easier by segregating and consuming the plutonium as it is created. The use of onsite reprocessing makes illicit diversion from within the process highly impractical. The combination of fast reactors and reprocessing is a promising option for reasons of safety, resource utilization, and proliferation resistance [6]."

"Reaping the full benefits of fast reactor technology will take a decade or more for a demonstration reactor, followed by buildup of a fleet of operating power stations. For now and in the intermediate-term future, the looming short-term energy shortage must be met by building improved, proven

(continued)

thermal-reactor power plants. To assure longer-term energy sustainability and security, the American Nuclear Society sees a need for cooperative international efforts with the goal of building a fast reactor demonstration unit with onsite reprocessing of spent fuel."

B.10.2 Thermal Neutron Reactor

Thermal reactors go through the same process as fast neutron reactors, but in a thermal reactor the process of obtaining plutonium is slower. These types of reactors use a neutron moderator to slow neutrons until they approach the average kinetic energy of the surrounding particles, that is, to reduce the speed of the neutrons to low velocity thermal neutrons. The nuclear cross section of uranium-235 for slow thermal neutrons is about 1000 b. For fast neutrons, it is in the order of 1 b. In a thermal reactor, the neutrons that undergo the reaction process have significantly lower electron-volt energy, so the neutrons are considered to be slower. A neutron's speed will determine its chances to interact with the nucleus of an atom; the slower its speed the bigger its fission cross section becomes and thus the higher its chance of interacting with the nucleus becomes (Fig. B.16).

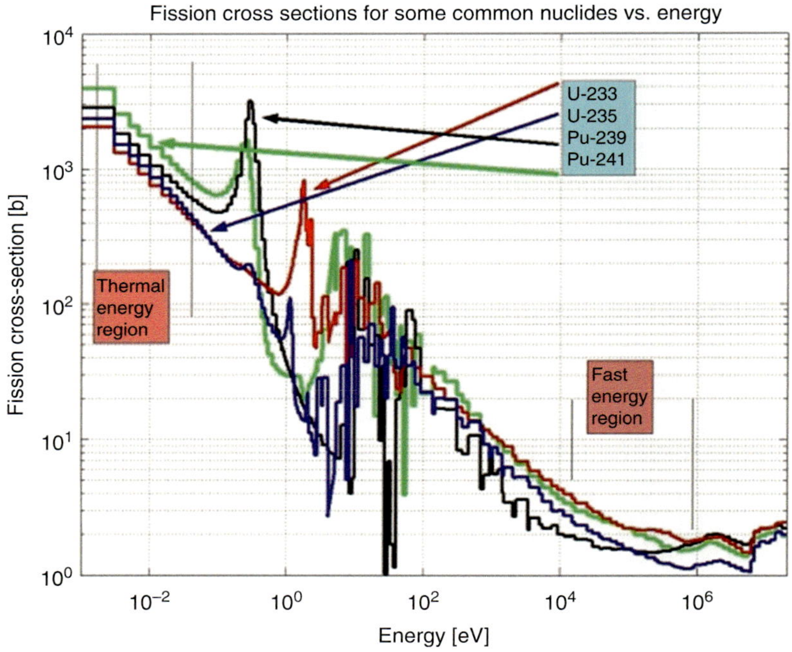

Fig. B.16 Fission cross section for some common nuclides vs. energy (Courtesy of TUDelft)

This figure gives the value of the fission cross section for some fissile isotopes. Note that both axes are logarithmic. The thermal and fast energy regions are indicated. For thermal energies, the fission cross section equals several thousand barns, at high energies the fission cross section is of the order of 1–10 b.

The fact that the fission cross section is rather large for low-energy neutrons has an important effect on the design of a nuclear reactor: in a reactor where the neutrons have a low energy, not much fissile material is required, because the probability of an interaction is very large. The lowest energy a neutron can have in a nuclear reactor is the energy at which it is in equilibrium with its environment. The movement of the neutron is then identical to the thermal movement of the atoms that constitute the reactor. The neutrons have slowed down from the high energy (2 MeV) where they are born to this equilibrium energy are called "thermal neutrons." The average energy of a neutron in thermal equilibrium is 0.025 eV—the neutron is slowed down over 9 decades, more than a billion times. Reactors in which most fissions are induced by thermal neutrons are called thermal reactors. Thermal reactors are by far the most widely used reactors in the world today. Most reactors use water, heavy water, or graphite as moderator. The reason for the choice of thermal reactors is a simple one: a thermal reactor requires a small amount of fuel to become critical, and thus the fuel is cheap [7].

B.10.3 Liquid Metal Fast Breeder Reactors (LMFBR)

The plutonium-239 breeder reactor is commonly called a fast breeder reactor, and the cooling and a liquid metal does heat transfer. The metals, which can accomplish this, are sodium and lithium, with sodium being the most abundant and most commonly used. The construction of the fast breeder requires a higher enrichment of U-235 than a light-water reactor, typically 15–30 %. The reactor fuel is surrounded by a "blanket" of non-fissionable U-238. No moderator is used in the breeder reactor since fast neutrons are more efficient in transmuting U-238 to Pu-239. At this concentration of U-235, the cross section for fission with fast neutrons is sufficient to sustain the chain-reaction. Using water as coolant would slow down the neutrons, but the use of liquid sodium avoids that moderation and provides a very efficient heat transfer medium (Fig. B.17).

The Super-Phoenix was the first large-scale breeder reactor. It was put into service in France in 1984. It ceased operation as a commercial power plant in 1997. Such a reactor can produce about 20 % more fuel than it consumes by the breeding reaction. Enough excess fuel is produced over about 20 years to fuel another such reactor. Optimum breeding allows about 75 % of the energy of the natural uranium to be used compared to 1 % in the standard light water reactor (Fig. B.18).

Under appropriate operating conditions, the neutrons given off by fission reactions can "breed" more fuel from otherwise non-fissionable isotopes. The most

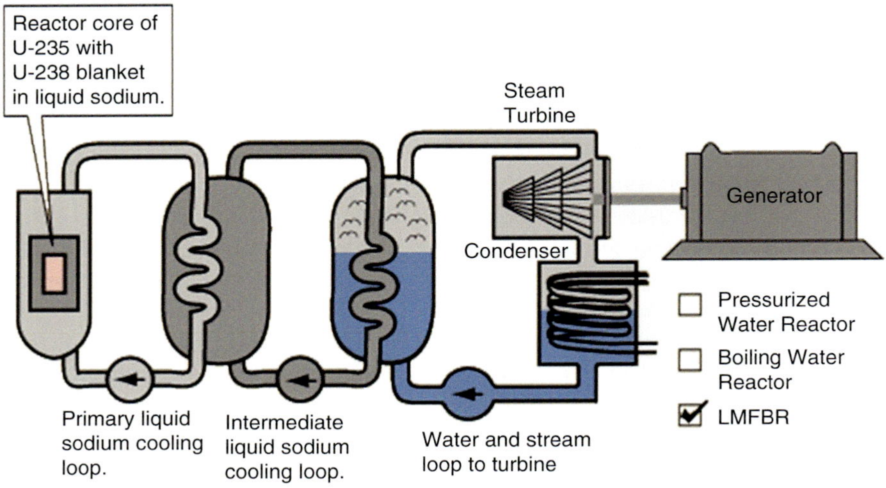

Fig. B.17 A typical layout of liquid metal fast breeder reactor

Fig. B.18 This is a photo of a model of the containment vessel of the Super-Phenix. It is displayed at the National Museum of Nuclear Science and Technology in Albuquerque, NM

common *breeding reaction* is that of plutonium-239 from non-fissionable uranium-238. The term "fast breeder" refers to the types of configurations, which can actually produce more fissionable fuel than they use, such as the LMFBR. This scenario is possible because the non-fissionable uranium-238 is 140 times more abundant than the fissionable U-235 and can be efficiently converted into Pu-239 by the neutrons from a fission chain reaction.

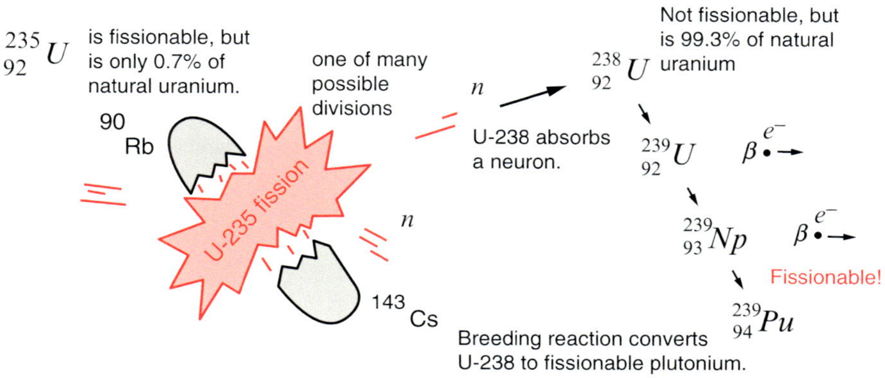

France has made the largest implementation of breeder reactors with its large Super-Phenix reactor (today is not in production line) and an intermediate scale reactor (BN-600) on the Caspian Sea for electric power and desalinization [27].

Fissionable breeding Plutonium-239 can be produced from non-fissionable uranium-238 by the reaction illustrated:

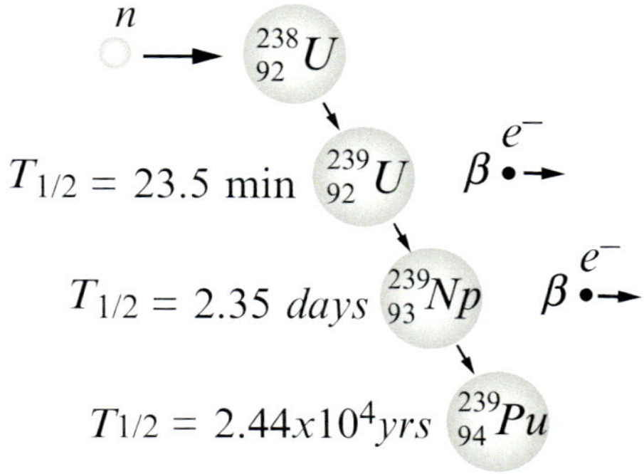

The concept of breading ratio of Plutonium-239 can be defined as follows. In the breeding of plutonium fuel in breeder reactors, an important concept is the breeding ratio, the amount of fissile plutonium-239 produced compared to the amount of fissionable fuel (like U-235) used to produce it. In the liquid-metal, fast-breeder reactor (LMFBR), the target-breeding ratio is 1.4 but the results achieved have been about 1.2. This is based on 2.4 neutrons produced per U-235 fission, with one neutron used to sustain the reaction.

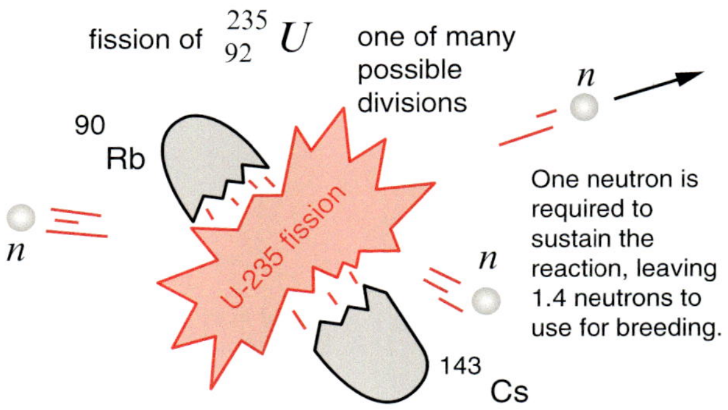

This particular fission path yields three neutrons, but the average neutron yield is 2.4 neutrons.

Liquid sodium is used as the coolant and heat-transfer medium in the LMFBR reactor. That immediately raised the question of safety since sodium metal is an extremely reactive chemical and burns on contact with air or water (sometimes explosively on contact with water). It is true that the liquid sodium must be protected from contact with air or water at all times, kept in a sealed system. However, it has been found that the safety issues are not significantly greater than those with high-pressure water and steam in the light-water reactors.

Sodium is a solid at room temperature but liquefies at 98 °C. It has a wide working temperature since it does not boil until 892 °C. That brackets the range of operating temperatures for the reactor so that it does not need to be pressurized as does a water-steam coolant system. It has a large *specific heat* so that it is an efficient heat-transfer fluid.

In practice, those reactors, which have used liquid metal coolants, have been fast-neutron reactors. The liquid metal coolant has a major advantage there because water as a coolant also moderates or slows down the neutrons. Such fast-neutron reactors require a higher degree of enrichment of the uranium fuel than do the water-moderated reactors [27].

B.11 Nuclear Fission Power Generation

Nuclear fission energy is today a competitive and mature low-carbon technology, operating at very high levels of safety. The installed nuclear electricity capacity in the Europe (EU), for example, is 132 GWe, which provides one-third of the EU's generated electricity [9, 10]. Most of the current designs are Light Water Reactors (LWR) of the second generation, capable of providing base-load electricity often with availability factors of over 90 %. There have been only a few new nuclear power plants connected to the grid in the last two decades, and as a result of decommissioning of old plants the total number of reactors in Europe has decreased. Nevertheless, electricity supply from nuclear has remained constant and the levelized cost has decreased owing to improved efficiency, power upgrade, and improved availability factor.

More recently, there has been a renewed interest in nuclear energy, referred to as "nuclear renaissance," mainly driven by concerns over climate change, security, and independence of supply and energy costs [8].

B.12 Generation IV Nuclear Energy Systems

Concerns over energy resource availability, climate change, air quality, and energy security suggest an important role for nuclear power in future energy supplies. While the current Generation II and III nuclear power plant designs provide a secure and low-cost electricity supply in many markets, further advances in nuclear energy system design can broaden the opportunities for the use of nuclear energy. To explore these opportunities, the US Department of Energy's Office of Nuclear Energy has engaged governments, industry, and the research community worldwide in a wide ranging discussion on the development of next-generation nuclear energy systems known as "Generation IV".

The goal of the Gen IV Nuclear Energy Systems is to address the fundamental research and development (R&D) issues necessary to establish the viability of next-generation nuclear energy system concepts to meet tomorrow's needs for clean and reliable electricity, and nontraditional applications of nuclear energy. Successfully addressing the fundamental Research and Development (R&D) issues will allow Gen IV concepts that excel in safety, sustainability, cost-effectiveness, and proliferation risk reduction to be considered for future commercial development and deployment by the private sector (Fig. B.19).

Gen IV reactor concepts are being developed to use advanced fuels, fashioned from recycled reactor fuel and capable of high-burnups. The corresponding fuel cycle strategies allow for efficient utilization of domestic uranium resources while minimizing waste. Reduction of proliferation risk and improvements in physical protection are being designed into Gen IV concepts to help thwart those who would target nuclear power plants for terrorist acts or use them improperly to develop materials for nuclear weapons. Gen IV concepts will feature advances in safety and reliability to improve public confidence in nuclear energy while providing enhanced investment protection for plant owners. Competitive life-cycle costs and acceptable financial risk are being factored into Gen IV concepts with high-efficiency electricity generation systems, modular construction, and shortened development schedules before plant start-up [11].

Gen IV is also an active participant in the International Project on Innovative Nuclear Reactors and Fuel Cycles (INPRO). INPRO was established in 2001 in

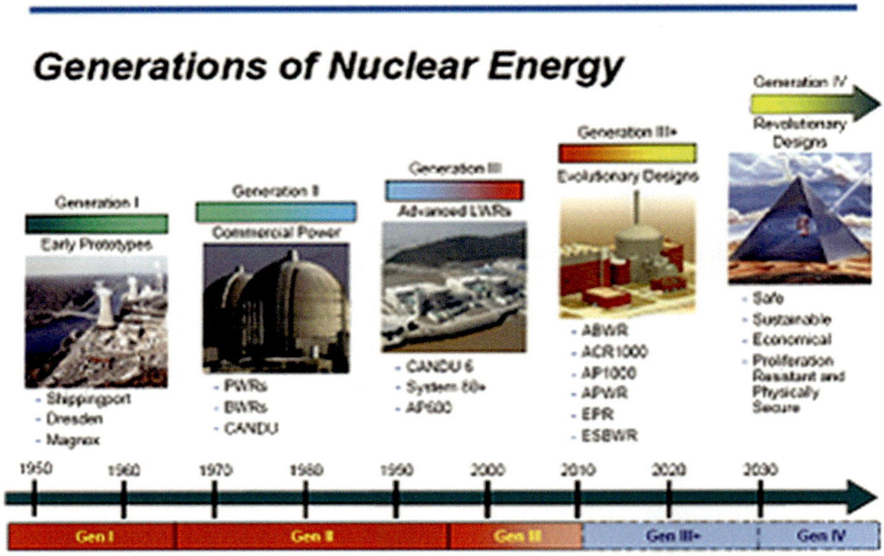

Fig. B.19 The evolution of nuclear power

response to a resolution by the IAEA General Conference to help to ensure that nuclear energy is available to contribute, in a sustainable manner, to meeting the energy needs of the twenty-first century and to bring together technology holders and users so that they can consider jointly the international and national actions required for achieving desired innovations in nuclear reactors and fuel cycles. INPRO provides a forum for discussion for experts and policy makers from industrialized and developing countries on all aspects of nuclear energy planning as well as on the development and deployment of innovative nuclear energy systems in the twenty-first century.

The Generation IV International Forum (GIF) was chartered in May 2001, to lead the collaborative efforts of the world's leading nuclear technology nations to develop the next generation of nuclear energy systems. The initial efforts of GIF resulted in the identification of the six most promising reactor concepts to be investigated by this international research community and are documented in the Generation IV Technology Roadmap. Thirteen members have signed the GIF Charter: Argentina, Brazil, Canada, People's Republic of China, Euratom, France, Japan, Republic of Korea, the Russian Federation, Republic of South Africa, Switzerland, the UK, and the USA. This unique international effort reached a major milestone on February 28, 2005, as five of the Forum's member countries (Canada, France, Japan, the UK, and the USA) signed the world's first multi-lateral agreement aimed at the international development of advanced nuclear energy systems—the Framework Agreement for International Collaboration on Research

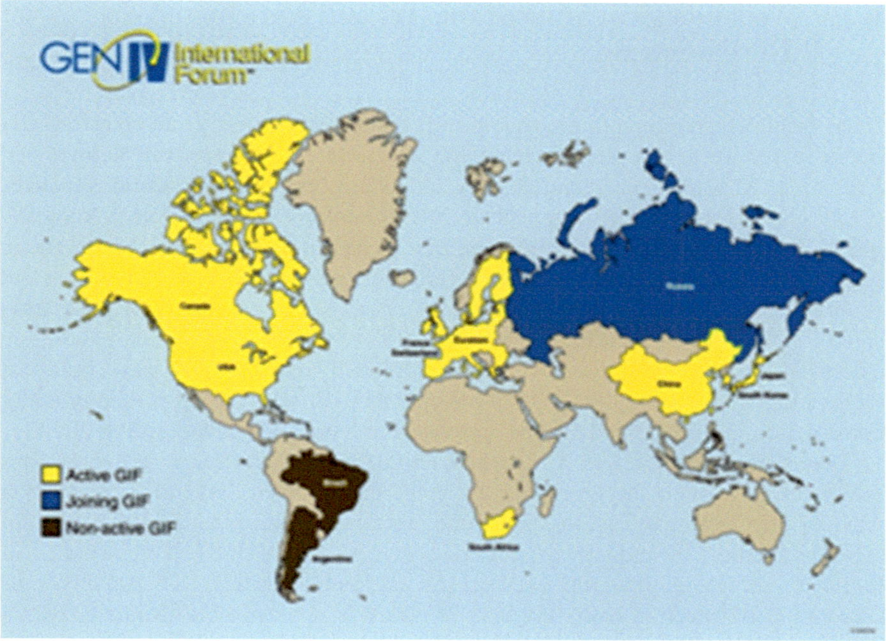

Fig. B.20 Map of member countries

and Development of Generation IV Nuclear Energy Systems. Subsequent signatories to the Framework Agreement included People's Republic of China, Euratom, Republic of Korea, Republic of South Africa, and Switzerland. The UK is a signatory of the Framework but is currently a non-active member. Argentina and Brazil have not ratified the Framework Agreement and are therefore considered non-active. The Russian Federation is working on the necessary approvals for its accession to the Framework [11] (Fig. B.20).

As detailed in its Charter and subsequent GIF Policy Statements, GIF is led by the Policy Group (PG), which is responsible for the overall coordination of GIF's Research and Development (R&D) collaboration, policy formation, and for interactions with other organizations. France with currently chairs the Policy Group vice chairs from the USA and Japan. An Experts Group and the Senior Industry Advisory Panel advises the Policy Group on (R&D) strategy, priorities, and methodology and on evaluating research plans for each Generation IV System. The Framework Agreement establishes two levels of implementing arrangements in order to conduct the joint (R&D). The first level consists of a System Arrangement for each Generation IV reactor concept directed by a System Steering Committee (SSC). Under each SSC, Project Arrangements are established with Project Management Boards to manage and implement the joint (R&D).

B.13 Technological State of the Art and Anticipated Developments

It has been demonstrated that Generation-II plants can be safely and economically operated for up to 60 years through the development of improved harmonized Plant-Life Management technologies and Plant License Extension practices (PLIM/PLEX) and that developments in fuel technologies can still lead to improvements in reactor performance [12]. The first Generation-III reactors, which are an evolution of thermal reactors with even further improved safety characteristics and economy, are now being built. In the coming decades, nuclear electricity generation should increase or at least maintain its current level by a combination of lifetime extension and power upgrades of Generation-II reactors and new build of Generation-III reactors. Two 1.6 GWe Generation-III reactors are presently under construction in Finland and France, targeted for connection to the grid in 2012.

The Finnish reactor was a First-of-a-Kind (FOAK) and the construction has suffered delays with the Overnight Cost increasing from 2000 to 3100 €/kWe, whereas the Overnight Cost for the second reactor in France is now 2400 €/kWe. In series production, the industry expects the cost to be 2000 ± 500 €/GWe, which is in line with recent international studies. An additional capacity of 100 GWe of Generation-III reactors over the next 25 years is a reasonable estimate, which would require an investment in the range of 200–280 billion Euros. The capital costs represent typically 60–70 % of the levelized cost for nuclear electricity, operation and maintenance 20–25 %, and fuel 10–15 %. The front-loaded cost profile means that the levelized cost is very sensitive to construction time and the financial schemes for the investment. Estimates in 2007 for UK resulted in a range of 31–44 £/MWh (37–53 €/MWh).

Though uranium is relatively abundant in the Earth's crust and oceans, estimates of natural reserves are always related to the cost of mineral extraction. As the price of uranium increases on world markets, the number of economically exploitable deposits also increases. The most recent estimates [17] identified 5.5 Million Tons of Uranium (MtU) that could be exploited below 130$/kg. The total amount of undiscovered resources (reasonably assured and speculative) available at an extraction cost below 130 $/kgU is estimated at 10.5 MtU. Unconventional resources, from which uranium is extracted as a by-product only, e.g., in phosphate production, lie between 7 and 22 MtU, and reserves in sea water are estimated to be 4 000 MtU. Japanese studies suggest that uranium from sea water can be extracted at 300€/kg [8]. At a conservative estimate, 25,000 t of the uranium are required to produce the fuel to generate 1000 TWhe in an open fuel cycle. The global electricity supplied by nuclear is 2600 TWhe, which means that the conventional resources below 130$/kgU at the current rate of consumption would last for at least 85 years with the already identified resources (5.5 MtU) and 246 years, if the undiscovered are also included (5.5 + 10.5 MtU). In addition to uranium, it is also possible to use thorium, which is three times more abundant in the Earth's crust, though would require different reactors and fuel cycles. Nonetheless, natural resources are plentiful and do not pose an immediate limiting factor for the development of nuclear energy.

However, in a scenario with a large expansion of nuclear energy, resources will become an issue much earlier, especially since new plants have at least a 60-year lifetime and utilities will need assurances when ordering new build that uranium supply can be maintained for the full period of operation. Eventually, known conventional reserves will all be earmarked for current plants or those under construction, and this could happen by the middle of this century. This underlines the need to develop the technology for a new generation, the so-called Generation-IV, of reactors and fuel cycles that are more sustainable. In particular, fast neutron breeder reactors could produce up to 100 times more energy from the same quantity of uranium than current designs and may significantly reduce the amount of ultimate radioactive waste for disposal.

Fast reactors convert non-fissile material (U-238) in the fuel into fissile material (Pu-239) during reactor operation so that the net amount of fissile material increases (breeding). After reprocessing of the spent fuel, the extracted fissile materials are then recycled as new fuels. Reduction of the radiotoxicity and heat load of the waste is achieved by separating some long-lived radionuclides, the minor actinides, which could then be "burned" in fast reactors or alternatively in Accelerator Driven Systems (ADS), through transmutation. The fast reactor concept has been demonstrated in research programs and national prototypes in the past, but further R&D is needed to make it commercially viable and to develop the designs in compliance with true Generation-IV criteria. Major issues involve new materials that can withstand higher temperatures, higher burn-ups and neutron doses, corrosive coolants; reactor designs that eliminate severe accidents; and the development of fuel cycles for waste minimization and elimination of proliferation risks. Fast Reactors are expected to be commercially available from 2040.

So far nuclear power has primarily been used to produce electricity, but it can also be used for process heat applications [13, 14]. Currently, LWRs are already being used to a limited extent for some lower temperature applications (200 °C), such as district heating and desalination of seawater. Existing designs of High-Temperature Reactors (HTR) that can reach 800 °C can be deployed in the coming decades and Very-High Temperature Reactors (VHTR) that can reach gas coolant temperatures beyond 1 000 °C are being studied as a Generation-IV concept for later deployment. Process heat applications include petroleum refinery applications (400 °C), recovery of oil from tar sands (600–700 °C), synthetic fuel from CO_2 and hydrogen (600–1000 °C), hydrogen production (600–1000 °C), and coal gasification (900–1200 °C). Small reactors that can be inherently safe and used to support specific high energy applications and often in remote areas form another very interesting application that are receiving more attention, in particular in the IAEA INPRO Initiative.

The management of radioactive waste, as either spent fuel or ultimate waste, depending on the national strategy, is a key issue for public acceptance of nuclear energy. There is scientific consensus that geological disposal is the only safe long-term solution for the management of ultimate waste. After a long period of intensive research and development coupled with in-depth political and social engagement, the world's first deep geological repositories for nuclear waste will be in operation in Sweden and Finland by 2020, with France following a few years later, demonstrating that practical solutions exist for the safe long-term management of hazardous waste from the operation of nuclear power plants. Though there will also be ultimate waste from Generation-IV fast reactor fuel cycles after reprocessing, the volumes and heat loads will be greatly reduced thereby facilitating disposal operations and optimizing the use of space in available geological repositories.

B.14 Next Generation Nuclear Plant (NGNP)

The Next Generation Nuclear Plant (NGNP) demonstration project forms the basis for an entirely new generation of advanced nuclear plants capable of meeting the Nation's emerging need for greenhouse-gas-free process heat and electricity. The NGNP is based on the Very-High-Temperature gas-cooled Reactor (VHTR) technology, which was determined to be the most promising for the USA in the medium term. The determination is documented as part of the Generation IV implementation strategy in a report submitted to Congress in 20031 following an extensive international technical evaluation effort. The VHTR technology incorporates substantive safety and operational enhancements over existing nuclear technologies. As required by the Energy Policy Act of 2005 (EPAct), the NGNP will be a prototype nuclear power plant, built at the Idaho National Laboratory (INL). Future commercial versions of the NGNP will meet or exceed the reliability, safety, proliferation-resistance, and economy of existing commercial nuclear plants [22].

It is envisioned that these advanced nuclear plants would be able to supply cost-competitive process heat that can be used to power a variety of energy intensive industries, such as the generation hydrogen, enhanced oil recovery, refineries, coal-to-liquids and coal-to-gas plants, chemical plants, and fertilizer plants [22].

The US Nuclear Regulatory Commission (NRC) is responsible for licensing and regulating the construction and operation of the NGNP. The EPAct authorizes the US Department of Energy (DOE) to build the NGNP at the Idaho National Laboratory and charges INL with responsibility for leading the project development. The project's completion depends on the collaborative efforts of DOE and its national laboratories, commercial industry participants, US universities, and international government agencies as well as successful licensing by the NRC. At present, and pending further evaluation as the NGNP proceeds through Phase 1 in cost-shared collaboration with industry as required by the EPAct, DOE has not made a final determination on whether the license applicant will be DOE or one or more entities that reflect a partnership between DOE and private sector firms [22].

Under the provisions of Section 644 of the EPAct, the Secretary of Energy and the Chairman of the Nuclear Regulatory Commission are to jointly submit to Congress a licensing strategy for the NGNP within 3 years of the enactment of the Act on August 8, 2005. This report addresses the requirement by outlining a NGNP licensing strategy jointly developed by the NRC and DOE. The scope of the document includes all four elements of the NGNP licensing strategy described in Section 644(b) of the EPAct:

of electricity,

1. A description of the ways in which current NRC light-water reactor (LWR) licensing requirements need to be adapted for the types of reactors considered for the project.
2. A description of the analytical tools that the NRC will need to develop in order to independently verify the NGNP design and its safety performance.
3. A description of other research or development activities that the NRC will need to conduct for the review of an NGNP license application.
4. A budget estimate associated with the licensing strategy.

DOE has determined that the NGNP nuclear reactor will be a very-high-temperature gas-cooled reactor (VHTR) for the production of electricity, process heat, and hydrogen. The VHTR can provide high-temperature process heat (up to 950 °C) that can be used as a substitute for the burning of fossil fuels for a wide range of commercial applications. Since the VHTR is a new and unproven reactor design, the NRC will need to adapt its licensing requirements and process, which have historically evolved around light-water reactor (LWR) designs, for licensing the NGNP nuclear reactor. Thus, Section 644 of the EPAct recognized the need for an alternative licensing strategy. This report provides the recommended NGNP licensing strategy, jointly developed by the NRC and DOE. As the technology matures, the government–industry partnership evolves, and input is provided by the general public, revisions to the strategy may be necessary and appropriate [22].

The report addresses the four elements of the licensing strategy set forth in Section 644(b) of the EPAct. These elements are summarized above [22].

B.15 Why We Need to Consider the Future Role of Nuclear Power Now

The following reasoning is some argument that shows why we need to consider the future role in design of new nuclear power plant:

1. Nuclear power has been part of the global energy need mix for the past five decades. Currently it provides about 18 % of the electricity we use in our homes and workplaces. For example, in the UK, about one-third of our emissions of carbon dioxide come from electricity generation [23]. The vast majority of those emissions come from coal and gas power plants.
2. Energy companies will need to invest in around 30–35 GW of new electricity generating capacity—as coal and nuclear plants retire—over the next two decades, with around two-thirds needed by 2020. This is equivalent to about

one-third of our existing capacity. The world needs a clear and stable regulatory framework to reduce uncertainty for business to help ensure sufficient and timely investment in technologies that contribute to our energy goals.

3. Of the capacity that is likely to close over the two decades, two thirds is from carbon intensive fossil fuel generation and about 10 GW is nuclear and therefore low carbon. So companies' decisions on the type of power stations they invest in to replace this capacity will have significant implications for the level of carbon emissions. As an illustration, if our existing nuclear power stations were all replaced with fossil fuel fired power stations, our emissions would be between 8 and 16 MtC (million tons of carbon) a year higher as a result (depending on the mix of gas and coal-fired power stations). This would be equivalent to about 30–60 % of the total carbon savings we project to achieve under our central scenario from all the measures we are bringing forward in the Energy White Paper [24]. Our gas demand would also be higher, at a time when we are becoming more dependent on imported sources of fossil fuels.
4. Electricity demand in the USA is expected to grow significantly in the future. Over the past decade, Americans used 17 % more electricity, but domestic capacity rose only 2.3 % (National Energy Policy, May 2001). Unless the USA significantly increases its generating capacity, the country will face an energy shortage that is projected to adversely affect our economy, our standard of living, and our national security. Coupled with this challenge is the need to improve our environment.
5. New nuclear power stations have long lead times. This time is necessary to secure the relevant regulatory and development consents, which must be obtained before construction can begin, and there is also a long construction period compared to other generating technologies. Our conservative assumption is that for the first new nuclear plant the pre-construction period would last around 8 years (to secure the necessary consents) and the construction period would last around 5 years. For subsequent plants, this is assumed to be 5 and 5 years, New nuclear power stations are therefore unlikely to make a significant contribution to the need for new capacity before 2020.
6. Even with our expectations that the share of renewable will grow, it is likely that fossil fuel generation will meet some of this need. However, beyond that date there are still significant amounts of new capacity needed; for example, in 2023 one-third or 3 GW of our nuclear capacity will still be operational, based on published lifetimes. Given the likely increase in fossil fuel generation before this date, it is important that much of this capacity is replaced with low carbon technologies. New nuclear power stations could make an important contribution, as outlined in this consultation document, to meeting our needs for low carbon electricity generation and energy security in this period and beyond to 2050. Because of the lead-times, without clarity now we will foreclose the opportunity for nuclear power.
7. The existing approach on new nuclear build was set out in 200311: "Nuclear power is currently an important source of carbon-free electricity. However, its

current economics make it an unattractive option for new, carbon-free generating capacity and there are also important issues of nuclear waste to be resolved. These issues include our legacy waste and continued waste arising from other sources. This white paper does not contain specific proposals for building new nuclear power stations. However, we do not rule out the possibility that at some point in the future new nuclear build might be necessary if we are to meet our carbon targets. Before any decision to proceed with the building of new nuclear power stations, there will need to be the fullest public consultation and the publication of a further white paper setting out our proposals."

8. Since 2003, there have been a number of developments, which have led the government to consider afresh the potential contribution of new nuclear power stations. Firstly, there has been significant progress in tackling the legacy waste issue:

    - We have technical solutions for waste disposal that scientific consensus and experience from abroad suggest could accommodate all types of wastes from existing and new nuclear power stations.
    - There is now an implementing body (the Nuclear Decommissioning Authority), with expertise in this area, and government is reconstituting the Committee on Radioactive Waste Management (CoRWM) in order to provide continued independent scrutiny and advice.
    - A framework for implementing long-term waste disposal in a geological repository will be consulted on in the coming months.

9. The government has also made progress in considering the issue of waste management in relation to potential new nuclear power stations:

    - This consultation provides the opportunity to discuss the ethical, intergenerational, and public acceptability issues associated with a decision to allow the private sector to invest in new nuclear power stations and generate new nuclear waste.
    - The government is developing specific proposals to protect the taxpayer. Under these proposals, private sector developers would meet the full decommissioning costs and full share of waste management costs. The proposals would be implemented in the event that we conclude that energy companies should be allowed to invest in new nuclear power stations. They would need to be in place before proposals for new power stations could go ahead.

10. Secondly, the high-level economic analysis of nuclear power, prepared for the Energy Review, concluded that under likely scenarios for gas and carbon prices and taking prudent estimates of nuclear costs, nuclear power would offer general economic benefit to the UK in terms of reduced carbon emissions and security of supply benefits [25]. Therefore, the government believes that it has a potential contribution to make, alongside other low-carbon generating technologies.

11. Thirdly, some energy companies have expressed a strong interest in investing in new nuclear power stations. They assess that new nuclear power stations could be an economically attractive low-carbon investment, which could help diversify their generation portfolios. Their renewed interest reflects assessments that with carbon being priced to reflect its impacts and gas prices likely to be higher than previously expected, the economics of new nuclear power stations are becoming more favorable.
12. Nuclear power stations have long lead times. If they are to be an option to replace the capacity closing over the next two decades, and in particular after 2020, a decision on whether allowing energy companies the option of investing in new nuclear power stations would be in the public interest needs to be taken now. Energy companies would need to begin their initial preparations in the near future in order to have a reasonable prospect of building new generation in this period. Not taking the public interest decision now would foreclose the option of new nuclear being one of our options for tackling climate change and achieving energy security.

B.16 The Generation IV Roadmap Project

As the Generation IV goals were being finalized, preparations were made to develop the Generation IV technology roadmap. The organization of the roadmap is shown in the Fig. B.21 below. The Roadmap Integration Team (RIT) is the executive group. Groups of international experts were organized to undertake identification and evaluation of candidate systems, and to define R&D to support them [26].

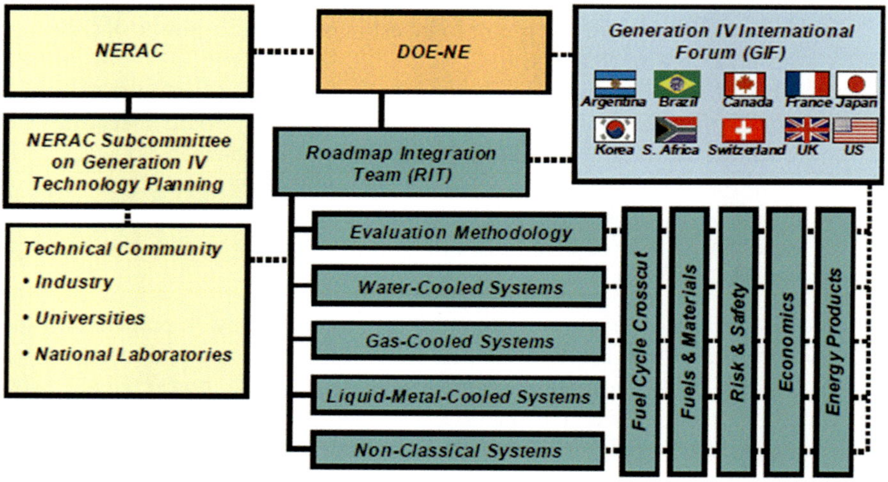

Fig. B.21 The roadmap organization

In a first step, an Evaluation Methodology Group was formed to develop a process to systematically evaluate the potential of proposed Generation IV nuclear energy systems to meet the Generation IV goals. A discussion of the Evaluation Methodology Group's evaluation methodology is included in this report. At the same time, a solicitation was issued worldwide, requesting that concept proponents submit information on nuclear energy systems that they believe could meet some or all of the Generation IV goals. Nearly 100 concepts and ideas were received from researchers in a dozen countries [26].

Technical Working Groups (TWGs) were formed—covering nuclear energy systems employing water-cooled, gas-cooled, liquid-metal-cooled, and nonclassical reactor concepts—to review the proposed systems and evaluate their potential using the tools developed by the Evaluation Methodology Group. Because of the large number of system concepts submitted, the TWGs collected their concepts into sets of concepts with similar attributes. The TWGs conducted an initial screening, *termed screening for potential*, to eliminate those concepts or concept sets that did not have reasonable potential for advancing the goals, or were too distant or technically infeasible [26].

A Fuel Cycle Crosscut Group (FCCG) was also formed at a very early stage to explore the impact of the choice of fuel cycle on major elements of sustainability—especially waste management and fuel utilization. Their members were equally drawn from the working groups, allowing them to compare their insights and findings directly. Later, other Crosscut Groups were formed covering economics, risk and safety, fuels and materials, and energy products. The Crosscut Groups reviewed the TWG reports for consistency in the technical evaluations and subject treatment, and continued to make recommendations regarding the scope and priority for crosscutting R&D in their subject areas. Finally, the TWGs and Crosscut Groups worked together to report on the R&D needs and priorities of the most promising concepts [26].

The international experts that contributed to this roadmap represented all ten GIF countries, the Organization for Economic Cooperation and Development Nuclear Energy Agency, the European Commission, and the International Atomic Energy Agency [26].

B.17 Licensing Strategy Components

A DOE and NRC working group was formed to develop the licensing strategy. This group conducted an in-depth analysis of LWR licensing process and technical requirements options, which was performed by the experienced senior staff of the two agencies. The methodology used in formulating the NGNP licensing strategy alternatives also included development of a phenomena identification and ranking table (PIRT) for a prototypical NGNP by subject matter experts in the nuclear field. The PIRT results assisted in the identification of key R&D needs. Based on the detailed analysis of these alternatives and balancing schedule considerations with licensing risk and other pertinent factors, the Secretary of Energy and the Commission concluded that the following NGNP licensing strategy provides the best opportunity for meeting the 2021 date for initial operation of a prototype NGNP, the details of which can be found in the NGNP report to Congress [22].

NGNP reactor technology will differ from that of commercial LWRs currently used for electric power generation. LWRs have a well-established framework of regulatory requirements, a technical basis for these requirements, and supporting regulatory guidance on acceptable approaches an applicant can take to show that NRC requirements are met. The NRC uses a Standard Review Plan to review licensing applications for these reactor designs. Additionally, the NRC has a well-established set of validated analytical codes and methods and a well-established infrastructure for conducting safety research needed to support its independent safety review of an LWR plant design and the technical adequacy of a licensing application.

New nuclear power plants can be licensed under either of two existing regulatory approaches. The first approach is the traditional "two-step" process described in Title 10, Part 50, "Domestic Licensing of Production and Utilization Facilities," of the *Code of Federal Regulations* (10 CFR Part 50), which requires both a Construction Permit (CP) and a separate Operating License (OL). The second approach is the new "one-step" licensing process described in 10 CFR Part 52, "Licenses, Certifications, and Approvals for Nuclear Power Plants," which incorporates a combined Construction and Operating License (COL). Both of these processes allow a deterministic or risk-informed performance-based approach to technical requirements.

Many of the regulatory requirements and supporting review guidance for LWRs are technology-neutral; that is, they are applicable to non-LWR designs as well as LWR designs. However, certain LWR requirements may not apply to the unique aspects of a VHTR design. Accordingly, in developing the NGNP licensing strategy, the NRC and DOE considered the various options available to the NRC staff for adapting current NRC LWR licensing requirements for the NGNP VHTR. These options related to legal, process, technical, research, and regulatory infrastructure matters and included an examination of historical licensing activities. These considerations led to the selection of a licensing strategy that would comply best with the considerations identified in the EPAct.

The licensing strategy outlined in this report is composed of two distinct aspects. The first aspect is a recommended approach for how the NRC will adapt the current LWR technical requirements to apply to a VHTR. The second aspect is a recommended licensing process alternative that identifies which of the procedural alternatives in the NRC regulations would be best for licensing the NGNP. To arrive at these recommendations, NRC and DOE evaluated a number of options and alternatives.

B.18 Market and Industry Status and Potentials

Europe plays a leading role in the development of nuclear energy and has 35 % of the globally installed capacity. The reactors in Europe have been in operation for 27 years on average. Current plans in most EU member countries are to extend their lifetime on a case by case basis beyond 40 years, and even beyond 60 years in some cases, in combination with power upgrades. The first two Generation-III reactors, European Pressurized-water Reactor (EPR) are currently being constructed.

The global growth of the nuclear energy can be measured by the increasing number of reactors [9, 10] (3 more in 2005 and 2006; 7 in 2007 and 10 in 2008), but with a strong concentration in Asia. Nevertheless a number of these reactors are of European design. There are presently four reactors under construction in Europe: the EPRs in Finland and France and two smaller reactors of Generation-II type (VVER 440) in Slovakia and with plans to build new reactors in France, Romania, Bulgaria, and Lithuania. Perhaps more importantly the UK has taken concrete steps toward new build with bidding beginning in 2009 from leading utilities, and Italy has declared that it intends to start a nuclear program with a target to produce 25 % of the electricity by 2030. The estimated maximum potential installed capacities of nuclear fission power for the EU-27 (150 GWe by 2020 and 200 GWe by 2030) appear more realistic than the baseline (115 GWe in 2020 and 100 GWe in 2030).

Programs to build fast reactor and high-temperature reactor demonstrators are being implemented in Russia and several Asian countries. Although these are not Generation-IV designs, transfer of knowledge and experience from operation will contribute significantly to future Generation-IV development. In Europe, a concerted effort is proposed in the form of a European Industrial Initiative in sustainable nuclear fission as part of the Community's SET-Plan. The EII has singled out the Sodium Fast Reactor (SFR) as its primary system with the basic design selected by 2012 and construction of a prototype of 250–600 MWe that is connected to the grid and operational by 2020.

In parallel, a gas- or lead-cooled fast reactor (GFR/LFR) will also be investigated. The selection of the alternative fast reactor technology is scheduled for 2012 on the basis of a current program of pre-conceptual design research. The reactor will be a 50–100 MWth demonstrator reactor that should also be in operation by 2020. The SFR prototype and LFR/GFR demonstrator will be complemented by a fuel fabrication workshop that should serve both systems, and by a range of new or refurbished supporting experimental facilities for qualification of safety systems, components, materials, and codes. A commercial deployment for a SFR reactor is expected from 2040 and for the alternative design a decade later.

High temperature reactors dedicated to the cogeneration of process heat for the production of synthetic fuels or industrial energy products could be available to meet market needs by 2025, which would trigger requirements to construct "first of a kind" demonstrators in the next few years. Indeed, such programs are currently being set up in some countries (USA, Japan, South Africa, and China). The key aspect is the demonstration of the coupling with the conventional industrial plant.

Supercritical water reactors and molten salt reactors, as well as accelerator driven subcritical systems dedicated to transmutation of nuclear waste, are currently being assessed in terms of feasibility and performance, though possible industrial applications have yet to be clearly identified.

B.19 Barriers

The high capital cost of nuclear energy in combination with uncertain long-term conditions constitutes a financial risk for utilities and investors. The lack of widespread support in the EU Member States may undermine the strength of EU industry for the development of new technologies. Harmonized regulations, codes, and standards at the EU-level would strengthen the competitiveness of Europe's nuclear sector and promote deployment of Generation-III technology in the near term. The industry, infrastructures, and services that support nuclear power has shrunk significantly during the last decades. This situation in Europe is not unique but it may pose a bottleneck for the deployment of reactors in the relatively near future. One example is large forgings needed for pressure vessel heads. World capacity is limited and even at the present new build construction rate there is a waiting list for the delivery of these components.

Public acceptance remains an important issue, but even though opinion is not very favorable in a number of Member States, there are signs that the mood is changing. Nevertheless, concerted efforts are still required, based on objective and open dialogue among all stakeholders. International cooperation currently exists at the level of research, and this is being facilitated in the area of Generation-IV systems by the Generation-IV International Forum (GIF). However, EU industry is facing stiff competition, especially in Asia where strong corporate support for R&D is putting industry in a better position to gain leadership in the near future. Another significant potential barrier for nuclear fission is the shortage of qualified engineers and scientists as a result of the lack of interest in nuclear careers during the 1990s and the reduced availability of specialist courses at universities. Preservation of nuclear knowledge remains a major issue, especially since most of the current generations of nuclear experts are nearing retirement.

B.20 Needs

The high initial capital investments and sensitive nature of the technology involved mean that renewed deployment of currently available nuclear technology can only take place in a stable (or, at least, predictable) regulatory, economic, and political environment. In June 2009, the EU established a common binding framework on nuclear safety with the adoption of the Council Directive establishing a Community framework for the safety of nuclear installations [20, 21]. This is the first binding EU legislation in this field.

In order to retain its leading position and to overcome bottlenecks in the supply chain, Europe also needs to reinvigorate the industrial supply chains supporting the nuclear sector. Apart from this overriding requirement for a clear European strategy on nuclear energy, a new research and innovation system is needed that can assure additional funding, especially for the development of Generation-IV technology. In this context the Sustainable Nuclear Energy Technology Platform [11] plays a key role. The timescales involved, and the fact that key political and strategic decisions are yet to be taken regarding this technology, mean that a significant part of this additional funding must be public. The launch of the European Sustainable Nuclear Industrial Initiative under the Community's SET-Plan, bringing together key industrial and R&D organizations, would be a very significant step toward the construction and operation of the necessary demonstrators and prototypes.

High temperature reactors based on existing technology can also be deployed in the near future with the aim of demonstrating the co-generation of process heat and the coupling with industrial processes. This would need to be built and funded through a European or International consortium, which should also include the process heat end-user industries. In the meantime, an enhanced research effort is needed to ensure Europe's leadership in sustainable nuclear energy technologies that include continuous innovation in LWRs, qualification and development of materials, closed fuel cycle with U-Pu multi-recycling, and (very) high temperature reactors and related fuel technology.

Breakthroughs are especially sought in the fields of materials to enhance safety, nuclear fuels, and fuel cycle processes. Additionally, there is a need for harmonization of European standards and a strategic planning of national and European research infrastructures for use by the European research community. The implementation of geological disposal of high-level waste is also being pursued as part of national waste management programs, though some countries are not as advanced as others. The new Implementing Geological Disposal Technology Platform, launched in November 2009, is coordinating the remaining necessary applied research in Europe leading up to the start of operation of the first geological repositories for high-level and long-lived waste around 2020 and will facilitate progress in and technology transfer with other national programs.

More effort is needed to inform and interact with the public and other stakeholders, and the education and training of a new generation of nuclear scientists and engineers and transfer of knowledge from the generation that designed and built reactors in the 1970s and 1980s needs urgent attention. The European Nuclear Energy Forum (ENEF) provides a unique platform for a broad open discussion on the role nuclear power plays today and could play in the low carbon economy of the future. ENEF analyses and discusses the opportunities (competitiveness, financing, grid, etc.) and risks (safety, waste) and need for education and training associated with the use of nuclear power and proposes effective ways to foster communication with and participation of the public.

B.21 Synergies with Other Sectors

Nuclear energy provides a very stable base-load electricity supply and can therefore work in synergy with renewable energies that are more intermittent. Nuclear energy should also contribute significantly to a low-carbon transport sector as high temperature applications can provide synthetic fuel and hydrogen, while generated electricity could provide a large share of the energy for electrical cars. Interactions are anticipated with activities in "Hydrogen Energy and Fuel Cells" through the potential of nuclear hydrogen production and with "grids" from the characteristics of nuclear electricity generation. With respect to basic materials research, there should be synergies with other applications, such as "Biofuels" and "Clean Coal," where materials are subjected to extreme environments. In addition, the opportunities for important common research with the fusion program, especially in the area of materials, need to be fully exploited. The European Energy Research Alliance under the SET-Plan is also expected to provide opportunities for synergies and collaborative work in the area of nuclear materials. In general, cross cutting research would benefit from more clearly defined channels of interaction, responsibilities, and increased flexibility regarding funding and programming.

References

1. F. Settle. Web Site: http://www.chemcases.com/nuclear/index.html
2. B. L. Cohen. (Jan. 1983). Breeder reactors: A renewable energy source. *American Journal of Physics*, *51*, 1.
3. A. Weinberg. The second fifty years of nuclear fission. *Proc. Special Symposium: 50 Years of Nuclear Fission in Review*, Ontario, Canada, June 5, 1989, Canadian Nuclear Society.
4. N. Seko. (Nov. 2003). Aquaculture of Uranium in seawater by a fabric-adsorbent submerged system. *Nuclear Technology*, *144*, 274.
5. Some Physics of Uranium. (December 2005). Retrieved from http://www.worldnuclear.org/education/phys.html
6. W. H. Hannum. (1997). The technology of the integral fast reactor and its associated fuel cycle. *Progress in Nuclear Energy*, *31*, 1.
7. TUDelft. Retrieved from http://www.iri.tudelft.nl/~rooijen/spectra_en.html
8. http://setis.ec.europa.eu/newsroom-items-folder/nuclear-fission-power-generation
9. World Nuclear Association Web site: http://www.world-nuclear.org/
10. IAEA Power Reactor Information System Web site: http://www.iaea.org/programmes/a2/
11. http://www.ne.doe.gov/geniv/neGenIV1.html
12. Sustainable Nuclear Energy Technology Platform, Strategic Research Agenda, May 2009, www.SNETP.eu
13. Nuclear Energy Outlook 2008, OECD/NEA Report No. 6348, 2008, Nuclear Energy Agency, Paris.
14. 2009 Update of the MIT 2003 Future of Nuclear Power, An Interdisciplinary MIT Study, 2003, Massachusetts Institute of Technology, Cambridge USA, May 2009 http://web.mit.edu/nuclearpower/pdf/nuclearpower-update2009.pdf

15. Projected Costs of Generating Electricity, 2005 Update, NEA/OECD, 2005
16. The Future of Nuclear Power – The Role of Nuclear Power in a Low Carbon UK Economy, Consultation Document, 2007, DTI, UK http://www.berr.gov.uk/files/file39197.pdf
17. Uranium 2007: Resources, Production and Demand. OECD Nuclear Energy Agency and the International Atomic Energy Agency, OECD 2008 NEA N 6345.
18. Strategic and Policy Issues Raised by the Transition from Thermal to Fast Nuclear Systems, 2009, OECD/NEA Report No. 6352
19. A Technology Roadmap for Generation IV Nuclear Energy Systems, 2002, GIF-002 - 00, Issued by the US DoE and the Generation IV International Forum http://www.gen-4.org/PDFs/GenIVRoadmap.pdf
20. Proposal for a COUNCIL DIRECTIVE (EURATOM) setting up a Community framework for nuclear safety COM(2008) 790/3, November 2008.
21. COUNCIL OF THE EUROPEAN UNION Legislative Acts and Other Instruments 10667/09, June 2009.
22. http://www.ne.doe.gov/pdfFiles/NGNP_reporttoCongress.pdf
23. Updated Emissions Projections, July 2006, DTI, http://www.dti.gov.uk/files/file31861.pdf
24. DTI: Energy White Paper, Meeting the Energy Challenge, http://www.dti.gov.uk/energy/whitepaper
25. DTI Analysis, http://www.dti.gov.uk/energy/whitepaper/consultations/nuclearpower2007
26. http://www.ne.doe.gov/geniv/documents/gen_iv_roadmap.pdf
27. http://hyperphysics.phy-astr.gsu.edu/hbase/nucene/fasbre.html
28. A. K. Raja, Amit Prakash Srivastava, Manish Dwivedi. (2006). Power Plant Engineering. New Delhi: New Age International (P) Limited.

Nuclear Systems Acronyms

Glossary of Nuclear Terms (US NRC)

AECL	Atomic Energy of Canada Ltd.
AECB	Atomic Energy Control Board
AESOP	Atomic Energy Simulation of Optimization (computer code)
AFCI	Advanced Fuel Cycle Initiative
AHTR	Advanced High Temperature Reactor
ASDV	Atmospheric Steam Discharge Valve
ASSERT	Advanced Solution of Sub-channel Equations in Reactor Thermal hydraulics (computer code)
ASTM	American Society for Testing Materials
BLC	Boiler Level Control
BLW	Boiling Light Water
BPC	Boiler Pressure Controller
CBA	Core Barrel Assembly
CCP	Critical Channel Power
CHF	Critical Heat Flux
CHP	Combined Heat and Power
CPR	Critical Power Ratio
CRL	Chalk River Laboratories
CRT	Cathode Ray Tube
CS	Core Structures
CSA	Canadian Standards Association
CSC	Core Structure Ceramics
CSDV	Condenser Steam Discharge Valve
CSNI	Canadian Standards for the Nuclear Industry

B. Zohuri, *Application of Compact Heat Exchangers For Combined Cycle Driven Efficiency In Next Generation Nuclear Power Plants*,
DOI 10.1007/978-3-319-23537-0

DBE	Design Base Earthquake
DCC	Digital Control Computer
DF-ET	Drift Flux-Equal Temperature
DF-UT	Drift Flux-Unequal Temperature
DHX	Dump Heat Exchanger
DNB	Departure from Nucleate Boiling
DRACS	Direct Reactor Auxiliary Cooling System
EBR-II	Experimental Breeder Reactor
ECC	Emergency Core Cooling
ECI	Emergency Core Injection
EFPH	Effective Full Power Hours
EVET	Equal Velocity Equal Temperature
EVUT	Equal Velocity-Unequal Temperature
EWS	Emergency Water Supply
FBR	Feed, Bleed, and Relief
FHSS	Fuel Handling and Storage System
FP	Full Power
FP	Fission Product
GFR	Gas-cooled Fast Reactor
GTHTR	Gas Turbine High Temperature Reactor
GT-MHR	Gas Turbine-Modular Helium Reactor
HEM	Homogeneous Equilibrium Model
HRSG	Heat Recovery Steam Generators
HTGR	High Temperature Gas-cooled Reactor
HTR	High Temperature Reactor
HTS	Heat Transport System
HWP	Heavy Water Plant
HX	Heat Exchanger
HYDNA	Hydraulic Network Analysis (extinct computer code)
I&C	Instrumentation and Control
IBIF	Intermittent Buoyancy Induced Flow
ICRP	International Commission on Radiological Protection
IGCC	Integrated Gasification Combined Cycle
IHX	Intermediate Heat Exchanger
LEU-TRISO	Low Enriched Uranium Triple-Coated Isotropic
LFR	Lead-alloy-cooled Fast Reactor
LMTD	Logarithmic Mean Temperature Difference
LOC	Loss of Coolant
LOC/LOECC	Loss of Coolant with Coincident Loss of Emergency Core Cooling
LOCA	Loss of Coolant Accident
LOP	Loss of Pumping
LOR	Loss of Regulation
LWR	Light Water Reactor
MCCR	Ministry of Corporate and Consumer Relations

MCS	Maintenance Cooling System
MHD	Magneto hydrodynamics
milli-k	Unit of reactivity for reactor physics
NERI	Nuclear Energy Research Initiative
NGNP	Next Generation Nuclear Plant
NPD	Nuclear Power Demonstration
NPSH	Net Positive Suction Head
NTU	Number of Transfer Units
NUCIRC	Nuclear Circuits (computer code)
OECD	Organization for Economic Co-operation and Development
OH	Ontario Hydro
PBMR	Pebble Bed Modular Reactor
PCS	Power Conversion System
PGSA	Pickering Generating Station A
PHTS	Primary Heat Transport System
PHW	Pressurized Heavy Water
PHWR	Pressurized Heavy Water Reactor
PRESCON2	Pressure Containment (computer code)
QA	Quality Assurance
R&M	Reliability and Maintainability
RAMA	Reactor Analysis Implicit Algorithm
RB	Reactor Building
RCS	Reactivity Control System
RIH	Reactor Inlet Header
ROH	Reactor Outlet Header
RSS	Reserve Shutdown System
RTD	Resistance Temperature Detectors
RU	Reactor Unit
SDM	Safety Design Matrices
SG	Steam Generator
SOPHT	Simulation of Primary Heat Transport (computer code)
SOX	Sarbanes Oxley
SRV	Safety Relief Valve
THTR	Thorium High Temperature Reactor
TMI	Three Mile Island
TOFFEA	Two Fluid Flow Equation Analysis (computer code)
TRIS	Triple-coated Isotropic
TRISO	Tristructural-Isotropic
UVUT	Unequal Velocity Unequal Temperature
VB	Vacuum Building
VC	Vacuum Chamber
VHTR	Very High Temperature Reactor
WRE	White-shell Research Establishment

Index

A
Absolute pressure, 11
Absolute zero, 16
Area density, 79

B
Brayton combined cycle (BCC), 162

C
CHEs with precision formed surfaces, 168
Classical thermodynamics, 7, 8
Clean-in-place (CIP), 170
Closed circuit, 86
Closed systems, 8
Combined cycle (CC)
    code, 5, 11
    efficiency, 87
    system, 5
Combined cycle gas turbine (CCGT), 86, 88
Compact heat exchangers (CHEs), 79, 167, 168
Computational fluid dynamics (CFD), 191
Control volume, 9
Cooling towers, 137
Cost of electricity (COE), 98
Counter flow, 151
Counterflow heat exchanger, 7
Cross flow, 151

D
Darcy friction factor, 221
Density, 10
Dynamic pressure, 221

E
Effectiveness method, 71, 193
Electricity, 17
Energy efficiency, performance and capacity (ECA) scheme, 168
Energy Information Administration (EIA), 17, 26
Energy technology list, 168
English (E) system, 5
Environmental program requirements (EPR), 95

F
Fanning friction factor, 221
First law of thermodynamics, 16

G
Gage pressure, 11
General electric (GE), 10
Generation IV, 21, 32–34, 46–51, 60, 89–91, 93

H
Head loss, 221
Heat exchangers (HXs), 3, 66, 68, 130, 135, 161
    counter flow, 69
    cross flow, 69

B. Zohuri, *Application of Compact Heat Exchangers For Combined Cycle Driven Efficiency In Next Generation Nuclear Power Plants*,
DOI 10.1007/978-3-319-23537-0

Heat exchangers (HXs) (*cont.*)
parallel flow, 69
Heat recovery steam generator (HRSG), 1, 3, 86, 88, 96, 98
Hydraulic diameter, 182

I
Independent power producers (IPPs), 97
Intensive variables, 10
Intermediate heat exchanger (IHX), 163
International Energy Agency (IEA), 26

K
Kelvins degree, 14
Kg oil equivalents (kgoe), 29

L
Laws of thermodynamics, 15
Logarithmic mean temperature difference (LMTD), 70, 142, 193

M
Macroscopic systems, 7
Minimum fluid capacity rate, 149
Mixed/unmixed flow, 69
Mixtures of pure substances, 10

N
National Energy Modeling System (NEMS), 18
Newtonian mechanics, 3
Next generation nuclear plant (NGNP), 60, 91, 125, 161, 176, 180, 223
Nuclear air-Brayton combined cycle (NACC), 162
Nuclear power plants, 139
Nuclear steam supply systems (NSSS), 46
Number of transfer units (NTU), 70, 149

O
Office of nuclear energy, 90
Open circuit, 86
Open systems, 8

P
Parallel flow, 127, 150
Plate heat exchangers, 168
Plate-finned heat exchangers, 168
Power conversion system (PCS), 60, 65, 66, 94
Pressurized water reactor (PWR), 1
Printed circuit heat exchanger (PCHE), 172
Proof stress (PS), 223
Pure substances, 10

Q
Quasi-equilibrium process, 8

R
Rankine cycle, 4
Rankine degree, 14
Ratio of stream capacity rates, 199
Return on investment (ROI), 31, 162
Reynolds number, 148, 159

S
Second law of thermodynamics, 16
Small modular reactors (SMR), 79
Specific volume, 11
Steady-state, 7
System International (SI), 5

T
T boundary condition, 213
Temperature, 13
Thermal energy, 2, 3, 6, 8, 16
Thermodynamic efficiency, 46
Thermosyphon, 177
Third law of thermodynamics, 16
Total cost of ownership (TCO), 31, 162
Transient analysis, 7

U
U.S. Department of Energy's Office of Science, 90
U.S. Nuclear Regulatory Commission, 90
Ultimate tensile stress (UTS), 223
United States, 1

V
Vacuum pressure, 12

Z
Zeroth law of thermodynamics, 16

GPSR Compliance

The European Union's (EU) General Product Safety Regulation (GPSR) is a set of rules that requires consumer products to be safe and our obligations to ensure this.

If you have any concerns about our products, you can contact us on ProductSafety@springernature.com

In case Publisher is established outside the EU, the EU authorized representative is:

Springer Nature Customer Service Center GmbH
Europaplatz 3
69115 Heidelberg, Germany

Batch number: 10615512

Printed by Printforce, the Netherlands